Introduction to Astronomy and Astrophysics

Arnold Hanslmeier

Introduction to Astronomy and Astrophysics

 Springer

Arnold Hanslmeier
Institute of Physics
University of Graz
Graz, Austria

ISBN 978-3-662-64639-7   ISBN 978-3-662-64637-3   (eBook)
https://doi.org/10.1007/978-3-662-64637-3

English translation of the 4th original German edition published by Springer-Verlag GmbH, DE, 2020
This book is a translation of the original German edition "Einführung in Astronomie und Astrophysik" by Hanslmeier, Arnold, published by Springer-Verlag GmbH, DE in 2020. The translation was done with the help of artificial intelligence (machine translation by the service DeepL.com). A subsequent human revision was done primarily in terms of content, so that the book will read stylistically differently from a conventional translation. Springer Nature works continuously to further the development of tools for the production of books and on the related technologies to support the authors.

This Springer imprint is published by the registered company Springer-Verlag GmbH, DE, part of Springer Nature.
The registered company address is: Heidelberger Platz 3, 14197 Berlin, Germany

Foreword

This book is now in its fourth edition. Modern astrophysics is now performed at almost all wavelengths; the entire electromagnetic spectrum, from gamma rays to radio waves, provides information about astronomical objects. We can observe up to a time when the universe was about 400,000 years old. The further we look into the depths of the universe, the more we observe into its past. The exploration of the early phases of the universe goes hand in hand with modern theories of physics, such as the unification of the four fundamental forces, or the attempt to unify quantum physics with the theory of relativity. New satellite missions have made it possible to obtain previously unimaginably accurate data on about 1% of all the stars in the Milky Way. The discovery of new planets outside the solar system has become routine. A sensation was the first direct observation of gravitational waves. This virtually opens a new window of observation for astrophysics.

Despite the great progress, however, there are still many unanswered questions. How did life develop on our planet? Is there life on other objects of the solar system or on exoplanets? Is our solar system unique? How can dark matter or dark energy be explained?

The book includes 17 chapters beginning with classical astronomy and a brief overview of the history of astronomy. After the description of astronomical observation instruments, the objects of the solar system are introduced, as well as the physics of the Sun. This is followed by chapters on state variables of stars, their structure and evolution. The next chapters deal with the description of interstellar matter and the structure of our Milky Way and galaxies in general. Then follows the chapter on cosmology in which also new physical theories are briefly outlined, such as loop gravity and string theories. The chapter on astrobiology goes into detail about the search for habitable planets. A special feature of this book is a brief introduction on common mathematical methods used in scientific practice. The programs are all given in the Python programming language, which is freely available and has now become enormously popular in astrophysics.

I would like to thank all my colleagues who contributed to this book. I would like to express special thanks to Springer Verlag for the excellent cooperation, especially with Ms. Meike Barth and Ms. Margit Maly. Many students of my lectures gave suggestions for improvement, as did numerous emails from readers.

I thank my partner Anita and my children for their patience and understanding.

Graz, Austria   Arnold Hanslmeier
December 2022

Contents

Introduction

Astronomy is one of the most rapidly developing scientific disciplines. This book is hereby in its fourth edition and has been updated in all chapters, and new sections have also been added. Again, our knowledge of our planetary system as well as exoplanets, the large-scale structure of the universe, dark energy, and dark matter developed particularly rapidly. The direct detection of gravitational waves was also a milestone.

This book deals with astronomy[1] and astrophysics. Why the distinction? In general, *astronomy* is the science of the stars. However, it also includes what is known as classical astronomy, which deals with planetary motions, spherical astronomy, coordinate systems, and so on. *Astrophysics* focuses mainly on the physics of objects, planets, stars, galaxies. Typical questions of astronomy/astrophysics are:

- Positioning: Where are we in the universe?
- Are we alone in the universe?
- Question about the origin: How and when did the universe come into being?
- Question about the future: How will the universe, the earth, our sun etc. develop further?

In contrast to all other natural sciences, astronomy does not allow direct experiments with the objects of research (stars, galaxies, universe). Due to the great distances, the only information we can evaluate from our objects is their radiation and their position. This is where astrophysics in particular comes in: What physical methods and laws can be used to explain or interpret this radiation?

Let us consider the *light of* a star. It has travelled several trillion kilometres to reach us. But we can disperse it through a prism or a grating. In physics, there is the well-known *Doppler effect*. When a source of radiation approaches the observer, the light appears

---

[1] $\alpha\sigma\tau\rho\text{o}\nu$ Greek star.

© Springer-Verlag GmbH Germany, part of Springer Nature 2023
A. Hanslmeier, *Introduction to Astronomy and Astrophysics*,
https://doi.org/10.1007/978-3-662-64637-3_1

blue-shifted, but when it is far away, it appears red-shifted. Therefore, by measuring the lines in the decomposed light (spectrum), we can determine whether the star is moving away from us or toward us. According to atomic theory, each chemical element emits very specific lines. By comparing the star's spectrum with spectra we know, we can determine the star's chemical composition. From the color of a star (a dark red glowing furnace is "cooler" than a bright red glowing one) we determine its temperature. Now it's up to the theorists. Experimental astrophysics provides the values for a star's composition, velocities, temperature, pressure, etc., and we can use simple equations to build a model of a star's structure and atmosphere.

The second piece of information about stars that is provided to us as passive viewers is their *position* in the sky. From this follows other important quantities such as distance. These measurements are made for many stars, and we are found to be part of a vast system, the Milky Way or Galaxy. Soon it was found that there are many galaxies and they are arranged in clusters. This then gives us the large-scale structure of the universe.

Accurate position measurements of stars or measurements of their position changes also led to ground breaking new discoveries during the last two decades. It has been proven that a super massive black hole with several million solar masses is located in the centre of our Milky Way system; dark, non-luminous matter has been detected as well as, for the first time, the existence of planets outside our solar system.

But what does astrophysics have to do with real life? Are there any applications of astrophysical findings at all? The above questions sound a bit abstract at first. What is the point?

It should be remembered that astronomy evolved from a very simple human need: division of time. All calendars are based on astronomical events: A day is one rotation of the Earth on its axis, a year is one revolution of the Earth around the Sun. The stars had something divine, immutable, for the ancient civilized peoples, and even Einstein, when he found solutions to his field equations that gave an unstable universe, believed that something could be wrong here, and introduced a mathematical trick to get the solution stable. Today, however, we know:

- The universe has an evolution.
- The universe is expanding.
- The universe emerged from a state of extremely high density and temperature about 13.7 billion years ago (big bang).
- Stars form and decay, becoming a white dwarf, or ending in a supernova explosion.

Astrophysics deals with objects of different expansions, densities and temperatures, as shown in Tables 1.1 and 1.2, and is therefore a science that deals with extreme physical conditions:

- extremely cool (near absolute zero, interstellar dust) or extremely hot (up to 100 million K in the interior of the star);

Table 1.1 Typical objects of astrophysics

Object	Density [g/cm^3]	Temperature [K]	Size [cm]
Sun	1.4 (mean)	$10^4 \ldots 10^7$	10^{11}
Solar system	10^{-23}	10^5	10^{15}
Gas nebula	$10^{-19} \ldots 10^{-16}$	$10^2 \ldots 10^4$	$10^{18} \ldots 10^{21}$
Galaxy	10^{-24}	$10^2 \ldots 10^4$	10^{23}
Galaxy cluster	10^{-27}	10^5	3×10^{24}
Universe	10^{-29}	$10^{-5} \ldots 10^9$	10^{29}

Table 1.2 Typical plasmas of astrophysics

Plasma	N_e [cm^{-3}]	Temperature T_e [K]	H [Gauss]
Interplanetary space	$1 \ldots 10^4$	$10^2 \ldots 10^3$	$10^{-6} \ldots 10^{-5}$
Sun, corona	$10^4 \ldots 10^8$	10^6	$10^{-5} \ldots 1$
stellar interior (normal)	10^{27}	$10^{7.5}$	–
Planetary nebulae	$10^3 \ldots 10^5$	$10^3 \ldots 10^4$	$10^{-4} \ldots 10^{-3}$
White dwarfs	10^{32}	10^7	10^6
Pulsars	10^{42} (center)		10^{12} (surface)
Intergalactic space	$\leq 10^{-5}$	$10^5 \ldots 10^6$	$\leq 10^{-6}$

- extremely thin (interstellar matter) and extremely dense plasma, crystallization states (e.g. in neutron stars);
- extremely high energies; formation of particles and elements at the big bang …

Essential insights and impetus for modern physics were gained from the study of the physics of such objects. Since astronomical instruments have to meet extreme requirements, this also resulted in industrially usable findings. One example of this is the material Zerodur for astronomical reflecting telescopes. Ceramic hobs, which are very common in kitchens today, can also be made from this material. By comparing the Earth's atmosphere with the dense CO_2 atmosphere of Venus, we learn what too high a greenhouse effect means.

Almost all the matter in space can be considered *plasma*. This is an ionized gas consisting of electrons, ions and neutral atoms. In Table 1.2 we give some typical values for the number of electrons per cubic centimeter, their temperature T_e and magnetic field strength (Gauss) of the plasmas in question.

New fields for astrophysics are opening up in astro particle physics and gravitational wave astronomy. We will discuss these in the corresponding chapters. The chapter on mathematical methods has been completely redesigned. Only the freely available programming language Python has been used, which is very common in astrophysics and in which there are countless additional packages.

Astrophysics always offers suitable research topics with which *Nobel Prizes* have been won. As an incentive for readers, we provide a list of Nobel Prize winners of recent years that are relevant here.

- 1967, H.A. Bethe, discoveries about thermonuclear energy production in stars.
- 1970, H. Alvén, Magnetohydrodynamics.
- 1974, M. Ryle, Aperture synthesis in radio astronomy.
- 1978, A. Hewish, A.A. Penzias, R.W. Wilson, Microwave background radiation.
- 1983, S. Chandrasekhar, W. Fowler, Stellar structure and stellar evolution.
- 1993, R.A. Hulse, J.H. Taylor, Double pulsars.
- 2002, R. Davis, M. Koshiba, R. Giacconi, Cosmic neutrinos.
- 2006, J.C. Mather, G.F. Smoot, Anisotropy of the cosmic background radiation.
- 2011, S. Perlmutter, B.P. Schmidt, A. Riess, Discovery of the accelerated expansion of the universe.
- 2017, R. Weiss, B. Barish, K. Thorne, Direct experimental discovery of gravitational waves produced by the collision of two black holes.
- 2019, M. Mayor, D. Queloz for the first discovery of an exoplanet around a Sun-like star, and J. Peebles for his contributions to cosmology.
- 2020, R. Penrose, R. Genzel, A. Ghez, Supermassive black holes in galaxies.

Furthermore, the following physicists have contributed significantly to the development of astronomy and received the Nobel Prize, but not for their contributions to astronomy:

- 1921, A. Einstein,
- 1938, E. Fermi,
- 1952, E.M. Purcell,
- 1964, C.H. Townes,
- 1980, J.W. Cronin.

Spherical Astronomy

2

To study the distribution of astronomical objects in the sky, it is necessary to use suitable coordinate *systems*, and by two coordinates the position of an object in the sky is unambiguously defined (compare longitude and latitude on earth). Depending on the problem, it is often convenient to use different coordinate systems. If one looks to the sky, then one can determine approximately in which *direction* the object was seen (S, W, N or E) and in which *altitude*. As a rule of thumb, an outstretched fist measures about 8° in the sky.

Spherical astronomy, together with celestial mechanics discussed later, is often referred to as *classical astronomy*. They experienced a renaissance in space travel. Exact position determinations of astronomical objects led to the discovery of white dwarfs and black holes (e.g. in the center of our Milky Way) and also to the discovery of extrasolar planetary systems.

We first turn to astronomical coordinate systems, then examine the influence of the Earth's atmosphere on the position of celestial objects, and finally questions of the calendar or time.

2.1 Coordinate Systems

We know coordinate systems from mathematics and geography. They are used to describe the position of objects. How to define coordinate systems in the sky, where is the zero point?

© Springer-Verlag GmbH Germany, part of Springer Nature 2023
A. Hanslmeier, *Introduction to Astronomy and Astrophysics*,
https://doi.org/10.1007/978-3-662-64637-3_2

2.1.1   Basic Principle for the Creation of Coordinate Systems

In general, only the position of a celestial object can be given or its radiation can be measured. For the indication of the position one thinks of the stars at an infinitely distant *celestial* sphere and then determine two spherical coordinates. The third coordinate would be the distance, which is not important here.

For each system one needs

- a base plane, a *base circle,* with associated Poles;
- a distance from the base circle, it is by a parallel circle *latitude* circles;
- *meridians*, where here again a starting point is to be defined *(zero meridian).* The distance of an object from this zero meridian then defines the second coordinate.

Analogue:geographic latitude (related to the base circle equator) and geographic longitude (related to the meridian through Greenwich).

2.1.2   Horizon System

This is the system of direct observation: in what celestial direction and at what altitude does one see a celestial object (Fig. 2.1)?

- Base circle: *Horizon,* the upper pole is called *zenith*, the lower pole *Nadir.*
- Longitude circles: Great circles through zenith, also called *Vertical* or circles of altitude. Zero meridian: Vertical by *south point,* this zero meridian is also called *meridian.*
- The coordinates in the horizon system are:
    - *Altitude h* above the horizon ($0°\dots90°$).
    - *Azimuth a*, angle between vertical through the object and meridian; counted from S through W, N, E. In radio astronomy also from N-E-S-W.

Fig. 2.1 Horizon system

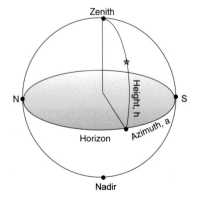

Fig. 2.2 Image rotation with a horizontally mounted telescope. On the left the moon at the rise, in the middle at the upper meridian transit, on the right at the set. Image: A. Hanslmeier

Very large telescopes and also amateur instruments are azimuthally mounted. Such mounts are simpler and cheaper, but there is a disadvantage: the altitude and azimuth are constantly changing. One must always track the telescope in two planes to follow the daily motion of a star. The image of an object seems to rotate over time (Fig. 2.2).

2.1.3   Equator System

Here we divide into fixed equatorial system and moving equatorial system.

Fixed Equatorial System
Serves to observation at telescope (Fig. 2.3).

- Base circle: *Celestial equator*, this is the projection of the earth's equator onto the celestial sphere. The north and south poles in the sky are an extension of the earth's axis. Coincidentally, a relatively bright star is currently located near the north pole, the *Pole Star.* There is no Pole Star in the southern hemisphere.
- Zero meridian: Meridian, passes through pole, zenith and south point.

Fig. 2.3 Equator system. t is
the hour angle, S is the south
point, α the right ascension, δ
the declination, Υ the vernal
equinox

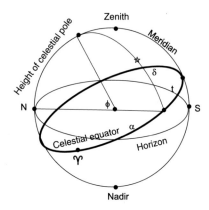

- The coordinates are: *Declination* δ: Distance from equator ($\pm 90°$); *Hour angle t:* Distance of the meridian of the meridian circle through the celestial object. Again, one counts from S-W-N-E.

Seen from the northern hemisphere of the Earth stars located north of the celestial equator are more than 12^h above the horizon (cf. Sun in the summer half-year). The *Altitude of the celestial pole* is given by the *geographic latitude* Φ. If the observer is at the earth's equator, then the pole is at the horizon (i.e. $\Phi = 0°$). If an observer is at the Earth's north pole, then Polaris, and thus the celestial north pole, is exactly perpendicular at the zenith, hence $\Phi = 90°$. Circumpolar *stars* go for a location of latitude Φ never below the horizon.

For the northern hemisphere of the Earth:

- $\delta = 0$, object 12^h above horizon.
- $\delta > 0$, object more than 12^h above horizon.
- $\delta < 0$, object less than 12^h above horizon.
- $\delta > 90° - \Phi$, Circumpolar stars.
- At the Earth's pole, all stars north of the equator are circumpolar.

This coordinate system has one major drawback: the hour angle passes through all values during the course of a day because of the rotation of the Earth.

The *culmination altitude* of a star is $h_{\max} = 90 - \Phi + \delta$. This is the highest point in the sky that a star (e.g., the sun) reaches.

The hour angle t is the time elapsed since the meridian passage of a star.

- $t = 0^h$ Star in *upper culmination*, is highest above horizon exactly in south;
- $t = 12^h$ Star in *lower culmination*.

To convert time to degrees use: $24^h = 360°$, corresponding to $1^h = 15°$ etc.

In the case of the *equatorial* or *parallactic mount* of a telescope, the hour axis is parallel to the Earth's axis, and therefore one needs only track in one plane.

$1^h \sim 15°$	$1° \sim 4^{min}$
$1^{min} \sim 15'$	$1' \sim 4^s$
$1^s \sim 15''$	$1'' \sim 0, .07^s$

Moving Equatorial System

Since the hour angle changes continuously, one defines the *moving coordinate system.*

- Base circle: celestial equator.
- Zero meridian: hour circle through the Spring point *Vernal point, vernal equinox.* This is defined as one of the points of intersection between the sun's orbit *(ecliptic)* and equator, namely the place of the sun at the beginning of spring.
- The coordinates are: δ *Declination*, α *Right Ascension*. Right ascension gives the distance of the hour circles by celestial object and vernal equinox. Important: α is counted according to the movement of the sun to the east.

Advantage Both coordinates are now fixed and suitable for the *cataloging* of the objects. One introduces still the *sidereal time* Θ which is the hour angle of the vernal equinox. The hour angle t of an object that has a right ascension α becomes:

$$t = \Theta - \alpha \qquad\qquad (2.1)$$

Most commonly used in astronomy are the coordinates right ascension α and declination δ. With the sidereal time Θ one obtains the hour angle t of an object at a given moment. The declination does not change with time. This determines the position of the celestial objects.

2.1.4 Ecliptic System

Here is the base circle the *Ecliptic* (Fig. 2.4), i.e. the apparent path of the sun in the sky (created by the motion of the earth around the sun), and one defines as coordinates the *ecliptic longitude* λ (counted from the vernal point) as well as the *ecliptic latitude* β. The ecliptic is inclined by 23.5° with respect to the equator. The declination of the sun at the beginning of summer is therefore 23.5°. Thus one can determine the *Inclination of the ecliptic*. In Table 2.1 all coordinates of the sun at the beginning of the seasons are given.

Fig. 2.4 Ecliptic system; ENP is the ecliptic north pole, pole refers to the celestial north pole. λ is the ecliptic longitude, β ecliptic latitude. The Sun's motion is, by definition, in the ecliptic plane. The moon and planets move near the ecliptic

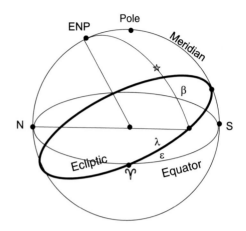

Table 2.1 Coordinates of the sun in the equator system or position of the sun in the zodiac

Date	Decl.	RA	Sun in
March 21, vernal equinox, equinox	0°	0^h	Aries
June 22, summer equinox, solstice	+23.5°	6^h	Cancer
Sep 23, autumn beg., equinox	0°	12^h	Libra
Dec 22, winter equinox, solstice	−23.5°	18^h	Capricorn

2.1.5 Galactic System

The base circle is defined by the *Milky Way plane* and the zero meridian passes through the *galactic center.* This leads to the coordinates b, galactic *latitude*, and l, *galactic longitude.* Galactic coordinates are used to study the distribution of stars and star clusters in our Milky Way (Galaxy).

2.1.6 Transformations of the Systems

Let us assume that at a certain time we observe an object at azimuth angle a and height h. How can one obtain from this the coordinates that lead to a cataloguing of the object? Or vice versa: one has given the coordinates of an object in a catalogue and wants to know in which cardinal direction and altitude the object can be found, or when it rises or sets.

Fig. 2.5 Spherical triangle
defined on a spherical surface.
Note that all quantities are
angles

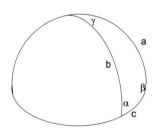

Fig. 2.6 Astronomical
spherical triangle

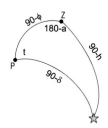

For this we need *Coordinate Transformations*. To derive the transformation formulas,
we use spherical triangles defined on a spherical surface (Fig. 2.5). In such a triangle the
sine and the *cosine theorem apply:*

$$\frac{\sin a}{\sin \alpha} = \frac{\sin b}{\sin \beta} = \frac{\sin c}{\sin \gamma} \qquad (2.2)$$

$$\cos a = \cos b \cos c + \sin b \sin c \cos \alpha \qquad (2.3)$$

$$\cos b = \cos a \cos c + \sin a \sin c \cos \beta \qquad (2.4)$$

$$\cos c = \cos a \cos b + \sin a \sin b \cos \gamma \qquad (2.5)$$

The so-called *astronomical triangle* (Fig. 2.6) is formed by superposition of the horizon
system with the equator system; its ends are defined by pole P, zenith Z and object; the
sides are $90° - \delta$, $90° - \Phi$, $z = 90° - h$. Furthermore one knows two angles t and $180° - a$
(Fig. 2.6).
   Using the formulas of spherical trigonometry, we find:

$$\sin z \sin a = \cos \delta \sin t \qquad (2.6)$$

$$\cos z = \sin \phi \sin \delta + \cos \Phi \cos \delta \cos t \qquad (2.7)$$

$$- \sin z \cos a = \cos \Phi \sin \delta - \sin \Phi \cos \delta \cos t \qquad (2.8)$$

respectively:

$$\cos\delta\sin t = \sin z \sin a \tag{2.9}$$

$$\sin\delta = \sin\Phi\cos z - \cos\Phi\sin z\cos a \tag{2.10}$$

$$\cos\delta\cos t = \cos\Phi\cos z + \sin\Phi\sin z\cos a \tag{2.11}$$

2.2   The Time

After the orientation in the sky, we now turn to the concept of time. How can we define time on the basis of astronomical events? When is the moon again at the same place in the sky after one revolution and why is it not the same moon phase? What problems arise when creating a calendar? How did the ancient Egyptians calculate the time of the flooding of the Nile and where does our calendar come from?

2.2.1   Definitions, Solar Time and Sidereal Time

There are several different concepts of time:

- *Sidereal time:* Hour angle of the vernal equinox.
- *Sidereal day:* Time between two meridian passes of the vernal equinox. The sidereal time is not suitable for everyday purposes because the sun moves and $\Theta = 0^h$ occurs at different times of the day.
- *True solar time:* Hour Angle of the true sun + 12^h (thus beginning of day at night). However, this is an uneven measure of time because the speed of the Sun changes (in northern hemisphere winter the Earth passes through the closest point to the Sun in its orbit and therefore moves faster) and the Sun travels along the ecliptic rather than the equator.
- *Mean solar time:* One defines a fictitious sun moving uniformly along the equator. The *Equation of time* is the difference between true time and mean time. There are two periods in the equation of time:
  - year-round period: occurs because of the changing speed of the Earth (in the northern hemisphere, the summer half-year lasts 186 days, while the winter half-year lasts only 179 days);
  - half-year period because of the projection effect on the equator.

    The maxima in the equation of time are on 14th May (+3.7 min) resp. 4th November (+16.4 min). The minima are on 12th February (−14.3 min) and on 26th July (−6.4 min).

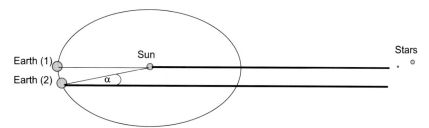

Fig. 2.7 Difference between synodic and sidereal rotation. After a sidereal rotation (Earth (1)→ Earth (2)) a star is again in the meridian, but the Sun is angularly α away

> Due to the orbit of the Earth around the Sun in a year, the Sun apparently moves around the entire sky in a year (360°), therefore per day about 1° eastwards, which corresponds to about 4 min of time.

In the rotation of the Earth, a distinction is made (see Fig. 2.7) between

- *synodic* rotation of the Earth, after a synodic rotation the Sun is again at the same position in the sky, and
- *sidereal* Rotation of the Earth, after a sidereal rotation a star is again at the same position in the sky.

This results in a difference of 4 min per day (in Fig. 2.7 by the angle α), which is 2 h within 1 month. If a star rises at the beginning of the month at 10 p.m., it will rise at the end of the month at 8 p.m. As the earth moves around the sun, after a sidereal rotation (from position 1 to position 2 in Fig. 2.7) the sun rotates by $\sim 360°/365 \sim 1° \sim 4^{\text{min}}$ and is not yet in the meridian. Thus the solar day is longer than the sidereal day.

24^{h} Solar time corresponds to $24^{\text{h}}3^{\text{min}}56.55^{\text{s}}$ sidereal time.

24^{h} Sidereal time corresponds to $23^{\text{h}}56^{\text{min}}4.09^{\text{s}}$ mean solar time.

Each longitude has its own solar time, and therefore the introduction of *Time Zones* is appropriate: In strips of about 15° length should be the civil Time equal.

> For example in central Europe, mean solar time = zone time for locations on 15° Longitude (l).

For $l < 15°$ the sun is later in the meridian, for $l > 15°$ earlier.

Table 2.2 Calculation of the sidereal time

$+14^{h}00^{min}$	Time in UT
$+ \sim 2^{min}$	Correction of 14 h to sidereal time
$4^{h}00^{min}$	Sidereal time at 0^{h} UT $\leftarrow$ Yearbook
$+1^{h}01^{min}47^{s}$	Conversion of geographic. Longitude to time
$19^{h}04^{min}$	Local sidereal time in Graz

Fig. 2.8 True noon calculated for Munich

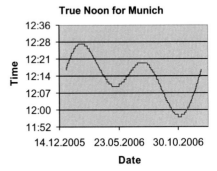

Examples:

- $l = 0°$ Greenwich = GMT (Western European Time). Mostly we use UTC= Universal time coordinated that corresponds to GMT.
- $l = 15°$ = CET (Central European Time).
- $l = 30°$ = OEZ (Eastern European Time).
- In the US four time zone are used: Eastern Time Zone (lUTC−5.00), Central Time Zone (UTC−6.00) Mountain Zone (UTC−7.00) Pacific Time Zone (UTC−8). During daylight saving e.g. the Eastern Standard Time (EST) changes zu EDT (Eastern Daylight time).

In Table 2.2 it is given how to calculate the sidereal time for Graz (length corresponds to $+1^{h}01^{min}47^{s}$) calculated for 15 o'clock CET (= 14.00 UT, Universal Time). In Fig. 2.8, the true noon for Munich is given.

For a place of longitude $l = 10°$ the meridian passage of the mean sun occurs at 20^{min} (because $1° \sim 4^{min}$) after true noon. In addition, there is the equation of time. This is e.g. on 12. February 14^{min} so the meridian passage there is only around $12^{h}34^{min}$ therefore relatively long afternoon.

Dateline: at $l = 180°$ If one travels to the east, one has 1 day twice, if one travels to the west, 1 day is skipped.

The *World Time* (Universal Time, UT) is the *Greenwich Mean Time*. It is usually given for astronomical events.

We now treat the unit of time *year*. Roughly speaking, a year is one revolution of the Earth around the Sun. However, there are different concepts of year:

- *Tropical year:* Orbit from vernal equinox to vernal equinox $365^{day}5^h48^{min}46^s = 365.24220^{day}$ mean solar time. **The tropical year determines the seasons, so a calendar should be adapted to it.** The beginning of the tropical year is calculated when for the mean sun $\alpha = 280°$ (annus fictus).
- *Sidereal year:* The time after which a star is again in the same position in the sky after one revolution of the Earth about the Sun. This is 20^{min} longer than the tropical year $365^{day}6^h9^{min}10^s = 365.25636^{day}$.
- *Anomalistic year:* Period from Perihelion (closest point of the earth's orbit to the sun) to next perihelion. The Earth's orbital ellipse is not fixed in space as a result of perturbations from the other planets, $365^{day}6^h13^{min}53^s = 365.25946^{day}$.

2.2.2 Calendar

With the settling down of the people, who practiced agriculture, a division of time, that is a *calendar* became important. The units of all calendars are essentially

- day (corresponds to one rotation of the Earth),
- month (roughly corresponds to one revolution of the Moon around the Earth) and
- year (corresponds to one revolution of the Earth around the Sun).

In principle a calendar should be adapted to the *tropical year*, since this determines the seasons.

In the old *Egypt* a pure *solar calendar* with a year of 365 days was used. The year had 12 months of 30 days, and then there were five extra days. Now, however, the actual solar year is about 1/4 day longer, and therefore the beginning of the year after 1461 years ($1461 = 365.25 \times 4$) once passed the the entire Egyptian year. The time of the *Nile floods* initially coincided with the heliacal rising of the *Sirius* (called Sothis in Egypt).[1] The Nile floods shifted by 1 day every 4 years, which became noticeable over time: The Nile floods no longer coincided with the heliacal rising of Sirius, because the length of the year was wrong. In 238 BC from Ptolemy III. in the *Decree of Canopus* the calendar was reformed:

The year has 365 days, and every 4 years there is an extra day, so 366 days.

In the year 48 BC *Julius Caesar* landed with his troops in Alexandria and in 46 BC he reformed the calender in Rome.

---

[1] This is understood to be the time at which Sirius after its conjunction with the sun can be seen again for the first time in the morning sky.

Julian Calendar

> The year has 365 days, and every 4 years there is a leap year of 366 days.

The length of the year is therefore $365^{\text{day}}6^{\text{h}}$ and is 11^{min} longer than the tropical year. This added up in the sixteenth century to a noticeable error of more than 10 days, and therefore Pope *Gregory XIII* instituted a reform, the Gregorian *calendar;* The 4th October 1582 was immediately followed by 15th October 1582.

Gregorian Calendar

> The year has 365 days, every 4 years is a leap year, but full centuries are only a leap year if they are divisible by 400 without remainder.

The year 2000 was a leap year, but the year 1900 was not according to the Gregorian calendar (it is divisible by four without remainder, but not by 400 without remainder). Thus, the error with respect to the tropical year is 1 day in 3300 years.

The length of the year in the Gregorian calendar is: $365+1/4-3/400 = 365.2425^{\text{day}} = 365^{\text{day}}5^{\text{h}}49^{\text{min}}12^{\text{s}}$.

Counting years was complicated at first, and we bring examples:

- After regents, "years after Diocletian".
- Ab urbe condita, a. u. c.:[2] after the foundation of Rome (753 BC according to the Varronian census. [mostly used] resp. 752 BC according to the Capitoline count); the year 1 a. u. c. = 753 BC. The death of Herod was 750 a. u. c. = 4 BC Note: 753 a .u. c.= 1 BC and 754 a. u. c. = 1 AD, there is no year 0.
- Olympiads: Beginning in 776 BC; 4 years constitute one Olympiad.
- Indiction: Roman numeral, a 15-year cycle derived from the division remainder (year +3)/15 (0 equals 15).
- Around 525 *Dionysius Exiguus* introduced the counting after Christ's birth (but made a mistake in fixing the year of Christ's birth). This counting was officially recognized by the popes only in 1491.

Note There is no year 0. The astronomical year 0 corresponds to the historical year 1 BC. Caesar died in 44 BC = −43, so his 2000-year death anniversary was 1957. In

---

[2] From the foundation of the city of Rome.

Table 2.3 Table for the calculation of the Easter date according to Gauss

J	m	n
1583–1699	22	2
1700–1799	23	3
1800–1899	23	4
1900–2099	24	5
2100–2199	24	6
2200–2299	25	0

the *Christian calendar*, the Christmas is always on December 25 and 26. At the *Council of Nicaea* 325 *Easter* was fixed on the first Sunday after the first full moon after the beginning of spring. This has historical reasons (Jesus was crucified on the Jewish Passover, but according to the Jewish calendar this always fell on the first full moon after the beginning of spring). The beginning of spring was fixed on 21 March. Simplified one can calculate the *Easter date* after the *Gauss' rule*. Let J be the year, m and n follow from Table 2.3. Further one calculates the year J divided by different values as given below and writes the remainder of the divisions as $a, b, \ldots$

$$J : 19 = a \qquad J : 4 = b \qquad J : 7 = c$$

$$(19a + m) : 30 = d \qquad (2b + 4c + 6d + n) : 7 = e$$

(always the remainders of the divisions).

The Easter Sunday then falls on the

$(22 + d + e)$. March resp. $(d + e - 9)$. April.

Borderline cases: instead of April 26 always April 19. instead of 25. April the 18. April $(d = 28, e = 6, a > 10)$.

The *seven-day week* is found for the first time in the *Jewish calendar*. The days were numbered except for the Sabbath (Saturday). The Jewish calendar is aligned only with the Moon.[3] Since the synodic revolution of the Moon (from full Moon to full Moon) is 29.5 days, the months are alternately 29 and 30 days long. To bring the Lunar calendar with the course of the sun, there is a common year with 12 months and a leap year with 13 months. There is the abbreviated common year of 353 days, the supernumerary common year of 355 days, the ordinary common year of 365 days, an abbreviated leap year of 383 days, an ordinary leap year of 384 days, and a supernumerary leap year of 385 days. The Jewish months are called Tishri, Marcheshan, Kisslev, Tevet, Shevat, Adar, Adar II (in the leap year), Nissan, Ijar, Siwan, Tammus, Av and Elul. Beginning of the Jewish world era is the year 3761 BC (creation of the world).

The months in the *Islamic calendar* begin with the first visibility of the crescent moon in the evening sky. 12 months are combined into a lunar year: Moharrem (30 days),

---

[3] This goes back to the Old Testament: "You made the moon to divide the year".

Safar (29 days), Rebî-el-awwel (30 days), Rebî-el-accher (29 days), Deschemâdi-el-awwel (30 days), Deschemâdi-el-accher (29 days), Redshab (30 days), Shabân (29 days), Ramadân (30 days), Shevwâl (29 days), Dsû'l-kade (30 days), Dsû'l-hedsche (29 days), leap year (30 days). The leap day was necessary to keep the beginnings of the months in line with the illumination of the crescent moon. Within an annual period of 30 years there are eleven leap years. The yearly counting takes place from the *Hedschra* (flight of the Prophet Muhammad from Mecca to Medina, 622 AD).

The old *Maya* possessed a cultic calendar *(Tzolkin)* with 260 days. Further there was however also the year *(Haab)* of 365 days. This was divided into 18 months of 20 days each. The last month had only 5 days. Beyond that there were interesting counts:

1 Kin = 1 day,
1 Uinal = 20 Kin = 20 days,
1 Tun = 18 Uinal = 360 days,
1 Katun = 20 Tun = 7200 days,
1 Baktun = 20 Katun = 144,000 days,

The starting point of the count is 11.8.114 BC. In the Mayan calendar this had the date 13.0.0.0 (13 Baktun, 0 Katun, 0 Tun,...). This date repeats every 1,872,000 days, thus on 21 December 2012, and according to some esoteric circles the world should end there once again.

In 1790, during construction work, the famous calendar stone of the *Aztecs* was found (now in the National Museum in Mexico City). The diameter is 4 m and the thickness is about 1 m. In the middle is the sun god, and then follows a ring, which is divided into 20 equal sections. The symbols are similar to the Mayan Tzolkin, except that here the 260-day period is called "tonalpohualli".

Also the *Chinese* knew a year with 12 months, with 29 or 30 days. The Chinese peasant calendar is called *nongli*, the civil calendar *gongli*. The day begins at midnight. The apparent path of the sun in the sky, *ecliptic*, was also known and called the yellow orbit. This was divided into 12 signs (similar to our signs of the zodiac): mouse, cow, tiger, rabbit, dragon, snake, horse, sheep, monkey, cock, dog, pig.

To facilitate chronological calculations and for observations and ephemerides, J. *Scaliger* (1592) introduced the *Julian day count* (JD). The Julian day begins at 12^h UT, and the beginning of the Julian day 0 was fixed at 1.1.4713 BC. On 01.01.2002 begins at 12^h UT the Julian day 2,452,276 begins. In the year 1975 the modified MJD Julian day count was introduced (MJD):

$$MJD = JD - 2,400,000 \tag{2.12}$$

Is a date given as year y, day d, month m, then the Julian date can be calculated using the following formula:

$$JD = (1461(y + 4800 + (m - 14)/12))/4 + (367(m - 2 - 12((m - 14)/12)))/12 - (3((y + 4900 + (m - 14)/12)/100))/4 + d - 32{,}075$$

2.2.3 The Star of Bethlehem

At this point we would like to briefly discuss the problematic nature of the Bible account of the Star of Bethlehem. The interpretation is very difficult and depends on the interpretation of the Bible passage Matthew 2:1–12 and 16. There we are told of the appearance of a bright star that stood still over the manger with the infant Jesus. Another clue is offered by the infanticide ordered by King Herod. Jupiter was considered the royal star of the Roman occupiers, Saturn the protector of the Jews.

The interpretation of the Star of Bethlehem is based on two conjunctions:

- Year 7/6 BC: Triple conjunction of the planets Jupiter and Saturn in the constellation Pisces.
- Year 3/2 BC: Conjunctions between Jupiter and Regulus and between Jupiter and Venus.

Possibly the three wise men from the East were astrologers from Babylon who interpreted the triple conjunction of the planet of the East with the planet of the West in the significant constellation of Pisces as a clear indication of the birth of the Saviour.

Nevertheless, there is a problem with the dating: Herod is said to have died in 4 BCE, and a lunar eclipse was observed shortly before. So the infanticide had to have occurred in the years before 4 BCE. Another lunar eclipse that would fit in would have been in 1 BCE. Then the star of Bethlehem would be a conjunction between Jupiter, Venus and Regulus.

A comet is out of the question, since comets were always considered to bring bad luck. For the year 5 BC, however, various historical sources report a comet visible for 70 days.

2.3 Star Positions

Why do we need exact star positions? From the analysis of the star positions we get important clues about the motion of the Earth and the solar system and we can determine the distance of nearby stars with the parallax method.

Fig. 2.9 Symbols of the signs
of the zodiac

♈ Aries	♌ Leo	♐ Sagittarius
♉ Taurus	♍ Virgo	♑ Capricornus
♊ Gemini	♎ Libra	♒ Aquarius
♋ Cancer	♏ Scorpio	♓ Pisces

2.3.1 Constellations and Zodiac

The night sky is divided into 88 constellations. The constellations of the northern sky are usually named after Greek mythology, while those of the southern sky are named by the seafarers. In astronomy, the Latin designation of the constellation is used and abbreviated with three letters. The 12 constellations of the *zodiac*, which lie along the ecliptic are: Aries (Ari), Taurus (Tau), Gemini (Gem), Cancer (Cnc), Leo (Leo), Virgo (Vir), Libra (Lib), Scorpius (Sco), Sagittarius (Sag), Capricornus (Cap), Aquarius (Aqr), Pisces (Pis). Zodiac symbols are shown in Fig. 2.9.

Star designation is done for bright stars with names (mostly of Arabic origin), e.g., Vega.[4] *Johannes Bayer* introduced the following designation in 1603: Greek letter (α for the brightest star, β for the second brightest in a constellation etc.) + constellation: The star α Aqr is the brightest star in the constellation Aquarius. Faint stars are given a number in a catalog.

The determination of the star positions is done by a telescope placed in the meridian (*meridian telescope*). The star's zenith distance is measured at the time of its meridian passage. Errors arise from the set-up (inclination error, axis not horizontal, azimuth error, axis not east-west) and from the instrument itself (collimation error if the optical axis is not perpendicular to the axis of rotation, division error of the circle for reading, eccentricity error if the center of rotation is not at the center, and zenith error if the zero points of the circles are wrong). For the *Fundamental stars* the coordinates are known exactly, and one has then:

On photographic plates and CCD images, the rectangular coordinates x, y of an unknown object (e.g. a newly discovered comet) can be transformed into spherical coordinates if there are at least three reference stars in the vicinity of the searched object whose positions must be known. Such determined star positions are called also relative locations. With the absolute locations one must consider the pole height as well as the obliquity of the ecliptic and the error of the clock. Further effects result from refraction in the earth's atmosphere ($\rightarrow$ refraction) as well as by the movement of the earth around the sun ($\rightarrow$ aberration) and the gyroscopic motion of the earth's axis ($\rightarrow$ precession).

---

[4] Means swooping eagle: an-nasr al-wa-qi.

Fig. 2.10 Snellius' *law of refraction*. A ray of light is refracted when passing from a medium with the refractive index n_1 into a medium with the refractive index n_2 where $n_2 > n_1$ is, refracted to the perpendicular

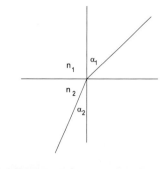

Fig. 2.11 Ray refraction in the Earth's atmosphere. It can be seen that the true position of a celestial object appears to be elevated especially near the horizon

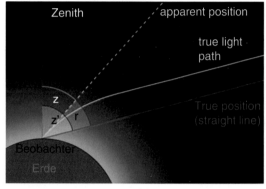

2.3.2 Refraction

The astronomical refraction is the refraction of rays in the Earth's atmosphere. The *Snellius' law* of refraction (Fig. 2.10) states:

$$\frac{\sin \alpha_1}{\sin \alpha_2} = \frac{n_2}{n_1} \tag{2.13}$$

$$n_i \sin \alpha_i = \text{const} \tag{2.14}$$

One thinks of the earth's atmosphere as consisting of many plane-parallel layers i are built up, for which the refractive index n_i amounts to.

Let z_b the observed zenith distance and R is the refraction (Fig. 2.11), then the following holds for the *true zenith distance* z (the stars appear lifted by the refraction):

$$z = z_b + R \qquad \sin(z_b + R) = n \sin z_b \tag{2.15}$$

Using the addition theorem and considering the small values for R holds:

$$\sin z_b \underbrace{\cos R}_{1} + \cos z_b \underbrace{\sin R}_{R} = n \sin z_b$$

From this follows:

$$R = (n - 1) \tan z_b \tag{2.16}$$

or as a rule of thumb: $R['] = \tan z_b$, when reckoned in radians. For a zenith distance of $10° R = 0.18'$, at a zenith distance $z = 90°$ (star on the horizon) already $R = 35'$ which corresponds approximately to the apparent diameter of the sun or moon.

> Therefore, when the sun touches the horizon, it is actually below the horizon.

The refraction R increases with air density (decreasing temperature and increasing pressure).

Scintillation is understood to mean rapid refraction changes due to turbulence in the Earth's atmosphere. There are the *directional* scintillation (trembling of a star) and the *brightness scintillation* (blinking of stars). Both effects are summarized under the term *Seeing*. One speaks of good observation conditions with a seeing below $1''$. The refractive index increases with decreasing wavelength (blue light is refracted more strongly than red light). The image of a star thus becomes a small spectrum: blue to the zenith, red to the horizon.

For $z = 70°$ you have: $\Delta R = R_{\text{blau}} - R_{\text{rot}} = 3.5''$.

To keep disturbances from the Earth's atmosphere small, modern observatories are usually built on high mountains in climatically favorable locations (Hawaii, Canary Islands, Chile, ...).

Nowadays observatories are built only after an extensive *site testing*. One looks for places with ideal observing conditions, stable air stratification, far away from terrestrial light sources (light pollution).

2.3.3 Aberration

Aberration is caused by the movement of the observer and the finite speed of the incoming light beam. The light then appears to come from a different direction.

Fig. 2.12 Aberration of
starlight

Analogy: If one walks through rain, one must hold the umbrella at an angle to the front.
If v is the velocity of the observer and c is the speed of light of the incoming beam
(Fig. 2.12) and $v \ll c$ and $v \perp c$ then:

$$\tan \alpha \sim \alpha = \frac{v}{c} \tag{2.17}$$

If v is not small vs. c, then holds:

$$\tan \alpha = \frac{v/c}{\sqrt{1 - (v/c)^2}} \tag{2.18}$$

Example: $v = 0.9\,c \rightarrow \alpha = 64°$.

The *annual aberration* arises by the motion of the earth around the sun. The light of a
star arrives with the speed of light $c = 300,000\,\text{km/s}$ the motion of the earth around the
sun occurs at $v = 30\,\text{km/s}$ therefore the ratio is $v/c = 10^{-4}$. If γ the angle between v and
c, then holds:

$$\alpha = \frac{10^{-4}}{\sin 1''} \sin \gamma = k \sin \gamma \qquad k = 20.50'' \tag{2.19}$$

k is the constant of the annual aberration. Therefore:

- Stars describe annual ellipses with major semi-axis k and minor semi-axis $k \sin \beta$ (β—
  ecliptic latitude of the star).
- At the pole of the ecliptic the stars describe a circle with radius k.
- For stars in the ecliptic plane the result is a straight line with length $2k$.

The *daily aberration* results from the Earth rotation.

$$v = v_{\text{eq}} \cos \Phi \qquad v_{\text{eq}} = 465\,\text{m/s} \tag{2.20}$$

Fig. 2.13 Daily parallax of a
star (only directly measurable
for the moon)

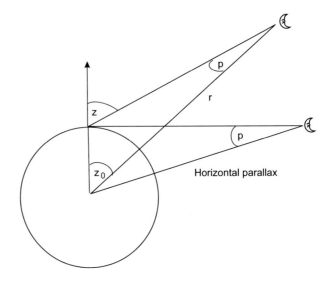

v_{eq} is the rotational speed of the earth at the equator. The aberration constant here is
$0.32''\cos\Phi$. So the culmination of a star (passing its highest point in the meridian) is
shifted by $0.021^s\cos\Phi\sec\delta$ the star passes too late through the meridian.

Due to the rotation of the Sun around the center of the Milky Way in 200 million years
there is also a *secular aberration* At the pole of the Milky Way the stars describe a circle
of about $2'$ Diameter in 200 million years.

2.3.4 Parallax

If one changes the observation place, the position of a nearer object shifts in relation to a
more distant one. Again, a distinction is made:

Daily parallax (Fig. 2.13): z is the zenith distance of a star at the observation site, z_o
the zenith distance at the center of the Earth; then (because of $a/\sin p = r/\sin(180 -
z)$; $\sin(180 - z) = \sin z$):

$$\sin p = \frac{a}{r}\sin z \tag{2.21}$$

a ... Earth's radius, r distance. The *Horizontal Parallax* applies to $z = 90°$. This is the
angle at which the Earth's radius appears when viewed from an object (Sun, Moon,...).
The horizontal equatorial parallax P is given by $\sin P = a_{eq}/r$. For the moon one has
$P = 57'$, i.e. almost two full moon sizes, for the sun $P = 8.79''$ so 1/400 of the moon.
Consequently, the sun is 400 times further away from us than the moon.

The *solar parallax* is not directly observable, because the sun is in the daytime sky; one
determines the solar distance by trigonometric distance determination of a minor planet,

Fig. 2.14 Annual parallax of a star relative to more distant stars. A nearby star moves through the Earth's annual orbit around the Sun; the magnitude of the parallax π depends on the distance of the star d ab

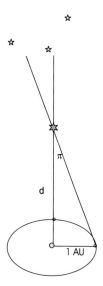

which comes much closer to the earth than the sun (e.g. Eros), and then apply Kepler's third law. For fixed stars, the diurnal parallax cannot be measured.

The *annual parallax* (Fig. 2.14) arises as a result of the Earth's motion around the Sun.

$$\sin \pi = \frac{a}{d} \sin \gamma \qquad (2.22)$$

a ... Earth's orbital radius, d ... distance of the star.

> The mean distance earth–sun is 150×10^6 km and is called the astronomical unit, AU.

The angle γ lies between the Earth and the star as seen from the Sun. If $\gamma = 90°$, the maximum value = angle at which the Earth's orbital radius appears from the star. The stars describe ellipses with the semi-axes π, $\pi \sin \beta$ where β is the ecliptic latitude of the star. The parallax $\pi = 1''$ corresponds to the distance from 206,265 Earth orbital radii = 30×10^{12} km = 3.26 Ly.[5] This distance serves as the unit of distance in astronomy and is called *Parsec*. Further: 1 kpc = 10^3 pc, 1 Mpc = 10^6 pc.

---

[5] 1 Ly = light year (Ly), is often given as a unit of distance in popular literature. It is the distance light travels within one year: 300,000 km/s $\times$ 86,400 s $\times$ 365.25 = 10 trillion km.

Fig. 2.15 The stars describe
an ellipse in the sky during the
year; if a star is in the ecliptic,
it describes a straight line, at
the pole of the ecliptic a circle

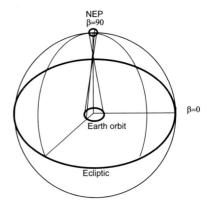

Stellar distances are given in units of parsecs. 1 pc $= 3.26$ lightyears.

$$\pi'' = \frac{1}{d[pc]} \tag{2.23}$$

The largest parallax (nearest fixed star) has been measured at Proxima Centauri: $\pi =$
$0.762'' \rightarrow d = 1.3\,\mathrm{pc} = 4.26\,\mathrm{Ly}$.

The aberration ellipses can be easily separated from the parallax ellipses (see
Fig. 2.15):

- They are around $90°$ out of phase with each other.
- The aberration ellipses are independent of the stellar distance.

The *secular parallax* arises due to the motion of the sun.

2.3.5 Precession, Nutation

Earth can be seen as a rotating *gyroscope*, which is exposed to the forces of the Sun and the
Moon. As a result, both the Earth's polar axis and the Earth's orbit itself (ecliptic) change
(Fig. 2.16).

A distinction is made between:

Lunisolar Precession By the attraction of the Moon and Sun to the equatorial bulge of
the Earth, a torque is produced which tries to straighten the Earth's axis, and the latter
responds by gyrating.

Fig. 2.16 Precession

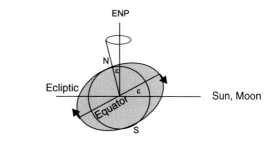

Fig. 2.17 Lunisolar and
planetary precession

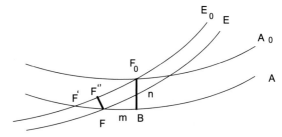

The Earth's axis describes a cone with 23.5° Opening around the pole of the ecliptic. The vernal equinox (which is one of the points of intersection between the ecliptic and the equator) therefore moves backwards (lat. praecedere, to go ahead) by the amount of $P_o = 50.3878''$ per year. This is the 20 min difference between the sidereal and the tropical year, resulting in an orbital period of 25,700 years. *(Platonic year)*. So in about 12,000 years. Vega become the Pole star (see Fig. 2.18). Since the discovery of precession by *Hipparchus* (130 BC) the vernal equinox has moved by more as a constellation further on. At that time it was located in the constellation Aries (hence the Aries sign for the vernal equinox), today it is located in the border region Pisces/Aquarius.

> Precession gives rise to the difference between the constellations and the signs of the zodiac (with which astrologers reckon).

Planetary Precession The influence of the planets causes the Earth's orbit to shift (Fig. 2.17), i.e. the pole of the ecliptic, this causes an additional shift of the vernal equinox by $P_{pl} = 0.1055''$ per year. Be A_0, E_0, F_0 equator, ecliptic resp. vernal point at the beginning of a year and A, E, F at the end of the year, then $P_o = F_o F'$ the lunisolar precession and $P_{pl} = F'F$ the precession by the planets. The effect of precession (per year) on the position of the stars follows from:

$$n = 20.0431''$$
(2.24)

$$m = 46.1244'' = 3.0750^s$$
(2.25)

Fig. 2.18 The precession of
the Earth's axis changes the
position of the celestial pole,
presently close to Polaris.
According to Tau Olunga

$$\Delta\alpha = m + n \tan\delta \sin\alpha \tag{2.26}$$

$$\Delta\delta = n \cos\alpha \tag{2.27}$$

Therefore, when specifying stellar coordinates, the specification of the *Equinox* is necessary. All these quantities are also time-dependent.

Nutation The Moon's orbit is by 5° with respect to the ecliptic plane. The sun exerts a torque on it, which tries to straighten the moon's orbit, and again a gyroscopic motion occurs, in which the points of intersection of the moon's orbit with the ecliptic plane (nodes) shift with a period of 18.6 years (Fig. 2.18). The displacement of the true pole therefore describes a nutation ellipse with semi-axes 9.21″ resp. 6.86″.

2.3.6 Star Catalogues

Now how are the positions of stars indicated?

- *Apparent position:* Observed Location, corrected for instrument error, refraction, and daily aberration.
- *true position:* apparent Location corrected for annual aberration and parallax.
- *Mean position:* true Location corrected for precession, nutation relative to an equinox...

In the catalogues you can always find the mean positions. Some catalogues are given with the years of the equinoxes. *Fundamental catalogs* (FK4, FK5) contain about 1500 stars with exact mean positions. AGK1...3: Catalogues of the Astronomical Society (AGK 1 equinox 1875, 200,000 stars; AGK 3 equinox 1950). By comparing the catalogues, the proper motions of the stars are obtained. The surveys contain many stars with lower precision (BD, *Bonn Survey*, 1855, 458,000 stars, *HD Henry Draper Catalogue*, 1900.0, 223,000 stars; SAO Smithsonian Astrophys. Obs., 25,000 stars), *POSS* (Palomar Sky Survey, photographic atlas, 879 fields taken with the Mt. Palomar telescope), *HIPPARCHOS* (catalogue for the HIPPARCHOS satellite, contains about 118,000 stars). The *Tycho catalog* contains about 1 million stars, but with lower precision than the HIPPARCHOS data. The *GSC* (Guide Star Catalogue) was made for the Hubble Space Telescope and contains data of stars from 9537 zones (each about 7.5°). It contains astrometric and photometric data of about 15 million stars (vers. 1.1, 1989).

We still want to cite here the GAIA (global astrometric interferometer for astrophysics) satellite mission, a successor to the HIPPARCOS mission. The astrometric accuracy is 200 times greater than HIPPARCOS, and 10,000 times more objects are studied. In total, about 1 billion objects, or about 1% of all the stars in the Galaxy, will be surveyed with micro arc second precision. The launch of this European Space Agency (ESA) probe took place on December 19, 2018, in French Guiana. The probe is located at the 1.5 million km Lagrange point L_2 (Fig. 2.19) where gravitational forces cancel and the probe can observe the sky undisturbed. In particular, one wanted to avoid temperature fluctuations, which would have affected the measurement accuracy.

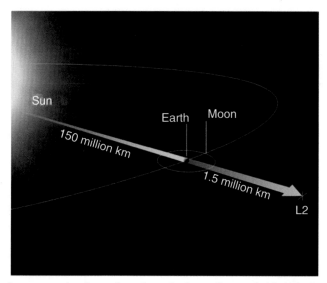

Fig. 2.19 The Lagrange point L_2, at four times the lunar distance behind Earth, where GAIA is located. NASA/ESA

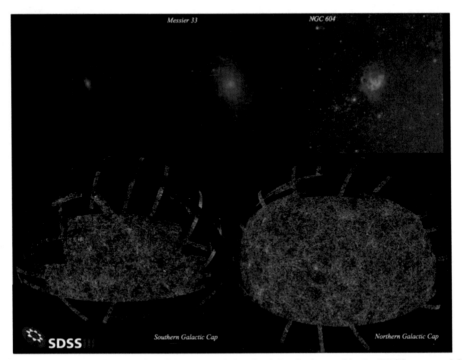

Fig. 2.20 Examples from the SDSS images and sketch of the sky imaged in the SDSS surveys. M. Blanton and the SDSS-III collaboration

A problem with such missions is the huge amount of data. It is expected that about 200,000 DVDs will be needed to store the data. The GAIA archive is accessible at **gea.esac.esa.int.** In addition to position data, one can also find velocity measurements of stars and many other objects. The GAIA data release 2 DR2 took place in April 2018.

A very widespread catalog today is the *SLOAN Digital Sky Survey* (SDSS). There are two versions: SDSS-I, 2000–2005; SDSS-II, 2005–2008, and SDSS-III has been in progress since 2008. The 2.5 m Apache Point telescope is used to compile the catalogue. The data set includes positions of about 500 million stars and spectra of nearly 2 million stars. In addition, more than 90, 000 galaxies and more than 100,000 quasars are found. In Fig. 2.20 shows an example.

The M-number stands for number in the *Messier catalog*, a catalog published by *Ch. Messier* in the years between 1764 and 1782 of more than 100 bright objects.

2.3.7 Light Deflection and Exoplanets

We want give an example of the application of very precise measurements of stellar positions; this is the deflection of light in the gravitational field of a mass *M*. The properties of space and time around a mass are described by the so-called *Schwarzschild metric* according to general relativity (Sect. 15.3). From this one can measure the deflection of light in a gravitational field derive:

$$\phi = 4\frac{GM}{R_0c^2}\ \text{rad} \qquad\qquad (2.28)$$

Here the light ray passes a mass M at distance R_0 from its center. If we insert the mass of the sun and the radius of the sun into the formula, we get $\phi = 1.7''$. By this amount the light ray of a star is deflected at the edge of the sun.

One discusses how to measure this in practice!

With this method, one can use so-called *Microlensing* finding planets outside our solar system *(exoplanets)*. So it is also possible to find planets of earth size in a distance of several kpc. An object in the foreground (acting as a lens) amplifies the brightness of a distant object (source). Amplification occurs when the source moves within the Einstein radius of the lens (Fig. 2.21). If D_L the distance between observer and lens, D_S the distance between observer and source, D_{LS} denote the distance between lens and source, then the

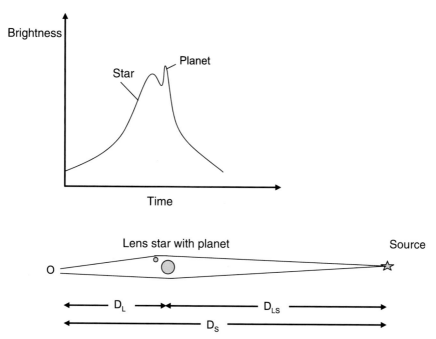

Fig. 2.21 Principle of microlensing. A source passes behind a lens (formed by the star) and is amplified by it. An additional effect occurs when a massive planet is in a suitable position (see light curve above)

Einstein radius applies:

$$\theta_E = \sqrt{\frac{4GM}{c^2}\frac{D_{LS}}{D_L D_S}}$$ (2.29)

If the lens now possesses a planet, then briefly the brightness can increase again. There are observing programs to monitor many stars after such microlensing events.

2.4   Determination of Time and Place

The determination of place and time used to be a very important practical application of astronomy. We will only briefly explain the principles here to illustrate how the location on earth (latitude and longitude) is determined from astronomical observations. Finally we will go into the modern navigation systems.

2.4.1   Latitude Φ

The latitude follows from the altitude of the pole or when the star is in the meridian: for stars south of the zenith:

$$\delta_1 + z_1 = \Phi$$

for stars north of the zenith:

$$\delta_2 - z_2 = \Phi$$

Further there is the method By *Horrobow-Talcott:* One observes two stars passing through the zenith one after the other with approximately the same zenith distance z through the zenith, one to the north, the other to the south:

$$\Phi = \frac{1}{2}(\delta_1 + \delta_2) + \frac{1}{2}(z_1 - z_2)$$

If the zenith distances are very similar, effects of refraction fall out. In 1888 *Küstner* determined with this method *Pole-height fluctuations.* These have a 1-year period (seasonal shifts due to snow load, etc.) and the so-called *Chandler period* of 14 months.

2.4.2   Time Determination

Time determination is done with the *Meridian circle*; at the meridian passage of a star its hour angle $t = 0$

$$\Theta = \alpha$$

Θ is the local sidereal time and α the right ascension of the star.
   The hour angle follows from:

$$\cos z = \sin \Phi \sin \delta + \cos \Phi \cos \delta \cos t$$

and the local sidereal time with knowledge of the right ascension α:

$$\Theta = t + \alpha$$

The oldest timekeepers were sundials, water clocks, and hourglasses. The Pendulum clock was invented around 1656 by *Huygens* the deviation was about 10 s per day . The absolute error of modern pendulum clocks is 3 s per year (3 s/a). Spring clocks exploit spring oscillations (Huygens 1675). *Quartz clocks* use periodic natural oscillations (piezoelectric effect, frequency 10^4 s^{-1}) of quartz crystals, which are controlled by an electric oscillating circuit. The absolute error of a very precise Quartz watch is 0.001 s/a. However, the oscillation of a quartz crystal is not constant over long periods of time. Even more accurate clocks are *Atomic clocks* based on the invariable natural oscillations of molecules. In the cesium atomic clock, parallel beams of Cs atoms emerge from a nozzle. They enter an electric field and are separated into high-energy and low-energy beams. The high-energy ones are deflected into a cavity resonator, where the Cs atoms are excited to frequency-stable natural oscillations. With quartz clocks and atomic clocks, time can be measured more accurately than with the *Earth's rotation*, which is variable. They have been used to detect irregular, periodic and secular changes in the Earth's rotation.

- Irregular fluctuations: due to mass shifts in the Earth's interior. Also strong earthquakes like the catastrophe of Fukoshima in 2011 have had an impact: This shifted the position of the Earth's axis by about 10 cm and shortened the Earth's rotation time by more than a microsecond.
- Periodic fluctuations: meteorological (year-round period, amplitude 22 ms), semi-annual period due to tidal forces of the sun; amplitude 10 ms.
- 13.8- and 27.6-day period with very small amplitude due to tidal forces of the Moon.

In May the Earth lags by 30 ms, in September it advances by 25 ms. Finally, there are the secular changes. These are mainly caused by tidal friction (chapter Celestial Mechanics).

All effects cause an increase in day length of 1 s in 60,000 years. These changes in the Earth's rotation make it necessary to define a constant measure of time.

One has the Ephemeris Time *Ephemeris Time, ET*, introduced as the length of the astronomical second at the beginning of the twentieth century. $1\,s = 1/31556925.9746$ of the tropical year. The difference between UT and ET for 1900 was 0. For 2006.0 it was already 64 s. TAI is the international atomic time.

Definition of second: $1\,s = 9,192,631,770$ oscillations of the ^{133}Cs during the transition to the ground state.

2.4.3 Modern Navigation Systems

The *Global Positioning System (GPS)* is a satellite-based navigation system, originally developed by the US Department of Defense for military applications and in operation since 1995. In this system, 24 satellites constantly emit signals, and from their transit time the GPS receiver can determine the position and altitude of its location. At a frequency of 1575.42 MHz, the C/A code is transmitted (*coarse acquisition*, for civilian applications). For more precise military applications there is the P/A code *(precision encrypted)*, which is encrypted. The measurement accuracy is better than 10 m in 90% of all cases.

By the end of 2019, the European satellite navigation system *GALILEO* became ready (for civilian use only). It uses 30 satellites, three of which are in reserve. Russia uses *GLONASS*.[6] China plans its own system (formerly known as *COMPASS*) designated *BeiDou*, BDS (Big Bear). The first of more than 35 satellites was launched into orbit in 2000. orbit brought (Fig. 2.22).

As we will see later, the accuracy of GPS navigation is determined by several factors, including the activity of the Sun (space weather). Therefore, one would like to predict the activity of the Sun as accurately as possible; here again arises a practical application of astrophysics, Space Weather an important branch of Solar Physics.

2.5   Further Literature

We give a small selection of recommended further literature.
Spherical Astronomy, R.M. Green, Cambridge Univ. Press, 1985
Practical Astronomy with your calculator, P. D. Smith, Cambridge Univ. Press, 1989
Orientation, navigation and time determination, G. Wolfschmidt, tredition, 2019

---

[6] Globalnaya Navigazionnaya Sputnikovaya Sistema.

Fig. 2.22 Launch of two Bei Dou satellites, Xichang Space Center, China, 2018. Sirf News

Astronomical navigation, W.F. Schmidt, Springer, 1996
Astronavigation, B. Schenk, Delius Klasing, 2018

Tasks

2.1 At which place on Earth are all the stars ever 12^h above or below the horizon?

Solution
Equator

2.2 At what hour angle is a star in the meridian?

Solution
At $t = 0$ and at $t = 12^h$.

2.3 (a) when is the sun at the vernal equinox and (b) when is sidereal time equal to civil time to a good approximation?

Solution
(a) at the beginning of spring, (b) at the beginning of autumn

2.4 The right ascension of Mars on March 21, 2012 is about 10^h40^{min}. When is the planet exactly in the south? Can you see the planet in the evening sky?

Solution

Since the right ascension of the Sun at the beginning of spring is 0^h Mars passes through the meridian about 10 h 40 min later than the Sun, that is, at 10:40 p.m. It is therefore easy to observe in the night sky.

2.5 When is the sun in the meridian for Munich on 11/4 (so when exactly is noon)?

Solution

We consider the longitude of Munich $= 11.5°$ east. This results in a difference of $3.5°$ compared to the $15°$ MEZ zone, and this corresponds to: $1° = 4$ min, i.e. the sun passes through the meridian 14 min later, i.e. at 12 o'clock 14 min. In addition, there is the equation of time, ZG $= 16.4$ min (true time—mean time $=$ ZG, mean time $=$ true time— ZG), i.e. true noon in Munich occurs at 11 o'clock 57 min (Fig. 2.8).

2.6 At what hour angle is a star located in the meridian?

Solution

At $t = 0$ and at $t = 12^h$.

2.7 When is a star of the right ascension α in the meridian?

Solution

If $\Theta = \alpha$ is.

2.8 Assume that the sun is in the vernal equinox. How many hours is it visible in Germany?

Solution

Exactly 12 h.

2.9 What is the maximum height of the sun in Graz at the beginning of summer?

Solution

The latitude of Graz is $\Phi = 47°$ so the altitude of the celestial equator is equal to $90° - \Phi$ and the maximum altitude of the sun is : $h_{\max} = 90° - \Phi + 23.5°$.

2.10 At what latitude is the sun at its zenith in the meridian at the beginning of summer?

Solution

At the tropic ($\Phi = 23.5°$).

2.11 Calculate the local sidereal time for a given date at 3 pm CET for Graz. The longitude of Graz is slightly more than 15°, so approx. 1^h difference compared to Greenwich. From a table one takes the sidereal time for Greenwich at 0^h UT (Universal Time) e.g. : 4^h.

Solution
The solution is given in Table 2.2.

2.12 Calculate the date of Easter for 1818.

Solution
22.3 (extreme value).

2.13 A star has the parallax of $0.1''$, one calculates the distance.

Solution
10 pc.

History of Astronomy

3

In this section we give an overview of the history of astronomy.

> Observation of the sky existed among all civilized peoples, and it certainly had practical purposes: calendars and navigation.

In addition, burial sites were aligned or buildings erected according to certain cardinal points. The first beginnings of the scientific method of measurement and experimentation can be found among the ancient Greeks.

3.1 Astronomy of Pre- and Early History

Astronomy played a role as early as the Paleolithic period in the alignment of graves and stone settings that served as calendars. In addition to the ancient cultures of the Near and Middle East, we also consider the ancient cultures of China and the Americas.

3.1.1 Stone Age

Since the *Neanderthal period* (about 60,000 years ago) man has buried his dead in graves. From these grave alignments, one can get indirect evidence of astronomical knowledge, which, however, was closely linked to religious ideas. From the *Paleolithic period* there are graves aligned according to the main cardinal points as well as cave paintings. (e.g. the cave paintings from *Lascaux*).

© Springer-Verlag GmbH Germany, part of Springer Nature 2023
A. Hanslmeier, *Introduction to Astronomy and Astrophysics*,
https://doi.org/10.1007/978-3-662-64637-3_3

Towards the end of the *Middle Stone Age* a westward orientation of the graves is found in Scandinavia (the deviation was a maximum of 3°). With the beginning of the Neolithic *period* people became sedentary and practiced agriculture and animal husbandry. The early graves from the Neolithic period show an orientation: the skeletons are oriented north-south, the eyes look towards the east (example Quedlingburg). This is characteristic for the Bell Beaker *Culture* (ca. 2600 BC–2200 BC). Typical for this culture is the burial:

- women: Head in the south, feet in the north,
- men: head in the north, feet in the south.

In both cases the skull faces east. This type of burial differs from that of the *Corded Ware* (ca. 2800 BC–2200 BC), where the orientation is east-west and the skull faces south. Likewise one finds here again a bipolar orientation, with the females the skull lies in the east, for men in the west (see Fig. 3.1).

In the European area, megalithic stone sites are known from the time before the birth of Christ (here especially the *stone setting* from *Stonehenge* in England, Fig. 3.2). The stones are arranged in a circle and form sight lines. For example, at Stonehenge, four specific stones form a rectangle whose sides indicated the sunrise at the beginning of winter and the beginning of summer, respectively. Stonehenge was built between 2200 and 1600 BC. A similar Setting is found in *Carnac* in France.

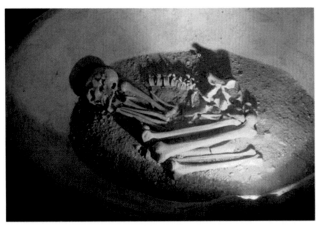

Fig. 3.1 During the Bell Beaker and Corded Ware cultures, the dead were buried in squatting positions. The orientation of the skeletons was different for men and women. The skull pointed either to the east or to the south. Mitterkirchen, Austria. W. Sauber

Fig. 3.2 The stone circle of Stonehenge

Such stone settings were therefore first and foremost *calendars* and thus for the people of the Stone Age the only aid to divide the year.

Without calendars, it would have been impossible for people to cultivate the land and settle down (sowing dates), which began around 10,000 BC. In addition, there was also the belief that ritual acts could be used to please the gods. This is the root of astrology.

The *Nebra* Sky Disk (Fig. 3.3), found near the town of Nebra in Saxony-Anhalt, has a mass of 2 kg, a diameter of about 32 cm and dates from around 1600 BC. There are 32 objects depicted, including the moon and sun and also the star cluster Pleiades.

3.1.2   Egypt, Mesopotamia

Around 2000 BC there were already flourishing cultures in the area of the rivers Nile, Euphrates and Tigris as well as Indus. The stars and their observation were an important part of the religious ideas. Thus astronomy developed together with astrology. However, astronomy became detached from astrology as observations became more and more exact.

Several Factors favored the development of astronomy in Egypt and Mesopotamia. First, the climate there was favorable for astronomical observations, second, the educated upper class had plenty of time for astronomical observations. Furthermore, the development of a written language was essential. This made it possible to record astronomical observations and pass them on to subsequent generations.

The calendar of the Egyptians has already been reported in the previous section. *Re* was considered the sun god and was often depicted as a hawk's head with a disc on

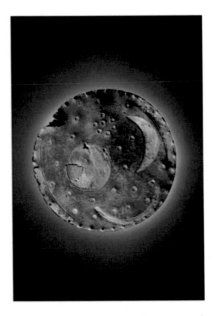

Fig. 3.3 Nebra sky disc

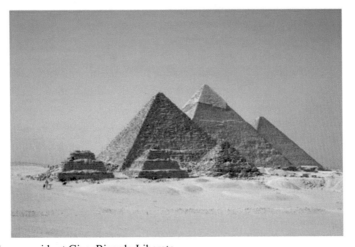

Fig. 3.4 The pyramids at Giza. Ricardo Liberato

top, the solar disc. The two Children of Re were *Nut* (sky goddess) and *Geb* (earth). The pyramids of Giza (Fig. 3.4) were built between 2700 and 2200 BC and are aligned N-S and E-W respectively. The entrance of the *Cheops pyramid* points directly to a specific star. A sarcophagus lid shows the Egyptian goddess Groove with many of the constellations known to us today, the *constellations of the zodiac* (e.g., Leo, Libra, Pisces, Taurus, and Gemini) (Fig. 3.5).

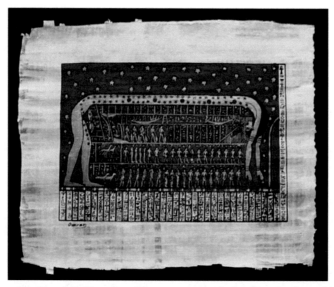

Fig. 3.5 The world view of the Egyptians. The stars are attached to Nut, the sky goddess, who spans the earth. The feet are in the west. The goddess swallows the sun in the evening and gives birth to it again in the morning (Source: Archive Vetter)

The beginning of *Babylonian astronomy* dates back to the third millennium BC. The mean duration between two equal lunar phases *(synodic month)* was determined to be 29.530641 days (end of the third century BC). Similarly, synodic orbital periods of the planets were known with great accuracy. The oldest datable observation of a total solar eclipse in ancient Babylon fell on 15.6.763 BC. The *Saros cycle* was already known. This says that eclipses repeat after 6585.3 days (that is 18 years 11 1/3 days or 18 years 10 1/3 days, depending on how many leap years lie in between).

Example: on 31.05.2003 at 5:30 o'clock there was a partial solar eclipse visible from central Europe. When there will be the next one according to the Saros cycle? Solution: 31.05.2003 + 18 years 10 1/3 days = 10 June 2021 about 11:30.

More details about the Saros cycle will follow in the Sect. 4.7 about eclipses.

3.1.3 China

Among other things, the first observation of very large sunspots (sometimes possible with the naked eye only when the sun is low in the sky) was recorded. Astronomers were court officials and had to predict astronomical events so that certain rituals could be performed. Like the Chaldeans, the Chinese recognized that *Solar eclipses* can only occur around the time of the new moon. Incidentally, they believed in ten suns, one for each day of the Chinese 10-day week. At the end of the 3rd millennium BC, the Astronomers *Hi* and

Ho were punished with death because they had not predicted a solar eclipse (on 13 Oct. 2128 BC). The Chinese believed that during an eclipse a dragon would try to devour the sun and that it had to be driven away by a loud drum roll. 613 BC. the *Halley's comet* was observed. Around the time of Christ's birth *Lio H* made an astronomical manual.

3.1.4 Central and North America

Here, too, observations have been handed down, and pyramids were aligned according to cardinal points. The orientation to the east embodies the resurrection, also birth, and the orientation to the west the end, the approaching death. On Feb. 15, 3379 BC, a lunar eclipse was observed and recorded. The zero point of the famous *Mayan calendar* was the 11. August 3114 BC. At Chichen *Itza* (on the peninsula Yucatan) were marked the points where the sun exactly rose in the east and set in the west. They knew relatively accurately the course of the moon and could also predict eclipses. The *Milky Way* was described as a world tree. The star clouds that make it up were called the Tree of Life, from which all life on Earth comes. The Maya knew the number 0 and used a vigesimal system (i.e. as a base 20: in the decimal system you have the sequence 10-100-1000 etc., in Vigesimal System 20-400-8000 etc.).

For the *Navajo Indians* was *Tsohanoai* sun god. The Inca considered *Inti* their sun god. He ordered them to build the capital *Cuzco*. The *Aztecs* believed in *Tonatiuh* as sun and thought that with each era a sun dies and a new one is born. According to them, they lived in the 5th era, so four suns had already passed. To the god *Huitzilopochtli,* sun and war, they offered human sacrifices. Figure 3.6 shows the famous calendar stone of the Aztecs.

Fig. 3.6 Calendar stone or sunstone. The four signs signify the four periods of destruction and reconstruction. In the center is the 5th sun, Tonatiuh. Museo Nac. de Antropologia, Mexico City

3.1.5 Old Europe

With the *Celts* the sun was also worshipped as the god *Lugh*. This name also occurs as Lleu and Lugos (the latter among the *Gauls*; the names of the cities of Lyon, Leyden, and Laon trace back to it). For the old *Teutons* solar and lunar eclipses were taken as a threat to the sun and moon by werewolves. Generally astronomical knowledge was not strongly developed here, which might be due to the worse weather conditions. The constellations known were the Big Dipper, Orion and the Pleiades. Of the planets, especially the morning and evening star (Venus) were known. The belief that the waxing moon was favourable for sowing and planting as well as for fertility also originates from the ancient Germanic tribes. Much about the ancient Germanic tribes is known from the Latin author *Tacitus*.

3.2 Astronomy of the Greeks

> In ancient Greece, for the first time, the fundamental method of the natural sciences was applied: experimentation.

One tries to gain new knowledge by measuring and not only by mere hypotheses, which are not testable. The astronomical knowledge of the ancient Greeks was amazing; they knew that the earth was a sphere and that the moon must be much closer than the sun. Even the circumference of the earth was calculated. How the simplest measurements could be used to determine such quantities without any tools is shown in this chapter.

3.2.1 Philosophical Considerations

The Greeks adopted the astronomical knowledge of the Babylonians; beyond this, however, they did not confine themselves to mere observations of the heavens, but for the first time put forward theories and hypotheses, and tried to explain regularities in the heavens by them. We give some examples.

At *Hesiod* (seventh century BC, is also called a peasant poet) the idea of a round earth disk is found, which is flowed around by the *Okeanos* . Above the earth, the sky is curved as a hemisphere; below it, there is also a hemisphere, called *Tartarus*. Outside the Celestial sphere is yawning void, the *Chaos*.

On the 28.5.585 BC *Thales* of Miletus predicted a solar eclipse, and this knowledge even influenced a battle. The opponents, not knowing about it, fled in fear. *Pythagoras* established theorems on trigonometry (Pythagorean theorem) in the late sixth century BC and made valuable contributions to positional astronomy. His thesis was that all celestial bodies revolve around the earth, which is at rest in the center of the universe. The earth

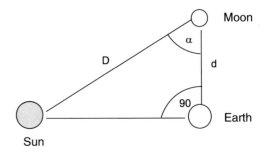

Fig. 3.7 Aristarchus' method
of determining the distance
ratio Earth-Sun to Earth-Moon.
The moon is in the first quarter
as seen from the earth, so it is
half moon

floats on water. According to *Anaximenes* air is the source of all being. The earth is a flat
plate supported by air.

Around 440 BC *Meton* determined by means of a gnomon (shadow staff) the solstices.
Around 400 BC *Democritus* speculated that the Milky Way consists of countless individual
stars, a very modern idea, which he could not prove, however.

According to *Anaxagoras*, the sun is a glowing mass of stone.

Aristotle (384–322 BC) held the idea that the Earth was a sphere. This could be
explained by the always circular shadow of the earth at the beginning and end of a total
lunar eclipse.

3.2.2 First Measurements

In addition to the above philosophical considerations, the first attempt was made in ancient
Greece to obtain astronomical knowledge through measurements.

Around 265 BC *Aristarchus* for the first time postulated a *heliocentric* world system.
In doing so, he also tried to determine distances, and he found for the relationship of the
Distance earth–moon to earth–sun: 1:19.[1] Aristarchus' method is explained in Fig. 3.7.
The Moon is in the first quarter, i.e. it has travelled one quarter of its orbit around the
Earth. For the ratio distance Earth-Moon (d) to distance Earth-Sun (D) we get:

$$\cos \alpha = \frac{d}{D} \tag{3.1}$$

The exact value for α (seen from the moon, the angle between the sun and the earth; in
reality very difficult to determine, since it depends on the determination of the time for the
moon in the first quarter) is $89°51'.4$.

---

[1] Correct value 1:400.

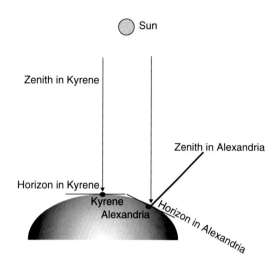

Fig. 3.8 The method of Eratosthenes for determining the circumference of the earth

Around 220 BC *Eratosthenes* estimated the *Earth's circumference*. .He investigated the fact that on a certain date of the year the sun in Kyrene *Cyrene* (also Syene, modern Aswan) can be seen from a deep well, but not from Alexandria, which lies more to the north (Fig. 3.8). For the sun to be seen from a deep well, it must be at the zenith. The distance between Syene and Alexandria is 770 km, and in Alexandria there was the Sun at the same time at 7.2° from the zenith.

From relation:

$$7.2°/360° = 770 \, \text{km}/U \tag{3.2}$$

the circumference of the earth follows U.

Around 150 BC *Hipparchus* discovered the *precession,* as he made new star charts and compared them with old ones. His star charts contained a total of 850 stars.

Around 140 AD *Ptolemy* (Fig. 3.9) summarized the astronomical knowledge of his time in the work "Mathematices syntaxeos biblia XIII", which was later adopted by the Arabs as the *Almagest* .

Ptolemy is considered to be the founder of the *geocentric world system.*

The earth rests at the center of the universe, and all planets, stars, and the sun and moon orbit it. The movements of these celestial bodies occur on circular orbits.

The complicated planetary orbit-loop movements, which an outer planet executes around the time of its opposition position, since the earth overtakes it on its further inward lying orbit, one explained oneself with the help of the *epicycle theory* (Fig. 3.9). The planet

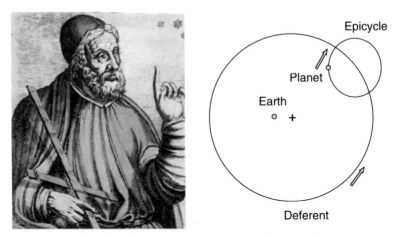

Fig. 3.9 Ptolemy, the founder of the geocentric world system. Right: The theory used to explain the loop orbits of the outer planets in the Epicycle theory geocentric system used *Epicycle theory* of the Ptolemy; the earth is not at the center (+) of Deferent

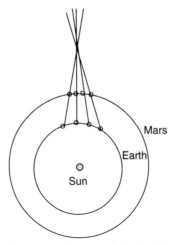

Fig. 3.10 The *looping orbits* of the outer planets result in the heliocentric worldview simply from the Earth orbiting faster around the Sun overtaking the outer planet orbiting slower around the Sun

runs on a circle called an epicycle. Its center moves on a second circle (deferent) and so on. In the Almagest we find the influence of Aristotelian philosophy based on dogmas ("Aristoteles dixit.").

In the heliocentric world system, the looping motion results from the Earth moving faster around the Sun than a planet orbiting outside the Earth's orbit (Fig. 3.10).

3.2.3 Navigation

Here we come briefly to another important aspect of astronomy, which was already of great importance to the peoples of antiquity: *Navigation.* In ancient times, people first preferred coastal waters, where landmarks and the plumb line allowed orientation. The Greeks had descriptions of coasts and harbours (Perploi). But since the fifth century BC there is evidence of deep-sea voyages by merchants, and it was clear that as soon as the coasts were out of sight, one had to rely on the stars as guides. *Thales of Miletus* is said to have written on nautical astronomy in the early sixth century BC. His pupil was *Anaximander,* who created the first celestial globe. *Cleostratos* defined the 12 parts of the *Zodiac*. For the measurement of the geographical latitude the position of the *polar star,* which at that time was even further away from the celestial pole (today it is only approx. 1° away. At that time the *Kochab* in the Little Dipper was known as the polar star) had to be estimated. For the determination of the geographical longitudes time measurements are necessary, which were however at that time too inaccurate.

3.3 Astronomy in the Middle Ages and Modern Times

In the Middle Ages there was a stagnation in the sciences. One referred mainly to the philosopher Aristotle and did not question reality. At the beginning of the modern age there was Copernicus, who advocated the heliocentric world system in his main work. Kepler later explained how the planets move, and Newton established the law of gravitation, which describes the forces of attraction between masses.

3.3.1 Astronomy of the Arabs

In the period between Ptolemy and Copernicus there were no new developments in the conception of the world picture, but the *Arabs* did develop observational techniques and mathematics. *Al Bettani*[2] (†929 AD) formulated the cosine theorem. *Abul Wefa* (940–998) created tables of the trigonometric functions.

Arabs brought ancient astronomical knowledge to the Occident.

From the *Indians* they adopted the number system as well as the representation of numbers (Arabic numerals). Likewise developed they the algebra (Al gebr).

---

[2] Known as Albategnius.

The most important book of Ptolemy, the Greeks called it *Megistos,* was translated by the Persian astronomer *Al Sufi* (903–986) (also called *Abd ar-Rahman as-Sufi,* or *Abd al-Rahman Abu al-Husain*) at the court in Isfahan, and since then the book is known as the famous Almagest. In his major work written in Arabic in 964 (Book of Fixed Stars[3]) Al Sufi mentions for the first time the great *Magellanic Cloud* and the *Andromeda Galaxy.* This still serves as a source for studying the proper motions of stars and for brightness measurements of variable stars.

Under caliph *Al Mamum*has the Earth was remeasured by *Muhammad Ibn Mu-sa-al-Khwarizmi* (c. 780–c. 850), who also solved linear and quadratic equations and published a book on the image of the earth.[4] *Ibn Junus* (†1009) discovered the phenomenon of astronomical *refraction.*

Ulugh Beg (1393–1449) created a very accurate catalogue of 994 stars at his observatory in *Samarkand* (built in 1428), thus expanding the Al Sufi catalogue.

3.3.2 Middle Ages

In the Middle Ages there was in Europe a stagnation of astronomy. Exclusively theologians dealt with it.

Ioannes de Sacrobosco (†1256) authored the first occidental textbook of astronomy. *Nicholas Cusanus* (†1464) advocated modern ideas: He spoke for the first time of the infinity of the universe.

Johannes Regiomontanus was born in Königsberg in 1436 (his real name was Johannes Müller); he was the most important astronomer of the late Middle Ages. In April 1450 he went to Vienna, where, among other things, he made the first weather records. He completed the work of his teacher *George of Peuerbach,* to retranslate the Almagest, and therefore went to Rome around 1460. In 1469 a new manual of astronomy was thus produced, and the calculations of Ptolemy were placed on a more precise mathematical basis. In 1467 Regiomontanus went to Pressburg (now Bratislava) to work on behalf of the king in the *Matthias I Corvinus* to build the largest library in the West, and in 1471 to Nuremberg, where he proved that comets were celestial phenomena. His accurate tables served Christopher Columbus as an aid to navigation. In 1475 he was summoned to Rome by Pope *Sixtus IV* to Rome to help reform the calendar, but he died in 1476.

---

[3] Kitab al-Kawatib al-Thabit al-Musawwar.

[4] Kitab surat al-ard.

3.3.3 Geocentric → Heliocentric

1453 it came to the conquest *Constantinople* by the Turks. As a result, many works of antiquity were brought to the West by Byzantine scholars. In 1492 the discovery of America took place, and a new age was heralded.

At that time, *Platonic way of thinking* replaced that of *Aristotle:* The process of knowledge consists in a progressive adaptation of our inner world of concepts and forms of thought.

Nicolaus Copernicus (1473–1543, Fig. 3.11) lived at the transition from medieval to modern times. *Martin Luther* (1483–1546), *Philipp Melanchthon* (1497–1560), *Christopher Columbus* (1446–1502), *Leonardo da Vinci* (1452–1519) and *Albrecht Dürer* (1471–1528) were his most famous contemporaries. Copernicus was born in Thorn in 1473 and lost his father as a boy. After studying mathematics, astronomy, and humanistic subjects at the University of Cracow, he was admitted to the cathedral chapter of Warmia at the instigation of his uncle (Bishop Luke Watzenrode) and was sent by his chapter brothers to Bologna, Padua, and Ferrara to study medicine and canon law. From 1512 Copernicus lived mostly in Frauenburg. There he was chancellor of the cathedral chapter, governor, bishopric administrator, deputy to the Prussian Diet, etc. Due to his humanistic education he was very interested in the ideas of antiquity. An essential motive for his astronomical studies was the exact determination of the length of the year. Between 1506 and 1514, in a writing known as the "Commentariolus", he made the first assumptions of a Central Position Sun in the solar system :

"The centre of the earth is not the centre of the world, but only of gravity and of the lunar orbit."

Fig. 3.11 Nicolaus Copernicus, the founder of the heliocentric world view

"What appears in the case of the Convertible Stars as a decline and advance, is not so of itself, but as seen from the earth. Their motion alone, therefore, is sufficient for many diverse phenomena in the heavens."

But his chief work, "De Revolutionibus Orbium Coelestium Libri VI," does not appear until the year of his death (1543), at the insistence of Cardinal Nicolaus von Schönberg. The interest of the Curia was also the reason why Copernicus dedicated the book to Pope *Clement VII*.

At first, the book was not widely read. Thirty-nine years after the death of Copernicus, due to his exact dates of observation, the calendar was reformed under Pope *Gregory XIII*. According to the ideas of Copernicus and the Church of the time, the revelatory content of the Bible was unaffected by the purely mathematical issue of whether the Earth or the Sun was at the center of the planetary system.

It was not until 1616 that the Copernican doctrine was condemned as contrary to the Bible, and it was not until 1822 that the ban was lifted again.

> Although Aristarchus already assumed the sun to be in the center, Copernicus is considered the founder of the heliocentric world system.

Tycho Brahe (1546–1601) advocated a mixed world system (Fig. 3.12). Around the earth run Sun and moon; around the sun run the other planets. Brahe was the last great unaided observer. In 1572 he discovered a *Supernova* at Cassiopeia constellation. In 1580 he completed an observatory called Uraniborg on the island of Hven. In 1599 he became an imperial court astronomer at *Rudolph II.* in Prague. Tycho Brahe was an excellent observer, the accuracy of his observations was 2′.

Already during his lifetime *Johannes Kepler* (1571–1630) moved to Prague and Tycho entrusted him with the task of an exact calculation of the orbit of Mars.

Fig. 3.12 World view of Tycho Brahe. The earth is at the center; the moon and sun revolve around the earth; the planets revolve around the sun

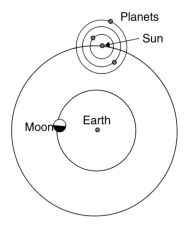

Fig. 3.13 Johannes Kepler, who explained the "how" of the planetary motions

3.3.4 Kepler, Galileo, Newton

Johannes Kepler was born on 27.12.1571 (Jul.) in Weil (Württemberg) (Fig. 3.13). His lifetime falls into the epoch of the *Renaissance,* in during which great upheavals occurred in the intellectual field. Because of his good school performance, Kepler received a scholarship from Duke Christoph, and he began studying theology, mathematics, and astronomy at the University of Tübingen in 1589. His teacher was *Mästlin,* and he familiarized him with the ideas of Copernicus. By his critical statements he offended against the dogmas and got no employment in Württemberg. Therefore he went to *Graz,* where where he worked as a landscape mathematician between 1594 and 1600. In 1597 he married Barbara Müller in Graz. Already at the age of 24 he wrote his first work "Mysterium Cosmographicum"(Fig. 3.14). In this work the earth's orbit is represented as measure of all things. "Let her circumscribe a dodecahedron, the sphere of Mars, inside lie an icosahedron, the sphere of Venus". In 1600 Kepler moved to *Prague* to Tycho Brahe, and he succeeded him after the latter's death (1601) as Rudolf II's astronomer and imperial mathematician respectively. The death of Tycho Brahe is one of the unsolved mysteries in the history of science; there are two theories: (i) poisoning of Brahe by Kepler. Kepler was desperate to get hold of Brahe's exact observations, but the latter did not release them. (ii) Tycho Brahe is said to have feasted extensively and died of a burst bladder. The fact is that Kepler, on the basis of Brahe's extremely accurate observations, came to the conclusion in 1605 that the orbit of Mars was an ellipse. Kepler's first two laws appeared in the "Astronomia Nova" in 1609. He learned of the use of the telescope in astronomy in 1611 and published a book on optics. In 1612 died *Rudolf II* and Kepler went as a landscape mathematician to *Linz,* after his wife had also died. There he married Susanne Reutinger in 1613. When he made preparations for the second wedding, he ordered wine barrels and developed a formula to calculate their volume. In 1615 his mother was accused of witchcraft. After a business dispute with the wife of a glazier, Ursula Reinhold, she is accused by the latter of having given her a bitter potion. Only after an interrogation under torture is she acquitted in 1621, whereupon, however, she soon dies.

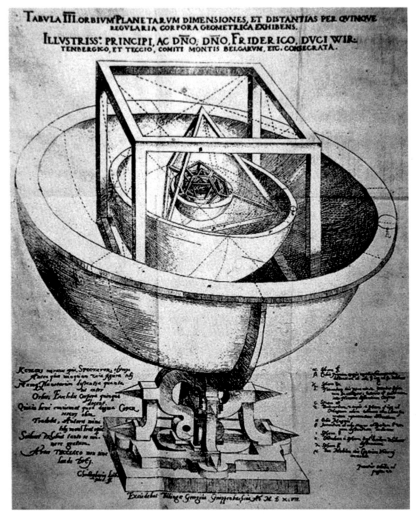

Fig. 3.14 Kepler's Mysterium Cosmographicum

In 1619 Kepler's third law appeared in the "Harmonices Mundi". The first five volumes deal with the concept of harmony in mathematics, the last three with applications of this concept to music, astronomy and astrology.

In 1618–1622 he published a seven-volume series on Copernican astronomy, and in 1627 appeared the *Rudolphine Tables,* which were the most accurate tables of planetary motions up to that time. In 1628 Kepler was in Ulm and Sagan in the service of *Wallenstein,* He worked mainly as an astrologer. In 1630 he travelled again to Linz, where he fell ill and died on 15 November.

Galileo Galilei (Fig. 3.15) was born in Pisa in 1564. In 1589 he became professor of mathematics in *Pisa,* 1592 in *Padua.* In 1609 he rebuilt a telescope (which was first used

Fig. 3.15 Galileo Galilei

in 1608 by the Dutch *Lippershey* (1570–1619)). His discoveries and observations were published in 1610 in the "Sidereus Nuncius" (Starry Messenger) and were for the epoch a sensation. He reports *sunspots, Jupiter's moons, phases of Venus* and *lunar mountains.*

In 1616, the Catholic Church declared the Copernican doctrines to be in error. They required Galileo to abandon them as well. In 1623, a friend of Galileo's became pope (Maffeo Barberini, *Urban VIII.*) Galileo immediately tried to get the decree lifted. He got permission to write a book about the Aristotelian and the Copernican theory, in which he was not allowed to take sides ("Dialogo"). This book appeared in Italian, and in it three scholars discuss the heliocentric and geocentric world systems. The representative of the geocentric world system, Simplicio, did very badly in it, the Pope charged Galileo with disobedience in 1633, and he was put under house arrest. Galileo died in 1642.

It was not until 1979(!) that the Vatican officially acknowledged that Galileo's ideas were correct.

Giordano Bruno (1548–1600) was burned at the stake in Rome in 1600 for his doctrine of the infinity of the world.

In the spirit of the *Aristotelians*, it was initially assumed that there was a difference in principle between celestial and terrestrial matter. Now came the rethinking, and it was assumed that earthly laws of nature could also be used to describe astronomical phenomena.

Isaac Newton (1643–1727) (Fig. 3.16) published in his main work "Philosophiae Naturalis Principia Mathematica" the law of gravitation and beside it the infinitesimal calculus. Newton's telescope also originated with him.

Newton was one of the few scientists to be honored during his lifetime. He became president of the Royal Society. Because of the publication of some data he clashed with the director of the Royal Observatory, *Flamsteed,* in conflict. Out of anger at having lost the case against Flamsteed (stolen data), he deleted every reference to Flamsteed from his

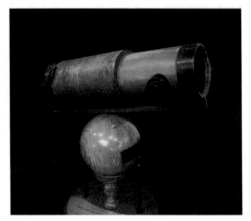

Fig. 3.16 Sir Isaac Newton and the telescope he constructed

works. With *Gottfried Wilhelm Leibniz* he came into dispute over the question who was the first to use the invented *infinitesimal calculus*. Many of the articles which appeared in defence of Newton were written by himself and published under the names of friends. By the Royal Society, of which Newton was president, Leibniz was made out to be a plagiarist.

- The final change from the geocentric to the heliocentric view of the world took place in 1543 with the publication of Copernicus' magnum opus.
- Kepler explains how the planets move.
- Newton explains the law by which the heavenly bodies move.

3.3.5   Celestial Mechanics

Carl Friedrich Gauss (1777–1855) (Fig. 3.17) developed in the field of astronomy the method of *orbit determination* and the *method of least squares*. At this time celestial mechanics became strongly emergent, and we mention here only such names as *Euler, Lagrange, Poincaré*, etc.

The first measurement of *trigonometric stellar parallaxes* arrive in the year 1838 *F. W. Bessel by (61 Cygni), F. G. W. Struve (Wega)* and *T. Henderson (Alpha Centauri)*. This was the final proof for the validity of the *of the heliocentric world system*.

> Star parallaxes can only be determined by the annual motion of the earth around the sun.

Fig. 3.17 Gauss and Caroline Herschel

Fig. 3.18 Famous depiction on a tapestry, Bayeux, eleventh century. It shows the conquest of England by the Normans as well as Halley's comet

Comets were of special interest. *Edmond Halley* (1656–1742) stated that the comets of the years 1531, 1607 and 1682 had to be the same comet, later named Halley's comet after him. He predicted its recurrence for 1759. Besides the representation of this comet in Fig. 3.18 that of the painter *Giotto* is very well known, who saw the comet around 1301 and depicted it as the Star of Bethlehem in his fresco "The Adoration of the Magi". During its last return in 1986, the comet was first studied by five space probes (including ESA's GIOTTO mission). The next return is expected in 2061.

3.4   Modern Astrophysics and Cosmology

At this point we bring only a few examples, because astronomy or astrophysics has developed extremely fast in the last centuries.

3.4.1   The Discovery of Further Objects in the Solar System

Already in ancient times there are reports about the five planets visible to the naked eye and their orbits (Mercury, Venus, Mars, Jupiter, Saturn). On March 13, 1781. *Caroline* and *William Herschel* discovered the planet *Uranus*. This planet is already visible as a star in binoculars and also shows a disk in relatively small telescopes.

The first observations of *Neptune* were reported by Galilei. On December 28, 1612, he observed Jupiter with its large moons. In his notes on it there is also an 8th magnitude star (the brightest stars are 1st magnitude, the faintest stars just visible to the naked eye are 6th magnitude). A month later, on 27 January 1613, he again observed these two objects, but thought that the faint star was an ordinary fixed star.

The astronomer *Lalande* observed Neptune between 8 and 10 May 1795. Also *Lamont* observed Neptune in 1845 and 1846, thinking it to be a fixed star. All these observations did not lead to the recognition of the true nature of the star and thus to the discovery of Neptune. The celestial mechanic *Delambre* calculated the positions of Uranus, and discrepancies were soon seen between the predicted values and the actual positions. 1841 *Adams* became interested in these discrepancies, as did the French celestial mechanic *Le Verrier.* Both made predictions of the position of the new planet. On 23.09.1846 *Johann Gottfried Galle* of the Berlin observatory received the information and observed the same night for the first time the Planet Neptune .[5]

Percival *Lowell* (1855–1916) established a private observatory in *Flagstaff,* Arizona, and was primarily interested in Mars observations. By 1905, due to advances in astronomical photography, there were marked improvements in position measurements of stars. At the Flagstaff Observatory, the search began for another planet predicted due to perturbations of Neptune's orbit (Uranus and Saturn also showed such perturbations). Later, the actual designer of the Flagstaff Observatory , *W. H. Pickering* (1858–1938), was concerned with the discovery of another planet.

On 3/13/1930 *C.W. Tombaugh* succeeded in making the discovery of Pluto. In doing so, he used the method of the *blink comparator.* Take two photographic plates which show the same celestial field separated by an interval of time. The two plates are superimposed. All the stars remain in position during the blink, a planet moving on makes itself known by "blinking". The discovery of Pluto coincides with Lowell's 75th birthday and the 149th anniversary of the discovery of Uranus.

---

[5] He always refused to be considered a discoverer, attributing the discovery of Neptune to Le Verrier.

However, there have been new discoveries in the solar system in recent times as well. There are many similar objects like Pluto, which strictly speaking should not be considered a planet at all, one speaks since 2006 of *dwarf planets*. Around 1990, objects of the so-called *Kuiper belt* were found, and our solar system is enveloped by the Oort's *cloud* (up to 100,000 times distance earth-sun). Kuiper belt and Oort's cloud are remnants from the formation of the solar system. Some Examples of Pluto like objects are *Quaoar* (discovered 2002, diameter 1260 km) or *Eris,* originally also as Xena designated (discovered 2003, diameter 2400 km, thus larger than Pluto), or Sedna (diameter 1700 km, discovered 2003). These objects are also called *trans-Neptunian objects*.

3.4.2 Astrophysics

The question of the nature of stars and their formation and evolution is also an essential element of astronomical research. In 1054, Chinese astronomers (and also the Indians) observed a bright star in the sky, which for some time was visible even during the day. In 1942, *Oort* and collaborators identified the *Crab Nebula* as a remnant of this Supernova explosion.

An essential source of information is the *light of* the stars. The speed of the propagation of the light was measured in 1676 by *Roemer* and he predicted that an eclipse of Jupiter's moon Io on 9 November 1676 would occur 10 min earlier than calculated. Earth and Jupiter are not always equidistant. If the Earth approaches Jupiter, the time between eclipses of Jupiter's moons becomes shorter than if it moves away from Jupiter (Fig. 3.19). In 1728 *Bradley* determined the speed of light from the aberration, and in 1849 *Fizeau* published the first measurement of the speed of light on Earth.

Astrophysics, i.e. the determination of the physical properties of a star (temperature, formation, evolution, . . .), began with spectroscopy in the nineteenth century. *Fraunhofer* made experiments with the decomposition of light around 1823. He discovered more than 500 dark lines in sunlight, the same in moonlight (reflected sunlight). However, the decomposed light of some bright stars showed other lines. In 1860 *Kirchhoff* published the fundamentals of radiation theory and the relationship between emission and absorption lines. By comparing the stellar spectra with lines of the elements in the laboratory, it became possible to determine the compositions of the *stellar atmospheres*. Another milestone in astronomy was the *photography*. In 1872 there were already first photographic star spectra *(Draper).Vogel* determined 1888 the first Radial Velocities of stars (by measuring the Doppler effect). Around 1900 established *Planck* the law of the spectral energy distribution of the radiation of a *black body* that stars radiate like one to a first approximation. Then, in 1913, one of the most important diagrams in astrophysics, named after its inventors *Hertzsprung-Russell diagram, HRD* was introduced. *Eddington* provided ground breaking work on the structure of stars (1916). *Hale* found around 1908 that sunspots are areas of strong magnetic fields.

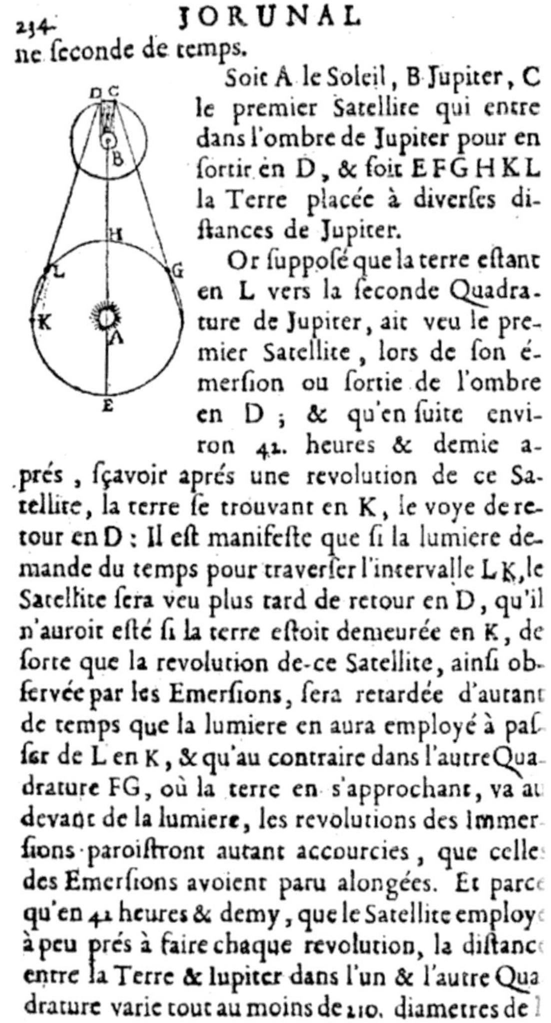

Fig. 3.19 Roemer's original
work on the measurement of
the speed of light

JORUNAL
ne ſeconde de temps.

Soit A le Soleil, B Jupiter, C
le premier Satellite qui entre
dans l'ombre de Jupiter pour en
ſortir en D, & ſoit E F G H K L
la Terre placée à diverſes di-
ſtances de Jupiter.

Or ſuppoſé que la terre eſtant
en L vers la ſeconde Quadra-
ture de Jupiter, ait veu le premier Satellite, lors de ſon é-
merſion ou ſortie de l'ombre
en D ; & qu'en ſuite envi-
ron 42. heures & demie a-
prés, ſçavoir aprés une revolution de ce Satellite, la terre ſe trouvant en K, le voye de retour en D : Il eſt manifeſte que ſi la lumiere de-
mande du temps pour traverſer l'intervalle L K, le
Satellite ſera veu plus tard de retour en D, qu'il
n'auroit eſté ſi la terre eſtoit demeurée en K, de
ſorte que la revolution de ce Satellite, ainſi ob-
ſervée par les Emerſions, ſera retardée d'autant
de temps que la lumiere en aura employé à paſ-
ſer de L en K, & qu'au contraire dans l'autre Qua-
drature FG, où la terre en s'approchant, va au
devant de la lumiere, les revolutions des Immer-
ſions paroiſtront autant accourcies, que celle
des Emerſions avoient paru alongées. Et parce
qu'en 42 heures & demy, que le Satellite employe
à peu prés à faire chaque revolution, la diſtance
entre la Terre & Iupiter dans l'un & l'autre Qua
drature varie tout au moins de 210. diametres de l

More supernova observations in our galaxy occurred in 1572 *(Brahe)*, in 1603 *(Kepler)*, and in 1987 *(Shelton* the large Magellanic cloud). *Chandrasekhar* stated in 1930, that there is an upper mass limit for white dwarfs. The first observed white dwarf was Sirius B (1862, *Clark*). *Zwicky* and *Baade* around 1933 for the first time assumed *Neutron stars* as supernova remnants. Supernovae are therefore not new stars, but stars at the end of their lives. In 1967 discovered *Bell* and *Hewish* radio pulses from a pulsar, and 1 year later *Gold* provided the explanation: Pulsars are rotating neutron stars.

Modern astrophysics is based on spectroscopy and the resulting analysis of stellar spectra.

3.4.3 The Universe

The first task was to find out the place of the solar system within the galaxy (Milky Way). In 1610 *Galilei* realized with his telescope that the *Milky Way* consists of many single stars. 1755 expressed *Kant* the assumption : the *Galaxy* is a rotating disk of stars, held together by gravity. In 1918 *Shapley,* found that Globular Clusters are arranged symmetrically not to the solar system, but to the center of the Milky Way. At that time, the debate began as to whether or not the many observed spiral nebulae were objects outside our system. 1923 determined *E. Hubble* for the first time the distance to the *Andromeda Galaxy* and it was clear that these objects were outside our system. A few years later Hubble discovered the expansion of the universe. In 1930, *Trumpler,* discovered that there is matter between the stars that attenuates the light, which he called *interstellar matter. Zwicky* was able to prove in 1933, with the help of the *Virial Theorem* that the Coma galaxy cluster must consist of much more matter than is visible. In 1963 *Schmidt* showed that *quasars* must be extremely distant objects, by identifying highly redshifted spectral lines in their spectra.

The *cosmic background radiation,* i.e. the remnants of the hot *Big Bang,* was discovered in 1965 by *Penzias* and *Wilson.* In 1987 it was found that galaxies within a radius of 200 million light years move with ours to the so-called *Great Attractor* and 2 years later the Great Wall was discovered. A *wall* only 15 million light-years thick but 500 million × 200-million-light-year layer of galaxies.

Great sensation was caused by the discovery around 1999 of the *accelerated expansion* of the universe which led to the concept of Dark Energy. In addition to the *Dark energy* exists also the *Dark Matter.* This concept explains the motion of the galaxies in a galaxy cluster as well as the rotation of the stars around the galactic center.

Alongside these discoveries played the development of theoretical physics, in particular relativity and quantum theory, a great role: 1905 *Einstein* introduced the *Special Theory of Relativity* with the famous formula $E = mc^2$ (energy and mass are equivalent). In 1915 Einstein published *General relativity* theory, which shows the connection between matter and space-time curvature. With relativity theory the perihelion movement of Mercury could be fully explained. *Le Verrier* found Mercury's perihelion movement that is caused by the influence of the planets but for a full explanation Einstein's relativity theory is necessary.

In the year 1919 *Eddington* undertook a solar eclipse expedition in order to measure the *deflection of light* by the Sun, which had been predicted by Einstein. *Zwicky* wrote already in 1937 about the existence of Gravitational *lensing.*

In 1982 *Taylor* and *Weisberg* determined the energy losses which result from the rotation of an observed Double quasar object. In 1967 *Wheeler* introduced the term *Black Hole*. Already in 1939 *Oppenheimer* and others showed that a gravitational collapse of a pressureless fluid can lead to a complete cutting off of it from the rest of the world. In 1974 *Steven Hawking* found that black holes radiate nevertheless although actually not even light can escape from them. In 2002, observations showed that the radio source *Sagittarius A* is a *supermassive Black Hole* at the center of our galaxy.

In 2015, for the first time, direct measurements of *Gravitational waves* emitted during the collapse of two massive black holes were made.

3.4.4   Concluding Remarks

We could mention here only the most essential discoveries. Further data can be found in the individual chapters, especially the results of modern satellite astronomy. It should be emphasized, however, that in all areas of astrophysics new findings and discoveries have also been made recently, be it the accelerated expansion of the universe and the mysterious dark energy associated with it, the *helioseismology,* with the help of which it is possible to look into the interior of the sun, new *dwarf planets* in the solar system and even planets outside our solar system, exoplanets. In addition, with the help of new supercomputers and computer clusters, more and more precise models of the structure and evolution of stars are being developed and even the dynamics of whole galaxies.

The discovery of gravitational waves opens a new window for astrophysics. New insights into the universe are also provided by the results of Astro particle physics. Here, elementary particles of cosmic origin are studied. Examples include neutrinos, which are produced in a supernova explosion, and high-energy cosmic ray particles. Other experiments look for candidates for dark matter, which may account for 25% of the matter in the universe. With the extremely small dimensions of the universe at the Big Bang, the validity of quantum physics and general relativity fails. Possible ways out of this dead end offer the *String theories* or *loop gravity.*

> The results of modern astrophysics are of great importance for theoretical physics. At the extremely high energies around the time of the Big Bang, more than 13.7 billion years ago, there was a unification of the four forces known today, of which gravity is by far the weakest force.

3.4.5 What is Matter?

In this section, we will briefly describe how our understanding of matter has evolved over time. As early as *Democritus* assumed that everything is composed of minute, indivisible particles. This was called atomism. Atom means indivisible. But scattering experiments in the previous century showed that atoms are made up of smaller units. The classical *Bohr model of* the atom assumes negatively charged electrons orbiting a positively charged nucleus composed of the positively charged protons and neutral neutrons. However, it turned out that the atomic world behaves differently from our everyday world. Electrons can only orbit around the nucleus in certain paths. Furthermore, it was found that while electrons are truly atomic, i.e. indivisible, protons and neutrons are made up of even smaller particles, quarks. String theories even assume a thread-like structure and not a point structure of these particles.

Important
Our view of the world is constantly evolving. The world at large:

- Earth at the center
- Sun at the center
- The sun is one of several 100 billion stars in the galaxy.
- There are several billion of galaxies.
- There are no distinguished points in the universe.
- The universe was formed about 13.7 billion years ago.

The world in miniature:

- Everything is made of atoms.
- Atoms are made of nucleus and orbiting electrons.
- Protons and neutrons are not elementary particles, but are made of quarks.
- Elementary particles may be elongated filaments, strings.
- At the high energies of the Big Bang, general relativity and quantum physics no longer fit together.

3.5 Further Literature

We give a small selection of recommended further reading.
The History of Astronomy: From Copernicus to Stephen Hawking, P. Aughton, National Geographic, 2009
A Brief History of Astronomy and Astrophysics, K.R. Lang, World Scientific, 2018.
The Discovery of the Universe, P. Murdin, Kosmos, 2014

Celestial Mechanics

4

The centuries after Newton were the golden age of celestial mechanics. While at first the focus was on describing the orbits of the planets, this was soon followed by precise orbital calculations. In 1682, the comet predicted by Halley appeared again on the sky. The last apparition was 1985/1986 (the comet has an orbital period of 76.2 years).

Today, the field of celestial mechanics has expanded to include the study of orbits in binary star systems, the dynamics of star clusters and galaxies, and stability considerations. Furthermore, it can be used to answer questions about the stability of planetary orbits. Many of the newly discovered extra solar planets are giant planets close to their Sun, which raises questions about the formation of these planets—they may have formed at a greater distance from the star and then migrated to it.

4.1 Moon and Planetary Orbits

Kepler's laws describe the motion of the planets. In this section we deal especially with the orbits of the Earth and the Moon. Because of the influence of the other planets, the orbital elements of the Earth are not constant, and the motion of the Moon around the Earth is extremely complicated because of the many perturbations. An important aspect is the tidal forces, especially those of the Moon on the Earth.

© Springer-Verlag GmbH Germany, part of Springer Nature 2023
A. Hanslmeier, *Introduction to Astronomy and Astrophysics*,
https://doi.org/10.1007/978-3-662-64637-3_4

4.1.1   Description of Planetary Orbits, Orbital Elements

Johannes Kepler (1571–1630) was the first to describe planetary orbits correct. The *1st Kepler's law* states:

Planetary orbits are ellipses with the Sun at the focal point.

To describe a planetary orbit and its position in relation to a reference plane (usually the ecliptic), the following set of *orbital elements is used:*

1. *a large orbital semi-axis, b* small orbital semi-axis. Let M be the center of an ellipse (Fig. 4.1) and F the focal point, then $\overline{MF} = c$ and

$$a^2 = b^2 + c^2$$

2. *Excentricity, e:* ϕ is of Angle of eccentricity, $c = ae$.
    Perihelion distance: Planet near sun (point P in Fig. 4.1):

$$\overline{PF} = a - c = a(1 - e) \tag{4.1}$$

   Aphelion distance: Planet far from the sun (point A in Fig. 4.1):

$$\overline{AF} = a + c = a(1 + e) \tag{4.2}$$

   By a, e is given the shape of the ellipse (Fig. 4.1).
        Currently, the Earth is at perihelion in early January and aphelion in early July.
3. Length of the *ascending node* Ω: is counted away from the vernal point (Fig. 4.2).
4. *Slope of the orbital plane* against the ecliptic i: $i < 90°$ Orbit is clockwise, $i > 90°$ Orbit is retrograde.

Fig. 4.1 Ellipse

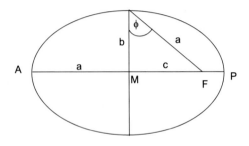

Fig. 4.2 Orbital elements.
Ecliptic is the Earth's orbital
plane, the orbital plane of a
planet is inclined to it

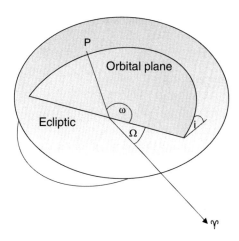

5. *Perihelion length* Ω: length of the perihelion in the orbit (counted from the Node line away). Usually the length of the perihelion is defined by the angle:

$$\tilde{\omega} = \Omega + \omega \qquad (4.3)$$

This angle is the sum of two angles in different planes. However, since the orbital inclinations of most planets are small, the summation is justified.

6. Location of the planet in the orbit: determined by the *Perihelion transit time T*.

We thus have six *orbital elements* (Fig. 4.2).

- $\Omega, i, \tilde{\omega}$ indicate the position of the ellipse in space, but change because of precession; one must therefore indicate the Epoch for which these orbital parameters were determined.
- a, e shape of the ellipse.
- T temporal position of the planet in its orbit.

Large values for e are found for Mercury (0.206) and Pluto (0.248). The orbital inclinations are, except for Mercury ($i = 7.0°$) and Pluto ($i = 17.1°$) small.

Let's calculate the perihelion and aphelion distances for the dwarf planet Pluto, $a = 39.88$ AU, $e = 0.25$, and compare the values with the large orbital semi-axis of Neptune ($a = 4509 \times 10^6$ km). Since 1 AU (astronomical unit, mean distance Earth–Sun) corresponds to 150×10^6 km, we obtain the result that Pluto's orbit passes inside that of Neptune for some time.

Fig. 4.3 The different
definitions of the anomalies: E
eccentric, v true anomaly; Pl
planetary location. The true
anomaly is computed away
from the elliptic focal point

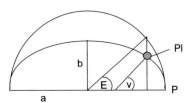

4.1.2   Ephemeris Calculation

Goal: Calculate planetary orbits and their positions (ephemeros Greek: for 1 day). Let r be
the *radius vector of* the planet and v the *true anomaly* (Fig. 4.3), then the following applies
(ellipse equation):

$$r = \frac{p}{1 + e \cos v} = \frac{a(1 - e^2)}{1 + e \cos v} \tag{4.4}$$

The *second Kepler law* states that the radius vector sweeps equal areas in equal
times.

Consider the ratio of elliptical area $ab\pi$ to orbital period U and set for $b = a\sqrt{1 - e^2}$:

$$\frac{\pi ab}{U} = \frac{\pi a^2 \sqrt{1 - e^2}}{U} \tag{4.5}$$

Further one still needs the *eccentric anomaly E* (Fig. 4.3) and the *mean anomaly M*,
which can be derived from the ellipse geometry:

$$M = \frac{2\pi T}{U} \tag{4.6}$$

The *Kepler equation* yields for given M and e the eccentric anomaly E:

$$E - e \sin E = M \tag{4.7}$$

In practice one does the calculations like this :

- One calculates M.
- From the Kepler equation then follow the eccentric anomaly E and the true anomaly v by means of:

$$\tan \frac{v}{2} = \sqrt{\frac{1+e}{1-e}} \tan \frac{E}{2} \tag{4.8}$$

- The radius vector is given by

$$r = a(1 - e \cos E) \tag{4.9}$$

Thus we obtain the position of a planet for each time t that has elapsed since its perihelion passage, i.e., the undisturbed orbit of a pure Kepler ellipse.

Here r, v the heliocentric orbital coordinates of the object, and by suitable transformations obtain from them the heliocentric ecliptic coordinates or the geocentric coordinates.

Earth's Orbit
The *mean distance to the sun* is $a = 149.6 \times 10^6$ km. This distance serves as a unit of distance, especially for calculations in the planetary system:

1 astronomical unit, AU = mean distance Earth-Sun.
The eccentricity of the Earth's orbit is $e = 0.0167$, the small orbital semi-axis is $b = 0.9998\,a$. From this follows:
Perihelion distance: $a(1 - e) = 147 \times 10^6$ km, velocity (formula (4.41)): 30.3 km/s
Aphelion distance: $a(1 + e) = 152 \times 10^6$ km, velocity: 29.3 km/s.

Currently, the Earth passes through its solar far side in July and solar nearside in early January. Therefore, in the northern hemisphere of the Earth, the summer is almost 8 days longer than the winter, because the Earth moves slightly slower around the Sun at greater solar distance.

As a result of disturbances from other planets, changes occur in these parameters and thus the vernal equinox and perihelion change:

- sidereal orbital period of perihelion: 111,270 years (right hand);
- sidereal orbital period of the vernal equinox: 25,800 years (retrograde);
- tropical orbital period of the perihelion: 20,900 years (right hand).

700 years ago, perihelion and winter solstice coincided; currently, Earth passes through perihelion on Jan. 2.

The motion of the earth is clockwise, seen from N. The vernal equinox moves backwards. The vernal equinox moves retrograde, the perihelion clockwise.

4.1.3 Apparent Planetary Orbits in the Sky

Seen from the earth one distinguishes apparent planetary orbits:

- *Lower (inner) planets* (Mercury, Venus): Orbit within the Earth's orbit (Fig. 4.4); important positions are the *lower conjunction* (planet between earth and sun); if the planet is to the left of the sun *(eastern elongation),* so it can be seen in the evening sky, if it stands to the right of the sun *(western elongation),* then it can be seen in the morning sky. It is farthest away from the earth in its orbit at its *upper conjunction.* In western elongation the planet Venus it is called morning star in eastern elongation Venus is found on the evening sky and called the evening star. Inner planets go through phases similar to our moon.
  - Lower conjunction: planet unlit, near Earth.
  - Upper conjunction: planet fully illuminated, but standing with the Sun in the daytime sky, at a distance from Earth.

    Immediately before and after the lower conjunction, the planet is seen illuminated as a narrow crescent. The same holds for Mercury.
- *Upper (outer) planets* (Mars, Jupiter, Saturn, Uranus, Neptune): Orbit outside the earth's orbit (Fig. 4.5).
  - In the Sun–Earth–Planet position (i.e. when the outer planet is exactly opposite the Sun and rises when the Sun sets and sets when the Sun rises) it is called *Opposition.* This is when the planet is closest to the Earth and shines brightest.

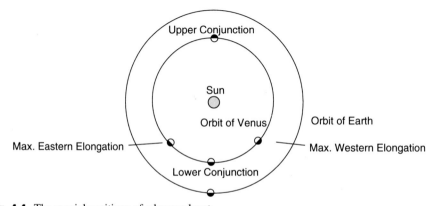

Fig. 4.4 The special positions of a lower planet

Fig. 4.5 The special positions of an upper planet

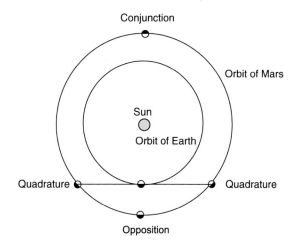

- At *conjunction* Sun, the planet is invisible with the Sun in the daytime sky and farthest from the Earth. Since the Earth moves faster around the Sun than the outer planets, they are overtaken by the Earth, which results in a *looping motion*.

Let J a sidereal year (orbit of the earth), U the sidereal orbit of a planet and S the synodic orbit of a planet (e.g. from opposition to opposition), then holds:

- The daily motion of the planet is 360°/U.
- The daily motion of the Earth: 360°/J.
- So the daily advance of the earth is. 360°/J–360°/U and identical with the daily synodic motion of the planet.

Therefore, one obtains:

$$\frac{360°}{J} - \frac{360°}{U} = \frac{360°}{S}$$

$$\frac{1}{U} = \frac{1}{J} - \frac{1}{S}$$

Thus, for upper (outer) planets:

$$U = \frac{SJ}{S - J} \tag{4.10}$$

and for lower planets:

$$U = \frac{SJ}{S + J} \tag{4.11}$$

Example

Mars; $S = 2.135$ years, from which follows $U = 1.881$ years. After 2.135 years Mars is in opposition, after 1.881 years at the same place in the sky. Therefore, if Mars was in opposition on 07/27/2018, then it must have been in opposition 2.135 years before that as well. It is easy to calculate where the planet will be at its next opposition in 2020 if it was in the constellation of western Capricorn at its opposition in 2018: $360 \times (1.881 - 2.135)° = 91.4°$ shifted east on the ecliptic. If you take the mean longitude for a constellation on the ecliptic to be 30° (12 signs of the zodiac!), then in 2020 it is shifted three constellations to the east, in the specific case in the eastern part of Pisces.

> Venus and Mercury are seen as either morning or evening stars in the sky, the outer planets are seen around opposition throughout the night.

4.1.4   Perihelion Rotation

Let's look at the planet closest to the Sun, *Mercury.* Besides the large solar mass the forces of the other planets are still acting. Already in the nineteenth century observed *Le Verrier* that Mercury's perihelion point (and thus orbit) is shifting, by $5.7''$ per year. If, however, one recalculates this value using the laws of Newtonian mechanics, one finds that it is too large ($0.43''$ per year). Only the Einstein's *general relativity* could explain this difference. According to general relativity, the perihelion of a planet orbiting a star of mass M on an orbit with the parameters a, e moves, after one revolution by the amount

$$\Delta\Theta = 2\pi \frac{GM}{c^2 a(1 - e^2)} \tag{4.12}$$

So the formula shows

- The larger the central mass M, the larger the perihelion rotation.
- The amount of perihelion rotation becomes smaller at larger distances from the central mass and at small orbital eccentricity.

> The *relativistic perihelion motion* of the Mercury is of the amount of $\Delta\Theta \approx 43''$ per century.

Table 4.1 Data of the moon orbit

Mean distance Earth–Moon	384,400 km
Smallest distance (perigee)	356,410 km
Largest distance (apogee)	406,740 km
Inclination of the orbit	$i = 5°8'$
Eccentricity	$e = 0.0549$
Sideric month (fixed star–fixed star)	27.321 days
Synodic month (phase–phase)	29.531 days
Tropical month (vernal equinox–spring equinox)	27.322 days
Anomalistic month (perigee–perigee)	27.555 days
Draconic month (node–node)	27.212 days

4.1.5 Lunar Orbit

The calculation of the Moon orbit (data in Table 4.1) is one of the most difficult celestial mechanics problems.

The sideric lunar orbit (orbit fixed star—fixed star), the sideric month, is equal to the rotation period of the of the Moon.

The moon has a *bound rotation,* we always see only the same half of the lunar sphere from the Earth.

We give the following perturbations of the Moon's orbit:

Great Inequality Due to the elliptical orbit of the Moon, its orbital velocities are different—near Earth it moves faster. The large inequality is the difference between the true Moon and the mean Moon. In addition, however, the following disturbing acceleration from the sun (mass $M_\odot$) applies:

$$a_\odot = 2GM_\odot \frac{R}{r^3} \tag{4.13}$$

R... distance moon-Earth; *r...* distance moon-sun. At full moon and new moon *(Syzygia)* you have a perturbation acceleration directed away from the Earth of the amount $a_\odot$. At first quarter and last quarter the acceleration is directed towards the earth, and with magnitude of $a_\odot/2$ and at 57.4° elongation it amounts to $a_\odot/\sqrt{2}$.

The radial component changes the distance of the moon, the tangential component changes the velocity of the moon, and the component perpendicular to the orbital plane tilts the moon's orbit.

The main perturbations of the lunar orbit are:

- *Evection:* perturbation of the large inequality due to different positions of the sun to the connecting line Perigee-Apogee*(apsidal line);* annual, amplitude $1°$.
- *Variation:* due to the tangential component there is an acceleration or deceleration; period: semi-monthly, amplitude: $39'$.
- *Annual inequality:* the radial component changes because of the eccentricity of the Earth's orbit. In aphelion the influence of the earth is stronger than in perihelion. This changes the orbital period of the moon by $10\,min$.
- *Secular acceleration:* the eccentricity of the Earth's orbit is changing. Today, the orbital period is $0.5\,s$ shorter than it was $2,000\,years$ ago.
- Rotation of the apsidal line: period $8.85\,years$.
- Rotation of the nodal line (caused by the perturbation component perpendicular to the orbit): The moon's orbit reacts to this with a precession motion. Period: $18.6\,years$ (retrograde).
- Change of orbital inclination: Period: $173\,days$.
- *Libration:* From the earth you can see more than half of the Moon because of:
  - Longitudinal libration: Rotation occurs at a constant angular velocity, but the orbital velocity around the Earth is variable due to the eccentricity of the Moon's orbit. Maximum effect: $\approx 8°$.
  - Latitudinal libration: The Moon's axis of rotation is not perpendicular to the orbital plane, so one sees once over the north pole and once over the south pole. Maximum effect about $1°$.
  - Parallactic libration: you look from different points from the earth or from one point at different times. Maximum effect: about $1°$.
  So, in total, you see 60% of the lunar surface from Earth.
- Earth-Moon interactions: the Sun and Moon cause lunisolar precession, and the Moon causes nutation. Earth and Moon move around the common center of gravity, which is about 3/4 Earth radii from the center of Earth. Furthermore the *tides* are to be mentioned.

4.1.6 Exoplanet Tidal Locking

Exoplanets are planets orbiting stars other than the Sun. Over the last two decades, more than 4000 such objects have been found. It is interesting for the search for life to detect planets in the so-called habitable zone. The habitable zone around a star defines the area where water can exist in liquid form on a planet that is located there. We will discuss this in detail in a separate chapter.

Many of the exoplanets that have been discovered are very close to their star, which is no wonder, since it is easier to find them there. If a planet is relatively close to its star, it can be forced by tidal interactions into a spin–orbit–resonance. This means that the planet rotates once around its own axis, while it moves once around its parent star. This effect is called *tidal locking*. The planet rotates relatively slowly and this of course has an effect on the irradiance on its surface. The planet closest to the Sun, *Mercury*, has the following resonance: Three rotations take the same amount of time as two orbits around the Sun:

- Orbital period 88 days
- Rotation time (length of day): 58 days and 15.5 h.

4.1.7 Tides

Consider a spherical body with radius R, which is attracted by a mass m at a distance $\mathbf{r}_0$. The tidal acceleration is the difference between the accelerations at the surface of the body (distance to mass m is $r_0 - R$) and at the centre of the body (distance to mass m in this case is r_0) (see Fig. 4.6) and thus the tidal acceleration is:

$$\mathbf{a}_{\text{tid}}(\mathbf{r}) = \frac{Gm}{|\mathbf{r}_0 - \mathbf{R}|^3}(\mathbf{r}_0 - \mathbf{R}) - \frac{Gm}{r_0^3}\mathbf{r}_0 \tag{4.14}$$

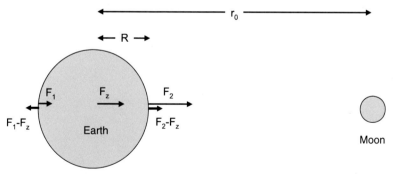

Fig. 4.6 Tidal force on the earth with radius R. The attraction of the moon, which is at a distance r_0 on the right, at the center of the earth is F_Z, at the side facing the moon F_2, at the side facing away from the moon F_1. It is valid $F_1 < F_Z < F_2$. If one forms the differences of the forces (e.g. $F_2 - F_Z$), one sees that on the side facing the moon as well as on the opposite side a resulting force (thick arrow) remains

One finds by series expansion: Let R is the radius of the earth and r_0 be the distance Earth–Moon, $r_0 \gg R$ then the difference in the acceleration exerted by the Moon at the Earth's surface and at the Earth's center:

$$
\begin{aligned}
a_{\text{tid}} &= \frac{GM_{\text{Moon}}}{(r_0 - R)^2} - \frac{GM_{\text{Moon}}}{r_0^2} \\[2mm]
&= \frac{GM_{\text{Moon}}}{r_0^2 - 2r_0 R + R^2} - \cdots \\[2mm]
&= \frac{GM_{\text{Moon}}}{r_0^2 \left(1 - 2R/r_0 + R^2/r_0^2\right)} - \cdots \qquad R^2/r_0^2 \to 0 \\[2mm]
&= \frac{GM_{\text{Moon}}}{r_0^2 \left(1 - 2R/r_0\right)} - \cdots
\end{aligned}
$$

and from

$$
(1+x)^p = 1 + px \qquad x \ll 1
$$

follows

$$
= \frac{GM_{\text{Moon}}}{r_0^2}(1 + 2R/r_0) - \frac{GM_{\text{Moon}}}{r_0^2} = \frac{2RGM_{\text{Moon}}}{r_0^3} \qquad (4.15)
$$

Let us consider separately the influence of the Sun and the Moon:

- The perturbative acceleration exerted by the moon:

$$
\Delta a_{\text{Moon}} = 2GM_{\text{Moon}} \frac{R}{r_{\text{Moon}}^3} \qquad (4.16)
$$

- The perturbative acceleration exerted by the Sun:

$$
\Delta a_{\text{Sun}} = 2GM_{\text{Sun}} \frac{R}{r_{\text{Sun}}^3} \qquad (4.17)
$$

Where r_{Moon} the distance Earth–Moon, r_{Sun} the distance Earth–Sun, and R is the radius of the earth.

The mass of the Moon is only about 1/80 of the mass of the Earth, which in turn is 1/330,000 of the mass of the Sun; nevertheless, the tidal effect of the Moon is about twice that of the Sun because of its proximity:

$$
\Delta a_{\text{moon}} / \Delta a_{\text{sun}} = 2.18
$$

The tide-mountain travels in one lunar day = 24^h50^{min} around the earth, i.e. every 12^h25^{min} is high tide at a certain place. At full moon resp. new moon, sun and moon work additive, and one speaks of a *Spring tide,* at half moon they affect subtractive *(neap tide).* Also the earth is deformed: By the moon by 21.4 cm, by the sun by 9.9 cm, and by a spring tide up to 1 m. The tidal range can be up to 15 m. Since the flood crest moves over the Earth's surface, *tidal friction* occurs. The Earth loses rotational energy E_{rot}.

$$E_{rot} = \frac{I}{2}\omega^2, \tag{4.18}$$

where I is the moment of inertia of the earth and ω is the angular velocity. If one considers the conservation of angular momentum $\mathbf{L}$ with

$$\mathbf{L} = I\omega \tag{4.19}$$

in the system earth–moon, then one sees that therefore the distance earth-moon must increase by 10 cm per year. The earth rotation duration, thus the day length, increases by 1/1000 s in 100 years. Therefore, about 370 million years ago, the year had 400 days of 22 h each.

The Moon also has a stabilizing effect on the Earth's axis, which currently is inclined around 23.5° to the normal to the ecliptic plane. Without the moon, the earth's axis could vary between 0 and 85° inclination. Mars has no large moons, and its rotational axis inclination can vary in the range of 0–60°. This could also have been the cause of large climate variations on that planet.

> The Moon stabilizes the orientation of the Earth's axis.

4.1.8 Comparison of Tidal Force of the Moon and Capillary Action in Plants

In this short section, we will estimate the influence of the Moon on biological processes. For the transport of water in living organisms we use the capillary action. Among other things, it is responsible for the fact that the water in the trees rises upwards (together with osmosis.[1]) For the capillary action of water in a glass tube whose radius is r, one finds the

---

[1] Theoretically, trees on earth can grow up to 130 m high, only then the osmotic pressure together with the capillary pressure is no longer sufficient to transport the water from the roots to the treetops.

following formula for the height of climb h of the of water:

$$h \approx 1.4 \times 10^{-5} \frac{m^2}{r} \tag{4.20}$$

Water rises in a glass tube with a radius of 0.01 mm to $h = 1.4$ m.

Let us consider the relationship of the tidal force of the Moon to this capillary force. The capillary force F_c is calculated from the mass of the liquid multiplied by the acceleration due to gravity, and is thus equal to the Weight of the liquid column:

$$F_c = mg = r^2 \pi h \rho g \tag{4.21}$$

The ratio of the tidal force of the moon F_G to F_c is then given by :

$$V = F_c/F_G = \frac{mg}{2GmM_{Moon}/r^3_{Moon}} = \frac{gr^3_{Moon}}{2GM_{Moon}} \tag{4.22}$$

The lunar mass is $M_{Moon} = 7.4 \times 10^{22}$ kg and the lunar distance $r_{Moon} = 3.8 \times 10^8$ m. In a thin capillary of $r = 0.01$ mm the ratio is $V \approx 5 \times 10^{13}$!

> Therefore, the influence of the Moon on capillary action in plants and animals is absolutely negligible.

4.2   Two-Body Problem

The two-body problem, i.e. the motion of two masses, can be solved analytically. As soon as more than two bodies are involved, the equations can only be solved approximately.

4.2.1   Definition of the Two-Body Problem

Definition: Let **F** be the force with which two masses attract each other, M, m be the masses of these two bodies, $\mathbf{r}/r$ the unit vector in the direction $\overline{Mm}$ and $G = 6.668 \times 10^{-11}$ Nm^2 kg^{-2} the gravitational constant, then holds:

$$\mathbf{F} = -G \frac{Mm}{r^2} \frac{\mathbf{r}}{r} \tag{4.23}$$

Fig. 4.7 Two-body problem. To the points M and m point to the corresponding vectors

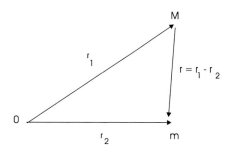

Let $\mathbf{r_1}$ is the distance from M to a point O and $\mathbf{r_2}$ be the distance from m to the point O, then the force of m relative to M (Fig. 4.7):

$$m\ddot{\mathbf{r}}_1 = -G\frac{Mm}{r^3}(\mathbf{r_2} - \mathbf{r_1}),$$

and for the force of M relative to m:

$$M\ddot{\mathbf{r}}_2 = -G\frac{Mm}{r^3}(\mathbf{r_1} - \mathbf{r_2}).$$

We divide the first equation by m and the second by M, subtract both equations and use $\mathbf{r} = \mathbf{r_2} - \mathbf{r_1}$: We get the equation of motion of the two-body problem:

$$\ddot{\mathbf{r}} = -G\frac{M+m}{r^3}\mathbf{r} \tag{4.24}$$

For $\mathbf{r} = (x, y, z)$ it is in component notation:

$$\ddot{x} = -G\frac{M+m}{r^3}x \tag{4.25}$$

and analogously for the other components. These are then three differential equations of second order. Thus six integrations are necessary for the solution, and one finds finds:

- 1st–3rd integral: Motion occurs in a plane.
- 4th integral: Area theorem (momentum theorem).
- 5th integral: Energy theorem.
- 6th integral: Path = conic section.

If $M \gg m$ then one can write $M + m \approx M$, and one has the equation of motion for a *One-body problem.*

Example

One-body problems are easy to solve and in many cases give at least good estimates: Let M = Earth mass and m is the mass of a satellite, then surely $m \ll M$.

4.2.2 Angular Momentum, Area Theorem

We investigate the vector product of the equation of motion (4.24) with $\mathbf{r}$. It follows that $\ddot{\mathbf{r}} \times \mathbf{r} = 0$ since on the right hand side $\mathbf{r} \times \mathbf{r}$ which is zero.

From $\ddot{\mathbf{r}} \times \mathbf{r} = \frac{d}{dt}(\dot{\mathbf{r}} \times \mathbf{r}) = 0$ follows:

$$\dot{\mathbf{r}} \times \mathbf{r} = \mathbf{v} \times \mathbf{r} = \text{const} = -\mathbf{N} \qquad (4.26)$$

Angular momentum = $m(\mathbf{r} \times \mathbf{v})$, so $\mathbf{N}$ is the angular momentum vector related to mass 1, and we have derived the law of conservation of angular momentum from the equation of motion. From the constancy of $\mathbf{N}$ it also follows that $\mathbf{r}$, $\mathbf{v}$ remain in the same plane.

> The motion in the two-body problem occurs in a fixed plane.

The change dF of the surface F is (Fig. 4.8):

$$dF = \frac{1}{2}(\mathbf{r} \times d\mathbf{r}) \qquad (4.27)$$

and for the derivation holds:

$$\dot{\mathbf{F}} = \frac{1}{2}(\mathbf{r} \times \dot{\mathbf{r}}) = \text{const} = 1/2\mathbf{N} \qquad (4.28)$$

Fig. 4.8 For the derivation of the surface velocity

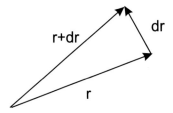

Thus we have that *second Kepler's law* is derived:

> The change of the area velocity is constant, in equal times equal areas are traversed by the radius vector (connecting line planet-sun).

4.2.3 Orbit Shape

To obtain the equation for the orbital, multiply the equation of motion vectorially by $\mathbf{N}$:

$$\mathbf{N} \times \ddot{\mathbf{r}} = -G\frac{M+m}{r^3}(\mathbf{N} \times \mathbf{r})$$

$$= -G\frac{M+m}{r^3}[(\mathbf{r} \times \dot{\mathbf{r}}) \times \mathbf{r}]$$

From

$$(\mathbf{a} \times \mathbf{b}) \times \mathbf{c} = (ac)\mathbf{b} - (bc)\mathbf{a}$$

follows:

$$\mathbf{N} \times \ddot{\mathbf{r}} = -G\frac{M+m}{r^3}(r^2\dot{\mathbf{r}} - (r\dot{r})\mathbf{r})$$

and because of

$$\frac{d}{dt}\left(\frac{\mathbf{r}}{r}\right) = \frac{r\dot{\mathbf{r}} - \dot{r}\mathbf{r}}{r^2}$$

you have:

$$\mathbf{N} \times \ddot{\mathbf{r}} = -G(M+m)\frac{d}{dt}\left(\frac{\mathbf{r}}{r}\right) \qquad \Big| \int$$

$$\mathbf{N} \times \dot{\mathbf{r}} = -G(M+m)\frac{\mathbf{r}}{r} - \mathbf{C}$$

$$(4.29)$$

Now multiply this expression scalar by r and note: $\dot{\mathbf{r}} \times \mathbf{r} = -\mathbf{N}$ so that $(\mathbf{N} \times \dot{\mathbf{r}})r = \mathbf{N}(\dot{\mathbf{r}} \times \mathbf{r}) = -N^2$ follows, and obtains:

$$N^2 = G(M+m)r + (\mathbf{r}\mathbf{C}).$$

With the quantities p and e:

$$p = \frac{N^2}{G(M+m)} \qquad e = \frac{|C|}{G(M+m)} \tag{4.30}$$

then follows the known equation:

$$r = \frac{p}{1 + e \cos v} \tag{4.31}$$

This equation describes a conic section (first Kepler's law!) e is the eccentricity. For a circle $e = 0$, for an ellipse $e < 1$ for a parabola $e = 1$ and a hyperbola $e > 1$. The pericentre distance for an ellipse is $a(1 - e)$ for a parabola $p/2$, the apocenter distance for an ellipse $a(1 + e)$.

4.2.4 Energy Theorem

The potential energy V is defined as:

$$V = -G\frac{Mm}{r} \tag{4.32}$$

This energy is released when the distance is increased by the amount of dr is. It holds:

$$\mathbf{F} = -\mathrm{grad}\,V \qquad \mathbf{F}d\mathbf{r} = -\mathrm{grad}\,V\,d\mathbf{r} \tag{4.33}$$

The potential of M at distance r (on $m = 1$) is then :

$$\tilde{V} = -GM/r \tag{4.34}$$

and the equation of motion is:

$$m\ddot{\mathbf{r}} = -\mathrm{grad}\,V. \tag{4.35}$$

This is scalar multiplied by $\dot{\mathbf{r}}$

$$m\ddot{\mathbf{r}}\,\dot{\mathbf{r}} = -\mathrm{grad}\,V\dot{\mathbf{r}} \qquad \frac{d}{dt}\left(\frac{1}{2}m\dot{\mathbf{r}}^2\right) = -\frac{dV}{dt}$$

Thus we have the energy theorem:

$$\frac{1}{2}mv^2 + V = \frac{1}{2}mv^2 - G\frac{Mm}{r} = \mathrm{const} \qquad E_{\mathrm{kin}} + E_{\mathrm{pot}} = E = \mathrm{const} \tag{4.36}$$

From this equation it follows that the velocity at perihelion (small r) must be greater than at aphelion.

By temporal averaging of the orbital motion (also valid for elliptical orbits) one obtains the *virial theorem:*

$$E_{kin} = -1/2 E_{pot} \tag{4.37}$$

Virial theorem: the time average of the kinetic energy is equal in magnitude to the time average of half the potential energy.

The case is especially simple for a *circular orbit:* here, quite simply, centrifugal force = attractive force:

$$\frac{mv^2}{r} = G\frac{Mm}{r^2} \tag{4.38}$$

From this we get the *circular* orbital velocity:

$$v_c = \sqrt{\frac{GM}{r}} \tag{4.39}$$

For a *parabolic trajectory* at infinity $E_{kin} = E_{pot} = 0$ and $mv^2/2 = GMm/r$ thus the velocity:

$$v_e = \sqrt{2}v_c \tag{4.40}$$

This is also called *escape velocity.* We therefore obtain for the relation between energy and orbital form:

- Ellipse $E_{kin} < E_{pot}$,
- Parabola $E_{kin} = E_{pot}$,
- Hyperbola $E_{kin} > E_{pot}$.

The energy GMm/r is required to move a mass m in the gravitational field of a mass M to infinity. The velocity at any point is for an elliptical orbit:

$$v^2 = GM \left(\frac{2}{r} - \frac{1}{a}\right) \tag{4.41}$$

Virial Theorem: Application to the Collapse of a Cloud

At this point, a brief application of the virial theorem. In equilibrium:

$$E_{\text{kin}} + 2E_{\text{pot}} = 0 \tag{4.42}$$

Let us examine at what point an interstellar gas cloud becomes unstable and collapses into a star.[2] The kinetic energy of the gas cloud (due to the motion of the gas particles) is:

$$E_{\text{kin}} = \frac{3}{2}NkT \tag{4.43}$$

k is the Boltzmann constant, T is the temperature, and N is the number of particles. The potential energy is

$$E_{\text{pot}} = \frac{3}{5}\frac{GM_c^2}{R_c} \tag{4.44}$$

Where M_c is the mass of the gas cloud, R_c is its radius. The condition for collapse is then:

$$2E_{\text{kin}} < |E_{\text{pot}}| \tag{4.45}$$

More about this in the chapter on star formation.

4.2.5   Third Kepler's Law

Let us consider again the orbital equation $r = p/(1 + e \cos v)$. If we put $v = 0$ then we have the distance at the pericenter and at $v = \pi$ is the distance at the apocenter. Therefore:

$$a = \frac{1}{2}\left(r_{\text{per}} + r_{\text{apo}}\right) = \frac{1}{2}\left(\frac{p}{1-e} + \frac{p}{1+e}\right) = \frac{p}{1-e^2}$$

and therefore:

$$1 - e^2 = \frac{p}{a} = \frac{N^2}{G(M+m)a}$$

The orbital time times the surface velocity gives the ellipse area:

$$U\frac{N}{2} = ab\pi = \pi a^2\sqrt{1-e^2} = \frac{\pi a^2 N}{\sqrt{G(M+m)a}}$$

---

[2] In reality, fragmentation occurs first.

and we get the strict form of *Kepler's third law:*

$$\frac{a^3}{U^2} = \frac{G}{4\pi^2}(M + m) \tag{4.46}$$

4.3 N-Body Problem

In the N-body problem, we study the motion of N bodies under the influence of their mutual gravitation. Once $N > 2$, there are no more analytical solutions, but we present a numerical integration method how to solve such problems. The restricted three-body problem provides a very good approximation for many cases and is treated separately. It has practical applications today, e.g. for the stationing of satellites at the Lagrangian points of the Earth-Sun system. Here the gravitational forces of the Earth and Sun cancel each other and satellites remain stable at these points.

4.3.1 The General N-Body Problem

Given are N-bodies; r_i is distance of mass m_i from the center of gravity, r_{ik} the distance between the masses m_i, m_k. The equation of motion for the mass m_i is then:

$$m_i \ddot{\mathbf{r}}_i = -\sum_{k=1}^{N} G m_i m_k \frac{\mathbf{r}_i - \mathbf{r}_k}{r_{ik}^3} \tag{4.47}$$

These are $3n$ 2nd order differential equations; one needs $6N$ integrations. However, there are only ten integrals:

• Center of gravity theorem: adding all equations of motion, is:

$$\sum m_i \ddot{\mathbf{r}}_i = 0, \tag{4.48}$$

since for each term $\mathbf{r}_k - \mathbf{r}_i$ a term $\mathbf{r}_i - \mathbf{r}_k$ exists. Integrating above equation twice, it follows:

$$\sum m_i \mathbf{r}_i = \mathbf{a}t + \mathbf{b}.$$

I. e. the center of gravity of a N-body system is either at rest or in rectilinear motion. In total, this gives six integrals.

- Angular momentum theorem: as in the two-body problem, we also obtain here:

$$\sum m_i \mathbf{r}_i \times \dot{\mathbf{r}}_i = \text{const} \tag{4.49}$$

and thus three more integrals.
- Energy theorem: The potential here is:

$$V = -G \sum_{i,k} \frac{m_i m_k}{r_{ik}} \tag{4.50}$$

and the equations of motion:

$$\sum m_i \ddot{\mathbf{r}}_i = \text{grad } V \tag{4.51}$$

Again one multiplies scalar by $\dot{\mathbf{r}}$ and obtains the law of conservation of energy.

> The law of conservation of energy is important in estimating the stability of a cluster of stars or a cluster of galaxies. In an N-particle system, individual particles can gain energy at the expense of others and escape the system.

Bruns and *Poincaré* showed that there are no further integrals besides the treated ones. So already a 3-body-problem cannot be solved analytically.

One can also give the virial theorem: If N bodies always remain in a finite volume, then the time-averaged kinetic or potential energy:

$$\overline{E_{\text{kin}}} = -\frac{1}{2}\overline{E_{\text{pot}}} \tag{4.52}$$

4.3.2   The General Three-Body Problem

Given three masses, m_1, m_2, m_3. The equations of motion are:

$$m_1 \ddot{\mathbf{r}}_1 = k^2 \left(\frac{m_1 m_2}{|\mathbf{r}_2 - \mathbf{r}_1|^3}(\mathbf{r}_2 - \mathbf{r}_1) + \frac{m_1 m_3}{|\mathbf{r}_3 - \mathbf{r}_1|^3}(\mathbf{r}_3 - \mathbf{r}_1) \right)$$

$$m_2 \ddot{\mathbf{r}}_2 = k^2 \left(\frac{m_1 m_2}{|\mathbf{r}_1 - \mathbf{r}_2|^3}(\mathbf{r}_1 - \mathbf{r}_2) + \frac{m_2 m_3}{|\mathbf{r}_3 - \mathbf{r}_2|^3}(\mathbf{r}_3 - \mathbf{r}_2) \right) \tag{4.53}$$

$$m_3 \ddot{\mathbf{r}}_3 = k^2 \left(\frac{m_1 m_3}{|\mathbf{r}_1 - \mathbf{r}_3|^3}(\mathbf{r}_1 - \mathbf{r}_3) + \frac{m_3 m_2}{|\mathbf{r}_2 - \mathbf{r}_3|^3}(\mathbf{r}_2 - \mathbf{r}_3) \right)$$

The kinteic energy becomes: $E = \frac{1}{2}(m_1 v_1^2 + m_2 v_2^2 + m_3 v_3^2)$. In this system of equations, as is still usual in celestial mechanics, the following system of units was used: We start from the third Kepler law and set the mass of the sun = 1:

$$T^2/a^3 = 4\pi^2/(k^2(1 + m_p)) \tag{4.54}$$

Where m_p is the mass of a planet. If now $T = 365.2564$ is days (goes back to Newton) and $a = 1$ and the mass of the planet = 0, then:

$$k = \frac{2\pi}{T} = 0.01720209895 \tag{4.55}$$

For the potential write:

$$V = -k^2 \left(\frac{m_1 m_2}{|\mathbf{r}_1 - \mathbf{r}_2|} + \frac{m_1 m_3}{|\mathbf{r}_1 - \mathbf{r}_3|} + \frac{m_2 m_3}{|\mathbf{r}_2 - \mathbf{r}_3|} \right) \tag{4.56}$$

For the barycenter: $m = m_1 + m_2 + m_3$, $m\mathbf{r} = m_1\mathbf{r}_1 + m_2\mathbf{r}_2 + m_3\mathbf{r}_3$. If one puts the zero point of the coordinate system in the center of gravity (barycenter), all $\mathbf{r}_i$ are replaced by $\mathbf{r}_i - \mathbf{r}$ and you still have the relations:

$$m_1\mathbf{r}_1 + m_2\mathbf{r}_2 + m_3\mathbf{r}_3 = \mathbf{0} \qquad m_1\dot{\mathbf{r}}_1 + m_2\dot{\mathbf{r}}_2 + m_3\dot{\mathbf{r}}_3 = \mathbf{0} \tag{4.57}$$

Further, one can introduce relative coordinates as: $\mathbf{q}_2 = \mathbf{r}_2 - \mathbf{r}_1$, $\mathbf{q}_3 = \mathbf{r}_3 - \mathbf{r}_1$. Then the equations of motion are:

$$\ddot{\mathbf{q}}_2 = -k^2 \frac{m_1 + m_2}{q_2^3} \mathbf{q}_2 + k^2 m_3 \left(\frac{\mathbf{q}_3 - \mathbf{q}_2}{|\mathbf{q}_3 - \mathbf{q}_2|^3} - \frac{\mathbf{q}_3}{q_3^3} \right) \tag{4.58}$$

$$\ddot{\mathbf{q}}_3 = -k^2 \frac{m_1 + m_3}{q_3^3} \mathbf{q}_3 + k^2 m_2 \left(\frac{\mathbf{q}_2 - \mathbf{q}_3}{|\mathbf{q}_2 - \mathbf{q}_3|^3} - \frac{\mathbf{q}_2}{q_2^3} \right) \tag{4.59}$$

$\mathbf{r}_1$ follows from the center of gravity theorem, since:

$$\mathbf{r}_1 = -\frac{1}{m}(m_2\mathbf{q}_2 + m_3\mathbf{q}_3) \tag{4.60}$$

4.3.3 Restricted Three-Body Problem

Two large masses orbit around the common center of gravity. In a restricted three-body problem , a third mass is not supposed to affect their motion, thus $m_3 \ll m_1, m_2$. From the general three-body problem

$$\ddot{\mathbf{r}}_i = \sum_{j=1, j \neq i}^{3} \frac{k^2 m_j}{|\mathbf{r}_j - \mathbf{r}_i|^3} (\mathbf{r}_j - \mathbf{r}_i), \qquad 1 \leq i \leq 3 \tag{4.61}$$

it follows for $m_3 = 0$:

$$\ddot{\mathbf{r}}_1 = \frac{k^2 m_2 (\mathbf{r}_2 - \mathbf{r}_1)}{|\mathbf{r}_2 - \mathbf{r}_1|^3} \tag{4.62}$$

$$\ddot{\mathbf{r}}_2 = \frac{k^2 m_1 (\mathbf{r}_1 - \mathbf{r}_2)}{|\mathbf{r}_1 - \mathbf{r}_2|^3} \tag{4.63}$$

$$\ddot{\mathbf{r}}_3 = \frac{k^2 m_1 (\mathbf{r}_1 - \mathbf{r}_3)}{|\mathbf{r}_1 - \mathbf{r}_3|^3} + \frac{k^2 m_2 (\mathbf{r}_2 - \mathbf{r}_3)}{|\mathbf{r}_2 - \mathbf{r}_3|^3} \tag{4.64}$$

Let $\mathbf{r} = \mathbf{r}_2 - \mathbf{r}_1$ and $\mu = m_2/(m_1 + m_2)$. Then holds:

$$\mathbf{r}_1 = -\mu \mathbf{r} \qquad \mathbf{r}_2 = (1 - \mu)\mathbf{r} \tag{4.65}$$

Now we choose a rotating coordinate system, which rotates with the same angular velocity n as the two primary masses m_1, m_2 around each other; $n = ka^{-3/2}\sqrt{m_1 + m_2}$. We assume that the two primary bodies move on a circular path with radius p. Then holds:

$$r = p = N^2/(k^2(m_1 + m_2)) \tag{4.66}$$

Finally we find the following equations of motion for so-called *circular restricted three-body problem*:

$$\ddot{x} - 2n\dot{y} - n^2 x = k^2 \left(m_1 \frac{\mu p + x}{\rho_1^3} + m_2 \frac{(x - (1 - \mu)p)}{\rho_2^3} \right) \tag{4.67}$$

$$\ddot{y} + 2n\dot{x} - n^2 y = -k^2 y \left(\frac{m_1}{\rho_1^3} + \frac{m_2}{\rho_2^3} \right) \tag{4.68}$$

Here $\mathbf{s} = (x, y, z)$ is the coordinate vector of the body m_3 and

$$\rho_1 = \sqrt{(\mu p + x)^2 + y^2} \qquad \rho_2 = \sqrt{((1 - \mu)p - x)^2 + y^2} \tag{4.69}$$

Fig. 4.9 Solutions of the constrained three-body problem

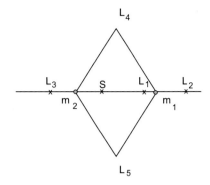

We find another integral the *Jacobi integral:*

$$n^2(x^2 + y^2) + 2k^2 \left(\frac{m_1}{\rho_1} + \frac{m_2}{\rho_2} \right) - \dot{x}^2 - \dot{y}^2 = J \tag{4.70}$$

Let us consider the motion of a spaceship in the Earth-Moon system. If we set $m_1 = 1, m_2 = 1/81.3$ as masses for earth and moon, respectively, we get $\mu = 1/82.3$. The motion of the spaceship then follows from the above equations of motion.

As stationary solutions one finds from the equation of motion the so-called *Libration points:* $\dot{x} = \dot{y} = 0, \ddot{x} = \ddot{y} = 0$:

$$\frac{n^2}{k^2} x = m_1 \frac{\mu p + x}{\rho_1^3} + m_2 \frac{x - (1 - \mu)p}{\rho_2^3} \tag{4.71}$$

$$\frac{n^2}{k^2} y = y \left(\frac{m_1}{\rho_1^3} + \frac{m_2}{\rho_2^3} \right) \tag{4.72}$$

This yields the following solutions (Fig. 4.9):

- Three stationary solutions for $y = 0$: One obtains the so-called libration points L_1, L_2, L_3: The arrangement of the masses looks like this:[3]
  $L_1 : M_2 - M_3 - M_1$
  $L_2 : M_2 - M_1 - M_3$
  $L_3 : M_3 - M_2 - M_1$
  The points L_1, L_2, L_3 are unstable. Objects will move away from these points due to small perturbations. The SOHO spacecraft, used for solar observation is located in the Sun-Earth system at the Lagrangian point L_1 1.5 million km from the Earth .

---

[3] m_3 is said to be located at the libration point L_i.

Fig. 4.10 The Trojan
asteroids. At L_4 we know 4188
objects, at L_5 2268 objects (as
of Jan. 2017)

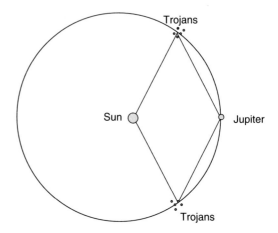

- Two stationary solutions with $y \neq 0$: Here the point masses form the vertices of an
  equilateral triangle, which is called the libration points L_4 with $y > 0$ and L_5 with $y <$
  0, respectively. As an example, consider the *Trojan* group of minor planets (Fig. 4.10),
  which reside at L_4 and L_5, respectively, and always form an equilateral triangle with
  the Sun and Jupiter. The first Trojan discovered was the object (588) Achilles and was
  found in 1906 by M. Wolf. In 1990 minor planets were also discovered at Mars: The
  minor planet Eureka has a relatively large orbital inclination, but is located at L_5 in the
  Mars-Sun system. Trojans were also found at Uranus and Neptune.

Hill's Surfaces: From the Jacobi integral follows considering that the velocity square must
always be positive ($\dot{x}^2 + \dot{y}^2 \geq 0$):

$$n^2(x^2 + y^2) + 2k^2 \left(\frac{m_1}{\rho_1} + \frac{m_2}{\rho_2} \right) \geq J \tag{4.73}$$

When results in Eq. (4.73) are zero, they are called zero velocity surfaces. These separate
possible whereabouts of a third mass from the forbidden zones where such a mass point
can never reside. One finds that m_3 can either reside:

- m_3 close to m_1,
- m_3 close to m_2,
- m_3 far outside the orbits of m_1, m_2.

Such studies are of interest for planets in binary star systems.

Let us consider the following example: consider the Earth-Moon system; where is the
neutral point at which the gravitational forces of the Earth and Moon cancel?

Solution: r_1 let be the distance of the neutral point from the earth, r_2 the distance from
the moon and r is the distance earth-moon, where $r = r_1 + r_2$. At the neutral point, the

forces cancel:

$$\frac{m_{Earth}G}{r_1^2} = \frac{m_{Moon}G}{(r-r_1)^2} \tag{4.74}$$

Thus, for the distance ratio, we find:

$$\frac{r_1^2}{(r-r_1)^2} = \frac{m_{Earth}}{m_{Moon}} = 81 \tag{4.75}$$

and from

$$\frac{r_1}{r-r_1} = 9 \rightarrow r_1 = 9r - 9r_1 \rightarrow r_1 = \frac{9}{10}r \tag{4.76}$$

The neutral point is therefore 346,000 km from the earth and 38,000 km from the moon remote.

4.3.4 The Runge-Kutta Method for Numerical Integration

This method was already established in 1895 by Carl Runge and later extended and improved by Heun and Kutta and is used for the integration of ordinary differential equations. If one knows a solution x_i at a time t_i then one can give the solution at a later point of time $x_{i+1} = x(t_i + h)$ by:

$$k_1 = hf(t_i, x_i) \tag{4.77}$$

$$k_2 = hf(t_i + \frac{h}{2}, x_i + \frac{k_1}{2}) \tag{4.78}$$

$$k_3 = hf(t_i + \frac{h}{2}, x_i + \frac{k_2}{2}) \tag{4.79}$$

$$k_4 = hf(t_i + h, x_i + k_3) \tag{4.80}$$

$$x_{i+1} = x_i + \frac{1}{6}(k_1 + 2k_2 + 2k_3 + k_4) \tag{4.81}$$

Let us take the following heliocentric equatorial coordinates of Jupiter:

$$\mathbf{r}(t_0) = \begin{pmatrix} -5.433317 \\ -0.406439 \\ -0.041833 \end{pmatrix} \quad \mathbf{v}(t_0) = \begin{pmatrix} 0.000441975 \\ -0.006598355 \\ -0.002839125 \end{pmatrix}$$

The units here are the astronomical unit AU for the longitude or for the velocity AU/day and the solar mass (= 1) for the masses.

The equation of motion is:

$$\ddot{\mathbf{r}} = -k^2 \frac{1+m}{r^3} \mathbf{r} \tag{4.82}$$

$m = 1/1047$ is the mass of Jupiter. If one chooses as time unit $N = 1000\,\text{days}$ and

$$\mathbf{v}(t_0) = \begin{pmatrix} 0.441975 \\ -6.598355 \\ -2.839125 \end{pmatrix}$$

and we set

$$\mathbf{x} = (x_1, ..., x_6) = (r_1, r_2, r_3, v_1, v_2, v_3)$$

then we get $N^2 = 10^6$:

$$
\begin{aligned}
\dot{x}_1 &= x_4 & \dot{x}_4 &= -10^6 k^2 \frac{1+m}{r^3} x_1 \\
\dot{x}_2 &= x_5 & \dot{x}_5 &= -10^6 k^2 \frac{1+m}{r^3} x_2 \\
\dot{x}_3 &= x_6 & \dot{x}_6 &= -10^6 k^2 \frac{1+m}{r^3} x_3
\end{aligned}
\tag{4.83}
$$

If we now set a step size of 50 days, $h_0 = 50$, then one obtains:

$$\mathbf{r}_1 = \begin{pmatrix} -5.398811 \\ -0.735176 \\ -0.183585 \end{pmatrix} \qquad \mathbf{v}_1 = \begin{pmatrix} 0.937583 \\ -6.546126 \\ -2.828813 \end{pmatrix}$$

As a test, one can test the constancy of the angular momentum, which for two arbitrary times t_0 and t_1 should be the same:

$$\mathbf{N}(t_0) = \mathbf{r}(t_0) \times \mathbf{v}(t_0) = \mathbf{r}(t_1) \times \mathbf{v}(t_1)$$

One can define a bound for the variation of the angular momentum. If this is violated during an integration step, then the step is repeated with a smaller interval h_0.

4.3.5 Stability

The Lagrange points $L_1 \ldots L_3$ are unstable, small perturbations destabilize the position of a mass m_3 from this position. L_4, L_5 are stable against small perturbations.

In general, a perturbation on a planet P becomes maximal when the sun, planet, and perturbation body m are in a line: The acceleration of m on P is:

$$b_P = Gm/r^2 \qquad (4.84)$$

The acceleration of m to the sun is:

$$b_\odot = Gm/(R+r)^2$$

This gives the perturbation to:

$$\Delta b = Gm \left(1/r^2 - 1/(R+r)^2 \right)$$

Only if R (distance from P to the sun) $\ll r$ is (distance from P to m), one can apply the following series expansion:

$$\frac{1}{(R+r)^2} = \frac{1}{r^2(\frac{R}{r}+1)^2} \qquad R/r << 1$$

$$\frac{1}{(1+\epsilon)^p} \approx 1 - p\epsilon \qquad (4.85)$$

thus:

$$\frac{1}{(R+r)^2} = \frac{1}{r^2} \frac{1}{(1+R/r)^2} = \frac{1}{r^2}(1 - 2R/r)$$

Thus, for the perturbation acceleration we obtain:

$$\Delta b = 2Gm \frac{R}{r^3} \qquad (4.86)$$

Setting the acceleration due to the Sun equal to 1, we obtain (mean values):
Perturbation of Venus on Earth (max): 1/37,000,
perturbation of Jupiter on Earth (opposition): 1/53,000,
perturbation of Jupiter on Saturn (max): 1/360.

This perturbation contains an orthogonal component acting perpendicular to the orbital plane, and the elements. Ω, i as well as a component acting in the orbital plane and a, e, ω influences.

4.4   Many-Particle Systems

Due to the enormous development of computing power, it is now possible to compute many-particle systems. Typical examples are:

- Star clusters, consisting of several 1000 to 10^6 Stars.
- Galaxy clusters, consisting of many galaxies.
- Galaxy clusters and distribution of dark matter.
- Gas clouds and their dynamics.
- Star formation in gas clouds.

For such systems, one can consider the individual particles, and solve differential equations for their motions. Often simplifications can be made that give faster results. We give two applications.

4.4.1   Virial Theorem and Distance of an Interstellar Gas Cloud

Consider a system of particles that do not collide with each other. Let the only interaction between the particles be gravitation. As an example, consider an interstellar gas cloud; collisions of particles are very unlikely because of low particle density. Virial theorem means for time average of kinetic and potential energy:

$$< E_{\text{kin}} > + \frac{1}{2} < E_{\text{pot}} > = 0 \tag{4.87}$$

We consider m_i particles with a velocity v_i.

$$< E_{\text{kin}} > = < \sum_i m_i v_i^2 / 2 > \tag{4.88}$$

If all particles have the same mass, then:

$$< E_{\text{kin}} > = M < V^2 > /2 \tag{4.89}$$

The potential energy is:

$$< E_{\text{pot}} > = G < \sum m_i m_j / r_{ij} > \tag{4.90}$$

respectively, if all the particles have the same mass:

$$< E_{\text{pot}} >= -GM^2/R \qquad (4.91)$$

Here, R is an averaged radius. For the case of homogeneous particle density, one can apply $(5/3)R$ and obtain

$$< V^2 >= \frac{3}{5}GM/R \qquad (4.92)$$

Now consider a gas cloud of neutral hydrogen. The velocity dispersion $< V^2 >$ is three-dimensional, but we measure only the radial component with the Doppler effect, i.e. the velocity component towards or away from us. It can be shown that there is the following relationship:

$$< V^2 >= 3\Delta V^2/(8 \ln(2) \qquad (4.93)$$

ΔV is the measurable radial velocity distribution of the gas cloud. The mass of the gas cloud is $\frac{4}{3}\pi R^3 \rho$, the mass density $\rho = N_{\text{HI}} m_H/2R$ where N_{HI} is the so-called column density. The column density denotes the density integrated (= summed up) along the line of sight and has the unit mass/area. It is responsible, for example, for the extinction (= weakening) of the radiation in the line of sight. It results under the assumption of the validity of the Virial Theorem for the radius of the gas cloud:

$$R = 15\Delta V^2/(16\pi \ln(2)GM_H N_H \qquad (4.94)$$

We obtain the true radius of the gas cloud from the observed radial velocity distribution and the column density. The distance then follows from comparing the true radius of the gas cloud with the easily measured angular extent.

4.4.2 Ergodic Behavior

We consider a many-particle system. This should behave like a thermodynamic system. We will get to know approaches for this, for example, in the chapter on cosmology; in the theory of the origin and structure of the universe, for example, whole galaxies are regarded as particles which behave similarly to the particles in a gas and can be described accordingly.

In our system, all possible states of the phase space are to be reached (molecular chaos). The phase space describes the set of all possible states of the system. If we consider the motions of many particles, we get a phase space with a very high dimension. In Hamiltonian mechanics, the phase space is the space of all locations and momentums of the particles. If there are N particles in the system (e.g. N stars in a star cluster), then the phase space is 6 N-dimensional, since each particle can be described by:

- 3 components of position
- 3 components of velocity.

A *trajectory* determines the time evolution of the system from a certain starting point.

> A system is called ergodic, if the trajectories fill the whole phase space.

Now we describe the mean behavior of the system. For an ergodic system, one can determine this behavior in two ways:

- One follows the evolution over a very long period of time $\rightarrow$ Time average.
- One averages over all possible states $\rightarrow$ Ensemble average.

This leads to another definition of ergodicity. A system is called ergodic if time averages and ensemble averages lead to the same result with probability 1. In a weakly ergodic system, only the expected values and the variances agree, higher order moments are neglected.

Example of an ergodic system: particle moving randomly in a closed volume. The volume determines the phase space. Over time, all points of the volume are reached. It does not matter whether one averages over time or space.

4.5   Spaceflight

How do you fly to the planets cheaply? On a direct way? What speeds must a satellite reach to overcome the Earth's gravitational pull or to travel in a circular orbit around it? How do rockets work and what are the possibilities of propulsion? These are the questions we will address in this chapter.

4.5.1   Escape Velocities

From the *law of conservation of energy*, we can estimate how high a particle of mass m can be thrown near the Earth; if the total energy is 0, then:

$$mv^2/2 = mgh \rightarrow h = v^2/(2\,g)$$

So this formula gives us the altitude h, which is a vertical upward throw with a velocity v.

A stone is thrown with a velocity of 14 m/s upwards. What height does it reach? Inserting it into the formula gives: 10 m.

If you throw a stone to greater heights, you have to take into account that the acceleration due to gravity g decreases with distance:

$$mv^2/2 - (GMm/R) = -GMm/(R+h).$$

From this also follows the *Escape velocity* from the earth to 11.2 km/s (one puts on the right side $h \to \infty$). In most space missions the satellites first arrive in a *Earth Orbit*. Further, rockets are fired to the east to take advantage of the Earth's *rotational speed* (at the equator 465 m/s).

In Table 4.2 the orbit data for Earth satellites at different altitudes are given. Note that satellites at altitudes below 200 km are strongly decelerated by the by the *earth's atmosphere*. Due to *radiation bursts* on the Sun, the upper Earth's atmosphere heats up and expands, causing low-Earth-orbiting satellites (LEOs, *low earth orbiting satellites*) to crash.

Table 4.3 shows the escape velocities for different objects.

Table 4.2 orbits of artificial earth satellites. $h = 0$ is of course hypothetical

h [km]	Orbital period	Velocity [km/s]
0	1^h24^{min}	7.91
100	1^h26^{min}	7.85
1000	1^h45^{min}	7.36
10,000	5^h48^{min}	4.94
35,000	23^h56^{min}	3.07

Table 4.3 Escape velocities satellite orbital velocity in km/s

Object	v_{esc}
Moon	2.4
Earth	11.2
Jupiter	60
Sun	620
White dwarf	7600
Neutron star	160,000
Black hole	c = 300,000

4.5.2   Rocket Formula

The *Rocket Formula* follows from the conservation of momentum: Let m, V, mV mass, velocity and momentum of the rocket. Let S be the outflow velocity of the propellants relative to the rocket. $V - S$ is the emitted propellant velocity relative to the space-fixed system.

We therefore get after a time dt:

- Mass of the outflowed gases: dm,
- momentum of the gas mass $dm(V - S)$,
- mass of the rocket $m - dm$,
- Velocity of the rocket $V + dV$,
- Momentum of the rocket: $(m - dm)(V + dV)$.

The *Law of Conservation of Momentum* then provides:

$$mV = (m - dm)(V + dV) + dm(V - S) \tag{4.95}$$

The first term on the right-hand side refers to the rocket, and the second term on the right-hand side refers to the outflowing gas. Thus:

$$mV = mV + mdV - Vdm - dmdV + Vdm - Sdm.$$

In this expression, $dmdV$ can be neglected (small 2nd order). One then obtains

$$dV = S\frac{dm}{m}$$
$$\Delta V = S \ln \frac{m_0}{m_B} \tag{4.96}$$

m_0 is the total mass at $t = 0$ and m_B the mass at burnout.

So if the rocket was launched at time $t = 0$ with $V = 0$ started, then m_0 is its starting mass and $\Delta V = V_B$ is the final velocity reached after burnout.

You can finally state the rocket formula like this: V_B/S is the velocity ratio and m_0/m_B is the mass ratio, then

$$\frac{V_B}{S} = \ln \frac{m_0}{m_B} \tag{4.97}$$

The *Efficiency* is given by the ratio of the kinetic energy of the rocket at burnout to the available energy of the propellants:

$$\eta = \frac{m_B V_B^2}{m_T S^2} = \frac{V_B^2/S^2}{(m_0/m_B) - 1} = \frac{[\ln(m_0/m_B)]^2}{(m_0/m_B) - 1} \tag{4.98}$$

Calculate the efficiency for various values of V_B/S and m_0/m_B!

The *escape velocity* is between 2 and 4 km/s for thermal propulsion. The maximum efficiency is given at a velocity ratio of 1.594 and a mass ratio of 4.93. Since the mass ratio cannot become arbitrarily large, a *multistage principle* is used, with first-stage rocket bodies being jettisoned (in the case of the Space Shuttle, it was the propellant tanks + solid boosters). If $m_0/m_B = \Pi(m_0)/(m_B)_i$, i.e. the product of the individual mass ratios, and $V_B = \sum_i \Delta V_i$ denote the velocities achieved in the individual stages, then:

$$V_B = S \ln \frac{m_0}{m_B} \tag{4.99}$$

Meanwhile, there are also private companies doing space travel, such as SpaceX with the Falcon 9 rocket (Fig. 4.12), which is two-stage and can carry a 228,000 kg payload to low-Earth orbit (LEO). The Falcon Heavy is expected to be able to carry a 16,800 kg payload to Mars, as well as about 63,800 kg to LEO and about 26,700 kg to geostationary orbit. SpaceX stands for Space Exploration and was founded in 2002 by entrepreneur E. Musk with the ambitious goal to colonize Mars. Compared to NASA's (American space agency) Apollo moon missions, these developments have the advantage of reusability of the first rocket stage, which saves enormous costs (Figs. 4.11 and 4.12). The landing is made on an autonomous spaceport drone ship. The Big Falcon Rocket, BFR, will be used to make a manned flight to Mars. The BFR is more than 100 m high and measures 9 m in diameter. Figure 4.13 shows a comparison between the different rockets.

A *thermal propulsion* uses the conversion of thermal energy of hot gases in a combustion chamber into kinetic energy of a gas jet. In this process:

$$S_{max} = \sqrt{\frac{2\kappa}{\kappa - 1} \frac{RT}{\mu}}, \tag{4.100}$$

where $\kappa = c_p/c_v$ and T is the temperature in the combustion chamber and μ is the molecular weight. κ is between 1.1 and 1.6 for real gas mixtures.

We see: Optimally, the temperature is high and the molecular weight is low. The problem is that you can only achieve a burning time of a few minutes with these. The highest performance is achieved with hydrogen, but storage of cryogenic liquid hydrogen is complicated, and it is therefore only suitable for launchers from the ground into Earth orbit.

Fig. 4.11 The Falcon rocket developed as part of the SpaceX project (Space X Photos)

With *ion propulsion* burning times of several months are possible. Ions are ejected which have been previously generated and then electromagnetically accelerated. The achievable velocities are 30–50 km/s (more than 10 times the velocity of conventional propellants), the ejection velocity is 50–100 km/s. The space probe Deep-Space 1 (launched in 1998) uses such a propulsion system, noble gas ions (xenon) are ejected at 50 km/s, and the energy needed for ionization is generated by solar cells. Although the thrust is only small, it can be maintained for several months, and therefore the achievable final speed is an order of magnitude higher than with chemical propulsion. The total xenon supply is 81.5 kg. Since the exhaust velocity is 10 times that of thermal propellants, only 1/10 the amount of propellant is required compared to conventional fuels. For telecommunication satellites, ion propulsion has become popular; not as the primary propulsion to reach orbit, but to compensate for drifts caused by interference from the Moon and Sun.

With the help of a *solar sail* the light pressure coming from the sun (which also deflects the tails of the comets) is used. The ionic propulsion system is a very powerful and efficient means of transporting masses up to the orbit of Jupiter at practically zero cost.

Fig. 4.12 The landed burned out rocket stage of Falcon (SpaceX Photos)

At the entry of a spacecraft into a planetary atmosphere, high heat is generated by friction, and one protects oneself against this by a so-called heat shield mostly in the form of a slowly evaporating layer. The evaporating material creates a thin layer of gas between the spacecraft and the atmosphere, which prevents too much heat from flowing to the spacecraft *(ablation cooling)*. The *Deceleration* in a planetary atmosphere is also exploited in aerodynamic braking, for example when a spacecraft is to be brought into orbit around a planet. On arrival, the spacecraft is too fast and flies tangentially through the top layers of the planet's atmosphere several times, slowing down each time.

For the calculation of the circular orbits of artificial satellites one starts from the *Kepler's third law* and takes into account that $m_{Sat} \ll m_{Earth}$ is. Then the orbital period is:

$$U = 2\pi\sqrt{\frac{r_k^3}{Gm_{Earth}}} = 84.491\left(\frac{r_k}{r_{Earth}}\right)^{3/2} \text{ [min]} \qquad (4.101)$$

A circular orbit (index k) was used in this calculation. If a satellite is 35,790 km above the Earth's surface, then it has an orbital period of 23^h and 56^{min}, i.e. one sidereal day, and if it is in the equatorial plane of the earth, then it is relative to the rotating Earth in rest *(geostationary orbit, GEO)*. Satellite missions to outer planets often follow very complicated orbits (Fig. 4.14).

Fig. 4.13 Comparison of the different rocket systems with a Boeing 747: BFR, Big Falcon Rockt, Apollo-Saturn C rocket, the BFR interplanetary spacecraft, and the lunar lander used in the Apollo missions (first manned lunar landing on July 21, 1969). Magnetostroy

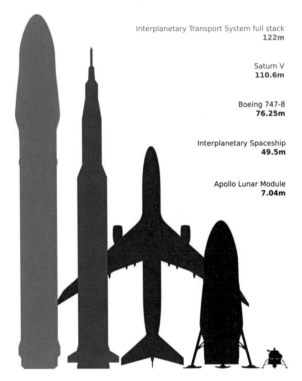

Interplanetary Transport System full stack
122m

Saturn V
110.6m

Boeing 747-8
76.25m

Interplanetary Spaceship
49.5m

Apollo Lunar Module
7.04m

Fig. 4.14 Orbit of the *Cassini mission* to Saturn. By several *gravity assists* the final orbit to Saturn is reached. Controversial was the close flyby on 18.08.1999 at the earth (1200 km distance), because on board of the spacecraft for the energy supply 30 kg Plutonium are

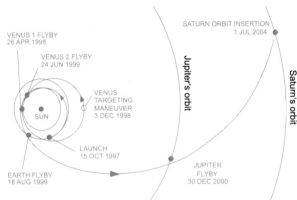

VENUS 1 FLYBY
26 APR 1998

VENUS 2 FLYBY
24 JUN 1999

VENUS TARGETING MANEUVER
3 DEC 1998

SUN

SATURN ORBIT INSERTION
1 JUL 2004

Jupiter's orbit

Saturn's orbit

LAUNCH
15 OCT 1997

EARTH FLYBY
18 AUG 1999

JUPITER FLYBY
30 DEC 2000

4.5.3   Satellite in Earth Orbit

The Earth is an ellipse Earth orbit with the eccentricity e and the major axes a_e and a_P. Where

$$a_p = a_e(1 - e^2)^{1/2} \qquad (4.102)$$

Let $\omega = 7.292115 \times 10^{-5}$ rad/s, $M = 5.977 \times 10^{27}$ g (Earth rotation and Earth mass, respectively) and r and ϕ radius and geocentric latitude of a point on the surface and V is the gravitational potential satisfying Laplace's equation, then the Geoid by a constant potential U:

$$U = V - \frac{1}{2}\omega^2 r^2 \cos^2 \phi \tag{4.103}$$

and

$$V = -\frac{GM}{r} \sum_{n=0}^{\infty} J_n \left(\frac{a_e}{r}\right)^n P_n(\theta) \tag{4.104}$$

$a_e = 6.378160 \times 10^8$ cm, $P_n(\theta)$ are *Legendre polynomials*. For example, one can express the Legendre polynomials by the formula of *Rodriguez*:

$$P_n(x) = \frac{1}{2^n n!} \frac{d^n}{dx^n} (x^2 - 1)^n \tag{4.105}$$

If the origin of the coordinate system is at the center of gravity, then $J_1 = 0$, $J_2 = 1$, and one obtains :

$$V \approx -\frac{GM}{r} + \frac{GMa_e^2 J_2}{2r^3}(3\sin^2 \phi - 1) \tag{4.106}$$

with $J_2 = 1.08270 \times 10^{-3}$. This can be derived from the change of the ascending node Ω or the change of the perigee ω of a satellite orbiting the Earth:

$$\frac{\partial \Omega}{\partial t} \approx -9.97 \left(\frac{a_e}{a}\right)^{3/2} (1 - e_0^2)^{-2} \cos i_0 \tag{4.107}$$

$$\frac{\partial \omega}{\partial t} \approx 4.98 \left(\frac{a_e}{a_0}\right)^{3/2} (1 - e_0^2)^{-2}(5\cos^2 i_0 - 1) \tag{4.108}$$

The units here are ° per day. This is also to illustrate how satellite orbits can be used to infer the shape of a planet.

4.5.4 Influences on Satellite Orbits

Satellite orbits are affected by:

- Residual atmosphere
- Solar pressure

- Flattening of the earth
- Triaxiality

Flattening The Earth is not an ideal sphere, but is slightly oblate at the poles. The length of the ascending node Ω and the angular distance of the perigee ω of the satellite orbit change.

The influence on Ω:

$$\dot{\Omega} = -9.9641 \left(\frac{R_{earth}}{a} \right)^{\frac{7}{2}} \frac{\cos i}{\left(1 - \varepsilon^2 \right)^2} \tag{4.109}$$

By choosing a suitable inclination, you get the satellite to move synchronously with respect to the sun:

$$\dot{\Omega} = \frac{360°}{364 \, \text{days}} \tag{4.110}$$

$\Rightarrow i = 98.1°$ (i.e., a nearly polar orbit)

Influence on ω:

$$\dot{\omega} = 4.98 \left(\frac{R_{earth}}{a} \right)^{\frac{7}{2}} \frac{5 \cos^2 i - 1}{\left(1 - \varepsilon^2 \right)^2} \tag{4.111}$$

By choosing a suitable *Inclination* one can also achieve that the perigee always lies over the same point:

$$\dot{\omega} = 0 \rightarrow 5 \cos^2 i - 1 = 0 \rightarrow i = 63° \tag{4.112}$$

Triaxiality Influences on major semi-axis a. The non-uniform mass distribution of the Earth causes different accelerations on the orbit.

Gravity Influences on inclination; if the satellite orbit is inclined relative to the sun's orbit (or moon's orbit) by the angle α inclined, then precession of the satellite orbit occurs with frequency:

$$\frac{di}{dt} = \frac{3}{4} \frac{\mu_{Sun}}{\mu_{Earth}} \left(\frac{\sqrt{R_{Earth}}}{R_{Sun}} \right) \sin \alpha \cos \alpha \tag{4.113}$$

Solar pressure Influences on eccentricity, the radiation pressure of solar radiation causes a continuous acceleration away from the sun. This depends on the area (solar panels) and the reflection factor.

Residual atmosphere: this interferes with the semi major axis a. Up to a height of about 500 km the residual atmosphere causes the strongest disturbance.

4.6   Resonances and Chaos in the Planetary System

Our planetary system initially gives the impression of being unchanging. But there are indeed chaotic phenomena. We will first briefly discuss the term chaos and then resonances in the solar system and in exoplanetary systems. These can have a stabilizing effect, but can also lead to planetary wanderings. In extreme cases, planets are catapulted out of their systems and then exist as *free floating* planets. We also show how the error increases with long-term integrations.

4.6.1   Chaos

Pierre Simon Laplace (1749–1827) investigated the *Stability* of the solar system by simplifying the equations and thus arriving at analytical solutions of the N-body problem.. Later, however, it turned out that these very terms can lead to *Chaos*. From *Laplace* originated the idea that if you know all the initial conditions (i.e., locations and velocities) of all the particles in the universe, you can completely predict the past and the future. This is known as Laplace's demon.

> Chaos theory showed that even completely deterministic systems can become chaotic.

Henri Poincaré (1854–1912) then found that chaos and resonance effects play a major role in the solar system.

Chaotic Behavior can be characterized as follows. for example: Two orbits (trajectories) with closely spaced initial conditions diverge exponentially. For example, one can show that an error in the eccentricity of the Earth's orbit of 10^{-n} rad produces solutions which after $n \times 10^7$ years, only yield meaningless values. Let the state of a system be described by n state variables x_i . A difference Δx_i then grows exponentially, and the *Liapunov exponents* λ_i result from:

$$\Delta x_i(t) = e^{\lambda_i t} \Delta x_i(t = 0) \tag{4.114}$$

If a $\lambda_i > 0$, then exponential growth results, the system becomes unpredictable.

A well-known example of this is the *butterfly effect.*

The flap of a butterfly's wings in Europe can cause a devastating hurricane in Florida.

A catastrophe can develop from an extremely insignificant small disturbance.

4.6.2   Resonances in the Solar System

The term *Resonance* is used when two periods are in an integer relationship. Very often this is true for orbital periods. The orbital period of Jupiter's moon *Io* is 1.769 days, that of the moon *Europa* is exactly twice that. Therefore it is a 2:1 resonance. Between *Mars* and *Jupiter* there is a belt containing thousands of *Small planets(asteroids).* Resonances can have a stabilizing effect or produce the opposite. Consider Jupiter, an asteroid, and the Sun, a classic restricted three-body problem. Conjunctions between an asteroid and Jupiter become dangerous when the asteroid is at aphelion. On the other hand, it is not dangerous if the asteroid is at aphelion but Jupiter is far away. In the *Saturnian system*, resonances are found between the moons *Mimas* and *Tethys, Enceladus* and *Dione* and *Titan* and *Hyperion.* Between *Neptune* and the dwarf planet *Pluto* there is a 3:2 resonance. In 1867, *D. Kirkwood* found gaps in the *asteroid belt* between Mars and Jupiter at 3:1 resonance ($a = 2.5$ AU) and 2:1 resonance ($a = 3.3$ AU) with Jupiter. Chaotic orbits are found at these locations, and asteroids can enter the inner solar system through slight perturbations, for example.

Other examples of resonances include *spin-orbit resonances: Mercury* is in a 3:2 spin-orbit resonance and our Moon in a 1:1 spin-orbit resonance, since it always shows us the same side. This is due to tidal effects (in the case of Mercury, due to the nearby, massive Sun).

4.6.3   Migration of Planets

The formation of a planetary system starts from kilometer sized *planetesimals* as the precursor of the large planets. These then coalesce into planets. Observation of extrasolar planets shows that chaotic processes often take place here. Resonant interactions of an already formed planet with planetesimals within its orbit can lead to two effects:

- The planetesimals are flung outwards. This could, for example, lead to the formation of the *Oort cloud* that develops our solar system.

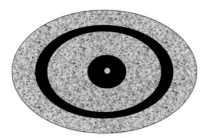

Fig. 4.15 Migration of planets. If a planet has more than 0.1 Jupiter masses, it can create a gap in the nebula of planetesimals and remain stable there. However, the viscosity in the disk pulls the planet towards the central star. In an area of about 0.1 AU around the central star, there is a zone that is free of particles due to the stellar magnetic field or stellar wind, and the planet can park there

- The planet itself continues to migrate to its parent star (timescales are relatively short, approx. 10^6 years). In this way one can explain exoplanets at short distances from their central star, whose mass is greater than that of Jupiter (Fig. 4.15).

So during the formation of our solar system, Jupiter could have moved inward by a few 0.1 AU and the planetesimals (mass a few Earth masses) moved outward, possibly also Neptune outward by a few AU.

4.6.4   Chaos in the Solar System

Our solar system was formed about 4.6 billion years ago by contraction of an interstellar gas cloud. Initially tiny dust grains formed into larger units, called *planetesimals* (a few kilometers in size), and from these the protoplanets. Our moon was formed when the early Earth collided with a Mars-sized protoplanet. Also the slow retrograde rotation of *Venus* suggests a collision in their early days.

The rotation of the two tiny *Pluto moons Nix* and *Hydra* is chaotic. This is due to the fact that Pluto has a relatively large moon, *Charon*. Saturn's moon *Hyperion* rotates chaotically as a result of interactions with Saturn's largest moon, Titan.

Long term calculations of the solar system show, that Mercury could possibly collide with Venus or even be ejected from the solar system. This could happen within the next 5 billion years. In making these calculations, one must take into account that the *error* of the calculated magnitude d increases as time t progresses according to the following law:

$$d(t) = d(0)10^{\frac{t}{10}} \tag{4.115}$$

$d(t)$ is the quantity at time t, $d(0)$ is the quantity at the starting time $t = 0$. Important: t is to be given in millions of years. So suppose we wanted to determine the value of a quantity

such as the Earth's major orbital semi-axis after 200 million years of calculation. We will assume that we can set the value to 10^{-9} m known accuracy. Then $d(200) = 10^{-9}10^{20} = 10^{11}$ m, which is almost as large as the great orbital semi-axis of the Earth itself.
The errors that inevitably arise with numerical methods increase greatly with time. Only special tricks can lead to reasonable numbers here!

4.7   Eclipses

Solar and lunar eclipses are among the most spectacular celestial events and evoke a great deal of media coverage. When do such eclipses occur, how can they be predicted?

4.7.1   Lunar Eclipses

Durinng a total lunar eclipse, the Moon dips completely into the Earth's umbra. The mean length of the *Shadow* of the Earth is 217 Earth radii and thus reaches far beyond the distance Earth-Moon (60 Earth radii). The angular radius is $41'$, the radius of the moon only $15,5'$. The angular radius of the penumbra is $75'$. The moon moves on about its own diameter per hour (thus $33'$). Therefore the totality of a lunar eclipse lasts up to 1.7 h. The total eclipse lasts up to 3.8 h, the penumbral phase up to 6 h. A *Penumbral Eclipse* is practically not visible to the naked eye. Important for a lunar eclipse is the refraction in the earth's atmosphere. The sunlight, which falls tangentially at the earth, is deflected by the double horizonat refraction (thus $70'$), and therefore the length of the actual shadow is only 40 Earth radii. Even during totality, therefore, the Moon is not completely darkened, but receives light (dark red to purple, blue light being more strongly absorbed than red). The brightness of the Moon during its total eclipse depends on the state of the Earth's upper atmosphere (which in turn is affected by solar activity). For example, lunar eclipses were very dark after the eruption of the volcano Krakatau, because during this time there were large masses of dust in the Earth's atmosphere. Lunar Eclipses can occur only at full moon, however as moon's orbit is inclined by $5°$ against the ecliptic, they only occur when the full moon is close to the node (intersection of the moon's orbit and the ecliptic) (Fig. 4.16).

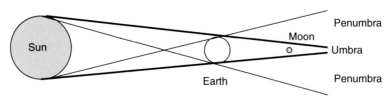

Fig. 4.16 Formation of a lunar eclipse (magnitudes are not to scale)

Lunar eclipses can be defined according to the *Danjon Scale*. The eclipses can be classified according to the eclipse scale, depending on how bright the moon appears during the total eclipse.

- L = 0: very dark; moon almost invisible during mid-totality.
- L = 1: dark; gray or brown; details on moon difficult to see.
- L = 2: deep red to rusty brown; very dark central shadow, edge of umbra bright.
- L = 3: brick red; umbra has bright yellow fringe.
- L = 4: very bright; moon appears copper red to orange. Core shadow has bluish fringe.

This classification is made with the naked eye or binoculars.

4.7.2 Solar Eclipses

Total solar eclipses occur at New Moon but only in the most favorable cases, the shadow of the moon reaches the Earth and then sweeps a small area (about 200 km). The Moon moves from west to east at 1 km/s, the rotation of the Earth is about 400 m/s, i.e. the *visibility zone Solar eclipse visibility zone* moves on the Earth (Fig. 4.17). The shadow arrives from the outside, first point of contact = eclipse at sunrise. Eclipses are favourable when the Sun is at aphelion (far from the Earth) and the Moon is at perigee (near the Earth). The maximum diameter of the umbra is 264 km, the velocity is about 0.5 km/s, and the maximum duration for a given location is thus $7^{\min}36^{s}$. Solar eclipses are 1.6 times more frequent than lunar eclipses, but for a given location on Earth, lunar eclipses are more frequent (a solar eclipse is visible on average only once every 400 years for a given location on Earth, while the number of lunar eclipses is more than 1000 times greater). In 1000 years there are about 1500 lunar eclipses and 2400 solar eclipses (Fig. 4.18). A graphical representation of all solar eclipses in the period 2001–2025 can be found in Fig. 4.19 (total solar eclipses) and Fig. 4.20 (annular solar eclipses).

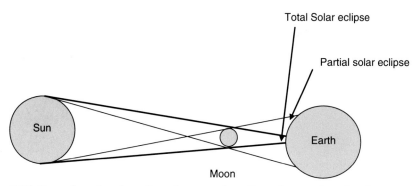

Fig. 4.17 Formation of a solar eclipse. In an annular eclipse, the moon's umbra does not reach the earth (the proportions are not to scale)

Fig. 4.18 Top: Partial lunar
eclipse. Bottom: Total solar
eclipse, diamond ring
phenomenON (the sun still
shines through moon valleys)
and prominence (A.
Hanslmeier, private
observatory)

Since 242 draconitic months correspond to 223 synodic months (= 18 years and 10.3
or 11.3 days, depending on how many leap years lie in between), eclipses repeat after this
time has elapsed *(Saros cycle)*.

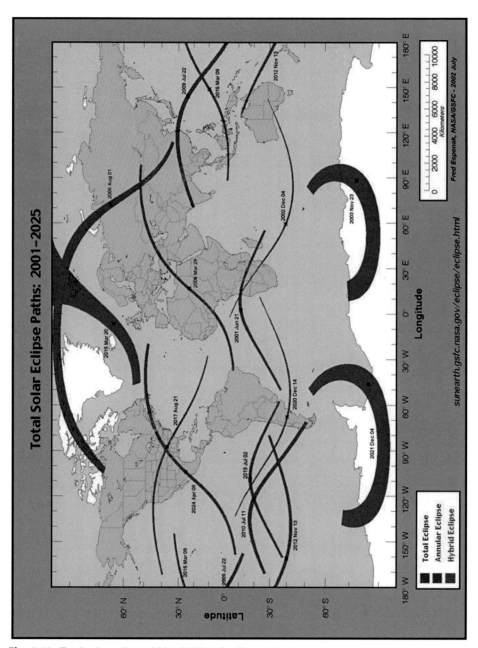

Fig. 4.19 Total solar eclipses 2001–2025 (After Espenak)

At the beginning of a Saros cycle eclipses always occur (mostly partially) at the polar regions and then move more and more to the S or N, and the degree of occultation also increases. At the middle of the cycle the eclipses have the longest duration and are

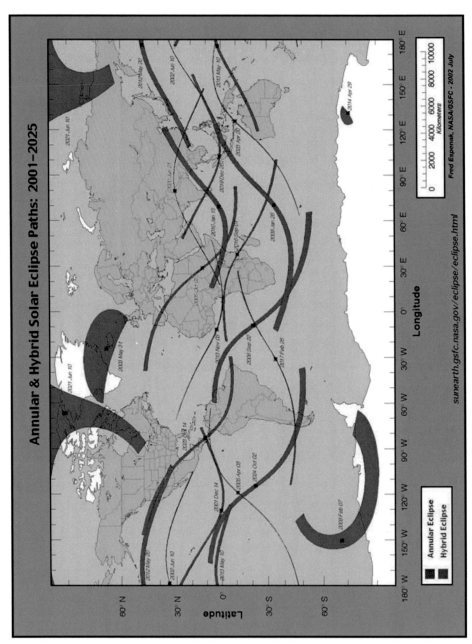

Fig. 4.20 Annular solar eclipses 2001–2025 (After Espenak)

observable around the equator. Currently there are 25 cycles that produce central eclipses and 14 cycles that do not yet produce central eclipses (i.e. are still at the beginning or end).

The total solar eclipse of 29.03.2006 (e.g. visible in southern Turkey) belongs to a cycle (number 139) which includes 71 eclipses in a time frame of 1262 years, of which 16 are partial, 43 total and 12 annular. The cycle began on 05/17/1501 and ends on 07/03/2763. The longest total solar eclipse of this cycle will occur on July 16, 2186.

However, for a given location on Earth, you have to wait three Saros cycles for an eclipse to repeat. The last total solar eclipse that could be seen from central Europe occurred on August 11, 1999.

Example
Consider the eclipse of 29.03.2006 (visible e.g. in southern Turkey). The next one will occur on April 8, 2024, but shifted W over North America. Previously, there was an eclipse visible in Indonesia on 03/18/1988.

Example
Estimate when the next eclipse will occur, belonging to the same cycle as the last total solar eclipse visible with us.
After three Saros cycles the eclipse will repeat itself approximately: thus 3×18 years + 3×11 days would result in a date of 13.09.2053, in fact an eclipse will take place on 12.09.2053, whereas seen from central Europe the sun will be eclipsed by about 50 % in the morning.

Since the Moon moves away from the Earth by 4 cm per year (tidal friction), it follows that after 600,000 years there will be only annular eclipses.

A partial lunar eclipse and a total solar eclipse can be found in Fig. 4.18.

Solar eclipses can only occur at new moon, lunar eclipses only at full moon. The moon must be in the ecliptic.

4.7.3 Planetary Transits

A planet is said to transit when it passes in front of the solar disk as seen from Earth. This concerns only the lower planets Mercury and Venus and is a rare event because of their orbital inclination. The last *transit of Venus* to be observed in Central Europe took place on 8 June 2004 and lasted about 6 h. The orbit of Venus inclined by 3.4° with respect to the orbital plane of the Earth. Venus transits occur at intervals of 8, 105, 5, 8, and 121.5 years. The transit of Venus on 06.06.2012 was visible from central Europe only in the last phases, as soon as the sun rose. In 1769 a transit of Venus was observed by *James Cook* in the

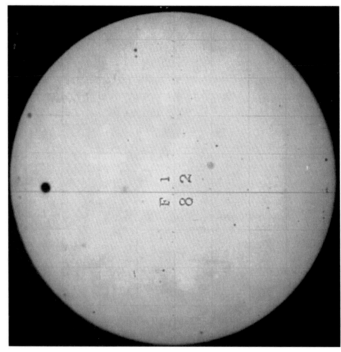

Fig. 4.21 Probably the first photographic image of the planet Venus in transit on 6 December 1882. US Naval Obs

Pacific Ocean and the Viennese court astronomer *Maximilian Hell* in Norway observed a transit of Venus. From this the *solar parallax* and the distance earth-sun-solar parallax are obtained. The next transit of Venus will occur on 11 December 2117 (Fig. 4.20).

A historical image of a Venus transit is given in Fig. 4.21. Venus can be seen in front of the solar disk above as a large black dot. Some sunspots can also be seen.

Example
Determination of the Venus parallax and solar distance. If you have two images of Venus during a transit, taken at the same time at two locations as far away as possible, then determine the distances of the planet from the center of the solar disk on the two images and get x_1, x_2. The Venus parallax is then $\pi_V = x_1 - x_2$.

Mercury transits occur 14 times per century. The last Transit was observed on November 11, 2019, and then the next one will not be seen until November 13, 2032.

Transits of extrasolar planets are also one of the most important detection methods to find these planets

If a planet close to the parent star passes in front of the star as seen from us, a small decrease in brightness occurs. By photometric measurements of the star HD 209458 1999 it was possible for the first time to measure the brightness decrease of the star due to a planetary transit with a 0.8 m telescope: The brightness drop was 1.58% or 0.017 magnitudes.

4.8   Further Literature

We give a small selection of recommended further literature.
Introduction to Celestial Mechanics, A. Guthmann, Spektrum, 2000
ABC of Celestial Mechanics, W. Broda, Curriculum, 2007
Ephemeris calculation step by step, D. Richter, Springer, 2017
Introduction to astrodynamics and ephemeris calculus, I. Klöckl, 2014
Celestial Calculations, J. K. Lawrence, MIT Press, 2019

Tasks

4.1 Calculate the perihelion and aphelion distances for the dwarf planet Pluto, $a = 39.88$ AU, $e = 0.25$, and compare the values with the major orbital semi-axis of Neptune ($a = 4509 \times 10^6$ km).

Solution
Hint: $r_{\text{perihel}} = a(1 - e)$

4.2 Calculate when perihelion and summer solstice coincide in the past and in the future. What effect might this have on the Earth's climate?

Solution
Coincidence of perihelion and summer solstice result in warmer summers.

4.3 Can Venus be seen in the morning or evening sky a few weeks after its upper conjunction?

Solution
Evening sky

4.4 Discuss for which objects the perihelion rotation could become very large.

Solution
For large masses M and small orbital semi-axes a.

4.5 Find the mass of Uranus in units of the mass of the Earth. The moon of Uranus, Miranda, orbits the planet in 1.4 days at a mean distance of 128,000 km.

Solution
Substitute this into Kepler's 3rd law and find a mass for Uranus of 14 Earth masses (the Earth mass is 5.9×10^{24} kg).

4.6 Given a comet with an orbital period of 7 years. What is the size of its major orbital semi-axis?

Solution
$a = 3.7$ AU.

4.7 How long does it take to travel to Jupiter ($a = 5.2$ AU) in an elliptical orbit?

Solution
The most economical orbit is one with $2a = 1$ AU $+ 5.2$ AU, therefore $a = 3.1$ AU. Using Kepler's third law $P^2 = a^3$ it follows: $P = 5.5$ years.

Astronomical Instruments

5

Before the light of astronomical objects reaches the Earth's surface and thus the telescope of an observatory or the receiver (CCD, photoplate, eye), it is exposed to numerous changes:

- intergalactic and interstellar medium cause reddening,
- Earth's atmosphere,
- telescope optics.

All this has to be taken into account when analyzing the light of an object by means of a detector (CCD, eye, ...). In this chapter we focus on astronomical instruments.

5.1 Telescopes

Telescopes are the most important observational instruments in astrophysics. What are the basic properties of a telescope? How to determine the magnification and what details can be seen on planets with it (resolving power).

5.1.1 Basic Properties

The basic principle of a telescope consists of: Lenses (refractor) or mirror (reflector) collect the light in a focal point, the *focus*.

© Springer-Verlag GmbH Germany, part of Springer Nature 2023
A. Hanslmeier, *Introduction to Astronomy and Astrophysics*,
https://doi.org/10.1007/978-3-662-64637-3_5

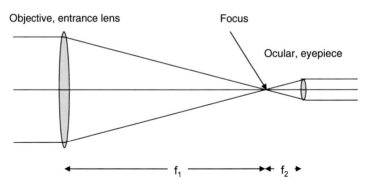

Fig. 5.1 Principle: refractor; f_1 Focal length of the objective, f_2 focal length of the eyepiece. The magnification is $V = f_1/f_2$

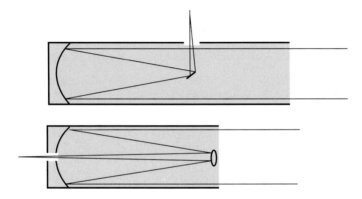

Fig. 5.2 Reflecting telescopes: Type mirror telescope Newton (top) and Cassegrain type (bottom)

With a lens (Fig. 5.1), an inverted image is produced, and the distance between the lens and the image is called *focal length*. Analogously, this is the case with a curved mirror (Fig. 5.2) (usually parabolically ground). Let f is the focal length and d is the diameter of the mirror or the objective, then one defines the focal ratio N:

$$N = f/d \qquad (5.1)$$

This determines the brightness of an image. The larger the focal ratio, the brighter the image of planar objects. In photography one uses the expression *aperture:* For example, if $f = 1000\,\text{mm}$ and $d = 200\,\text{mm}$, then one speaks of aperture 5 or $f/5$. At $f/8$ you have to expose four times as long as at $f/4$, because the light is divided over four times the area.

Let d_0 the distance of the object to the lens (mirror) and d_i the distance of the image to the lens (mirror), then:

$$1/d_0 + 1/d_i = 1/f \qquad (5.2)$$

How *large* does an image of an extended object appear? If the moon, whose apparent diameter in the sky is 0.5° gets mapped to a size of 10 cm, then the scale would be 0.05° cm. The formula is obtained:

$$s = 0.01745 f \tag{5.3}$$

Thereby s in units of f per degree and is as *plate scale*. The factor 0.01745 is the number of rad per degree.

Suppose a focal length of 90 cm is given. So to get 1° to be imaged on a CCD chip, a chip size of $0.0175 \times 90 = 1.57$ cm would be necessary!

Another simplified formula for calculating the plate scale is:

$$\text{plate scale } [''/\text{pixel}] = \frac{206.3 \times \text{pixel } [\mu\text{m}]}{\text{focal length } [\text{mm}]} \tag{5.4}$$

Thereby means *Pixel* the size of a pixel element on a CCD chip (CCD section). To enable the resolution of an image detail on the chip, two pixels are required (this is known as *Nyquist theorem*).

The *exit pupil* is defined as the product of eyepiece focal length and focal ratio. For a telescope with an eyepiece of 10 mm focal length and a focal ratio of 1:10, the exit pupil is 1 mm.

Telescopes collect light. The *light gathering power, LGP*, of a telescope depends on the area $d^2 \pi / 4$ of the lens.

Let's compare the light-gathering power of the human eye to the light-gathering power of a 50-cm lens. The pupil the human eye is about 0.5 cm. $\rightarrow LGP = (50/0.5)^2 = 10,000$

The diameter of the *human eye pupil* changes with age (Table 5.1).

The *Magnification V* is defined as follows: If F is the focal length of the objective (mirror or lens), and f is the focal length of the eyepiece through which the image produced by the objective is viewed:

$$V = F/f \tag{5.5}$$

Thus, for example, if one observes an object with a magnification of 200 times, then it has an angular diameter 200 times larger than if one observes it with the unaided eye.

As a simple rule of thumb, you can also remember: The diameter of the solar image in centimeters is given by the focal length in meters. So a telescope with 50 m focal length produces a 50 cm solar image.

Table 5.1 Diameter of the human eye pupil

Age	10	20	30	40	50	60	70	80 years
Diam.	8	8	7	6	5	4	3	2.3 mm

The *minimum magnification,* at which one can observe without edge shading, is found by:
Entrance pupil[1] eye pupil.

A very important parameter of a telescope is the *resolving power,* which determines at
which angle two objects can still be seen separately or which fine structures can be seen
on planets, moon, sun, galaxies etc..

The resolving power is:

$$\Theta_{min} = 206,265\lambda/d \ ['']$$ (5.6)

The number 206,265 is the number of arc seconds in 1 rad $= 3600 \times 180/\pi$. λ is the
wavelength at which one observes and d is the lens diameter. Since the objective (mirror or
lens) is circular, one must multiply Eq. (5.6) must be multiplied by 1.22, because additional
diffraction effects occur.

Example
Resolving power of a 10-cm telescope at a wavelength of 500 nm.

$$\Theta_{min} = 1.22 \times 206,265 \times 500 \times 10^{-9}/0.1 = 1.25''$$

So if there are two stars more than 1.25″ distant, one can still see them separated with
a 10-cm telescope. The exact theory of the resolving power of a telescope shows:

- The lens produces a diffraction figure whereby, with good objectives, 84 % of the light
  is in the central image.
- The further minima according to the diffraction-theory result from the zeros of the
  Bessel-functions. The central diffraction-disk is also called Airy-disk. For the intensity
  measured from the centre at the distance results r:

$$I(r) = \left(\frac{J_1(r)}{r}\right)^2$$ (5.7)

Thereby J_1 the Bessel functions are of the first kind:

$$J_n(x) = \sum_{r=0}^{\infty} \frac{(-1)^r (\frac{x}{2})^{2r+n}}{\Gamma(n+r+1)r!}$$ (5.8)

The Bessel functions are solutions of the Bessel differential equation:

$$x^2\frac{d^2y}{dx^2} + x\frac{dy}{dx} + (x^2 - n^2)y = 0$$ (5.9)

---

[1] The entrance pupil is equal to the aperture of the objective (mirror).

Fig. 5.3 Radial variation of intensity behind a circular aperture. With good objectives, 84 % of the light is in the central image, which has a radius of $\alpha = 1.22\,\lambda/d$ has

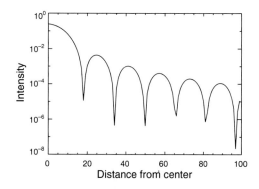

In Fig. 5.3 are the solutions of (5.7) are plotted.

- The first minimum occurs at $1.22\,\lambda/d$

> The most important parameter of a telescope is its resolving power, which is given by the telescope aperture.

5.1.2 Seeing and Large Telescopes

Due to turbulence in the *Earth's atmosphere* natural limits to the resolving power of a telescope have to be considered. The twinkling of stars is caused by turbulence in the Earth's atmosphere. If larger air masses of the Earth's atmosphere move against each other, the image or parts of it also move (in the case of extended objects such as the Sun), known as *Blurring*. The effects of the Earth's atmosphere on the quality of astronomical observations are described under the term *seeing*. Since seeing is usually in the range of 1 arcsec the actual resolving power of a 10-cm telescope is hardly worse than that of a 5-m telescope such as the Mt. Palomar telescope.

Why then do we build giant telescopes (Table 5.2) with diameters up to more than 10 m? On the one hand, the larger the telescope diameter, the higher the light gathering power, so fainter objects can be detected. On the other hand, the seeing can be kept low by:

- Site testing: One tests very carefully the observing conditions before deciding on the location of a new telescope. The observing conditions and thus the seeing are clearly better on high mountains or mountains on small Islands.
- Adaptive optics: The image is corrected by deforming the mirror.

Table 5.2 Large earthbound telescopes

Aperture [m]	Height [m]	Name/location
10,4	2500	GRANTECAN La Palma, Spain
10.0	4145	W.M. Keck, Mauna Kea, USA
≈ 10	1750	SALT, South African Astr. Obs.
8.4	3170	LBT, Mt. Graham, AZ
8.3	4100	Subaru, Hawaii, Nat. Obs. of Japan
8.2	2600	ESO-VLT, Paranal, Chile
8.1	4100	Gemini North, Mauna Kea
8.1	2700	Gemini South, Chile
6.00	2070	BTA6, Bolshoi, Mt. Pastukov, Russia
5.08	1706	Hale, Mt. Palomar, USA

Fig. 5.4 Achromat

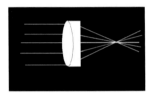

5.1.3 Imaging Errors

The price of a telescope depends on the correction of aberrations. We discuss the most important ones here:

Already with an axis-parallel incident light bundle occur:

- *Chromatic aberration:* The focal length depends on the refractive index n of the glass, which in turn depends on the wavelength. Short wavelength light is refracted more strongly than long wavelength light, hence $f_{blue} < f_{red}$. The focal point of the blue light component is closer to the objective. The images appear with a coloured edge. Therefore one uses as objective *Achromats,* a compound lens with different $n(\lambda)$ e.g. crown glass and flint glass. Depending on the type of glass, one speaks of a photographically or visually corrected lens (Fig. 5.4).

  Important: This error does not occur in a reflecting telescope, since the reflection is independent of the wavelength.
- *Spherical aberration:* The outer areas of a lens (mirror) have a shorter focal length than the zones further inside (Fig. 5.5). Correction: *Aplanat,* this is understood to be a compound lens; in the case of a mirror, this defect is circumvented by turning it into a *paraboloid* grinding.

Fig. 5.5 Spherical aberration; edge rays are more strongly refracted (their focal point is at (1))

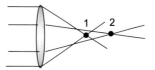

With oblique incidence of the light beams additionally occurs:

- *Astigmatism:* For oblique incident light rays, the lens is not a circle but an ellipse. The image is therefore not formed in one point, but in two focal planes.
- *Image field curvature:* Figure not in one plane.
- *Coma:* Outside of the optical axis, the stars appear "comet-shaped".
- *Distortion:* The stars appear drawn in a pincushion or barrel shape.

According to the *Rayleigh-Strehl criterion* there are the following tolerances:

- Focus: $|\Delta| < 2N^2\lambda$, where N the focal ratio f/d is.
- *Spherical aberration:* $D/N^3 < 512\lambda$. If this is satisfied, a spherical mirror need not be parabolically ground.
- Coma: $D/N^2 \tan w < 19.2\lambda$, w angle against the major axis.
- Astigmatism: $D/N \tan^2 w < 2.7\lambda$.

5.1.4 Telescope Types

The *refractors* or refracting telescopes (Fig. 5.1) usually consist of a two-lens achromat, and the focal ratio is between 1:15 and 1:20 → relatively small field of view. The largest refracting telescope in the world is located at the Yerkes Observatory; it was completed in 1897 and has an objective diameter of 1.02 m. Larger refractors can practically no longer be manufactured technically, since

- the manufacture of the objective, which must be as free from defects as possible, is extremely costly, and
- the distribution of the weight becomes unfavorable with a large refractor, since most of the weight (due to the large heavy lens) is concentrated at the front of the objective.

That is why telescopes with larger apertures—but also many telescopes in the amateur sector—are built as reflecting *telescopes* (*reflectors,* Fig. 5.2). These have the advantages

- no chromatic aberration,
- no spherical aberration,
- inexpensive, only the surface must be exact ($\lambda/10$-criterion).

There are different systems of reflectors, depending on the type of beam path:

- *Primary focus:* Designates only the focus of the main mirror (primary mirror). Directly used only with very large telescopes.
- *Newtonian* focus (Fig. 5.2 above): Through a 45°-plane mirror the light is guided laterally out of the main tube, the focus is close to the main tube. Advantage: very cheap, Disadvantage: Many cheap reflectors in the amateur sector are built this way, but it results in an unfamiliar telescope view and difficulties in sighting objects.
- *Cassegrain focus* (Fig. 5.2 bottom): the secondary mirror is located in front of the primary focus and is hyperbolically ground. The light is reflected back towards the primary mirror and passes through a central hole through it to the eyepiece.

  Thus, one observes directly behind the pierced primary mirror, which makes sighting much easier. In addition, the focal length is extended by the parabolically ground secondary mirror → Compact telescopes of large focal length. Such telescopes are offered in different variations for the upper amateur range.
- *Coudé system:* by a deflection mirror, the beam path is directed into the tracking axis of the telescope (hour axis), and one gets a spatially fixed position of the focus, no matter in which direction the telescope points. This makes it easier to mount heavy post-focus instruments like cameras, CCD, spectrographs etc. there.
- *Schmidt mirror:* By a spherical primary mirror the focal surface is curved. → Correction: aspherical correction plate at the telescope aperture. Advantage: large field of view. In Tautenburg there is the largest Schmidt mirror, the diameter of the correction plate is 1.34 m, that of the mirror is 2.00 m, the focal length is 4.0 m. This gives a very large field of view of 2.4° × 2.4° (Fig. 5.6).

Fig. 5.6 ESO's very large telescope very. Four 8-m telescopes are connected to form one telescope via interfeometry

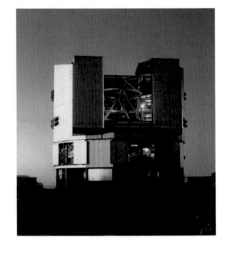

5.1.5 Telescope Mounts

The best optical system is worthless if the mount is poor. It should make it possible to point the telescope precisely at any point in the sky. A distinction is made between:

- *Azimuthal mount:* The telescope is moved around two axes, vertical and horizontal. To compensate for the daily movement of the celestial objects, one must move the telescope in both axes simultaneously. At the zenith, you have a singular point. In the age of computer controls, this is no longer a problem. The large radio telescopes as well as large optical telescopes, but also more and more amateur telescopes are tracked this way; mostly as fork mount (Figs. 5.7, 5.8 and 5.9).
- *Parallactic mount* (also called equatorial mount): Here the hour axis is parallel to the earth axis and the declination axis is perpendicular to it. The Earth's rotation can thus be compensated by rotation about a single axis (Fig. 5.10).

  There are different types of mounts: German mount, knee column mount, fork mount, frame mount.
- *Coelostat:* Large Solar telescopes are permanently mounted (often in a vacuum tank, to avoid heating of the optical elements by the intense solar radiation). Light is directed into the entrance window of a telescope by two mirrors, coelostats (Fig. 5.11).

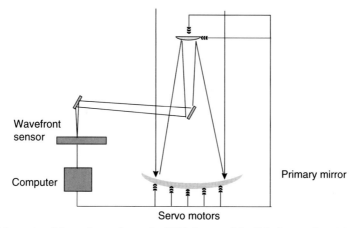

Fig. 5.7 Schematic of the active optics at the VLT. A part of the light is transferred from the beam path to a wavefront *sensor,* which measures the disturbance of the wavefront by the atmosphere. The data is then passed to servo motors which deform the partial mirrors accordingly

Fig. 5.8 Big Bear Solar Observatory in California, USA. The water surface has a calming effect on air turbulence

Fig. 5.9 The telescopes on Mauna Kea in Hawaii, USA. The two Keck telescopes (back right of image center) are twins; each has a mirror consisting of 36 segments (thickness only 7.5 cm) with a total diameter of 10 m. The segments are made of Zerodur (low thermal expansion). The weight of the glass is about 15 t; the mount is azimuthal. The total weight of the moving parts is about 300 t

Fig. 5.10 Example of a parallactic mount: refractor of the University Observatory Vienna. The rotation around the two axes is sketched (After a photograph by F. Kerschbaum)

In a good telescope, the mount must also be very stable and mechanically sound in order to compensate for the movement of the stars in the sky due to the rotation of the Earth.

5.1.6 Robotic Telescopes

A robotic telescope can do astronomical observations completely automatically. The following components must work together:

- Automatic dome control
- Control of weather conditions
- Control and positioning of the telescope
- Control of the CCD camera

Examples of robotic telescopes are ATF (Automated Telescope Facility, Univ. Iowa), LINEAR (Lincoln Near-Earth Asteroid Research), Catalina Sky Survey, Spacewatch etc. Typical research projects for robotic telescopes include supernova searches, gamma ray bursts (GRB), monitoring active galactic nuclei (AGNs), minor planets—in particular of the NEOs (Near Earth Orbits, i.e. Asteroids, their orbit is in near proximity of the Earth's orbit).

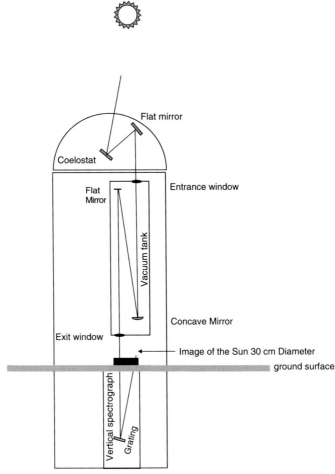

Fig. 5.11 Sketch of the beam path of a vacuum tower telescope for solar observation. The light is reflected into a vertical tower, which is evacuated to avoid heating. To obtain the largest possible image of the Sun, a long focal length is necessary

5.2   Modern Optical Telescopes

5.2.1   Modern Earthbound Telescopes

The Keck Observatory Keck telescope (Fig. 5.9) comprises two individual telescopes. One can either operate the two 10-m telescopes individually or connect them together as an interferometer. The telescopes were built according to new technologies: The 10-m mirrors consist of 36 individual segments, each 1.8 m in diameter and only 75 mm thick. Each mirror is controlled separately to minimize seeing effects. This gives a resolving power of 0.3″. The telescope is installed on Mauna Kea in Hawaii.

The European Southern Observatory ESO: this organization consists of 15 member states, which have realized major telescope projects, has built the NTT built (New Technology Telescope. The 3.6 m mirror can be deformed: As a reference, one observes the image of a bright star. As soon as this is altered by the Earth's atmosphere, the image can be corrected by means of servo motors, and therefore by means of this so-called *active optics* eliminate the seeing to a large extent. By actuators a narrow segment of the mirrors is moved and one tries to keep a point-like light source as point-like as possible by deformation of the mirror (up to 1000 times/s). As a point source one uses so-called Guiding Stars.

In the Chilean Atacama Desert, ESO has built the VLT (Very Large Telescope) at Paranal (Fig. 5.6). It consists of four telescopes with a diameter of 8.2 m, each of the telescopes can be operated either individually or interferometrically, i.e., all four telescopes are connected together (interferometry). In interference, waves must be superimposed coherently. The coherence length is important. This is the maximum difference in light path that two light beams coming from the same source may have in order to produce an interference pattern. Laser light has a very large coherence length of millimeters to kilometers, visible light only in the μm range. The coherence length depends on

- wavelength,
- spectral bandwidth.

Interferometry is more difficult in visual light than in IR and radio because of the smaller coherence length. The principle of how adaptive optics can be used to improve seeing is shown in Fig. 5.7.

An ambitious project at ESO is the ELT, Extremely Large Telescope, scheduled for completion around 2024 (Fig. 5.12). The construction costs will be about 1.1 billion EUR. The main mirror diameter is just under 40 m. The primary mirror consists of more than 900

Fig. 5.12 Sketch of the ELT and comparison with the Arc de Triomphe in Paris or the four VLT units (Image: ESO)

hexagonal segments, each 1.45 m in diameter. The secondary mirror has a diameter of 6 m, and from a tertiary mirror the light comes onto a 2.5 m mirror equipped with 5000 actuators to compensate for the blurring caused by the air turbulence of the Earth's atmosphere.

In general: Large telescope projects can hardly be financed by individual states, which is why the ESO, for example, was founded.

The Gran Telescopio de Canarias (GRANTECAN, GTC) was 90% built by Spain and consists of 36 segments that make up a 10.4 m mirror. It was ceremonially opened on the island of La Palma at the Observatorio Roque de los Muchachos in 2009. First instruments at the GTC are OSIRIS (camera and spectrometer for light between 0.365 and 1.05 μm wavelength) and CanariCam (camera and spectrometer for the mid-infrared between 8 and 25 μm wavelength).

The telescopes discussed so far work mainly in visible light and near infrared (IR).

Special telescopes are required for solar observation (Figs. 5.7 and 5.10). The 4-m Daniel K. Inoue Solar Telescope, DKIST, was built in Hawaii and became operational in 2021 (Fig. 5.13). Other solar telescopes are discussed in the chapter on the Sun.

5.2.2 The Hubble Space Telescope and Other Projects

On April 24, 1990, the American Space Shuttle Discovery transported the Hubble Space Telescope (HST) into space, which was released by astronauts the following day. It was originally supposed to have entered orbit in 1982, but the launch was postponed until 1986. Due to the Challenger disaster (all astronauts died in the explosion in the space shuttle shortly after the launch on 28.01.1986) the launch was then postponed to 1990,

Fig. 5.13 The D.K. Inouye Solar Telescope, in Hawaii (Image: NSF)

Fig. 5.14 The space telescope (Hubble Space Telescope). You can see the two Stretched out solar cell arrays as well as the open protective flap. NASA/ESA

Primary Mirror

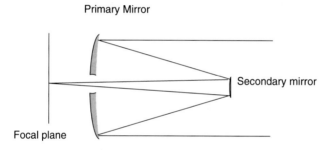

Secondary mirror

Focal plane

Fig. 5.15 Optical schematic of the HST

which was about three decades after the first considerations regarding a space telescope. The HST is so far the most expensive astronomical experiment with a cost of more than US\$ 2.0 billion (Fig. 5.14).

The first images revealed that the 2.4 m primary mirror (Fig. 5.15) was incorrectly ground, with the deviation from the edge to the center being 2.5 μm (cf. $\lambda/10$-criterion!). This led to a strong spherical aberration. The image of a point-like star was thus distributed over 0.7 arc s instead of the required 0.1 arc s. It was not until 1993 that an additional optical system (Fig. 5.16) to compensate for this error, and a new camera was also installed. In its present state, 70% of the starlight falls within 0.1 arc s. In 1997, a new service mission took place.

The successor to the Hubble telescope (Fig. 5.17) is the James Webb telescope, which was launched in Dec 2021. *James Webb Telescope* (JWST). It has a gold coated mirror of 6.5 m size (because observtaions are made in the IR). This mirror consists of 18 hexagonal segments which only unfold in space. The mirrors are made of the very light element beryllium. It will be possible to observe the very strongly red-shifted light of early galaxies

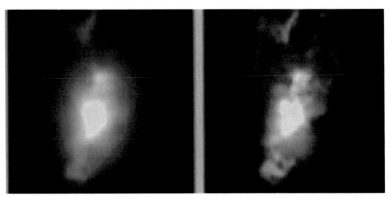

Fig. 5.16 With the help of the COSTAR system, a significant image improvement of the HST was achieved in 1993

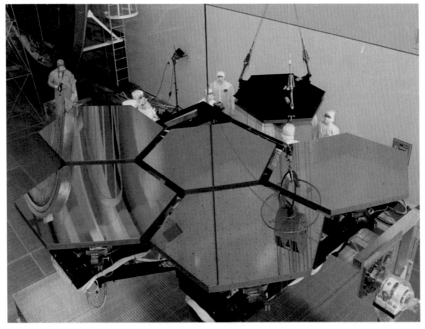

Fig. 5.17 Individual segments of the JWST being tested for cold suitability (Source: NASA/MSFC/Emmett Givens)

(i.e. in the IR). The telescope is located at the Lagrange point L2. This point is always behind the Earth in the Sun-Earth system. The total costs are at almost 10 billion US$.

Special satellite projects (e.g., for solar observation) are discussed in the appropriate chapters. Although large Earth-based telescopes are becoming very expensive to build, the cost is still lower than for space telescopes and, most importantly, necessary service missing are simpler. Active optics can be used to obtain observation quality comparable to that of space telescopes.

5.3   Detectors

The light captured in the telescope is observed or recorded with a detector. Only very rarely one observes directly visually at the telescope. We discuss the characteristics of detectors. Mostly one uses:

- Photography,
- photomultiplier,
- CCD.

5.3.1   Human Eye and Photography

In astronomy one uses mostly *photographic plates* instead of film.[2] Photographic plates are flat glass plates, on which the photographic emulsion is applied (e.g. bromide silver, AgBr; grain size $20 \,\mu$m). Often the plates are also subjected to special treatment to make them more sensitive to light. The *Quantum yield* is defined as Q:

$$Q(\lambda) = \text{number of photons detected/number of photons captured} \qquad (5.10)$$

Examples of quantum yield:
human eye: at $550 \,$nm , $Q = 0.01$,
photographic plate: between 440 and $650 \,$nm (visible region), the quantum yield is also $Q = 0.01$, but there is a *accumulating effect:* the longer the *exposure time,* the greater is the quantum yield caused by the action of light.

The relationship between incident light quantity and darkening is the empirically to be determined *density curve.* The density curve is linear only in a certain range.

The *human eye* has two light-sensitive receptors:

- *Cones:* Color sensitive but only slightly sensitive to light,
- *Rods:* rods very sensitive to light, not very sensitive to color.

The cones are mainly located at the point of the retina where a fixed object is imaged *(fovea),* the rods outside. Therefore, faint luminous or dimly illuminated objects are best seen by so-called *indirect vision,* i.e. one does not look directly at the object, but somewhat

---

[2] The first photograph was made by *Nicéphore* 1826.

beside it. This also explains why faint gas nebulae can only be seen in black and white by
eye in a telescope.

When observing with the naked eye, you can't see colors with a telescope because
of the faintness of most objects.

Adaptation is the process by which the eye becomes accustomed to darkness. The pupil
diameter adapts in a few seconds, the change in sensitivity of the retina, on the other hand,
goes hand in hand with the regeneration of the *visual purple (Rhodopsin)* and needs up to
30 min. The resolving power of the human eye t is about $1'$.

With the *photocells* one uses the *photoelectric effect*. As soon as light hits certain
materials, an electron is released and these electrons generate a measurable current. In
the case of a *Photomultiplier* one released electron generates a lot of electrons (up to 10^5).
In photocells, the quantum yield is between 10% and 20%.

5.3.2   CCD

A *charged coupled device (CCD)* it is a small chip (about $1\,\mathrm{cm}^2$) consisting of a large
number of light-sensitive pixels (typical sizes today are 4000 by 4000 pixels). Each of
these pixels can be compared to a photocell that accumulates charges. These charges are
then read out line by line by means of complicated electronics and result in the image. The
great advantage is the almost 100% quantum yield. In addition, the image is immediately
available in digital format and can be processed further, placed on the Internet, etc. One
hour exposure of a CCD chip with a 60 cm telescope yields images of equally faint objects
as 1 h photographic exposure with a 5 m telescope!

If you count photons of a signal, this number will fluctuate, due to *noise* from the
detector, interference from the Earth's atmosphere, and other effects. Let σ_m the standard
deviation from the mean value of the counted photons and $<N>$ the mean number of
photons, then the *Signal to Noise Ratio* given by:

$$S/N =<N>/\sigma_m \tag{5.11}$$

The larger this number, the better the quality of the observations. If one assumes *Poisson
distribution* is assumed, then

$$\sigma_m =<N>^{1/2} \tag{5.12}$$

and thus

$$S/N = < N >^{1/2} \qquad (5.13)$$

Therefore, if one counts 10^4 photons, then the S/N ratio is 100. If f denotes the photon flux falling on the detector and t the exposure time (integration time) and Q the quantum yield, then holds:

$$< N >= Qft \qquad S/N = (Qft)^{1/2} \qquad (5.14)$$

5.3.3 CMOS

CMOS is for *complementary metal oxide semiconductor*. The elements are produced in large numbers for computer chips and are therefore very cheap. In principle, it is a complete camera on a chip, i.e. all the necessary transistors, circuits, etc. are already present. This simplifies the readout: You can read out parts of a CMOS and even individual pixels (this is not possible with CCDs). You can find CMOS chips in all digital cameras or video cameras. Currently CCDs are still used for high requirements, but the cheaper CMOS technology has developed a lot.

5.3.4 Back-Illuminated Sensor

Normal CCDs have a detection probability for a single photon of about 60%. Using special circuitry, this can be increased to 90% for back-illuminated CCDs. This technique is used commercially in cameras for object protection, and also in astronomy, and is strongly on the rise.

> Photoplate, CCD, CMOS possess the advantage over the human eye that they accumulate photons; the longer the object is exposed, the fainter details become visible.

5.3.5 Speckle Interferometry

The seeing is usually larger than $1''$. With larger telescopes, the resolving power can be increased by very short exposures (0.1 ... 0.01 s), because the seeing is frozen during these short exposures: The *turbulence elements of* the Earth's atmosphere are in the range of 10–20 cm and the fluctuation time is in the range of 1/100 s. In this way one obtains individual structural elements *(speckles)* of the order of the resolving power of the telescope. To improve the signal-to-noise ratio, several 100 images are superimposed (stacking).

With this technique it was possible to directly determine stellar diameters for the first time, e.g. the star α Cyg, *Deneb,* has 145 times the radius of the sun.

5.3.6 Image Correction

As already discussed, the quality of astronomical observations is significantly affected by the air turbulence in the Earth's atmosphere. With the *Shack-Hartmann-sensor* the telescope image is split into many sub-apertures by an array of small lenses. One calculates the shift *(tilt)* of the optical wavefronts at these sub-apertures, and from the *tilt corrections* then results in a much sharper image. This is the simplest form of adaptive optics.[3] However, a suitable reference star (guiding stars) is always required. The signal is then passed to a deformable mirror for correction. With *multiconjugate adaptive optics* several deformable mirrors are used. As an alternative to a *guiding star* one can also use a laser guiding star (LGS). Rayleigh guide stars use lasers near the UV range, and the beams are backscattered at 15–25 km altitude. Sodium guide lasers emit light at 589 nm, which excites Na atoms. The advantage is a larger field of view by a factor of 5–10. This technique should be used with all telescopic systems above 8-m size.

5.4 Non-Optical Telescopes

Here we discuss telescopes that do not work in the optical range. Earthbound one can also receive radio signals from astronomical objects. For the UV or X-ray range, however, one must use space telescopes, since this radiation is absorbed by the Earth's atmosphere.

5.4.1 Radio Telescopes

In 1930 *Jansky* on behalf of the Bell Telephone Company interfered with transoceanic radiotelephony. In the process he discovered a source of substantial disturbance. This, as we now know, is the center of the Galaxy. Then, around 1940, the astronomer *Reber* systematically examined the sky for radio sources and found the *Sun* to be a radio source, which was later confirmed by *Hey* .

The *Radio Window* (transmission range) of the Earth's atmosphere lies in the range of wavelengths between 30 m and 1 mm.

---

[3] So-called *tip tilt correction.*

A radio *telescope* consists of a surface collecting radio waves (reflector), which concentrates them again in a focal point.

The *reflector* is made of metal, the requirements for the accuracy of the "mirror" depend on the wavelength of the radiation to be examined. The unevenness u must not exceed the maximum wavelength.

$$u \approx \frac{\lambda}{20} \tag{5.15}$$

Above a certain wavelength, one can also use mesh wire as a receiver.

The radio receiver converts the incoming radio waves into voltage, and this is recorded. These measurements are then plotted in a contour figure or false color. Since the *resolving power* of a telescope depends on the resolving power of the observed wavelength, it is very low in the radio range because of the long wavelengths. Visible light has a wavelength of 500 nm, radio waves of about meters, so the resolving power in the radio range is several orders of magnitude worse than in the visible range (as a rule of thumb 10^5). To obtain the same resolving power in the radio range as for the 5 m Hale telescope in the optical, the radio mirror would have to be 500 km in diameter. That's why you connect several radio telescopes together—*interferometer.* Let's consider two antennas that are a distance L apart from each other (Fig. 5.18). Let L be an integer multiple of the wavelength at which one observes:

$$L = n\lambda \tag{5.16}$$

Consider a radiation source O at a large distance from the two antennas. The waves from this source travel different paths P_1, P_2, where $P_1 > P_2$ is and P' is given by $P_1 = P + P'$. In order to obtain *constructive interference*, must hold:

$$P' = L \sin \Theta = m\lambda \tag{5.17}$$

Fig. 5.18 About the interference condition

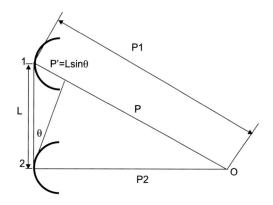

The interference condition is thus:

$$\sin \Theta = m\lambda/L \qquad (5.18)$$

The rotation of the earth changes Θ. If $m = 1$ one writes $\Theta = \Theta_f$:

$$\sin \Theta_f = \lambda/L = \lambda/(n\lambda) = 1/n \qquad (5.19)$$

since Θ_f is small in this case: $\sin \Theta_f \sim \Theta_f$ and one has

$$\Theta_f = 1/n \text{ rad} \qquad (5.20)$$

Where n is the number of wavelengths λ at the baseline L.

With an *interferometer* consisting of two antennas, one can only obtain a separation of points in the sky in the same direction in which the baseline is aligned. Now, to get a complete sky section with better resolution, one does aperture synthesis: The object is rotated, either by observing it as it moves across the sky each day, or by using the interferometers arranged in a form, such as with the VLA *(Very Large Array)* in New Mexico, where the antennas are arranged in a Y-shape configuration. With the VLBI*(very long baseline interferometry)* one takes the diameter of the Earth (about 12,000 km) as the maximum base, and for the example given above we obtain $\Theta_f = 3 \times 10^{-3''}$.

The largest free-moving parabolic antenna is the 100 m radio telescope at *Effelsberg* (near Bonn, Fig. 5.19), where the radio range i$\lambda > 2$ cm is observable. In a valley in *Arecibo,* Puerto Rico, there was a permanently mounted 305 m reflector which was used for observations at $\lambda > 30$ cm but the telescope crashed in 2020. In the Chinese province of Guishou, the 500-m radio telescope FAST (**F**ive hundred meter **A**perture **S**pherical radio **T**elescope, Fig. 5.20) was constructed. The telescope was first put into test operation in

Fig. 5.19 The 100 m radio telescope in Effelsberg

Fig. 5.20 The 500-m radio telescope FAST in China (Photo: CSN)

Fig. 5.21 Elements of the LOFAR network

2016. More than 9000 inhabitants within a radius of 5 km were relocated to minimize the impact from other sources of electromagnetic interference. The telescope operates in the frequency range 70 MHz–3 GHz (later upgraded to 8 GHz).

Very often the unit Jansky is used in radio astronomy:

$$1Jy = 10^{-19} \, \mathrm{erg \, s^{-1} m^{-2} \, Hz^{-1}} = 10^{-26} \, \mathrm{Wm^{-2} \, Hz^{-1}} \tag{5.21}$$

An important large-scale project in the field of radio astronomy is LOFAR (Low Frequency Array). The array consists of 25 000 plain radio sensors in the Dutch and northern German regions, and one can observe in the frequency range between 10 and 200 MHz (Fig. 5.21).

The ALMA project *(Atacama Large Millimeter Array)* consists of an array of 7 m and 12 m radio telescopes, scattered over several square kilometres in the Chilean Atacama Desert, and opened in 2013 (see Fig. 5.22). Observations are made in the millimeter and

Fig. 5.22 Some ALMA-antennas, in the background the Milky Way of the southern sky (ESO)

submillimeter range, which is only made possible by the location of the instrument at about 5000 m altitude. The final configuration consists of 66 antennas with a diameter of 12 m each. In total, this allows an interferometer baseline of 16 km. Observations can be made in the radio window of the Earth's atmosphere up to a frequency of 1 THz. The radio window in the Earth's atmosphere extends from 15 MHz to 100 GHz (i.e. 20 m to a few mm). ALMA will be used to observe extremely cool objects.

5.4.2 Infrared Telescopes

Due to the carbon dioxide contained in the Earth's atmosphere (CO_2) and the water vapor (H_2O) one can examine in the infrared only the ranges 2–25, 30–40 as well as 350–450 μm with earthbound observatories, the other wavelength ranges are absorbed by these gases. Infrared observatories are located in extremely dry zones where the content of H_2O-vapor is low.

The infrared range is divided into:

- Near Infrared (NIR): 700 nm–4 μm
- Mid IR: 4–40 μm
- Far IR: 40–300 μm

In astrophysics, filter designations are also very often used: I (around 1.05 μm), J (around 1.25 μm), H (around 1.65 μm), K (around 2.2 μm), L (around 3.45 μm), M (around 4.7 μm), N (10 μm), Q (20 μm). *Infrared detectors* are available for a good 30 years.

A bolometer consists of a germanium chip the size of a pinhead, cooled to 2 K. When IR radiation falls on it, it heats up and the electrical resistance increases.

Why do we observe the sky in the IR? IR radiation is less affected by dust between the stars, and objects with a temperature below 3000 K emit mostly IR radiation. Furthermore, IR measurements can be made during daytime, because sunlight is hardly scattered in the IR.

Lead sulfide cells (PbS) as well as indium antimonide cells (InSb) are used as detectors. The problem with IR observations is the heat sources, and so one uses a *chopper:* The source and the sky background are observed in quick succession and subtracted in order to better filter out the signal from the source. Cryostats (liquid nitrogen or He) are used to cool the detectors. In 1983 the IR satellite IRAS scanned the sky (between 12 and 100 μm about 300,000 sources), a little later COBE measured the microwave background, in 1995 the Infrared Space Observatory ISO was launched. The Hubble Space Telescope was made IR-capable in 1997 with the NICMOS. Since 2003, the Spitzer Space Telescope has been observing the sky in the IR. In 2009, the coolant was depleted (it cooled to 2 K), and the mission is now limited, as the maximum temperature is 31 K. Other IR telescopes are *IRAS* (start 1983), *ISO* (launched 1995) and *Herschel* (launch 2009)l. The Herschel telescope has a mirror diameter of 3.5 m and is the largest mirror in space until the launch of the James Webb telescope. Herschel is located at the Lagrange point L2 (Fig. 5.23).

The 2 Micron All Sky Survey, 2MASS contains more than 500 million sources in an area from 1.25–2.17 μm.[4]

The IRAS catalogue contains many IR objects, stars and galaxies.

5.4.3 X-Ray Telescopes

The X-ray range extends from 0.1 to 500 keV, i.e. 12 nm to 2.5 pm. Radiation below 2 keV is called soft X-ray. The first cosmic X-ray source to be discovered was the corona of the Sun in 1949 (with a V2 rocket). In 1962 Giacconi and co-workers discovered the X-ray binary star Scorpius X-1.

X-ray telescopes can only be used on satellites because of the absorption of X-rays in the Earth's atmosphere. With perpendicular incidence, the reflectance in the X-ray region is almost zero, so focusing instruments can only be built with grazing incidence (Wolter telescope[5]). There is a reflection at an elongated paraboloid (only 1–2° angles of incidence), one has a total reflection of X-rays at grazing incidence (Fig. 5.24). As detectors one uses proportional counters, scintillation counters, Cerenkov counters, semiconductors. Important X-ray satellites were:

---

[4] Conducted with automated 1.3-m telescopes in Arizona and Chile.

[5] *Wolter,* 1952.

Fig. 5.23 The Herschel telescope (Source: NASA)

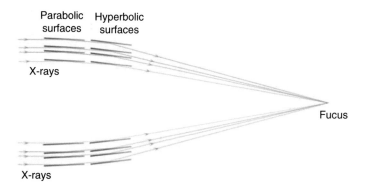

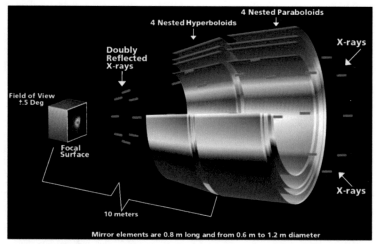

Fig. 5.24 Principle of a Wolter X-ray telescope. Bottom: Sketch of the CHANDRA telescope. The mirrors have to be extremely flat and plane ground (analogue: If the earth surface is equal to the mirror surface, the tolerance is max. 2 m elevation) (Source: NASA)

UHURU, EINSTEIN, ROSAT (about 100,000 sources), CHANDRA, XMM Newton. CHANDRA[6] was the previously largest research satellite that a space Shuttle transported.

5.5 Spectroscopy

Spectroscopy is about the decomposition of light. Important astrophysical information can be obtained from the analysis of the spectrum or spectral lines: Temperatures, composition, etc.

---

[6] Named after the astrophysicist *Chandrasekhar* (1910–1995), 1983 Nobel Prize in Physics.

Fig. 5.25 Prism

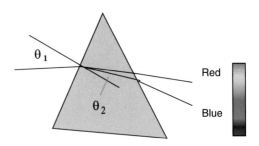

5.5.1   General Information About Spectroscopy

White light exists from all wavelengths, whereby the *wavelength* determines the *color* of the light:

- Red light is at 600 nm,
- blue light is at 400 nm.

Spectrum
The intensity is plotted as a function of wavelength.

There are two ways to decompose white light:

- *Prism:* At a glass prism (Fig. 5.25) there is a refraction of the light after the *Snellius' law of refraction:*

$$n_a \sin \Theta_1 = n_g \sin \Theta_2 \tag{5.22}$$

n_a...*Refractive index* for air[7] ≈ 1; n_g...refractive index for glass, this depends on the wavelength λ of the light. It holds:

$$n_g \approx 1/\lambda^2 \tag{5.23}$$

So, short wavelength light (blue, smaller wavelength) is refracted more than long wavelength light (red, larger wavelength). The refractive index for water, for example, is 1.33. If Φ is the deflection of the light, then for the *angular dispersion:*

$$\frac{d\Phi}{d\lambda} \approx \frac{dn}{d\lambda} \approx \frac{1}{\lambda^3} \tag{5.24}$$

---

[7] n_a indicates the ratio of the speed of light in vacuum to the speed of light in air.

In astronomy one rarely uses prism spectrographs because on the one hand the production is more complicated (the light goes through glass, this must be absolutely faultless), on the other hand the dispersion is not linear.

- *Grating:* Either as a reflection grating or as a transmission grating. The grating constant d is the (constant) distance between two grooves, ϕ_i the incidence angle and ϕ_e the exit angle; then the *optical path difference* between incoming and outgoing wavefront becomes:

$$\Delta = \Delta_1 + \Delta_2 = d \sin \phi_i + d \sin \phi_e \tag{5.25}$$

Amplification occurs if this is a multiple of the wavelength:

$$\Delta = \Delta_1 + \Delta_2 = d \sin \phi_i + d \sin \phi_e = m\lambda \tag{5.26}$$

It is easy to construct the directions: Draw a semicircle with radius d and take the direction of incidence ϕ_i. The reflection (zero order) results simply from reflection. Then one applies from O to the left and to the right λ and to obtain the first order and so on. The maximum number of *orders* is:

$$m_{\max} = 2d/\lambda \tag{5.27}$$

The *angular dispersion* of a grating is obtained by differentiating the grating equation ($\phi_e = $ const):

$$\frac{d\phi_i}{d\lambda} = \frac{m}{d \cos \phi_i} \tag{5.28}$$

It is *independent* of the wavelength. If N is the number of grooves, $N = D/d$, D is the total width of the diffraction grating, then we obtain maxima for the path differences $n\lambda$. The *Resolving power* is defined as:

$$A = \frac{\lambda}{\Delta\lambda} = mN \tag{5.29}$$

Blaze grating: using a particular groove shape one can direct the main intensity in a particular direction. Let N be the grating normal, B the normal to the mirror surface (blaze normal), and Θ the blaze angle. The blaze effect occurs when the incident and outgoing rays are aligned with respect to the blaze normal, B fulfilling the reflection law:

$$1/2(\phi_i + \phi_e) = \Theta$$

One speaks of *Echelle grating,* when $\Theta = 45°$ is; the narrow side of the blaze acts as a mirror. Echelle gratings yield very high dispersion, and light entry is very oblique, i.e., the grating is illuminated at an oblique angle.

Consider a grating with 500 grooves/mm; it follows that $d = 2 \times 10^{-3}$ mm. Observing at 500 nm gives: $m_{max} = 8$. The orders overlap, therefore a predisperser is used.: $m_k \sin \lambda_k = m_i \sin \lambda_i$; thus λ_1 in the first order corresponds to $\lambda_1/2$ in the 2nd order, etc.

We use the relation $m_i \lambda_i = m_k \lambda_k$: 600 nm, 2nd order = 400 nm ($2 \times 600 = 3 \times \lambda_k$), 3rd order 300 nm ... ; 500 nm in 1st order corresponds to 512.8 nm in 40th order ...

At the *Littrow set-up,* the light enters in direction B, therefore one has $\phi_e = \phi_a = \Theta$, and the formula is obtained:

$$m\lambda = 2d \sin \Theta$$

Given a grating with 73 grooves/mm; $\Theta = 64.5°$. Then one has at $\lambda = 500$ nm from the above formula we have a Blaze effect in the 49th order.

Usable orders are therefore between 50 and 100 for echelle gratings. To separate the overlapping orders, one places a low dispersion grating perpendicular to the echelle grating axis.

Grating production used to be carried out by means of a scribing technique, whereby a diamond was used to scribe onto a thin metal layer which had been vapour-deposited onto a glass body. Spindle errors of the grating scribers result in a superimposed grating and cause symmetrically arranged additional lines, the so-called grating Ghosts. Nowadays gratings are mostly produced holographically, a broadened laser beam is split into two coherent beams which produce a spatial interference system. This implies a short fabrication time and no grating ghosts. Figure 5.26 shows a typical arrangement of a grating spectrograph.

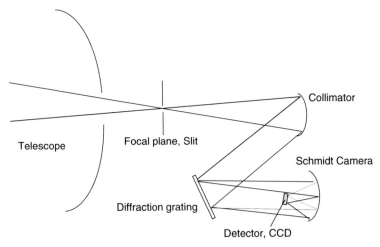

Fig. 5.26 Principle of a grating spectrograph

5.5.2 Types of Spectrographs

We distinguish three types of spectrographs. High spectral resolution is provided by slit spectrographs and Fourier transform spectrometers.

1. Objective grating: placed in front of the objective; a small spectrum is produced from each star. However, the dispersion is only a few 10 to 100 nm/mm. Often used to classify several stars simultaneously, e.g. in a star cluster.
2. Slit Spectrograph: Light coming from a telescope is focused in the entrance slit, brought parallel to the grating by means of a collimator lens etc.
3. Fourier transform spectrometer: One obtains an interferogram by scanning the path difference in a two-bundle interference. From this one gets the spectrum by means of a Fourier transformation.

5.6 Radiation and Spectrum

5.6.1 The Electromagnetic Spectrum

Electromagnetic waves were for the first time by *J.C. Maxwell* described in his equations (he worked on these four equations between 1861 and 1864). Their propagation is not dependent on a medium, they also propagate in a vacuum, with the oscillating electric and magnetic fields perpendicular to each other. If the electric field always oscillates in a certain plane, the wave is called *polarized.* Let us consider the two components of a wave vector **E**:

$$E_x = E_{x1} \sin(kz - \omega t - \delta_1) \tag{5.30}$$

$$E_y = E_{y1} \sin(kz - \omega t - \delta_2) \tag{5.31}$$

Waves are polarized when $\delta_1 - \delta_2$ is constant.

Light is electromagnetic radiation in the range 390 nm–720 nm. The following units are used:

$1\,\text{nm} = 10^{-9}\,\text{m}$; $1\,\mu\text{m} = 10^{-6}\,\text{m}$; $1\,\text{Å} = 10^{-10}\,\text{m} = 0.1\,\text{nm}$. The conversion from wavelength to frequency is done with:

$$c = \lambda \nu \tag{5.32}$$

and ν is given in Hertz (Hz), 1 Hz = 1 oscillation per second.

In Table 5.3 an overview is given.

Table 5.3 The
electromagnetic spectrum

Wavelength	Energy or frequency	Radiation type
10^{-6} nm	1240 MeV	Gamma radiation
10^{-2} nm	124 keV	X-rays (X)
1.0 nm	1.24 keV	UV
$\approx 1\,\mu$m	1.24 eV	Visible light
$10\,\mu$m $-$ 1 mm	0.124 eV $-$ 0.0012 eV	IR
10 mm	30,000 MHz	Microwaves (radar)
10 cm	3000 MHz	UHF
100 cm	300 MHz	FM
10 m	30 MHz	Shortwave
1000 m	300 kHz	Longwaves

The relationship between energy and frequency is defined by the formula

$$E = h\nu \tag{5.33}$$

Thereby $h = 6.626 \times 10^{-34}$ Js is the Planck's constant. Formula (5.33) characterizes the energy of a quantum of light.

5.6.2 Thermal Radiation

The radiation of a body (stars, sun, planet) can be described simplified by the radiation of a black body. A black body is defined as follows: All electromagnetic radiation incident on it is completely absorbed by the black body.[8]

This is an idealization, but in astrophysics very often a very good approximation.

Stars radiate like black bodies to a good approximation.

The cosmic background radiation corresponds to that of a black body with a temperature of 2.7 K.

Example

The temperature of a body determines at which wavelength the maximum of its radiant power lies. The radiant power at room temperature is 460 W/m², our eye is not sensitive

---

[8] 1860, *Kirchhoff.*

to radiation of this wavelength, but you can feel the heat. Sun: $T = 5800\,\mathrm{K}$, has a radiant power of $64\,\mathrm{MW/m^2} \rightarrow$ visible radiation.

For a black body the *Planck's law of radiation* describes the intensity of radiation of a black body at temperature T in the unit $\mathrm{W\,m^{-2}\,sr^{-1}\,Hz^{-1}}$ Is:

$$I(v, T) = \frac{2hv^3}{c^2} \frac{1}{e^{\frac{hv}{kT}} - 1} \tag{5.34}$$

$\rightarrow$ Thus, the intensity of radiation from a black body depends on T and of v, the frequency. Because of $c = \lambda v$ one obtains the intensity as a function of wavelength and temperature of a black body:

$$I(\lambda, T) = \frac{2hc^2}{\lambda^5} \frac{1}{e^{\frac{hc}{k\lambda T}} - 1} \tag{5.35}$$

The radiation over all wavelengths in $\mathrm{Wm^{-2}}$ is obtained by integrating Planck's law of radiation over all wavelengths $\rightarrow$ *Stefan-Boltzmann law*:

$$E = \sigma T^4 \qquad \sigma = 5.67 \times 10^{-8} \frac{\mathrm{W}}{\mathrm{m^2 K^4}} \tag{5.36}$$

The maximum of the radiation of a black body shifts to shorter wave length $\rightarrow$ *Wien's displacement law* (Fig. 5.27):

$$T\lambda_{\mathrm{max}} = 0.002897\,\mathrm{mK} \tag{5.37}$$

Besides of this thermal radiation there is the so called *non-thermal* radiation like Synchrotron radiation. Here, charged particles are accelerated in the magnetic field, which leads to radiation.

Fig. 5.27 Planck's law of radiation: The higher the temperature of a black body, the greater the radiation power at all wavelengths. The maximum of radiation shifts to short wavelengths, Wien's law. Blue stars are therefore hotter than red ones ($\lambda_{\mathrm{blue}} < \lambda_{\mathrm{red}}$)

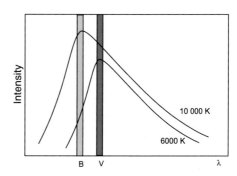

5.6.3 Emission and Absorption Lines

The emergence of emission or absorption lines in the spectrum can be described simplest by the *Bohr's atomic model.*

In an atom, the electron can only exist in certain quantized energy states, which are defined by the integer value of n, the *principal quantum number,* . This thus characterizes the various possible *orbits* (shells) of the electron and thus simultaneously its energy states, and the radius of the electron orbit is proportional to n^2. On the innermost orbit ($n = 1$) the electron has the lowest energy. The *Ionization energy* is the energy that must be applied to release the electron from the sphere of influence of the nucleus. For the *Hydrogen atom* (electron in ground state on $n = 1$) it amounts to 13.6 eV (1312 kJ/mol); $1\,eV = 1.6 \times 10^{-19}\,J$ is the kinetic energy that an electron receives when passing through a voltage difference of 1 V in vacuum. Beyond the ionization limit, the electron (mass m_e) can assume any kinetic energy.

It holds:

$$E_{Tot} = \frac{m_e v^2}{2} - \frac{e^2}{4\pi \epsilon_0 r}$$
(5.38)

This is the total energy, which is made up of

- kinetic energy and
- potential energy (given by the *Coulomb potential*) components.

e...elementary charge, ϵ_0...Dielectric constant.

Let us consider an electron about a nucleus. The condition for the equilibrium of forces is: the outward centrifugal force is equal to the electric attraction force (between the electron and the nucleus of the atom):

$$\frac{m_e v^2}{r} = \frac{1}{4\pi \epsilon_0} \frac{e^2}{r^2}$$
(5.39)

The two *Bohr's postulates* are:

- An atomic system has stationary (nonradiative) states with certain discrete energy values.
- An atomic system can only change its energy by transitioning from one stationary state to another stationary state. If the transition is associated with emission or absorption of radiation, its frequency is related to the energy change by the frequency condition:

$$E = h\nu$$
(5.40)

Furthermore, it holds that the orbital velocity is a multiple of h ($\hbar = h/2\pi$):

$$v = \frac{n\hbar}{m_e r} \tag{5.41}$$

With the condition for the equilibrium of forces (5.39) and substitute v from (5.41) we obtain:

$$r = \frac{h^2 \epsilon_0}{m_e e^2 \pi} n^2 \qquad n = 1, 2, 3 \ldots \tag{5.42}$$

If one sets $n = 1$ in a hydrogen atom, then we obtain Bohr's Atomic radius $a_0 = 0.529 \times 10^{-10}$ m. Electrons with $n = 1$ are called electrons of the K-shell, with $n = 2$ L-shell, with $n = 3$ M-shell etc.

At the transition from n_2 to n_1 an energy of the amount

$$\Delta E = 2.179 \times 10^{-18} \left(\frac{1}{n_1^2} - \frac{1}{n_2^2} \right) \text{ J} \tag{5.43}$$

is released. An electron lifted to a higher shell by a an incoming photon with energy $h\nu$ falls back to an energetically lower shell, and the energy released in the process is radiated as a quantum of light. One sees an *emission line* in the spectrum. On the other hand, absorption of a quantum of light from a continuous spectrum occurs when the energy of the quantum of light is just equal to the difference in energy between the shell in which the electron is located and a higher shell into which the electron is lifted. One sees then an *absorption line*.

One has for hydrogen (Fig. 5.28):

- Lyman series: (all lines in UV light)
  - Absorption lines: Transitions from $n = 1$ at $n = 2, 3, 4, \ldots$.
  - Emission lines (energy is released): transitions from $n = \ldots 4, 3, 2$ at $n = 1$.
- Balmer series (all lines in visible light):
  - Absorption lines: Transitions from $n = 2$ on $n = 3, 4, 5, \ldots$.
  - Emission lines (energy is released): transitions from $n = \ldots 5, 4, 3$ on $n = 2$.
- Paschen series (all lines in the IR region):
  - Absorption lines: Transitions from $n = 3$ on $n = 4, 5, 6, \ldots$.
  - Emission lines (energy is released): transitions from $n = \ldots 6, 5, 4$ on $n = 3$.

Often the following relations are useful:

$$1 \text{ J} = 6.24 \times 10^{18} \text{ eV} \qquad \lambda \, [\text{nm}] = \frac{1240}{\chi \, [\text{eV}]} \tag{5.44}$$

Fig. 5.28 Transitions in the
hydrogen atom. $n = 1$
corresponds to $-13.6\,$eV,
$n = 2$ corresponds to $-3.4\,$eV,
$n = 3$ equals $-1.5\,$eV, $n = 4$
equals $-0.85\,$eV. To ionize an
atom of $n = 113.6\,$eV of
energy are therefore necessary

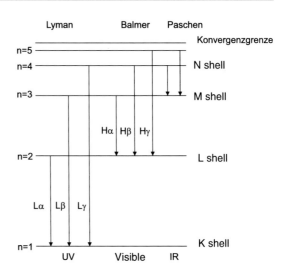

Here χ denotes the energy difference between two levels. If one knows this in the unit
[eV], then follows from (5.44) the wavelength of the transition in [nm].

> Because of the discrete energy states of the electrons bound to an atom, the spectral
> lines are formed.

5.6.4 Polarized Light

Light waves are *transverse waves,* oscillation is perpendicular to the direction of propagation. The *Wave Vector* points in the direction of propagation, the amplitude vector is
perpendicular to it. Thus, in three-dimensional space, the rotation about the wave vector is
indeterminate.

Polarization: Amplitude vector points in a preferred direction.

There are three types of polarization:

- Linear polarization: amplitude vector always points in certain direction.
- Circular polarization: amplitude vector rotates at constant angular velocity as wave
  propagates.
- Elliptical polarization: Amplitude vector rotates around wave vector and also changes
  its magnitude periodically.

Light is normally not polarized. Polarization occurs e.g. by *reflection* of light[9] or *Rayleigh scattering* (This is the scattering of electromagnetic waves by particles whose diameter d is small compared to the wavelength λ of the radiation. The scattering cross section is proportional to λ^{-4}; this explains the blue color of the sky, because blue light is scattered more than red light).

For the description one uses the Stokes vector (P denotes the power when the polarizer is oriented at a certain angle. There are right circular (RC) and left circular (LC) polarizations):

$$I = P_{0°} + P_{90°} \tag{5.45}$$

$$Q = P_{0°} - P_{90°} \tag{5.46}$$

$$U = P_{45°} - P_{135°} \tag{5.47}$$

$$V = P_{RC} + P_{LC} \tag{5.48}$$

If the light is completely polarized, then:

$$I^2 = Q^2 + U^2 + V^2 \tag{5.49}$$

If the light is unpolarized, then $Q = U = V = 0$. The degree of polarization p is:

$$p = \frac{\sqrt{Q^2 + U^2 + V^2}}{I^2} \tag{5.50}$$

Write the Stokes vector for the following cases: (a) linear, horizontally polarized, (b) linear, +45° polarized.

Solution:

$$(a) \begin{pmatrix} 1 \\ 1 \\ 0 \\ 0 \end{pmatrix} \qquad (b) \begin{pmatrix} 1 \\ 0 \\ 1 \\ 0 \end{pmatrix}$$

Polarized light can be produced by the following processes in the universe:

- Emission processes: Cyclotron, synchrotron radiation non-thermal bremsstrahlung.
- Scattering from non-spherical particles: e.g. accretion disks, interstellar dust.

---

[9] With reflection in the Brewster angle complete polarization.

Fig. 5.29 Zeeman effect at (a)
magnetic field = 0, (b)
magnetic field longitudinal,
one observes two oppositely
circularly polarized σ
components. (c) magnetic field
transversal in direction to the
observer. One has two σ- and
an unshifted π component

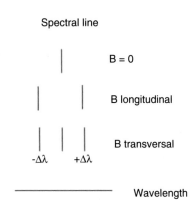

- Vacuum polarization[10] and birefringence due to extremely strong magnetic fields. Quantum mechanics predicts that there are virtual particle-antiparticle pairs in vacuum. If one brings a charge into vacuum, then an electric force acts on these particle-antiparticle pairs. The virtual particles separate from the antiparticles, and the charge on the nucleus is amplified immediately at the nucleus, and attenuated further away from it.

By combining spectroscopic measurements and polarimetry, called spectropolarimetry, it has been possible to detect magnetic fields of the strength of the Sun's magnetic field on other stars.

5.6.5 Magnetic Fields and Radiation

In this section we first consider the splitting of spectral lines in magnetic fields found by *Zeeman*.[11] In addition to splitting, polarization is also observed.

- If no magnetic field $\rightarrow$ unshifted absorption line.
- If the magnetic field is transverse to the direction of observation $\rightarrow$ one unshifted and two components shifted by the amount of $\Delta\lambda$; *transverse Zeeman effect*.
- If the magnetic field lies in the direction of propagation $\rightarrow$ line split into two components only; *longitudinal Zeeman effect* (Fig. 5.29).

---

[10] First detection 1990.

[11] *P. Zeeman,* 1896.

For the amount of splitting $\Delta\lambda$ one obtains (λ in cm and B in Gauss):

$$\Delta\lambda = 4.67 \times 10^{-5} B g \lambda^2 \text{ [cm]} \tag{5.51}$$

Thereby g, the *Landé factor* is known from quantum mechanics. Lines with Landé factor $g = 0$ therefore do not split by a magnetic field. B denotes the magnetic flux density (induction, usually simply called magnetic field strength). Important is also the polarization. If one looks longitudinally, i.e. along the field, then one sees only two components, which are shifted by $\Delta\lambda$ and are circularly polarized to the left and right, respectively. Looking transversely to the direction of the magnetic field, there is one unshifted linearly polarized (π) and two circularly polarized (σ), shifted by the amount $\pm\Delta\lambda$ components of the line.

In the case of very strong magnetic fields one also speaks of the *Paschen-Back effect*.

The Zeeman effect can be derived in a semi-classical simple way. Circular motion of an electron around the atomic nucleus with velocity v and radius r leads to a current

$$I = -e \cdot \frac{v}{2\pi r} \tag{5.52}$$

This generates a magnetic moment

$$\mu = I\mathbf{A} \tag{5.53}$$

The vector $\mathbf{A}$ is perpendicular to the surface enclosed by the circular orbit. The orbital angular momentum of the electron is :

$$\mathbf{l} = \mathbf{r} \times \mathbf{p} = m_e \cdot r \cdot v \cdot \hat{\mathbf{n}} \tag{5.54}$$

therefore

$$\mu_l = -\frac{e}{2m_e} \cdot \mathbf{l} \tag{5.55}$$

The potential energy in a magnetic field is

$$E_{\text{pot}} = -\mu_l \cdot \mathbf{B} = \frac{e}{2m_e} \cdot \mathbf{l} \cdot \mathbf{B} \tag{5.56}$$

m—magnetic quantum number, μ_B—Bohr's magneton:

$$\mu_B = \frac{e}{2m_e}\hbar = 9.274078 \times 10^{-24} \text{ A m}^2 \tag{5.57}$$

The *anomalous Zeeman effect* still takes into account the spin of the electron. The Landé factor is:

$$g_J = 1 + \frac{J(J+1) + S(S+1) - L(L+1)}{2J(J+1)} \tag{5.58}$$

Where J the total angular momentum, S the spin, and L the orbital angular momentum of the atom. If $B = 0$ then the M_J states are in the same energy. This degeneracy is cancelled if an external magnetic field B is present. By the way, if one sets $S = 0$ in the expression for the Landé factor, one gets the normal Zeeman effect again.

It is important to note that the splitting also depends on the wavelength itself. Thus, infrared lines split more strongly than lines in the visible range.

Another important effect is the *Hanle effect*, which leads to a rotation of the plane of polarization.

5.6.6 Einstein Coefficients

As we have seen, spectral lines are formed by transitions between discrete energy states.

- Emission: described by the so-called *Einstein coefficient* A_{ji}, where $j > i$ for the states holds.
- Absorption: B_{ij}.
- *Einstein* showed that holds:

$$B_{ij} = \frac{c^3}{8\pi h \nu_{ji}^3} \frac{g_j}{g_i} A_{ji} \tag{5.59}$$

g_i, g_j are the *statistical weights* (degeneracy) of the states j, i.

Very often one uses also the oscillator strength:

$$f_{ij} = \frac{4\pi \epsilon_0 m_e}{e^2 \pi} h \nu_{ji} B_{ij} \tag{5.60}$$

And it holds $g_j f_{ij} = -g_i f_{ji}$

- Stimulated emission: A photon with suitable emission wavelength initiates the emission of photons$\rightarrow$ coherent emitted photons. An example is the laser,[12] in astrophysics one often has to deal with masers.[13]

---

[12] Light Amplification by Stimulated Emission of Radiation.

[13] Microwave Amplification by Stimulated Emission of Radiation.

5.6.7 Coherence

Two waves are called *coherent* if they can cause temporally constant interference phenomena → temporally invariant phase difference.

Let us consider two waves $E(\mathbf{r_A}, t)$, $E(\mathbf{r_B}, t)$, then one defines a cross-correlation function

$$\Gamma_{AB}(\tau) = < E(\mathbf{r_A}, t)E^*(\mathbf{r_B}, t) >= \lim_{T \to \infty} \frac{1}{T} \int_{-T/2}^{T/2} E(\mathbf{r_A}, t)E^*(\mathbf{r_A}, t + \tau)dt \qquad (5.61)$$

The contrast function for spatiotemporal coherence

$$K_{AB}(\tau) = \frac{2\Gamma_{AB}(\tau)}{\Gamma_{AA}(0) + \Gamma_{BB}(0)} \qquad (5.62)$$

has a value between 0 and 1. If $K_{AB}(\tau) = 1$, we have a full coherence.

The Michelson interferometer can be used to measure temporal coherence. Light waves are not infinitely long sinusoidal wave trains. There is only a fixed phase relationship between light waves whose photons have left a source at different times if the time difference is very small. This results in a maximum time difference at which stationary interference phenomena can still be observed. One of the two light beams travels a slightly longer path than the other.

5.7 Further Literature

We give a small selection of recommended further reading.
Space Telescopes, N. English, Springer, 2016
Telescopes and Techniques, C.R. Kitchin, Springer, 2012
Experiencing the Night Sky, A. Hanslmeier, Springer, 2017
The Astrophotgraphy Manual, C. Woodhouse, Taylor and Francis, 2017
Astrophotography, K. Seidel, Rheinwerk, 2019

Tasks

5.1 Determine the focal ratio of a telescope with 150 mm aperture and 750 mm focal length.

Solution
$750/150 = 5$ or $f/5$.

5.2 Given a $f/15$ telescope with a 60 cm mirror. What is the plate scale?

Solution

$f/d = 15 \qquad f = 15 \times 60\,\text{cm} = 900\,\text{cm} \rightarrow s = 0.01745 \times 900\,\text{cm} = 15.7\,\text{cm}/°$

5.3 Given a telescope with $f = 1000$ mm. For observation, a camera with a pixel size of 5 μm is used. What resolution can be obtained with this configuration?

Solution

Substituting into the formula gives a plate scale of about $1''$/pixel. According to the Nyquist theorem, one can therefore resolve a maximum of $2''$ (the resolving power of the telescope has not been considered yet!). Consider how to increase this resolution!

5.4 Suppose a 5-m telescope with seeing conditions of about $1''$. What are the advantages of such a telescope compared to a 10 cm telescope which achieves a resolution of 1 arc s under similar conditions?

Solution

In this case, although seeing limits resolving power, a 5-m telescope has much greater light-gathering power than a 10-cm telescope

5.5 Let the focal ratio of a telescope be $N = 3.2$ and one observes at 500 nm; one calculates the focal tolerance!

Solution

This gives a focal tolerance of $\Delta = 20\lambda = 0.01$ mm.

5.6 By what amount does the light yield increase with the 42 m ELT compared to the Keck telescope (10 m)?

Solution

Compare the areas! The light yield is proportional to the area. The telescope could be put into operation around 2025.

5.7 Given a radio interferometer with $L = 21$ km and one observes at a wavelength of 21 cm; one calculates the resolving power!

Solution

$\Theta_f = 21/(21 \times 10^5) = 10^{-5}\,\text{rad} = 2''$. That means, with an interferometer of the base length 21 km, one gets a resolution of $2''$ which is close to that of optical telescopes.

5.8 Grating with 500 grooves/mm; $D = 100$ mm; calculate the resolving power to the 4th order.

Solution
$A = 4 \times 500 \times 100 = 200,000$; for $\lambda = 500$ nm one then has $\Delta\lambda = \lambda/A = 0.25$ nm.

5.9 Calculate at which wavelength the following objects have the maximum of their radiation: (a) the sun ($T = 5800$ K), (b) a red giant ($T = 3000$ K), (c) an A0 star with $T = 10,000$ K.

Solution

a) 510 nm, so in the green range
b) 970 nm, i.e. in the IR,
c) 290 nm, i.e. at the blue/UV boundary.

5.10 Calculate the wavelength of the hydrogen line Hα which results from the transition from $n = 3$ to $n = 2$.

Solution
Using Eq. (5.43) one finds $\Delta E = 3.026 \times 10^{-19}$ J $\approx 1..888$ eV and from this $\lambda \approx 656$ nm.

Physics of the Solar System Bodies

6

In this section we bring briefly the most essential results of the study of the bodies of the solar system. The sun itself is treated separately. The study of the atmospheres of other planets and their development and structure enables conclusions to be drawn about our Earth; the formation of our planetary system can be better understood by comparison with extrasolar planetary systems.

6.1   Overview

The dominant role of the Sun around which the other bodies of the solar system orbit is evident from their physical parameters such as mass, diameter, etc.

6.1.1   Sun and Planets

The dominant body in the solar system is the *Sun*. Let us consider some of its physical parameters

- Radius $R_\odot = 696,000$ km, this corresponds to 109 Earth radii.
- Mass $M_\odot = 1.989 \times 10^{30}$ kg, this corresponds to 333,000 Earth masses.
- Gravity acceleration at the surface $g_\odot = 274 \, \text{m/s}^2$, this corresponds to 28 times the acceleration due to gravity.

The Sun contains 99.8% of the mass of the entire solar system.

© Springer-Verlag GmbH Germany, part of Springer Nature 2023
A. Hanslmeier, *Introduction to Astronomy and Astrophysics*,
https://doi.org/10.1007/978-3-662-64637-3_6

Furthermore, the solar system includes the known eight large *planets*. Since the decision of the IAU[1] in 2006, Pluto no longer counts as a large planet. The planets have diameters between some 10^3 km up to 10^5 km; they move around the Sun in nearly circular orbits, all of which lie almost exactly in the plane of the ecliptic, and the total mass of all the planets together is 448 Earth masses. Enumerating the planets according to their distance from the sun, we have Mercury, Venus, Earth, Mars, Jupiter, Saturn, Uranus and Neptune. The newly defined group of the *dwarf planets* include Pluto and other objects (Sedna, Ceres, and others).

The *asteroids* or minor planets (also called Planetoids) are, with a few exceptions, smaller than 100 km, and most are located between the orbits of the planets Mars and Jupiter, but there are also some that cross the Earth's orbit, for example. Outside the orbit of Neptune lie the objects of the *Kuiper belt,* to which also Pluto belongs.

The *Moons* or satellites are companions of the major planets (Mercury and Venus have no moon), dwarf planets (Pluto has five moons, the last two discovered in 2012), and asteroids.

Many *Comets* are located in the *Oort's comet cloud.* This envelops the solar system spherically symmetrically. Disturbances cause comets to enter the interior of the solar system, where evaporating gases produce the comet tail. They are small in diameter (less than 100 km).

As the last component of the solar system there is still the *interplanetary matter* (meteoroids, gas and dust). In good conditions, the matter distributed along the ecliptic can be seen after sunset or before sunrise as a *Zodiacal light* when dark enough.

6.1.2   A Model of the Solar System

Consider a $1:10^9$ scale Solar System Model (Table 6.1). In this model, the 12-mm Earth would orbit the 1.4-m Sun at a distance of 150 m. The nearest star (apart from the Sun) would be about 40,000 km away (equivalent to the circumference of the Earth).

Table 6.2 gives a summary of the most important properties of the planets.

> Units of distance in the solar system are often expressed in *astronomical unit, AU*).
> This is the average distance from the Earth to the Sun (150 million km).

---

[1] International Astronomical Union.

Table 6.1 Model of the solar system, scale $1:10^9$

Object	Diameter	Distance from sun
Sun	1.4 m	
Mercury	5 mm	60 m
Venus	12 mm	110 m
Earth	12 mm	150 m
Mars	7 mm	230 m
Jupiter	14 cm	800 m
Saturn	12 cm	1.5 km
Uranus	5 cm	3 km
Neptune	5 cm	4.5 km
Pluto	2 mm	6 km

Table 6.2 The most important properties of the planets and of Pluto; d—distance from the Sun

Planet	$d[10^6$ km]	Orbital time	Rotation time	Equator. inclination [^{o}]
Mercury	57.9	87.9d	58.65 d	0
Venus	108.2	224.7	243.01 d	2.01
Earth	149.6	1.00 a	23 h56 min	23.5
Mars	227.9	1.88	24 h 37 min	24
Jupiter	779	11.87	9 h 50 min–9 h 56 min	3
Saturn	1432	29.63	10 h 14 min–10 h 39 min	24
Uranus	2888	84.66	17 h 06 min	98
Neptune	4509	165.49	15 h 48 min	29
Pluto	5966	251.86	6.3 d	122.5

6.1.3 The Solar System Seen from Outside

From the nearest star (α Cen) the solar system would appear as follows: The Sun would have an apparent magnitude of $0.^m4$, so it would be a bright star. Earth would have a brightness of $23.^m4$ and would be only $0.''76$ from the Sun. The largest planet in the solar system, Jupiter, would have a brightness of $22.^m0$ and would be $3.''94$ from the sun. Theoretically, Earth and Jupiter would be detectable with the largest telescopes from our neighboring star, however they would be too close to the Sun.

Planets do not shine themselves, so they are very faint and are also outshone by their parent star. For this reason, the search for planets outside our solar system *(extrasolar planets)* is very difficult. So far, they can usually only be detected indirectly, for example by the motion of the center of gravity of the parent star when a large, massive planet is near it or when a transit of a planet in front of its host star occurs.

A completely new method of detecting extra solar planets is provided by the *microlensing effect*. As is shown in the Cosmology section, the path of a beam of light bends in the presence of gravitational fields. Microlensing events can be observed by a change in the

Table 6.3 The major planets. D—Equatorial diameter, v_e—escape velocity, gravity is the surface gravitational acceleration

Planet	D[km]	M[M_{Earth}]	ρ[g/cm^3]	Gravity Earth = 1	v_e [km/s]
Mercury	4878	0.055	5.43	0.4	4,25
Venus	12,104	0.815	5.24	0.9	10.4
Earth	12,756	1.0	5.52	1.0	11.2
Mars	6794	0.107	3.93	0.4	5.02
Jupiter	142,796	317.8	1.33	2.4	57.6
Saturn	120,000	95.15	0.70	0.9	33.4
Uranus	50,800	14.56	1.27	0.9	20.6
Neptune	48,600	17.20	1.71	1.2	23.7

brightness of an object as it moves into the light path of a background source. If planets are present with this object, additional changes in brightness occur. For example, evaluations of an event in 1999 showed that the gravitational lensing in question was a binary star, with the masses of the two components behaving as 4:1 and being only 1.8 AU apart. Furthermore, a planet of three Jupiter masses is suspected at a distance of 7 AU.

There are two groups of planets in our solar system:

The *terrestrial planets* are Mercury, Venus, Earth and Mars. They have a relatively high density and a solid surface.

The *Giant planets* (sometimes called Jovian planets) are Jupiter, Saturn, Uranus, and Neptune. They are gas planets with no solid surface. Sometimes Uranus and Neptune are also called ice planets (have a large core of ice).

The most important physical properties of the planets are given in Table 6.3.

> Planets around other stars are only directly observable in exceptional cases because of the large difference in brightness and proximity to the star.

6.2   Properties of the Planets

In this section, we briefly cover how to derive the most important properties of the planets, most of which are already known from Earth-based observations.

6.2.1 Rotation Period

Planet rotation determination in some cases simply follows from the observation of permanent surface phenomena. This works quite well for Mars and Jupiter, however,

Fig. 6.1 Inclinations of the rotational axes of the planets as well as Pluto. Note the inclination of the rotation axis of Uranus. Planets in order as seen from the Sun. The sizes are not to scale. © Photobucket.com

e.g. Venus has a dense cloud cover and thus not reveal any surface details. One can also determine planet rotation from the *Doppler shift* of the reflected *Fraunhofer lines* of the solar spectrum or from the absorption lines originating from the planet's atmosphere itself. In the case of the nearest planets, the *Radar technology* apply: Annular zones around the center of the planet disk are distinguished according to the transit time differences of the radar waves. In 1964/1965 the 300 m parabolic antenna at Arecibo on Puerto Rico was used for the first time to determine the rotation of the planets Venus and Mercury.

Figure 6.1 shows the inclinations of the *rotational axes* of the planets. Mercury, Venus, and Jupiter have the least inclinations. Venus, however, rotates retrograde, i.e. in opposite sense of its solar orbit. The inclination of Uranus is extreme, as well as the inclination of the rotation axis of the dwarf planet Pluto.

6.2.2 Mass Distribution

The Mass distribution in the interior of the planet, i.e. the planetary structure, can be determined by:

- Measurement of *flattening,* due to rotation,
- gyroscopic motions in the gravitational field as well as precise determination of the gravitational potential by means of artificial satellites in orbit; from this one deduces the internal structure of a planet.

On Earth, Mars and the Moon, *seismic measurements* are possible, from the analysis the inner structure follows. In the Table 6.4 the proportion of the core of the planets in % of the total mass is given for the Earth-like planets and the Earth's moon.

6.2.3 Albedo

Albedo describes the ratio of the sunlight reflected or scattered in all directions to the incident light. Its strength and wavelength dependence provide clues to the nature of the planet's surface.

Table 6.4 Terrestrial planets
and earth's moon: proportion
of the core in the total mass

Planet	Core%
Mercury	42
Venus	12
Earth	16
Earth moon	4
Mars	9

The *Bond's Albedo* is defined by:

$$A = pq \tag{6.1}$$

Let r be the sun-planet distance in AU, Δ the distance between earth and planet in AU, ρ the diameter of the major semimajor axis of the planet's orbit in AU, α the *Phase angle* (= angle between the sun and the earth as seen from the planet), $\phi(\alpha)$ the phase law, p is the ratio of planetary brightness at $\alpha = 0$ to a perfectly diffuse surface, then holds:

$$\log p = 0.4(m_\odot - m_{\text{planet}}) + 2\log(r\Delta/\rho) \tag{6.2}$$

Whereby m is the apparent magnitude of the object.

The magnitude q is determined by the law of reflection:

$$q = 2\int_0^\pi \phi(\alpha)\sin\alpha \, d\alpha \tag{6.3}$$

According to Lambert's law $q = 1.50$, according to the law of *Lommel-Seeliger:* $q = 1.64$. Here the luminosities are given in *Magnitudes* [m]. Brightest stars are 1st magnitude stars, faintest stars just visible to the naked eye are 6th magnitude (Sect. 8.1).

One then has the following laws:

- Mercury: $p = 0.093$, $q = 0.65$, $A = 0.060$ (Moon 0.070). At its greatest easterly elongation $r\Delta = 0.357$, and the visual magnitude of Mercury is :

$$m_V = -0.21 + 5\log r\Delta + 3.82 \times 10^{-2}\alpha - 3.37 \times 10^{-4}\alpha^2 \ldots. \tag{6.4}$$

- At Jupiter the dependence on phase angle is already very small (why?): $p = 0.37$, $q = 1.10$, $A = 0.41$. At its opposition one has:

$$m_V = -9.1 + 5\log r\Delta + 0.015\alpha \tag{6.5}$$

In both formulas m_V stands for the brightness measured in the visual.

6.2.4   Spectrum

With spectroscopic observations, gases in the planetary atmospheres can be detected by absorption bands. The terrestrial H_2O-, CO_2- and O_3 bands can be bypassed from satellites. The *Intrinsic radiation* of a information about its temperature in the atmosphere.

6.2.5   Global Energy Budget

The irradiance from the Sun at Earth distance (1 AU) is referred to as the *Solar constant S*.

$$S = 1.37 \, \text{kW} \, \text{m}^{-2} \tag{6.6}$$

On an area of $1 \, \text{m}^2$ we obtain on the earth a radiant power of this amount with perpendicular incidence of solar radiation and without absorption in the earth's atmosphere.

If there is a planet at distance r, then its solar constant (amount of power from the Sun to $1 \, \text{m}^2$):

$$S(r) = S \left(\frac{r}{1 \, \text{AU}} \right)^{-2} \tag{6.7}$$

A planet with radius R takes from this the amount $\pi R^2 (1 - A) S(r)$ on. The mean albedo of the earth is $A \approx 0.3$. This albedo depends essentially on the mean the cloud fraction (clouds: Aldebo = 0.5).

If the radiation takes place according to the *Stefan-Boltzmann law* for a black body, is valid for the total power:

$$4\pi R^2 \times \sigma T^4, \qquad \sigma = 5.67 \times 10^{-8} \, \text{W} \, \text{m}^{-2} \, K^{-4} \tag{6.8}$$

Further, one has to consider the internal heat flux. Consider Q coming from internal heat sources of a planet. Such heat sources can be (i) radioactive decay, (ii) residual heat from the time of the formation of the planet (but small planets cool down quickly) and (iii) so-called tidal heating, caused by strong tidal forces (plays for some moons of the large Planet a role).

Now we make a balance:

Irradiation + Q = radiation:

$$\pi R^2 (1 - A) S(r) + 4\pi R^2 Q = 4\pi R^2 \sigma T^4 \tag{6.9}$$

For the Earth Q is due to the decay of the radioactive elements and amounts to $0.06\,\mathrm{W\,m^{-2}}$. Infrared measurements show that the planets Jupiter, Saturn, and Neptune have thermal radiation that exceeds the absorbed Solar radiation around by a factor of 1.9 and 3.5 and 2.4 respectively.

Actual temperatures of planets differ from values derived above, because planets rotate, furthermore atmospheric currents occur, as well as due to *greenhouse effect,* heating of individual layers (e.g. *Ozone layer* of the earth's atmosphere) etc.

6.2.6 Hydrostatic Equilibrium

We consider a volume element at distance r from the center of a planet with base area dA and height dr. Let its mass be $\rho(r)d\,A\,dr$, and for the mass within r holds:

$$M(r) = \int_0^r \rho(r')4\pi r'^2 dr' \qquad \frac{dm(r)}{dr} = 4\pi r^2 \rho(r) \qquad (6.10)$$

The gravitational acceleration by $M(r)$ is:

$$g(r) = \frac{GM(r)}{r^2} \qquad (6.11)$$

In the area of the considered volume element the pressure changes p um:

$$-dpd\,A = \rho(r)d\,Adrg(r) \qquad (6.12)$$

we have thus the relation: force = mass $(\rho(r)d\,Adr) \times$ acceleration $(g(r))$.

This gives us the condition for the *hydrostatic equilibrium* of a planet:

$$\frac{dp}{dr} = -g(r)\rho(r) \qquad (6.13)$$

Special case: homogeneous sphere with $\rho(r) = \bar{\rho} = $ const. Estimation for the central pressure p_c: M is the total mass, and it holds: $M(r)/M = (r/R)^3$.

$$p_c = \int_R^0 \frac{dp}{dr}dr = \bar{\rho}\int_0^R \frac{GM}{r^2}\left(\frac{r}{R}\right)^3 dr = \frac{1}{2}\bar{\rho}\frac{GM}{R} \qquad (6.14)$$

This estimation gives for the center of the earth : $p_c = 1.7 \times 10^{11}\mathrm{Pa}$. This value is lower than the actual one by a factor of 2.

> The hydrostatic equilibrium condition is also a basic equation of stellar structure.

To build a planet, one still needs an equation of state of the form $p = p(\rho, T,$ chemical composition).

6.2.7 Stability of a Satellite, Roche Limit

Here the question is, how close e.g. a moon can come to its mother planet before it is torn apart by the tidal forces. Let there be a central body (planet) with mass M, radius R, and an average density $\bar{\rho}$ and a satellite with the parameters M_S, R_S, $\bar{\rho}_S$. We think of the satellite as consisting of two halves $M_S/2$ which attract each other at a distance R_S:

$$F \approx G\frac{M_S M_S}{4R_S^2} \tag{6.15}$$

So this force holds the two halves of the satellite together.

The *tidal force* acts to the Satellite to be torn apart. Consider the tidal acceleration exerted by the Earth on the Moon. At lower culmination, this is equal to the difference between the gravitational acceleration due to the Earth's attracting mass M at the lunar center (distance Earth-Moon r) and the acceleration acting on the lunar surface (distance $r - R_S$). The tidal acceleration is therefore:

$$b_G = \frac{GM}{r^2} - \frac{GM}{(r - R_S)^2} \sim \frac{2GM}{r^3}R_S \tag{6.16}$$

In our case, the tidal force of M on M_S is then:

$$2GMM_S R_S/r^3 \tag{6.17}$$

The condition for the *Stability* of a satellite is therefore:

$$G\frac{M_S M_S}{4R_S^2} \geq k2G\frac{MM_S}{r^3}R_S \tag{6.18}$$

Where k is a constant of order 1. Because of $M_S = (4\pi/3)\bar{\rho}_S R_S^3$ and for the planet with $M = (4\pi/3)\bar{\rho}R^3$ follows:

$$\frac{r}{R} \geq (4k)^{1/3}\left(\frac{\bar{\rho}}{\bar{\rho}_S}\right)^{1/3} \tag{6.19}$$

Roche has 1850 showed that for the stability of a satellite it holds:

$$\frac{r}{R} \geq 2.44 \left(\frac{\bar{\rho}}{\bar{\rho}_S} \right)^{1/3} \tag{6.20}$$

→ A larger satellite, having the same density as its parent planet, must not come nearer to it than 2.44 planetary radii, otherwise it will be torn apart by the tidal forces of the parent planet.

→ In the case of smaller satellites, the cohesive forces also become effective, and one gets $r/R = 1.4$.

6.2.8 Planetary Atmospheres

If the expansion of a planet's atmosphere is small compared to the planet's radius R, then the gravitational acceleration can be assumed to be constant: $g = GM/R^2$. The altitude is $h = r - R$. At hydrostatic equilibrium

$$\frac{dp}{dh} = -g\rho(h) \tag{6.21}$$

We use the equation of state:

$$p = \rho \frac{kT}{\bar{\mu} m_u} = \rho \mathfrak{R} T / M = nkT \tag{6.22}$$

Where $k = 1.38 \times 10^{-23}\,\mathrm{J\,K^{-1}}$ is the Boltzmann constant, $\bar{\mu}$ the mean molecular weight, and $\mathfrak{R} = 8.31\,\mathrm{J\,K^{-1}\,mol^{-1}}$ the Gas Constant and $m_u = 1.66 \times 10^{-27}\,\mathrm{kg}$. From this we calculate ρ and put this into the condition for hydrostatic equilibrium.

This leads to:

$$\frac{dp}{p} = -\frac{g\bar{\mu} m_u}{kT} dh = -\frac{dh}{H} \tag{6.23}$$

where here the *Equivalent Height (scale height)* H is introduced:

$$H = \frac{kT}{g\bar{\mu} m_u} \tag{6.24}$$

If we assume H is constant, so follows the *barometric altitude formula:*

$$\ln p - \ln p_0 = -h/H \qquad p = p_0 e^{-h/H} \tag{6.25}$$

Here p_0 is the pressure at the bottom (h = 0).

Now we investigate a *convective atmosphere:* Hot matter rises adiabatically (without heat exchange) upwards and cooled matter downwards. Then the adiabatic equation:

$$T \approx p^{1-(1/\gamma)} \tag{6.26}$$

Where $\gamma = c_p/c_V$, the ratio of the specific heat capacities at constant pressure (c_p) or constant volume (c_v). If we differentiate this logarithmically to h, it follows:

$$\frac{1}{T}\frac{dT}{dh} = \left(1 - \frac{1}{\gamma}\right)\frac{1}{p}\frac{dp}{dh} \tag{6.27}$$

Here we insert the hydrostatic equation as well as the condition $c_p - c_v = k/\bar{\mu}m_u$ and get for the *adiabatic temperature gradient:*

$$\frac{dT}{dh} = -\frac{g}{c_p} \tag{6.28}$$

The lowest layer of the Earth's atmosphere, which is about 12 km high, the *Troposphere,* is convective. Since $g = 9.81\,\text{m/s}^2$, $c_p = 1005\,\text{J kg}^{-1}\,\text{K}^{-1}$ we get a gradient of $9.8\,\text{K km}^{-1}$. This is only true for dry air. For humid air *latent heat,* released during condensation will reduce the gradient by half. So we get an average temperature decrease with altitude in the Earth's atmosphere of $6.5\,\text{K km}^{-1}$.

Finally, on the question of when a planet can have its own atmosphere. We consider molecules of mass m of a gas of temperature T. From the kinetic theory of gases, the most probable velocity depends on the temperature and the mass of the particle:

$$v_{\text{th}} = \sqrt{\frac{2kT}{m}} \tag{6.29}$$

A molecule of velocity v can escape from a planet of mass M if holds:

$$v^2/2 \geq GM/R \tag{6.30}$$

A planet can hold an atmosphere only if holds:

- Its mass is sufficiently large;
- due to the temperature at the surface v_{therm} sufficiently small.

Therefore, the Moon (too low mass) and Mercury (too low mass, due to the proximity to the Sun, the thermal velocity of the particles is too high) cannot hold an atmosphere.

The outermost layers of a planetary atmosphere are called the also *exosphere.* Here the density of the gas particles is low, only few collisions take place. *Ionisation processes* (solar UV radiation) produce charged particles. Their motion is determined by the planetary magnetic fields.

6.3   Earth and Moon

6.3.1   Structure of the Earth

The Figure of the earth is, in consequence of its rotation, an ellipsoid with:

Equatorial radius a =6378.1 km.
   Polar radius b =6356.8 km.

The Flattening is:

$$\frac{a - b}{a} = 1/300 \tag{6.31}$$

The average density of the Earth is $\bar{\rho} = 5520 \, \text{kg/m}^3$. The density of the rocks at the earth's surface is $2800 \, \text{kg/m}^3$ but near the center, where the metals are located, is $13{,}000 \, \text{kg/m}^3$. The interior of the earth is shell-like constructed:

- *Earth's crust:* $3300 \, \text{kg/m}^3$, it is divided into: (a) *Lithosphere,* which consists of granitic rocks and reaches a depth of 35 km under the continents, the basalts under the ocean floors reach only 5 km deep; (b)*Hydrosphere:* 70% of Earth's surface is water.
- *Mantle:* The crust is floating in large blocks on the Earth's mantle, where silicates such as olivine occur. Density values range between 3400 and $5500 \, \text{kg/m}^3$. The mantle extends to a depth of 2900 km.
- Liquid outer *Earth's core:* about 2000 km thick, Fe, Ni.
- Solid inner core with radius 1300 km, Fe, Ni.

The theory of *Plate tectonics* was developed around 1960. Even before that a drift of the continents was suspected (A. *Wegener,* "The Origin of the Continents", 1915). 225 million years ago there was only one supercontinent, *Pangaea.* The distribution of the continents determines the ocean currents, and these in turn exert a significant influence on climate. For example the Mediterranean Sea evaporated several times during the last 6 million years, as can be seen from the massive salt deposits. The Strait of Gibraltar was closed several times by tectonic shifts. During the *Würm Ice Age* (115,00–10,000 years before today) the sea level was 120 m lower, and the upper Adriatic was land, and many Greek islands were connected to Anatolia. Plate tectonics is driven by convection in the Earth's mantle.

Due to the propagation of *earthquake waves (seismology)* we can study the Earth's interior. Earthquake waves occur in two forms:

- as longitudinal compression waves *(p-waves)*,
- as transverse waves *(s-waves)*.

Only the p-waves can pass through a liquid zone. You don't see s-waves going through the core because the outer core of the Earth is liquid. The Earth's core is predominantly Ni and Fe and should rotate faster than the outer regions (The Earth's core rotates once in 900 years more than the outer regions, super-rotation).

In the *oceanic deep-sea trenches* lava flows upwards and pushes the plates away from each other. The present plates were formed about 200 million years ago by the breakup of the great supercontinent. Continental drift is a few centimeters a year. The deep ocean basins are the youngest areas of the earth's surface because lava is constantly flowing up there.

The *age* of the Earth as well as the earth's rocks can be determined from the *radioactive decay*. Let n be the number of radioactive atoms, λ the decay constant:

$$-dn = \lambda n dt \qquad dn/n = -\lambda dt \qquad n = n_0 \exp(-\lambda t) \tag{6.32}$$

The *Half-life* τ indicates when the number of particles has decreased to half the original number $n = n_0/2$.

$$\tau = (ln2)/\lambda = 0.693/\lambda \qquad n/n_0 = \exp(-0.693t/\tau) \tag{6.33}$$

From this you get the radioactive age (t_h is the half-life):

$$\frac{n(t)}{n_0} = 2^{-\frac{t}{t_h}} \tag{6.34}$$

The rate of radioactive decay is given in the unit *Becquerel*:

1 Bq = 1 decay/second.

Some values for half-lives: ^{14}C (5730 years), ^{239}Pu (24,000 years), ^{235}U (800 million years), ^{238}U (4.5 billion years), ^{232}Th (14.5 billion years), ^{40}K (1.2 billion years).

One observes an increase in temperature with depth. The *geothermal depth gradation* is about $30K\,km^{-1}$. This temperature increase is due to the heat generation of radioactive substances such as ^{238}U, ^{232}Th, ^{40}K as well as by the slow outward heat transport due to heat conduction and convection of the magma. In the Earth's core the temperature is $\leq 10,000\,K$.

6.3.2   Geological and Biological Evolution

At the time of the *Precambrian* (more than 590 million years ago) there were three *Mountain building phases* (Laurentian, Algonquian as well as Assynitic mountain building). The oldest traces of life are the stromatolites and cyanobacteria (age approx 3.5×10^9 years). At the time of the Cambrian (590−500 million years from today) the algae time (Eophytic) began, the oldest vertebrates come from the time of the Ordovician (505−438 million years before today). In the Silurian (438−408 million years before today) there was the Caledonian mountain building, trilobites as well as oldest vascular plants. In the Devonian (408−360 million years before today) the Armoured fish emerges. Towards the end of the Carboniferous (360−286 million years before today) the Variscan mountain building begins (as well as vascular spore plants), and at the beginning of the Permian age (286−248 million years before today) there are the first reptiles. The dinosaur age begins with the Triassic (248−213 million years before today), the first gymnosperms appear. At the beginning of the Jurassic period (213−144 million years before today) one finds the first mammals. The oldest birds come from the Cretaceous period (144−65 million years before today), also the oldest angiosperms appear now. At the beginning of the Tertiary period (65−2 million years before today) the dinosaurs died out and the Alpidic mountain building began. The first humans are found at beginning the Quaternary period (two million years ago).

6.3.3   Earth's Magnetic Field

Our Earth has a magnetic field which resembles a *Dipole field* (dipole means two poles, thus one magnetic north and south pole): The magnetic flux density at the equator is 0.31 Gauss. The magnetic moment **M** points in the direction of the dipole axis, and by gradient operation one obtains the vector of *magnetic flux density* **B**.

$$\mathbf{B} = -\operatorname{grad}\frac{\mathbf{M.r}}{r^3} = -\operatorname{grad}\frac{M \sin \lambda}{r^2} \tag{6.35}$$

Here λ is the magnetic latitude, i.e. at the magnetic poles $\lambda = \pm 90°$. The axis of the earth's magnetic field is inclined by $12°$ against the rotation axis of the earth. The earth's magnetic field is maintained by a self-exciting *dynamo process*. Paleomagnetic studies showed that the Earth's magnetic field changes direction randomly in periods from 10^4 to 10^5 years. It is observed today that the field is slowly decreasing.

At the *Magnetopause*, the Earth's magnetosphere encounters a stream of particles coming from the Sun, the *solar wind*. On the side facing the Sun, the Earth's magnetic field is compressed and extends only about ten Earth radii (Fig. 6.2).

On the side facing away from the sun, the magnetic field expands to form a *tail* (engl. Magnetotail) off. In the magnetopause, the incoming protons and electrons emitted by

Fig. 6.2 Sketch of the Earth's magnetic field compressed on the side facing the Sun. Sun and Earth are not at the correct scale (Source: NASA)

the Sun are deflected (movement transverse to the field lines is not possible, only along the field lines). Some of them do get into the magnetosphere, and these particles become trapped in the Van Allen *radiation belts*. The *inner belt* extends between one and two Earth radii and contains protons with an energy of 50 MeV and electrons with 30 MeV. Then follows a gap, and between three and four Earth radii is the *outer belt*, where less energetic electrons and protons are present. The inner belt is relatively stable, the outer depends on the activity of the Sun and can vary greatly. Particles trapped in the belts make a spiral motion around the magnetic field lines, bouncing back and forth between the so-called magnetic mirror points with periods between 0.1 and 3 s. The particles of the inner belt can be trapped by the magnetic field lines. Particles of the inner belt can interact with the atmosphere, causing then *aurorae* (northern lights, southern lights). These normally appear at height of about 100 km within earth-atmosphere. Here the *magnetic reconnection* plays an important role. Areas with opposite magnetic fields come together, and the field lines collapse and combine into new combinations. Such processes take place in the Earth's magnetic tail at a distance of about 100 Earth radii. If the solar wind adds enough energy, then the field lines overstretch and reconnection occurs at distances as small as 15 Earth radii. The field collapses, electrons enter the atmosphere, and the auroras appear.

In Table 6.5 the data of the planets with magnetospheres given. Note the huge dimensions of Jupiter's magnetic field! The magnetic moment is given in units related to magnetic moment of the earth = 1, the distance of the magnetopause in multiples of the respective planetary radius.

Table 6.5 Comparison of magnetospheres of some planets. Distance Magn. means distance of the magnetopause in units of the planet radius from its surface

Planet	Distance to Sun	Magnet. moment ME (Earth=1)	Solar wind-pressure	Distance magn.
Mercury	0.4 AU	4×10^{-4}	20 nPa	1.5
Earth	1.0	1.0	3.0 nPa	10
Jupiter	5.2	1.8×10^4	0.1 nPa	70
Saturn	9.5	580	30 pPa	21
Uranus	19.2	50	8 pPa	27
Neptune	30.1	24	3 pPa	26

The force **F** which acts on a particle of the charge q is the *Lorentz force*:

$$\mathbf{F} = q(\mathbf{v} \times \mathbf{B}) \tag{6.36}$$

Here the electric field **E** was set equal to zero. This simplification is often made in astrophysics.

The particles move in a homogeneous magnetic field on a circular path where the equilibrium between centrifugal force and Lorentz force gives the radius of this gyration motion:

$$mv^2/r = qvB \tag{6.37}$$

Particles of cosmic rays originating from outside the solar system interact with the Earth's high atmosphere. Thereby the high energetic cosmic protons produce free neutrons, which decay into protons p and electrons e^- and neutrinos $\bar{v}$.

$$n \rightarrow p + e^- + \bar{v} \tag{6.38}$$

In a magnetic field, particles of positive and negative charge move in different directions. The particles produced in this way accumulate in the radiation belts. There they have a relatively long lifetime due to their small effective cross-section. A 20-MeV proton at an altitude of 2000 km has a lifetime of about one year. The particles move very fast around the earth (one orbit takes about 2 min). The particles of the radiation belt cause damage to space missions.

With the help of the NASA-operated imager mission, the faint UV glow of the relatively cool plasma around the Earth was seen in August 2000 (Fig. 6.3). The ring of aurorae (northern lights) is clearly visible. as a small faint luminous circular ring in the center. The particles are trapped by the Earth's magnetic field. The satellite was above Earth's north pole at the time the image was taken, and the Sun was outside the image at the upper right edge. Note also the shadow of the Earth (towards the lower left).

Fig. 6.3 Luminous plasma in the Earth's magnetosphere; the small oval ring is the region of the aurorae (here around the magnetic South Pole)

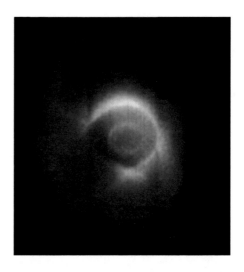

Without the Earth's magnetic field, we would be without protection against the energetic charged particles from the cosmos and life would not be possible on the Earth's surface.

6.3.4   Earth's Atmosphere

The Earth's original atmosphere consisted of H_2, He and hydrogen compounds. Through volcanism and degassing processes the atmosphere was then enriched with H_2O, CO_2, N_2, Ar. The Table 6.6 shows the composition of the today's atmosphere near the ground.

Table 6.6 Composition of the earth's atmosphere

Element	Volume percent
Nitrogen N_2	78.08
Oxygen O_2	20.95
Argon Ar	0.934
Carbon dioxide CO_2	0.033
Neon Ne	0.0018
Helium He	0.00052
Methane CH_4	0.00015
Water vapor H_2O	$\approx 10^{-4}$

Fig. 6.4 Course of the
temperature in the earth's
atmosphere

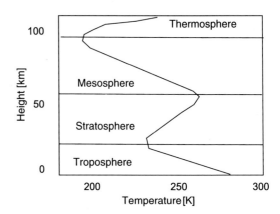

The Earth's atmosphere is divided into the following layers (Fig. 6.4):

- *Troposphere:* reaches up to the tropopause in 15 km height. Up to this height the temperature decreases to about −50 °C down.
- *Stratosphere:* Up to the stratopause (about 50 km height) the temperature increases slightly to −20 °C.
- *Mesosphere:* The temperature decreases again, at 90 km altitude you have another temperature minimum of about −100 °C
- *Ionosphere, Exosphere:* Here the temperature increases in the range of the thermosphere (90–250 km) up to 1500 K (at the base of the exosphere). At these heights (from 700 km) molecules escape into space. Ionization also increases strongly—this is also referred to as the ionosphere.

In the region of the Earth's atmosphere, electromagnetic radiation arriving from outside is absorbed, scattered, or refracted, depending on the wavelength. Radiation from most wavelength regions of the electromagnetic spectrum is absorbed, except for two windows where the atmosphere is transparent to electromagnetic radiation:

- *optical window:* from 290 nm (near UV) to 1 μm (near infrared),
- *radio window:* 20 MHz to 300 GHz.

We can therefore study the radiation of stars and galaxies only n these wavelength ranges from Earth-based observing stations. In the ozone layer (10–40 km) radiation from 300 nm to 210 nm is absorbed (UV range). In the infrared range (1 μm up to 1 mm) the radiation is absorbed by the molecules N_2, O_2, CO_2, H_2O . At wavelengths greater than 15 m (below 20 MHz) there is reflection,

In the visible light range (410–650 nm) it comes to

- Refraction,
- scattering,

- extinction,
- dispersion.

In the case of scattering of light, one must consider: (a) the wavelength of the light, (b) the size L of the scattering particle. When light is scattered by molecules in the earth's atmosphere, the intensity of the scattered light is given by the *Rayleigh scattering law:*

$$I_{\text{scatter}} \propto \frac{1}{\lambda^4} \tag{6.39}$$

Light of shorter wavelengths is therefore scattered more strongly, and that is why the sky is blue. If the sun or a star is low on the horizon, it appears reddened.

If you have dust particles of size $1\,\mu\text{m}$, there is Extinction, whereby here the following scattering law applies:

$$I_{\text{scatter}} \propto \frac{1}{\lambda} \tag{6.40}$$

And if $L \gg \lambda$ is (water droplets in clouds), then the scattering is independent of the wavelength—therefore clouds appear as white.

Light is also refracted in the Earth's atmosphere, and density in homogeneities cause stars to twinkle *(Seeing)*.

All planets with atmospheres have *Weather.* This can be understood as the circulation in the atmosphere. Weather patterns are driven by solar radiation, which heats the surface of the planet. Other factors include the rotation of the planet and seasonal changes. At Earth, oceans and ocean currents play an important role. Heat is transported towards cooler regions.

The Earth's global circulation system is sketched in Fig. 6.5 .At the equator heated air rises, near the ground colder air flows towards the equator. Due to the rotation of the earth *(Coriolis force)*, there is a rightward deflection (northeast winds) in the northern hemisphere of the Earth, and a leftward deflection in the southern hemisphere. At the pole, high air masses sink. At about 30° latitude, the air masses that have risen from the equator have cooled down to such an extent that they sink. Air masses flowing from the poles towards the equator rise up to 60° latitude.

Climate refers to effects of the atmosphere that change over decades or centuries. A hot summer alone does not indicate a change in climate. Climate thus contrasts with random changing weather conditions from year to year. Modern agriculture is very sensitive to temperature changes. A consistent drop in temperature of only $2\,^{\circ}\text{C}$ would cut U.S. and Canadian grain production in half. During the ice ages, large-scale glaciations and temperature decreases occurred in the northern hemisphere. In 1920 *Milankovich* indicated that the tilt of the Earth's axis was changing, which may have led to the *Ice ages* led to the climate change. We now know that there have been also global climate changes on Mars.

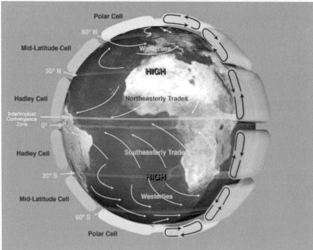

Fig. 6.5 Top: storm low. Bottom: the complex flow system in Earth's atmosphere, global circulation (Photocredit: sealevel.jpl.nasa.gov)

The first living beings emerged 3.5 billion years ago (*stromatolites*, still exist today, they are actually biogenic sedimentary rocks). However, there have only been species-rich fossils for 600 million years.

Two billion year old rocks do not yet contain oxygen (O_2), although plants have already released oxygen through photosynthesis here. Probably this oxygen immediately reacted with rocks of the earth's crust. After this time the atmosphere was enriched with O_2 oxygen, and the *ozone layer* formed. By the formation of ozone, O_3, life-hostile UV-Radiation was absorbed. The *Chapman reactions* describe the *photolysis* (splitting by electromagnetic radiation) of the O_2 by short-wave UV ($\lambda < 240$ nm) and the formation of O_3 and the destruction of the O_3 by longer wavelength photons ($\lambda > 900$ nm). There is an equilibrium:

$$3O_2 \rightleftharpoons 2O_3 \tag{6.41}$$

Only when the ozone layer was sufficiently strong could living beings leave the protective oceans. Most of the CO_2 is now bound in the sediments in the form of carbonates or in the fossil fuels coal and petroleum.

The *greenhouse effect,* which is mainly due to the content of CO_2 and H_2O has caused the global temperature to be $33\,°C$ higher than would be expected from solar radiation. Modern industrial society, however, increases the amount of CO_2 by burning fossil fuels and by cutting down tropical rainforests. Within 1900–2000, the amount of CO_2 increased by 25% and is increasing by 0.5% each year.

It is interesting to note that throughout the history of the earth, there have always been *Mass Extinction* of animal species. Half of all animal species became extinct within a short period 65 million years ago *(KT event)*. This is thought to have been caused by the impact of an asteroid and the dust particles kicked up by the impact caused global cooling. The dinosaurs died out 65 million years ago, and the triumph of mammals began. The mass of the asteroid was a trillion tons, the diameter was only about 10 km, and it is believed to have impacted in the Peninsula area of *Yucatan* (Mexico). That it was an asteroid is indicated by the unusually high Ir content in sediments from this period. The heavy metal *Iridium* sank into the deeper interior of planets like the Earth when the Earth was still liquid. Such a *Differentiation process* could not take place with smaller bodies like asteroids. The crater was 250 km in diameter, and the explosive force was equivalent to that of 5 billion Hiroshima bombs. The material hurled into the atmosphere spread all over the world, and the sun was darkened for several months. Further, there was acid rain and large-scale fires. Almost certainly there was also a period of intense volcanism at that time.

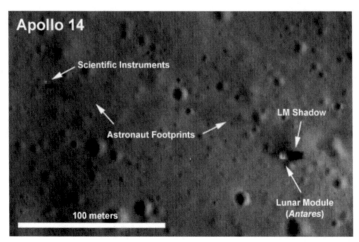

Fig. 6.6 Image of the Apollo 14 mission landing site taken from a lunar orbiter. From Earth, even with the largest telescopes, the lander modules left on the lunar surface cannot be seen (NASA)

6.3.5   The Moon-General Porperties

The mass of the Moon is only 1/80 of Earth's mass and its equatorial diameter is 3476 km; therefore, the Moon's surface gravity is too low to sustain an atmosphere. Between 1968 and 1972, nine manned spacecraft were launched to the Moon as part of the *US Apollo program* , and twelve astronauts have walked on its surface. The Russian space probe Luna 9 made a soft landing on the moon back in 1966. On July 21, 1969, the astronaut *N. Armstrong* became the first man to walk on the moon. The cost of the American manned lunar landing program was approximately 100 US for each American, spread over ten years.

Figure 6.6 shows the Apollo 14 landing site made by a probe orbiting the Moon.

A general map of the Moon is given in Fig. 6.7.

The Moon has an average density of only $3.3 \, \text{g/cm}^3$, that is, it is composed mainly of silicate rocks, and there is no Fe or other metals, and no water or volatile elements. The moon appears to be composed of the same material as the earth's crust. Seismometer records showed that there is no metal core, the moon is cold and solid inside. There are barely moonquakes.

The Early telescope observers gave the lunar formations fanciful names such as *Mare Imbrium* (Sea of Rain). The surface, which has numerous craters (Fig. 6.8), is called *Terrae* (Highlander). It consists of bright silicate rocks, anorthosites. The highlands are the oldest areas on the moon. The Craters are impact craters from early in the formation of the solar system. The dark large plains with only a few craters are called the *Maria* (seas). They make up 17% of the lunar surface, most of them are on the side of the moon facing the Earth. These are huge impact basins, which were formed by asteroid impacts and then

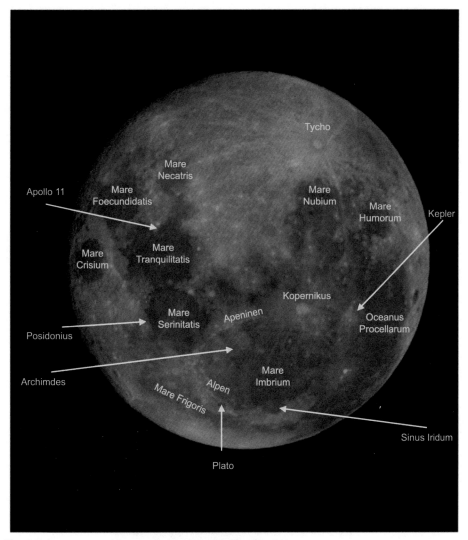

Fig. 6.7 Lunar map with some designations (A. Hanslmeier)

flooded with lava. They are composed of basalt (similar to the composition of Earth's oceanic crust). These areas are younger than the highlands by about a billion years.

The lunar mountains, mostly named after terrestrial mountain ranges (Alps. . .), are also the result of impacts, and occur at the edges of the great maria. The surface of the Moon is covered with a layer of dust created by the impacts. The astronauts sank several inches deep, leaving footprints that will be admired for many millions of years to come (Figs. 6.9 and 6.6).

Fig. 6.8 Lunar landscape (Apollo 11, taken in 1969)

Fig. 6.9 First excursion of a human being on the moon. (1969, Apollo 11)

The lunar night, which lasts two Earth weeks, is very cold: $-173\,°C$, since there is no protective atmosphere, the surface is highly porous and thus cools easily. During the lunar day the temperature is $120\,°C$.

The number of craters a given area has can be used to infer its age if a planet or moon has low internal activity or erosion. The rate of crater formation has been constant for several billion years, that is, the number of craters is proportional to the length of time the surface has been exposed to impacts. During the last 3.8 billion years, the rate of crater formation has been similar to today (age of oldest maria, lunar oceans). However, more than 3.8 billion years ago, the impact rate was much higher than today. There are ten times as many craters on the older terrae as in the maria. If the impact rate were the same, then the highlands would have an age of 38 billion. years. On the moon, therefore, a phase of intense Bombardments took place more than 3.8 billion years ago. Similar results are found for Mercury and for Jupiter's and Saturn's moons. For the other planets, but also e.g. for Saturn's moon Titan, erosion plays a role.

> On Earth, a crater impact of 1 km diameter occurs on average every 10^4 years. Craters of a few 10 km are formed every few million years; a 100 km crater every 50 million years.

Important for future moon missions is the (almost certain) discovery of water ice near the two poles. This could have been deposited by comet impacts. How can water be detected? The moon has neither a magnetic field nor an atmosphere. Therefore particles of cosmic rays can fully impact and produce fast neutrons. With a *Neutron spectrometer* of the Lunar Prospector in 1998 slow neutrons were detected near the two poles, these are slowed down by collisions with protons—an indication of water ice in the polar regions which are permanently in shadow. The ice volume at the North Pole is estimated at 10,000–50,000 km^3 for the North Pole. For comparison: the ice masses of Antarctica are estimated at $29 \times 10^6 \, km^3$, Greenland about 1/10 of this value. For the South Pole 5000–20,000 km^3. However, the ice deposits at the South Pole are doubted, since no water-containing cloud was registered near the South Pole during the controlled impact of the Prospector probe, 1999.

One source of hydroxyl ions on the Moon could be protons (H^+) transported by the solar wind. The first Indian lunar satellite, Chandrayaan-1, as well as NASA's Moon Mineralogy Mapper (M3) satellite, detected traces of water or hydroxyl molecules near the lunar surface. Other measurements also showed that water is found on the Moon, above a latitude of 10, most of the water is located in at the polar regions.

6.3.6   Origin of the Moon

Capture hypothesis: The moon had been formed somewhere elsewhere in the solar system and then came close to the Earth and was captured. This is conceivable in principle, but a close encounter between the Moon and Earth would be more likely to result in a collision, or the Moon would accelerate so much as to preclude another encounter. It is known from the Apollo missions that oxygen isotopes occur in lunar rocks in similar proportions to terrestrial ones.

Secession hypothesis: Already advocated by Darwin (1845–1912). When the Earth formed, it rotated very rapidly, a bulge formed at the equator from which the material for the Moon came loose. This would explain why the Moon is composed mainly of materials similar to those of the Earth's mantle. The Earth would rotate on its axis in just two hours at that time, but this is unlikely. The Apollo missions did show some similarities in the compositions of the Moon's rocks and the Earth's crust, but there are also marked differences, with hardly any potassium or sodium compounds on the Moon.

Double planet: Moon and Earth formed almost at the same time from the protoplanetary gas and dust cloud. Problem: Moon has only a very small metallic core.

Collision hypothesis: Was set up in 1975 by *Hartmann* and *Davis*. The young Earth was struck by a celestial body about the size of Mars, which knocked out the material from which the Moon was then formed. This can easily explain why there are hardly any metals in the moon's core: The core of the impacting celestial body got stuck in the Earth's core. The moon was formed from the silicate parts of both celestial bodies. The different ratio of iron to magnesium oxide in the Earth and Moon can be explained in this way: The lunar rock is mostly material from the impacting body. The angular momentum problem of the Earth-Moon system can also be explained: The impacting body was of Martian size, hit the Earth sideways, and thus increased the Earth's rotational velocity to its present value.

In a simulation calculation, it was also shown that without the Moon, the Earth's axis would undergo much larger tilt fluctuations (from 15 to 30°), which would make our Earth's climate become extremely unstable. The tilt of the Earth's axis essentially determines the seasons, and a greater inclination would have resulted in greater contrasts between the seasons.

> Our moon was formed by Earth's collision with a Mars-sized planet in the early days of the solar system.

6.3.7   The Interior of the Moon

The interior of the earth can be studied by evaluating the propagation of earthquake waves. Seismometers have been positioned on the lunar surface, but they have registered practically no moonquakes. The only tremors recorded came from

- from meteorite impacts,
- tidal forces from the Earth to the Moon.

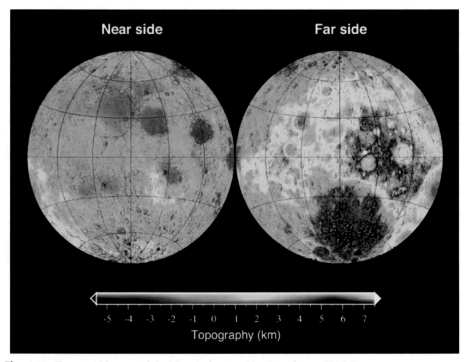

Fig. 6.10 Topographic map of the Moon's front and back surfaces. (NASA)

So the Moon is geologically inactive and has a cold core, while the Earth's core is about 6800 K hot. The moon does not have an iron core, which is already evident from its relatively low density of 3300 kg/m^3. That is why there is no magnetic field.

The surface of the Moon is different: On the side facing Earth, you find the large lunar seas, which were filled in with heavier basaltic lava after asteroid impacts. The lunar crust is thick here, the mass somewhat heavier, and therefore this side faces Earth (Fig. 6.10).

6.3.8 The Far Side of the Moon

From Earth, we see only one hemisphere of the Moon because the Moon exhibits a 1:-1 spin-orbit resonance. The Moon rotates once on its axis in the same amount of time it takes to orbit the Earth. In fact, we see a little more than 50% of the Moon's surface because its rotation is uniform, but its orbit around the Earth is not. The Moon's orbit is elliptical; it moves faster near Earth (perigee) than far away (apogee). In addition, the Moon's axis of rotation is slightly tilted, and we still have to take into account the parallax effect on the Earth. So we see nearly 60% of the moon's surface, this effect is also called *libration*. The back side of the Moon shows significantly fewer large basaltic plains (maria) than the front, and also lacks the lunar mountains formed by large impacts on the front. In 1959, a

Fig. 6.11 Release of a lunar rover at the far side of the Moon. Chang'e 4 Mission.(Chinese Space Agency)

Russian probe first radioed images of the Moon's far side to Earth. In January 2019, the Chinese space probe Chang'e 4 succeeded for the first time in taking a soft landing on the Moon's far side (Fig. 6.11).

The far side of the Moon would be an ideal base for scientific experiments and observatories.

6.4   Mercury and Venus

The planet closest to the Sun, Mercury, is difficult to observe from Earth. However, space missions have brought surprising results and Mercury is the target of further missions. Venus is considered Earth's sister in size, but is the planet with the most extreme greenhouse effect.

6.4.1   Mercury: Basic Data

The apparent diameter of the planet disk as seen from Earth is between $4.5''$ and $13''$, the maximum brightness is $-1.^{m}9$. Like Venus, Mercury appears as a morning and evening star.

Similar like our Moon, Mercury has a surface littered with craters (Fig. 6.12) as well as no permanent atmosphere. Its orbit has a high eccentricity of 0.26, so its distance from the Sun varies between 46 million and 70 million km. The Orbit is tilted 7° from the

Fig. 6.12 Mercury (Image: Messenger probe)

ecliptic plane. Its mass is 1/18 that of Earth and its diameter is only 4878 km, making it the smallest planet in the solar system. Its density is 5.4 g/cm^3 relatively high, which means that it consists mainly of metals. The metallic Fe-Ni core makes up 60% of its mass, and the diameter of the core is 3500 km. The outer rocky crust makes up only 700 km. Mercury has a weak magnetic field, which suggests a partially metal core (cf. dynamo process on Earth). The field strength is only 0.0035–0.007 Gauss, or about 1% of the surface field strength of Earth. Nevertheless, Mercury's field is stronger than the fields found at Mars and Venus. The planet is also surrounded by a thin shell of He gas; the planet's magnetic field traps He nuclei originating from the solar wind. As on Earth, the field lines are compressed by the pressure of charged particles from the solar wind on the side facing the Sun.

In 1985 one has spectroscopically detected an extremely thin Na atmosphere. The extremely thin *sodium atmosphere* of Mercury produces an exosphere, which is pronounced like a comet tail behind Mercury. The particles of the exosphere are created by bombardment of Mercury's surface with solar wind particles and micro meteorites.

6.4.2 The Rotation of Mercury

The *rotation period* is 58.85 days, this corresponds to 2/3 of its *orbital period* around the sun (88 days). This coupling between orbital angular momentum of Mercury and spin angular momentum comes from the tidal effect of the Sun. The rotation of Mercury has been determined from radar observations, since no details on Mercury's surface are discernible from Earth-based observations. If radar signals are sent to a rotating planet, then they are reflected there. The planet can be thought of as consisting of two halves:

one half is moving towards the observer, so the radar signal reflected there is blue-shifted, the other half is moving away from the observer, so the radar signal reflected there is red-shifted. Overall, the reflected radar signal thus appears broadened, and from the broadening one can infer the rotation speed. Mercury is the planet with the largest temperature contrasts between day and night: 700 K during the day, 100 K at night.

6.4.3   The Surface of Mercury

In 1974 the US probe Mariner 10 passed Mercury at a distance of only 9500 km and sent images to Earth with a resolution of 150 m. The surface of Mercury is covered with craters, the most conspicuous being the Caloris Basin *Caloris* Basin, 1300 km in diameter. This was created about 3.8 billion years ago by an impact body over 100 km in size. Gravity at Mercury's surface is twice that on the Moon, and therefore material ejected from a primary impact crater will cover only 1/6 the area on Mercury that it would on the Moon. Thus, the secondary craters are closer to the primary craters. The large basins resemble the maria on the Moon, and there is evidence of lava flooding. Mercury shows no tectonic activity. However, there are furrows that could be from slight compression of the crust. In 1992, signs of *Ice* below the surface (due to increased radar reflection) at the poles were detected.

The crater density on Mercury's surface is greater than that on the Moon. This indicates a very old surface, and there is no braking atmosphere.

Why does Mercury have such a thin crust and such a massive core? During the formation phase of the planet there were many impacts, with much of the crust being thrown away.

In August 2004, the spacecraft *Messenger* was launched, the and it entered in a Mercury orbit since 2011 due to a braking maneuver. The orbit lies between 200 and 15,000 km altitude above the Mercury surface. Color images (see Fig. 6.13) of Mercury's surface is obtained by means of MDIS (Mercury Dual Imaging System). The Camera MDIS works in visible light and near infrared (up to 1.1 μm). With the Gamma Ray and Neutron Spectrometer (GRNS) can man detect elements such as O, Si, S, or H on the surface of the planet. The Gamma Ray Spectrometer measures gamma rays produced by either galactic cosmic ray bombardment (O, S, Si, Fe, and H) or natural radioactive decay (K, Th, and U) down to a soil depth of about 10 cm. The neutron spectrometer measures low-energy neutrons. These are produced when cosmic rays are decelerated by hydrogen-rich material. With this we are able to explore the upper 40 cm of the planet's surface.

Because of its proximity to the Sun, it was thought that there were few light elements on Mercury's surface, yet sulfur and other elements have been found. Mercury's magnetic field is strongly distorted. The center is shifted 480 km to the north. Therefore, solar wind particles can increasingly penetrate above Mercury's magnetic south pole.

At the end of 2018, the ESA mission *BepiColombo* to Mercury was launched. After several swing-by maneuvers at Earth and Venus and also Mercury, it will become an artificial Mercury satellite in 2025.

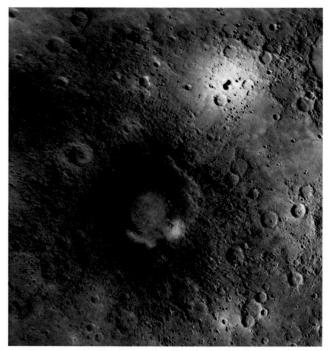

Fig. 6.13 Color-enhanced image of Mercury's surface (Messenger spacecraft, 2011). From the colors, one can infer the composition and evolution of the surface details

6.4.4 Venus: Basic Data

The apparent diameter of the planet's disk ranges from 9.7″ (upper conjunction) and 66″ (lower conjunction). The greatest brightness is $-4.^{m}6$. From all planets in the solar system, Venus and Earth are closest (to within 42 million km). The orbit is almost circular, and the planet is 108 million km from the Sun. Venus sometimes appears as an evening star and sometimes as a morning star, and is the brightest object in the sky after the Sun and Moon. Except that Venus shows phases like the Moon (and also Mercury), the planet is featureless even in large telescopes due to its dense cloud cover. Similar to Mercury, it was possible to determine the rotation period of Venus from radar observations and a peculiarity became evident: The sidereal period is 243.08 days, and Venus rotates retrograde (i.e., from east to west; thus, on Venus, the Sun rises in the west). The Rotation period is about 19 days longer than the sidereal orbital period of 224.7 days around the Sun, and a Venusian day thus lasts 116.75 Earth days. The reason for this slow and retrograde rotation could have been a collision with an asteroid in the early days of the formation of the solar system. Furthermore there are *resonance effects:* At each upper and lower conjunction, Venus shows the same side to Earth.

The *distance* Venus-Earth can be determined very precisely with radar measurements. At the upper conjunction of Venus (just before or after), the radar signal passes through the Sun's gravitational field—space is curved according to Einstein's theory of general relativity. Therefore the transit time of the radar signals is 200 μs longer than according to the classical calculation *(Shapiro Effect)*.

The first European Venus space probe was *Venus Express* (since April 2006 until the end of 2014).

6.4.5 Surface of Venus

Radar measurements with the telescope in Arecibo show details on the *Venus surface* with up to 20 km resolution. The *Atmosphere* consists for the most part of CO_2 surface has temperatures of 460 °C and the pressure is 90 times that of the Earth's atmosphere. Radio radiation with wavelengths of a few centimeters can penetrate these clouds. In 1978, the US spacecraft Pioneer and the Soviet *Venera probes* sent images of the surface. Venera 9 and 10 (Fig. 6.14) landed softly on the surface in 1975 and delivered the first television pictures. Better pictures were taken in 1982 by Venera 13 and 14 (colour pictures). The hard conditions on the surface of Venus ($T = 740$ K, $p = 90$ bar) only allowed measurements of one to two hours. Since the sun does not shine directly through the dense clouds, the surface of the planet is about as illuminated as the Earth on a heavily clouded

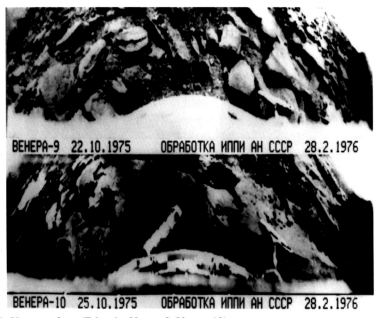

Fig. 6.14 Venus surface. (Taken by Venera 9, Venera 10)

Fig. 6.15 Surface of Venus (Image Magellan)

day, but there is a strong reddish tinge to the *Venusian sky,* because the atmosphere scatters blue very strongly. 70% of the surface is plains, 20% depressions up to 2000 m deep, and 10% continental highlands *(Ishtar Terra, Aphrodite Terra)*. The Ishtar continent, formed by uplift processes, and its high *Maxwell mountains* remind of the highlands of Tibet with the Himalayan massif. Crustal compression is also likely to have occurred here, and there is evidence for mantle convection of Venus.

In 1990 the probe *Magellan* the scanned the surface with the *Synthetic aperture radar method* and detected details up to 120 m in diameter (Fig. 6.15). The surface of Venus appears to be relatively young: 100 million years to 1 billion years old. Terraced volcanic calderas and extensive lava flows are found—but no evidence of plate tectonics as on Earth. Many impact craters have been flooded by lava.

There are hardly any winds on the surface (only in the higher Venusian atmosphere). Nevertheless, there are also signs of deposits caused by winds. Many impact craters are strongly asymmetric—this is explained by the influence of the dense Venusian atmosphere on the trajectory of the impacting meteors. Erosion valleys resembling river beds are also found. At the high temperatures liquid water on the surface can be excluded. Here one assumes thin lava flows (Fig. 6.16).

Venera 13 and 14 explored the surface of Venus by irradiating rock samples. From the X-ray fluorescence radiation, they were able to detect rocks similar to basalts.

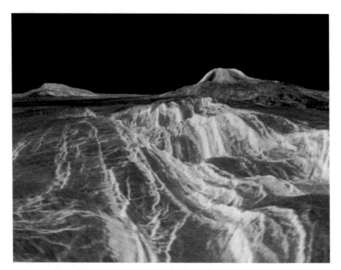

Fig. 6.16 The volcano Gula Mons on Venus (height 3 km). On the left you can see the Sif Mons volcano, which has a diameter of 300 km and a height of 2 km (Image: Venus-Magellan)

Table 6.7 Comparison of the atmospheres of Venus, Earth and Mars. Volume fractions in percent

Gas	Venus	Earth	Mars
CO_2	96.5	0.03	95.3
N_2	3.5	78.1	2.7
Ar	0.006	0.93	1.6
O_2	0.003	21.0	0.15
Ne	0.001	0.002	0.0003

6.4.6   Atmosphere of Venus

Table 6.7 shows the comparison of the atmospheres of Venus, Earth and Mars.

The main constituent of the atmosphere of Venus, CO_2, was first detected spectroscopically by Earth-based observations in 1932.

The pressure of the Venus atmosphere on the surface is equivalent to the pressure at 1000 m ocean depth on Earth. There is a *Troposphere* that extends to an altitude of 50 km. Within this, as on Earth, the gas below is heated and circulates slowly, rising at the equatorial region and falling down at the poles. Since Venus rotates very slowly, this convection current is very steady. The *clouds* are located at an altitude of more than 30–60 km and consist of H_2SO_4 droplets. This sulfuric acid forms from the reaction of SO_2—which is formed by volcanic outgassing—and H_2O. On earth the SO_2 is washed out by precipitation, which does not exist on Venus. Measurements with the space probe *VEGA* (USSR, 1985. showed that at a height of 53 km room temperature prevails at a pressure of 0.5 bar. The high surface temperature can be explained by the *greenhouse effect*. The atmosphere of Venus contains much more CO_2 than the Earth's atmosphere. The diffuse sunlight heats the surface, and the CO_2 reflects the resulting IR radiation from the ground. Thus, the surface continues to warm until it emits enough heat to reach equilibrium with the incoming sunlight. In the uppermost cloud layers (60–70 km height) the UV radiation of the sun splits atmospheric SO_2 into molecular components, and these radicals undergo various chemical reactions until they form sulphuric acid droplets with the water droplets, which have also been split. These sink downward, collide with other droplets, and below the clouds they decay again into SO_2 and H_2O and the process begins again.

Winds on Venus are governed by an east-west circulation that extends around the entire planet, in the uppermost cloud layer at 360 km/h. Cloud formations rotate around the planet in only four Earth days. Thus, Venus' atmosphere rotates much more rapidly than the planet. Earth's atmosphere rotates in sync with Earth. The solar radiation drives the atmospheric circulation on Venus, flow patterns form in N-S direction *(Hadley cells)*.

Surface temperatures on Venus change very little, even between day and night. The lower atmospheric layer has a very large thermal inertia, similar to the oceans on Earth, and stores large amounts of heat. The most important difference between the atmospheres of Venus and Earth is as follows:

- Earth: cold below, hot above;
- Venus: hot at the bottom, cold at the top, this is called the *Cryosphere.*
- Solar wind: Because the magnetic field of Venus is very weak, this directly affects the atmosphere. On the side facing the Sun, neutral atoms are ionized by the high-energy UV radiation and carried away by the solar wind.

ESA's Venus Express mission has detected *Ozone* in the atmosphere of Venus. The occultation of a star by Venus was observed with the spacecraft. Ozone in the atmosphere of Venus reduced the UV content of the star occulted by Venus (cf. Absorption of UV radiation by the ozone layer in the Earth's atmosphere). Ozone in the atmosphere of Venus is formed by the splitting of CO_2 molecules by UV sunlight. The split molecules are also transported to the night side of Venus' atmosphere, where oxygen atoms react with each other to form either O_2 or ozone, O_3. The ozone layer is located at an altitude of 100 km. The discovery of ozone in the Venusian atmosphere is very interesting for astrobiology, because until now it was believed that ozone O_3 that ozone was a biomarker, since atmospheric free oxygen could only be formed by biological processes. Oxygen on Earth was formed by photosynthesis of Plants, by which O_2 is released.

> Venus is the planet with the most extreme greenhouse effect.

Through interaction with the *solar wind* (charged particles continuously repelled by the Sun), Venus loses charged particles of its upper atmosphere. In Fig. 6.17 we can see the influence of the solar wind to the shape of Venus' ionosphere. Very low solar wind density

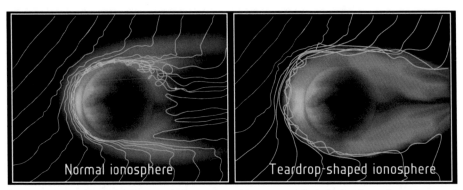

Fig. 6.17 The ionosphere of Venus is not protected by a magnetic field. Its shape depends strongly on the solar wind. The figure on the right shows a comet-like ionosphere at very low solar wind in 2010 (Venus Express, ESA/Wei)

in 2010 revealed a comet-like structure, which is also thought to be present in exoplanets close to the stars. The yellow lines indicate the solar magnetic field. The solar wind density in the figure on the right was about 0.1 particles per cubic centimeter, which is 1/50th of the normal density. The state of extremely low solar wind density lasted about 18 h.

6.4.7   Venus and Climate Change on Earth

The Climate Change on Earth is in all the media, and there are numerous proposals to mitigate climate warming. One whimsical idea has been to introduce droplets of sulfuric acid into Earth's high atmosphere. That way, less solar radiation reaches the Earth's surface.

Sulfuric acid clouds form at altitudes between 50 and 70 km in Venus' atmosphere, as mentioned above. SO_2 combines with H_2O. In 2008, the Venus Express orbiter detected a SO_2 layer at an altitude of about 100 km. This sulfur dioxide is formed by fission from the sulfuric acid clouds further down. If an attempt were made to stop global warming by the process described above, a similar effect to that seen on Venus could occur in the Earth's atmosphere. Thus this measure would be pointless, apart from other possible dangers.

This example is meant to illustrate the importance of comparative planetary research and how findings about the atmospheres of other planets can be applied to Earth's atmosphere.

Was Venus formerly a *habitable planet?* Water is one of the most important elements for life. There's a big difference between Earth and Venus here: If we were to combine the total present *Water resources* on both planets Venus and Earth and distribute them evenly over their entire surfaces, then the thickness of the water layer would be

- Earth: 3 km,
- Venus: 3 cm.

Venus is therefore extremely dry at present.

Nevertheless, a few billion years ago Venus could have had a similar amount of *water* as on Earth today. Due to the more intense solar radiation on Venus H_2O split into H_2 and O_2, the lighter hydrogen escaped into space, and oxygen rapidly combined with other elements. Venus Express measurements show that this process is still occurring today: a ratio of 2 H to 1 O atoms is found, so a water molecule was split. Some of the water on Venus could also have been formed by *Comet impacts* brought there.

So liquid water in the form of oceans could only persist on Venus for a short time, if at all. Water is also an important sink for CO_2. This greenhouse gas is reduced by the oceans on Earth.

Also important for astrobiology is *Lightning*. Electrical discharges were detected on Venus. These cause molecules to split and new compounds can form. Electrical discharges have only been found on four planets in our solar system: Earth, Venus, Jupiter and Saturn.

6.5 Mars

Mars is the most Earth-like planet in the solar system. Few other planets have fascinated mankind as much as Mars. There are several reasons for this. Mars strongly changes its brightness and color; it appears reddish at high brightness and it has long been considered a candidate for life in the solar system. All major space-faring nations therefore undertook missions to the red planet.

6.5.1 Mars: General Data

Due to the large orbital eccentricity, we only give the extreme values of the apparent diameter or opposition brightness (for near-Earth oppositions): The diameter of the planet's disk at the time of its opposition is 25.7″, at the time of the greatest Earth distance only 3.5″. The maximum Opposition brightness is $-2.^{m}91$.

Observations from Earth reveal surface structures on the planet's disk, as the surface is barely obscured by clouds. One can see details up to 100 km in size from Earth with telescopes, similar to the details seen with the naked eye on the lunar surface. In 1877, the astronomer *Schiaparelli* observed the Martian canals, *canali*, but these soon proved to be optical illusions. The *Rotation Period* Mars is $24^h37^m23^s$. The *Rotation axis* has a tilt of 25°, and therefore Mars has *Seasons* similar to Earth. You can see *Polar caps* as well as changing fuzzy dark areas depending on the season. The *mass* of Mars is 1/9 of the Earth's mass. The density is $3.9\,g/cm^3$. The planet has only an extremely weak *Magnetic field*. Besides there are magnetic field concentrations in small areas.

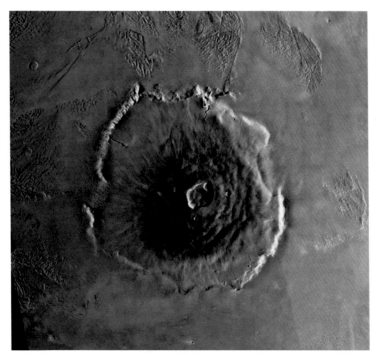

Fig. 6.18 Olympus Mons, the largest volcano in the solar system (NASA)

In 1965, a US probe visited Mars for the first time, *(Mariner 4),* and the photographs revealed details more like the surface of the Moon. In 1971, for the first time, *Mariner 9* space probe orbited another planet and photographed details up to 1 km in size. *Viking 1* landed softly on the surface in 1976 and transmitted images and data until 1982 (Fig. 6.18).

6.5.2   Martian Surface

The total surface of Mars (Fig. 6.19) roughly equals the area of all Earth's continents. About half of the planet consists of highlands dotted with craters, most of which lie in the southern hemisphere. The lava plains, which are about 4 km lower, are found in the northern half. *Hellas* is a huge impact basin 1800 km in diameter.

The *Tharsis region* is about 10 km high and contains four volcanoes that rise another 15 km. The largest volcano in the solar system is *Olympus Mons* (Fig. 6.18) (500 km diameter, 25 km high). Some volcanoes show impact craters on their slopes, so must be about 1 billion years old. Olympus Mons has very few impact craters, and its surface can therefore only be at most 1 million years old, there is also evidence of lava flows that still are much younger. There are numerous canyons on Mars, the most impressive being the

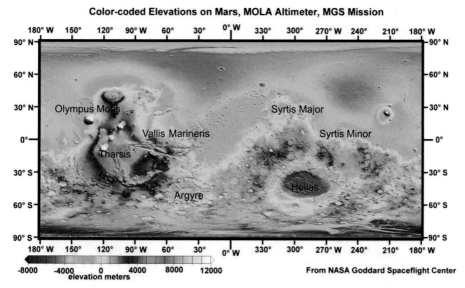

Color-coded Elevations on Mars, MOLA Altimeter, MGS Mission

Fig. 6.19 Map of Mars (Source: NASA)

Vallis Marineris (5000 km extended, 7 km deep and 100 km wide). These are tectonic rifts. The polar cap was also investigated, Fig. 6.20.

Viking 1 landed on 20 July 1976 in a 3 billion year old plain. At the landing site there are many 1 m large rock debris, and dune-like deposits of sand are found. The landing site of Viking 2 is similar, the landing took place on 1 September 1976. Wind speeds up to 100 km/h were measured. The weather stations on board measured larger changes in the Martian atmosphere than in the Earth's atmosphere. In summer, the maximum temperature was −33 °C and at the end of the night of −83 °C. Viking 2 photographed during the winter on the Martian soil. X-ray fluorescence spectrometers were used to determine the composition of the Martian samples. The composition differs significantly from that of terrestrial rocks: The Martian soil consists of basic rocks rich in Mg and Fe.

In 1997, a Mars probe landed softly on Mars and, with the help of a small robotic car *(Sojourner)* it was also possible to make a short trip on Mars. This mission, called "Pathfinder," provided 90 days of data to Earth. Sojourner worked 12 times as long as its planned lifetime (7 days). The landing took place on 04.07.1997, and 16,000 images were obtained from the lander and 550 from the rover. The mass of the lander was 264 kg, that of the rover 11 kg.

In 2004, the two Mars rovers Spirit (until 2011) and Opportunity started to be in operation (Fig. 6.21). Opportunity covered a distance of more than 40 km on the surface of Mars by 2015. The Mars orbiter Odyssey has been in Mars orbit since October 2001. The European mission Mars Express has been in Mars orbit since Christmas 2003, and the lander Beagle 2 was unfortunately lost.

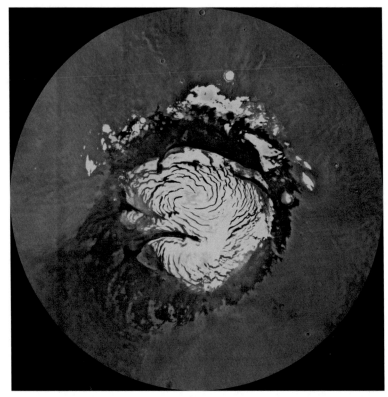

Fig. 6.20 The northern polar cap of Mars (NASA, Viking 1)

Fig. 6.21 The Mars rover Opportunity photographed the landing site. The large white braking parachute can be seen to the right of the lander module (NASA)

Details of a Martian crater partially covered with ice can be found in Fig. 6.22.

The SNC meteorite found *SNC meteorites* found in the Antarctica are only about 1.3 billion years old and may have come from Mars because of their composition (Moon

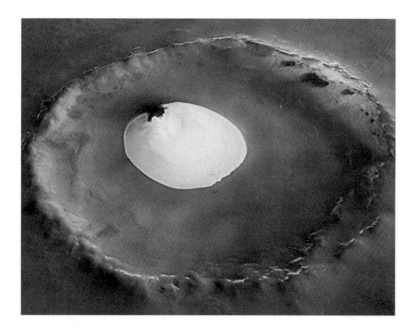

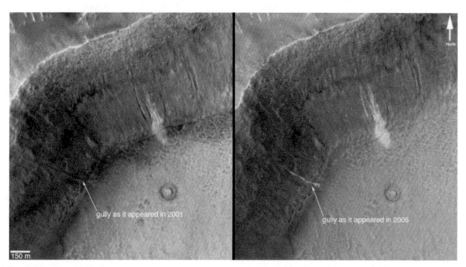

Fig. 6.22 Top: Martian crater, 30 km, covered with water ice. The image resolution is 21 m per pixel point. ESA/DLR/FU Berlin (G. Neukum); bottom: formation of so-called gullies, a system of outflows within two images taken five years apart (Image credit: NASA/JPL/Malin Space Science Systems)

is ruled out because of different composition, Venus because of its dense atmosphere and higher gravity as well). Gas bubbles were found in the SNC meteorites, whose composition resembles that of the Martian atmosphere. This gas was trapped during the impact of a body on Mars and subsequent ejection of the material or heating.

The *Mars interior* will be investigated with a seismometer (SEIS experiment on board the Mars Insight Mission, launched 2018). This will be used to investigate frequencies in the range 0.05 mHz to 50 Hz. Previous data show that Mars has a core that is liquid in the outer region (core size between 1500 and 1800 km). The presumed Martian earthquakes are expected to be 100 times stronger than on the Moon and are mainly due to thermal tensions.

6.5.3 Mars Atmosphere

The Surface pressure is only 0.007 bar, less than 1% of Earth's surface pressure, and 95% of Mars' atmosphere is composed of CO_2. There are several types of *Clouds* in the Martian atmosphere:

- Dust clouds: these are formed by swirling dust from the Martian soil; they are driven into the atmosphere by winds, and can cover much of the planet's surface.
- Clouds of water ice: These form around mountains, much like on Earth.
- Clouds of carbon dioxide: Form veils of dry ice.

The *polar ice caps* are made up of two components. Seasonally changing polar ice caps consist of CO_2-ice. The southern permanent cap is about 350 km in diameter, the northern about 1000 km, and this consists of frozen H_2O. One therefore finds large water reserves in the polar caps. Up to 80° latitude, sediment deposits are found in both hemispheres due to the expanding polar caps. All this points to *Climate changes* on Mars with periods of a few 10 000 years. At present, no water can exist in liquid form on Mars. However, one finds evidence of large water flows in the past. Perhaps there is frozen water beneath the Martian floor that has been thawed by volcanic activity. This still gives rise to speculation about possible lower life forms on Mars. However, experiments by the Viking probes (looking for signs of metabolism) turned out negative.

Numerical calculations showed that the inclination of the rotation axis of Mars can change by up to 60° which could have a significant effect on the climate. This would increase the seasonal contrast between north and south. At high and mid latitudes, summer temperatures would then rise above the freezing point of water. Released CO_2 from the melting polar cap increases the density in the atmosphere, the greenhouse effect takes effect, and large amounts of surface water can form. Add to this chemical reactions form salts and limestone (calcium carbonate), and so the amount of carbon decreases again, and the greenhouse effect is reduced. As soon as the inclination of the axis decreases, the planet cools down, dry ice and snow fall, and Mars is again in its cold state. The last warm period may have been 300 million years ago. In late Martian history, warm periods may have lasted only about 1 million years. In earlier Martian history, however, they may have been longer, and lower forms of life may have evolved, perhaps surviving in sheltered areas to the present day.

> Mars is the planet with the largest climate changes. The search for life has been fruitless so far, but will continue.

6.5.4 Mars: Terraforming?

One finds many similarities between Mars and Earth:

- Polar caps
- Atmosphere with clouds
- Seasons
- Weather with storms

However, there are also differences:

- Mars: no permanent magnetic field; particles of the solar wind reach the Martian surface almost unhindered
- Atmosphere very thin on Mars
- At present, water in liquid form is not possible on Mars.

Very often *Terraforming* of Mars is discussed. To make Mars habitable, its atmosphere would have to be made much denser, which would warm the surface of Mars. At the same time, however, the atmosphere must not be destroyed by processes of splitting molecules (e.g. CO_2, H_2O). It is assumed that enough frozen or bonded CO_2 exists in the Martian soil or at the poles to increase the pressure in the Martian atmosphere to 300 mbar. This would raise the temperature enough to melt ice at the poles and in the Martian soil, allowing plant life. There are other proposals for Martian terraforming. However, in the best-case scenario, it would take several thousand years before Mars would be habitable for humans.

The Mars rover *Curiosity* landed on Mars in August 2012 to conduct extensive chemical soil sample analyses and determined whether there was any water in the past on Mars (Fig. 6.23). On average, the rover can travel 30 m per hour on the Martian surface, overcoming obstacles up to 75 cm in height. Energy is supplied by decaying radioisotopes (plutonium) and lasts for one Mars year (687 Earth days). In total, between 5 and 20 km should be covered on the Martian surface. An example of an area of highly probable

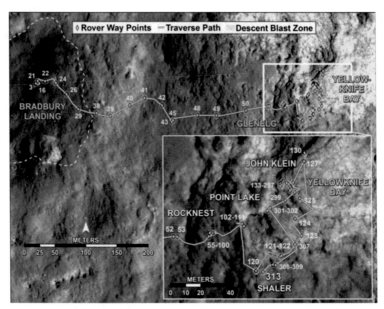

Fig. 6.23 The distance traveled by the Mars rover Cursiosity. The numbers indicate the Martian days since landing (Image credit: NASA/JPL/Malin Space Science Systems)

sedimentary deposits is given in Fig. 6.24. Figure 6.25 shows sediment deposits against a mountain in the background.

6.5.5 Martian Moons

In the year 1877 *A. Hall* first saw the two tiny Moons of Mars Phobos and Deimos.

The orbital period of Phobos is only 7^h39^m; it is thus shorter than the planet's rotation period. The moons are irregularly shaped and are about 20 and 12 km in size, respectively. Both resemble a triaxial ellipsoid:

- Phobos: $27 \times 21 \times 19$ km,
- Deimos: $15 \times \times 11$ km.

Phobos is only 2.8 Mars radii from the surface of Mars, Deimos 7 Mars radii (orbital period 30.3^h). Deimos moves W to E and very slowly, staying above the horizon for almost three Martian days, passing through all phases several times.

The Martian moon Phobos is very close to the Roche boundary and will be torn apart, or crash into Mars, within the next 100 million years (Fig. 6.26).

Fig. 6.24 Details of the Martian surface (NASA/Curiosity)

Fig. 6.25 Details of the Martian surface; the layered deposits in the foreground indicate formerly flowing water (NASA/Curiosity)

6.6 Jupiter and Saturn

Jupiter is the largest planet in the solar system. The properties of Jupiter and those of the other gas planet Saturn are fundamentally different from those of the terrestrial planets. Saturn is especially known among the planets for its bright ring system. But also the exploration of its moons showed unexpected results; for example, water geysers were found on one of its moons.

Fig. 6.26 Phobos (NASA)

6.6.1 Jupiter: General Properties

The largest planet of the solar system is due to its fast *rotation* noticeably flattened:
    polar diameter: 133,700 km
    Equatorial diameter: 142,984 km.
    Jupiter rotates differentially, so it does not rotate like a rigid body:

- At the equator, the rotation period is 9 h 50 min 30 s (System I).
- At the polar regions 9 h 55 min 41 s (system II).

The apparent diameter of the planetary disk at the time of its opposition is 49″ at the time of its greatest distance from Earth only 30″. The opposition brightness is $-2.^{m}9$ and it is the brightest planet in the sky after Venus (only rarely at near-Earth oppositions Mars appears brighter). Jupiter has about 1/1000 the mass of the Sun, or 318 Earth masses, but has 2.5 times the mass of all the other planets in the solar system combined. The center of gravity of the Sun-Jupiter system is 1.068 solar radii, $R_{\odot}$, outside the sun's center.

Jupiter is 778 million km from the Sun, and it takes 12 Earth years to orbit the Sun. Its mean density is 1.3 g/cm^3. Its axis is inclined only 3° from the orbital plane, and therefore there are no seasonal changes.

6.6.2 Space Missions to Jupiter

The first spacecraft explored Jupiter in 1974/1975 (*Pioneer* 10, 11, launched 1972/1973). *Voyager 1* reached Jupiter in 1979 and subsequently investigated Saturn (1980). *Voyager*

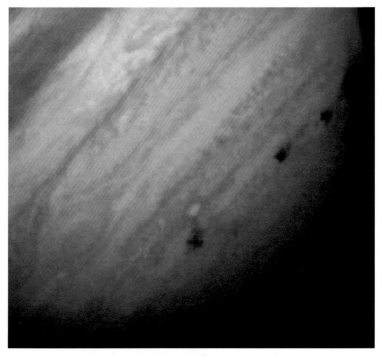

Fig. 6.27 After the impact of the fragments of Comets Shoemaker-Levy in July 1994 in Jupiter's atmosphere, visible traces appeared (HST image)

2 reached Jupiter a few months later and studied Saturn in 1981, Uranus in 1986 and Neptune in 1989.

The space probe *Galileo* orbited Jupiter from 1995 to 2003, and on its flight in 1994 was able to observe the impact of comet Shoemaker Levy on Jupiter (Fig. 6.27). Upon its arrival, a descent capsule separated in December and plunged into Jupiter's atmosphere. At a depth of 160 km it was finally destroyed by the external pressure, and the last reading was 22 bar at a temperature of 150 °C. The Galileo probe itself was then crashed in 2003 to avoid hitting Jupiter's moon Europa (Contamination with terrestrial bacteria). The space probe *Cassini* passed Jupiter in 2000/2001.

In July 2016, the Jupiter Polar Orbiter (JUNO) spacecraft entered orbit around Jupiter (it was launched in 2011). The polar orbit of this probe allowed the use of solar cells for power supply. Figure 6.28 shows a close-up of Jupiter's south polar region.

6.6.3   Structure of the Giant Planets

The internal structure of the giant planets (Table 6.8) differs in principle from that of terrestrial planets:

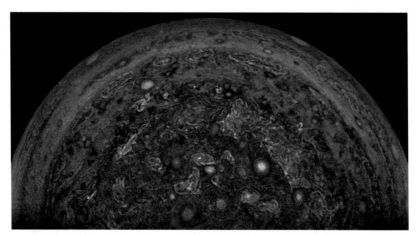

Fig. 6.28 Jupiter's south pole photographed from about 100,000 km away by the Juno spacecraft in February 2017. Note the many cloud swirls (NASA)

Table 6.8 Comparison of the structure of the giant planets; distance from the center in 1000 km

Planet	Molecular H	Metallic H	Ice	Rock
Jupiter	71–59	59–14	14–7	0–7
Saturn	60–30	30–16	16–8	0–8
Uranus	26–18		18–8	0–8
Neptune	25–20		20–10	0–10

- They are surrounded by a dense atmosphere.
- A few 1000 km below the surface, the pressure becomes so high that hydrogen changes into its liquid state. Even deeper H behaves like a metal. Most of Jupiter is composed of metallic H. Metallic hydrogen is formed when hydrogen is subjected to high pressure. A lattice of protons (distance smaller than Bohr's radius) forms as well as free electrons.
- The core of the giant planets consists of rock and ice (up to about 20 Earth masses).

A strong internal heat source generates 4×10^{17} W, equivalent to the amount the planet receives from the Sun. Jupiter's atmosphere is therefore a mixture between a normal planetary atmosphere and a stellar atmosphere that is heated from below. This involves a cooling process in the interior: Jupiter is about 20,000–30,000 K hot at its core.

6.6.4   Jupiter Atmosphere

In the spectrum of reflected sunlight, absorption lines of gaseous methane and ammonia can be detected. In fact, however, Jupiter's atmosphere consists mainly of hydrogen and has a composition similar to that of the Sun. Furthermore, nitrogen, carbon, water vapour and hydrogen sulphide have been detected. The colors in the atmosphere come from different forms of molecular sulfur (S_n, n is an integer natural number), this forms brown and yellow particles. The highest clouds glow red, below them white and then brown clouds, and still further below then blue clouds. The main constituents of the clouds are ammonia at the top, then ammonium hydrosulphide and water at the bottom. The NH_3 cloud cover marks the upper limit of the convective troposphere, the temperature there is 140 K. Within the troposphere, the temperature increases downward. At a pressure of 10 bar liquid water as well as ice crystals are assumed to exist.

Due to rapid rotation, the N-S flow in the atmosphere becomes less important and an east-west flow is formed. On Jupiter, a pattern of dark bands and bright zones is seen extending parallel to the equator. Gas flows upward in the bright zones, culminating in the white NH_3-Clouds. In the dark bands the cooler atmosphere moves downwards; here there are fewer NH_3 clouds, and one can see the NH_4SH-clouds.

One also finds large oval-shaped, very persistent regions of high pressure, of which the *Great Red Spot* (GRF, Fig. 6.29) which has an extent of 30,000 km is the best known and has been observed for 300 years. In it there is a counter clock wise rotation in six days. On Earth, a hurricane (which here, however, is a depression) has a lifetime of a few weeks, as it loses energy through friction with the land masses. Jupiter does not have a solid surface, and therefore such storms can be very long-lived. In March 2003 we observed the formation of another "red spot". *(Red Spot Jr.).* This structure appeared as an oval BA as early as 2000, but it didn't turn reddish until 2006. The extent was about 1/2 that of the

Fig. 6.29 The Great Red Spot in Jupiter's atmosphere taken by the Juno spacecraft (NASA)

Fig. 6.30 Sequence of Jupiter images taken on October 22 and 23, 2000. Clearly seen is how the Great Red Spot (GRF) continues to move due to Jupiter's rotation. The clouds in the GRF rotate counter clock wise. The equatorial zone is currently bright, suggesting high clouds. Twenty years ago, the equatorial zone consisted of dark bands (Images: Cassini mission, NASA)

Great Red Spot. It has since disappeared again, and the Great Red Spot also appears to be undergoing a major change. A sequence of Jupiter images clearly shows how the cloud formations change within a few hours because of Jupiter's rapid rotation (Fig. 6.30).

It is believed that there is a 70-year climate cycle on Jupiter. According to this, there was a decrease in cyclone activity on Jupiter until about 2011; now activity should be increasing again. The last minimum in cyclone activity dates back to 1939. The turbulent south polar region of Jupiter is shown in Fig. 6.28.

6.6.5 Magnetosphere of Jupiter

At In 1950 Jupiter's Radio emission was measured as a synchrotron radiation (the intensity increases with increasing wavelength; it is caused by charges accelerated in a magnetic field, which radiate). This radiation was shown to originate from an area outside the planet, and therefore charged particles must be moving around Jupiter, similar to the Van Allen belts in the Earth's magnetosphere. Jupiter's field is much stronger than Earth's magnetic field. The magnetic field axis is inclined by $10°$ from the axis of rotation and 18,000 km from the centre and has the opposite polarity to the Earth's magnetic field. If Jupiter's magnetic field could be seen in the sky from Earth with the naked eye, it would be three times the size of the Moon. The magnetic field of Jupiter is compressed by the solar wind similar to the earth on the side facing the sun to 6×10^6 km, while on the side facing away from the sun it rises up to 700×10^6 km, i.e. almost as far as Saturn's orbit.

The sources of ions in Jupiter's magnetosphere are:

- Protons and electrons from the solar wind,
- particles from the atmosphere,
- particles from Jupiter's inner moons. These may either have been knocked out by the infall of energetic ions, or come from the active volcanoes of the moon *Io*, that is, S and O ions. Io loses about 10 tons per second. These ions create a Plasmatorus.

In 1992 the space probe Ulysses explored Jupiter's magnetic field and was ejected from the ecliptic plane by a *Gravity Assist* out of the ecliptic plane and then returned to the interior of the solar system to study the Sun's polar regions.

6.6.6 Jupiter's Rings and Moons

The Jupiter's ring system, consisting of tiny black dust particles, was first photographed by Voyager 1 in 1979. The fine dust particles will spiral down on Jupiter in a few 10^4 years. The dust particles are charged by Jupiter's magnetic field and slowed down by collisions. By absorption and re-emission of radiation they also lose orbital angular momentum (*Poynting-Robertson effect*). The material of the rings comes from dust released by the impact of small meteorites on Jupiter's innermost moons: Adrastea and Metis, for example, produce the main ring. The moons Thebe and Amalthea feed the two fainter Gossamer rings. An extremely thin outer ring circles Jupiter retrograde.

Jupiter has 79 natural satellites (Fig. 6.31) (As of 2019). In 1610 *Galileo*, discovered the four largest moons; these are called *Galilean moons*: Io, Europa, Ganymede, and Callisto. The two moons, Europa and Io, are about the size of our moon, Ganymede and Callisto are larger than Mercury. Jupiter's other moons, however, are much smaller.

The surface of *Callisto* is littered with impact craters. It is an icy moon, and the craters look shallower than those of Mercury or the Moon. The ice crust heats up to 130 K, and

Fig. 6.31 The large red spot in Jupiter's atmosphere and, by comparison, the Galilean moons (NASA)

this causes the ice to have less resistance, making the impact craters shallower. Callisto is the Jupiter moon with the oldest surface. *Ganymede* (Fig. 6.35) is the largest moon in the solar system, and a billion years after its formation, there has been a restructuring of its surface (striking groove patterns), which otherwise appears rather featureless.

At *Io* (Fig. 6.32) there are active volcanoes, emitting sulfur or sulfur dioxide. There are also snowfalls of sulfur dioxide. A hot spot has also been found on Io, extending 200 km, where there is a temperature of 300 K. The cause of volcanism is heating due to tidal action. Io is at the same distance from Jupiter as the Earth's moon is from the Earth, but Jupiter's mass is 300 times that of the Earth, and therefore Io is virtually kneaded.

Jupiter's moon *Europa* (see also Fig. 6.33) is a white glowing sphere covered by ice with dark stripes, which may be cracks in this ice shell. Perhaps internal parts of this ice shell had melted as a result of heating from radioactivity, and the cracks formed during solidification. It is possible that there is liquid water underneath the ice shell even now, and this would make this moon of Jupiter a candidate for life. This ocean is a result of Jupiter's strong tidal heating. The surface temperature of Europa is only about $-150\,°$C. Since very few impact craters are seen on Europa, the age of the surface (Fig. 6.34) is estimated to be about 30 million years old. Measurements from space probes have shown that Europa has a weak changing magnetic field. This is strong evidence for a saline ocean beneath the surface.

Fig. 6.32 Jupiter's volcano-covered moon Io (NASA)

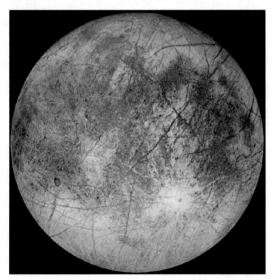

Fig. 6.33 Europa, the moon of Jupiter covered by a thick ice shell. The many grooves are clearly visible, but there are hardly any impact craters

In 2023, the mission *JUICE* (JUpiter ICy Moon Explorer, ESA project) will be launched to Jupiter, reaching the Jupiter system in 2030. After several approaches to Europa and Callisto, the probe will enter to orbit Ganymede in 2032.

Also *Ganymede* has a magnetic field and a salty ocean lying under an ice crust (Fig. 6.35).

Fig. 6.34 Details of the surface of Jupiter's moon Europa (NASA)

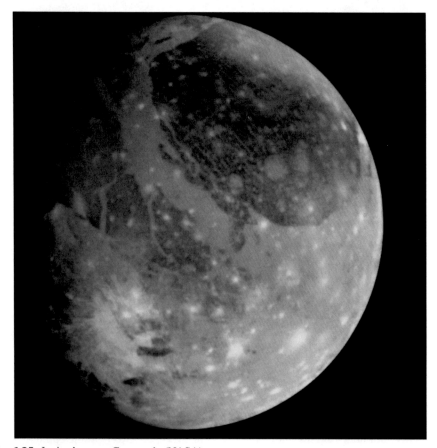

Fig. 6.35 Jupiter's moon Ganymede (NASA)

The Galilean moons of Jupiter are extremely interesting objects for *astrobiology.* Altogether, Jupiter's moons (with J1, J2, ... is also referred to as) are classified as:

- Four innermost moons: small irregular bodies, except for *Amalthea* (240 km diameter) no structures detectable.
- Galilean moons: distance 6–26 Jupiter radii, they are inside Jupiter's magnetosphere.
- Four moons of the middle group: the largest are Himalia (170 km) and Elare (80 km), distance 155–164 Jupiter radii, orbital inclinations up to 29°, eccentricities 0.13–0.21.
- Outermost group: 10–30 km diameter. 290–332 Jupiter radii away, very large orbital inclinations (147–163°) and high eccentricities. Probably captured minor planets.

> Jupiter's moon Europa is a candidate for life in the solar system because of its ocean beneath the ice crust.

6.6.7   Saturn: Basic Data

Saturn (Fig. 6.36) was already known to ancient civilizations: the diameter of the planet's disk at the time of its opposition is 20.1″, at the time of the greatest distance from Earth only 14.5″. The opposition brightness is $0.^m43$.

The ring planet requires 30 years to orbit the Sun and is on average 1 427 million km from the Sun. It has 95 times the mass of the Earth and the lowest density of all planets, with only 0.7 g/cm³. The rotation period at the equator is 10^h40^m and at the poles 10^h39^m. The causes a strong *Flattening:* equatorial diameter: 120,536 km, pole diameter 107,812 km.

Fig. 6.36 Saturn. You can see very clearly the broad Cassini division and the fine Encke div.sion on the outside (NASA)

The *axis of rotation* is around 27° inclined, i.e. one observes seasonal effects.

The *Structure* of Saturn resembles that of Jupiter and has been shown in Table 6.8 outlines. Its internal energy source is only half that of Jupiter. It is assumed that in the liquid H-mantle the heavier He-drops sink downwards and in this way gravitational energy is released. This does not happen with Jupiter because it is warmer than Saturn. This is supported by the finding that Saturn contains only half the He that Jupiter has.

The temperatures in the *Atmosphere* is slightly lower than Jupiter's, and the atmosphere is more extended as a result of its lower gravity. There are fewer large storms than on Jupiter, but their occurrence here varies seasonally. Every 30 years an eruption of spots is observed in the equatorial region (most recently in 1990). The strength of its magnetic field is 0.2 Gauss, and the magnetic field axis is only 1° inclined with respect to the axis of rotation. The yellowish-brown cloud cover contains mainly ammonia crystals.

6.6.8   Saturn's Rings

The Saturn's rings have been known for a long time, and they orbit Saturn in the equatorial plane, which is inclined by 27° inclined to the orbital plane of the planet. We therefore look at one side of the rings for 15 years. The brightest rings are the A ring (radius 136,780 km), the B ring, and the innermost C ring (only 12,900 km from Saturn's surface). The brightest ring is the B ring, and its total mass is equivalent to that of a satellite 300 km in diameter. The A and B rings are separated by the *Cassini division* (discovered in 1675 by *Cassini*). Although the rings are 70,000 km wide, they are only 20 m thick. The rings consist of countless ice particles about the size of tennis balls. Voyager showed a D-ring further inside and an F-ring outside the A-ring. This resembles the rings of Uranus and Neptune. One recognizes innumerable single rings on satellite photographs. Important for structure of rings resp. occurrence of gaps are *Resonances*. One speaks of a resonance when two objects have orbital periods that are in an integer ratio to each other (Figs. 6.36 and 6.37).

Cassini division: a particle at the inner edge would have half the orbital period of the satellite Mimas. The outer sharp edge of the A ring is in 7:6 resonance with the Satellite Janus and Epimetheus. Spiral density waves are also observed in the rings. New theories explain these spoke-like structures by electrical discharges. Density waves are also important for explaining the spiral structure of galaxies.

The found blue glow of the E-ring is interpreted like this: The ring is made of ice crystals, which derived from geysers of the moon *Enceladus*.

6.6.9   Saturn's Moons

As of end of 2022 83 moons of Saturn are known. The Saturn's moon *Titan* (Fig. 6.38) is the second largest moon in the solar system after Jupiter's moon Ganymede, with a diameter of 5150 km, and has 1.9 lunar masses. The density is 1.9 g/cm³. It was discovered

Fig. 6.37   Details of Saturn's ring system (NASA)

Fig. 6.38   Titan (NASA/JPL/Space Science Institute)

in 1655 by *Huygens*. In 1944, an atmosphere was discovered on Titan. The space probe Voyager passed Titan at a distance of only 4000 km and was covered by Titan as seen from Earth. During this occultation, the radio signals had to pass through various layers of Titan's atmosphere *(Radiooccultation),* so that it could be probed. The main component could be N_2 CH_4 and Ar make up only a few percent. Furthermore, one found HCN which is a basic building block for DNA. The low clouds reach up to 10 km altitude and consist of CH_4. This gas plays a similar role on Titan as H_2O on Earth. Titan's surface temperature is 90 K, and it is thought that there are lakes of liquid methane here. The difference in temperature between the poles and the equator is less than 3 K. There may also be continents of water ice.

In June 2004, the spacecraft *Cassini* after seven years of flight, entered into the Saturn system. 14 January 2005 marked a high point in European space travel: the probe detached from Cassini *Huygens* landed softly on Saturn's moon Titan and transmitted images of its surface to Earth. Instead of the oceans that had been suspected, sand dunes up to 150 m high were found. They consist of fine particles, possibly frozen ethane. Furthermore volcanoes are suspected on Titan, but not fire volcanoes like on Venus, Mars or Earth, but *Cryovolcanoes.*

Radar measurements show lakes of methane at the North Pole, some of which are evaporating. The seasons on Titan last seven Earth years each!

> Titan is the only moon in the solar system with a dense atmosphere and organic compounds are also found.

Cryovolcanoes (ice volcanoes) were also found on *Enceladus.* In the case of cryovolcanoes, substances such as methane, carbon dioxide, ammonia and frozen water are melted in the interior of a moon by heating (e.g. tidal forces) and penetrate to the surface. Images taken by the space probe *Cassini* (launch 1997, reached Saturn 2004) suggest that liquid water is present in chambers just below the surface of Enceladus (Fig. 6.39). This is ejected in fountains up to 500 km high. In 2008, Cassini passed Enceladus at a distance of only 52 km. The ejected material was analyzed during a stellar occultation, and traces of organic material were also found.

> Thus Enceladus, which is only 500 km across, is another candidate for life in the solar system.

Figure 6.40 shows the moon Tethys. After Titan are *Japetus* and *Rhea* the largest moons of Saturn (both about 1500 km in diameter). Japetus is 60 Saturn radii away from Saturn, Rhea only nine. Between the two moons are Titan and Hyperion. Japetus has a density of only 1.1 g/cm^3. It has a bright and a dark hemisphere. It is believed that impacting micrometeorites on *Phoebe* release dark particles that rain upon Japetus.

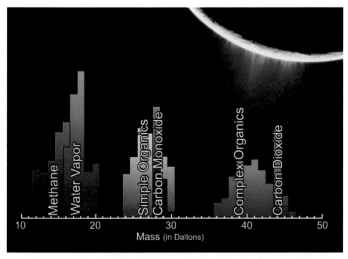

Fig. 6.39 Geyser-like fountains ejected from the surface of Enceladus. Below is the chemical analysis. The Cassini spacecraft flew through the fountain (plume) in March 2008. A dalton is the term used in the United States for the atomic mass unit, 1 Da = 1 amu = 1.66×10^{-27} kg (Credit: Cassini Mission)

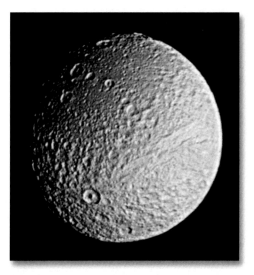

Fig. 6.40 Saturn's moon *Tethys*. Image taken by Voyager 2 from a distance of 282,000 km. The rift system visible on the right may have developed as Tethys expanded (NASA/Cassini)

Fig. 6.41 Saturn's F ring with the shepherd moons Pandora (outside) and Prometheus (© NASA/JPL/Space Science Institute)

The two moons Janus and Epimetheus (Fig. 6.41) move in nearly identical orbits. Every four years they approach each other and then exchange orbits.

6.7   Uranus and Neptune

The two outermost planets of the solar system are sometimes described as ice giants(as opposed to the gas giants Jupiter and Saturn).

6.7.1   Discovery of Uranus and Neptune

Uranus was discovered on 13 March 1781 by *W. Herschel*. At the time of its opposition the diameter of the planet disk reaches $4''$ and the apparent brightness is $5.^{m}5$ so it can be seen with binoculars if you know its exact position in the sky.

Uranus takes 84 years to orbit the Sun, while Neptune takes 165 years.

Neptune was discovered by *Galle* after predictions from perturbations of the orbit of Uranus on September 23, 1846. At the time of its opposition, the diameter of the planet's disk reaches about $2.3''$ and the visual magnitude is $7.^{m}8$. With a diameter of 49,258 km, Neptune is the fourth largest planet (Uranus: 51,118 km), and in terms of mass it is the third largest.

Both planets have approximately 15 times Earth's mass and densities of $1.2\,g/cm^3$ (Uranus) and $1.6\,g/cm^3$ (Neptune). Neptune's rotation axis is inclined by $29°$ but that of Uranus is inclined by $98°$. Each pole of Uranus is exposed to the Sun for about 40 years. The cause of this unusual tilt may have been a collision with a large planet in the early stages of the solar system. The rotation period for both planets is about 17^h. Uranus rotates retrograde. Both planets lack metallic hydrogen in their interiors. Neptune has an internal energy source, and although it is farther from the Sun than Uranus, both planets have the same surface temperatures. Uranus appeared featureless at the time of the Voyager

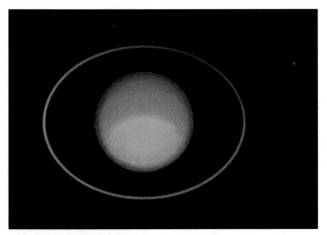

Fig. 6.42 Uranus with ring (Image: HST)

2 flyby (1986). Instead of NH_3clouds (which are found on Jupiter and Saturn), here they found CH_4-clouds. Since it has no internal heat sources, there is no convection, and the atmosphere is very stable. Neptune also has CH_4-clouds, and at the upper limit of the troposphere the temperature is 70 K at a pressure of 1.5 bar. The atmosphere is clear, and the scattered sunlight causes the blue coloration of the planet Neptune. Furthermore, there are strong east-west winds on Neptune with wind speeds up to 2100 km/h. Both planets have a magnetosphere, the field strength is 0.3 Gauss (Uranus) resp. 0.2 Gauss (Neptune), the inclination of the magnetic field axis to the rotation axis is 60° (Uranus) resp. 55° (Neptune).

In addition to satellite images, modern large telescopes (Fig. 6.42) can be also used to observe details on the two planets from Earth (e.g., Keck telescope). Long-lived large storms were found in the atmosphere of Uranus.

6.7.2   Rings and Satellites of Uranus and Neptune

The *rings* of Uranus were only discovered by chance in 1977: the planet was occulting a star. Shortly before the disappearance of the star one observed repeated light attenuations, which one interpreted as attenuation by four rings. Today 13 rings and 27 moons are known (end of 2006). The brightest ring is called ϵ-ring. The ring particles consist of dark chunks up to 10 m in diameter. In 2005 with the Hubble Space Telescope Uranus moon *Mab* was discovered and this moon appears to dissolve from meteorite impacts, providing material for new rings (fine ice crystals). The five largest moons range in size from 500 to 1600 km. The ring system of Neptune is still little explored, there seem to be thickenings—by gravitation of the moon *Galatea.* The rings were named after *Adams, Le Verrier, Galle, Lassel* and *Arago* .

Fig. 6.43 Neptune's moon
Triton (NASA/Voyager 2)

Neptune has 13 moons. The Neptune moon *Triton* (Fig. 6.43) was found 17 days after the discovery of Neptune. It has an atmosphere as well as volcanism. The diameter is 2720 km and the density 2.1 g/cm^3. It is probably consists of 75% rock and 25% water. The surface temperature is between 35 and 40 K. The atmosphere consists of N_2-steam. The lava (cryovolcanism) here consists of H_2O and NH_3. The distance to Neptune is about 350,000 km. It orbits Neptune in five days 21 h retrograde (i.e., opposite the direction of rotation), and the orbital inclination to the equatorial plane of Neptune is 156°. It is thought that it will be torn apart in about 100 million years due to the large tidal forces, and the particles will then give Neptune a spectacular ring system.

Triton has an extremely thin atmosphere consisting of 99% nitrogen, with a pressure of only 1/70,000 of that of Earth's atmosphere. Solar radiation causes convection currents and geyser-like eruptions at the surface. Similar to the dwarf planet Pluto, Triton's surface is 55% frozen nitrogen, 35% water ice, and 20% CO_2-covered by ice. Due to the strong inclination of the rotation axis, the poles are alternately illuminated by the sun, similar to the planet Uranus.

The *Cantaloupe terrain* (Fig. 6.44) consists of craters flooded with dirty ice. However, these craters, about 30 km in size, are unlikely to have been formed by impacts, but rather by Cryovolcanism.

In 2006, at the Lagrange point L_4 four Neptune Trojans were found. At L_5 no objects have been found yet.

Jupiter and Saturn (about 10 times Earth's size) are also called gas giants, Uranus and Neptune (about 4 times Earth's size) are called ice giants.

Fig. 6.44 The cantaloupe
terrain on Triton
(NASA/Voyager 2)

6.8   Dwarf Planets and Asteroids

Since the decision of the International Astronomical Union in the summer of 2006, there
a new group of objects was defined in the solar system: the *dwarf planets* Pluto is also
counted as one of them.

Dwarf planets are defined as follows: Their masses are enough to make their shape
approximately spherical, but unlike planets, they do not have their orbits cleared of other
objects.

6.8.1   Pluto

Pluto is very similar to Triton. It was detected on February 18, 1930 by *C. W. Tombaugh*.
Its orbit is around 17° inclined against the ecliptic, and it has a large orbital eccentricity
($e = 0.24$). Its distance from the Sun therefore varies between 4.4 billion and 7.3 billion
km and its magnitude between magnitudes 13.6 and over 16. It was inside Neptune's orbit
until February 11, 1999, and has been outside since then, reentering inside Neptune's orbit
on April 5, 2231. To orbit the Sun Pluto requires 248 years. In 1978, the Pluto moon
Charon was discovered. This moon moves retrograde around Pluto (distance is only 20,000
km) and has a diameter of 1200 km. The diameter of Pluto is 2200 km (so it is smaller
than our moon). The density is 2.1 g/cm^3 and from the high reflectivity of the surface one
assumes there frozen CH$_4$ as well as NH$_3$ with surface temperatures at 50 K (aphelion)
and 60 K (perihelion). It has an atmosphere of CH$_4$ and N$_2$. The measured Bond's albedo
is 0.14 for Pluto (0.31 for Earth). The axis of rotation is inclined by 122°, so it rotates
retrograde. The mass is only 0.0021 Earth masses. Pluto rotates around its own axis in
6.3 days. The HST was used to discover two tiny Pluto moons, Nix and Hydra, in 2005

Fig. 6.45 Pluto with its
moons Charon, Nix, and Hydra
(HST telescope image)

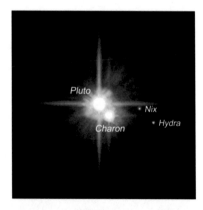

Fig. 6.46 The first time Pluto
was studied at close range by
the New Horizons probe in
2015 (NASA)

(estimated diameters between 40 and 160 km). In 2011 and 2012, respectively, another
tiny satellite of Pluto was discovered (Fig. 6.45).

Pluto has a very thin predominantly nitrogen atmosphere.

Pluto and Charon are in a Synchronous rotation, caused by the tidal effect; both objects
always show the same side to each other.

In 2006 the probe *New Horizons* was sent to the Pluto system; it arrived there in 2015.
In 2007, the probe flew by Jupiter and took data. For example, the eruption of a volcano
on *Io* was observed (Fig. 6.46).

Pluto was reached in the summer of 2015 (Figs. 6.47 and 6.48). New Horizons flew past
this dwarf planet at a distance of 12,500 km. Because of the large distance Earth–Pluto the
data could not be transmitted directly, but had to be stored temporarily on an 8 Gb memory.

Fig. 6.47 Ice-covered mountains on Pluto's surface (New Horizons/NASA)

Fig. 6.48 Eruption of a volcano in February 2007 on Jupiter's moon Io, observed by the New Horizons probe en route to Pluto (NASA)

The data transfer took place between 05/09/2015 and 25/10/2016 and thus the data transfer took more than a year.

Outside of Neptune's orbit, there is the *Kuiper belt,* which consists of many thousands of objects (asteroids, comet nuclei). Pluto is one of the brightest objects of them. Also the moon Triton captured by Neptune might have been a member of this belt.

Kuiper Belt objects will be other targets of the New Horizons mission.

6.8.2   Ceres and Other Dwarf Planets

Ceres was discovered in 1801 by *G. Piazzi* and is the largest object in the *Asteroid belt* between Mars and Jupiter. The orbital semimajor axis is 2.77 AU, the sidereal orbital

Fig. 6.49 Ceres with two bright spots (NASA/Dawn mission)

period 4.6 years. The mean equatorial diameter is 975 km. This dwarf planet was observed by the Dawn spacecraft in 2015 from a distance of 13,000 km. Ceres has about 1/4 of the total mass of all objects in the asteroid belt. The bright spots found are probably salt deposits (Fig. 6.49). Ceres has a rocky core surrounded by a mantle of minerals and water ice. On the surface is a dusty thin crust. During the radioactive decay of the aluminum isotope ^{26}Al could have formed a mantle of liquid water in the early days of the solar system, except for the outer crust which is about 10 km thick. With the IR telescope *Herschel* one could detect *Water Vapor* around Ceres, a water release of about 6 kg/s takes place at two locations on the surface. The water release is greatest when Ceres is close to the Sun in its elliptical orbit. Aliphatic *Carbon compounds* on Ceres have also been detected.

Further dwarf planets are the objects *Sedna* ($\approx$1400 km, strongly eccentric orbit, perihelion distance 76 AU, apohelion distance 900 AU, orbital period around Sun 10,787 years, strongly reddish color). Another object is *Quaoar*, diameter $\approx$1250 km, semimajor axis 43.5 AU, with the 8-m Subaru telescope in 2004 crystalline water ice was detected on its surface—an indication of internal water sources kept liquid by radioactive decay heating. *Eris* (also as Xena larger than Pluto) has a diameter of 2400 km, perihelion distance 37.8 AU, apogee distance 97.5 AU.

In August 2001, the European Southern Observatory (ESO) discovered a minor planet even larger than Ceres; the object was given the provisional designation 2001 KX76 and has a diameter between 1200 and 1400 km. The mean distance of this object from the Sun is more than 1.5 times that of Neptune. Its size corresponds to that of the Pluto moon Charon, and the object belongs to the Kuiper belt.

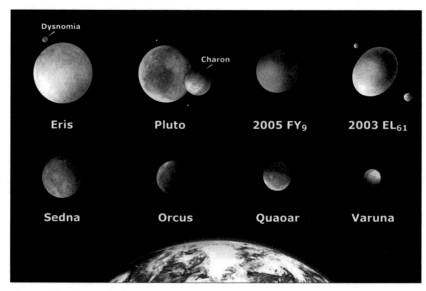

Fig. 6.50 Known dwarf planets and size comparison with Earth (NASA sketch)

In 2008 the dwarf planet *Makemake* (diameter 1500 km, it also has a moon) was found in the Kuiper belt, as well as *Haumea* which should be strongly oblate because of the rapid rotation (equator diameter 2200 km, pole diameter only about 1100 km). In the asteroid belt is likely to be located next to Ceres Object the object *Hygiea* (diameter 430 km) (Fig. 6.50).

> The dwarf planets Pluto (thin atmosphere) and Ceres (water ice) showed unexpected details.

6.8.3   Asteroids: Naming and Types

We now come to the asteroids or minor planets. Their exact number can only be estimated, but is certainly several 100,000 objects. Their arrangement in belts is striking. Some asteroids also cross the Earth's orbit and could collide with the Earth at some point.

After the discovery of *Ceres* in 1801, 300 objects were already known around 1890 with orbits between Mars and Jupiter, all smaller than 1000 km (Figs. 6.49 and 6.50).

The *Designation* of newly discovered asteroids is done today with a combination of letters. The first letter marks the month half of the discovery (thus 24 letters, A−Y, without I), and then follows after the order of the discovery the second letter A−Z (without I). Ex:

Table 6.9 The four Galilean moons of Jupiter

Name	Diameter [km]	Mass Earth moon = 1	Density [g/cm^3]	Reflection [%]
Callisto	4820	1.5	1.8	20
Ganymede	5270	2.0	1.9	40
Europa	3130	0.7	3.0	70
Io	3640	1.2	3.5	60
Earth's moon	3476	1.0	3.3	12

Table 6.10 Data of the largest minor planets

Name	Discovery	Major semimajor axis [AU]	Diam. [km]	Class
Pallas	1802	2.77	540	C
Vesta	1807	2.36	510	–
Hygeia	1849	3.14	410	C
Interamnia	1910	3.06	310	C
Davida	1903	3.18	310	C
Cybele	1861	3.43	280	C
Europe	1868	3.10	280	C
Sylvia	1866	3.48	275	C
Juno	1804	2.67	265	S
Psyche	1852	2.92	265	M

2006 AB is the second object (letter B) found in the first half of January (letter A) 2006. If these combinations are not sufficient, simply write a number (Table 6.9).

About 300,000 asteroids are known. 14 asteroids (or minor planets or planetoids) are larger than 250 km (Table 6.10). The *total mass* of all asteroids but is much smaller than that of the Moon, so they were not formed by the breakup of a large planet. The *Asteroid Belt* is located between 2.2 and 3.3 AU, and accordingly the orbital periods around the sun are between 3.3 and 6 years. 75% of all minor planets are located in this belt, nevertheless the average distance between them is more than 1 million km. It is therefore safe to cross the asteroid belt with a space probe.

> Most asteroids are arranged in belts (e.g. between Mars and Jupiter), some also cross the Earth's orbit.

Regarding asteroid group composition, asteroids are divided into three *groups*:

- C asteroids: carbon-rich, e.g. *Ceres* (now, however, defined as a dwarf planet) or *Pallas*.
- S asteroids: rocky; there are no dark carbon compounds, and thus higher albeda (reflectivity 16%, like Earth's moon); consist of silicate compounds.

- M asteroids: *Psyche* is the largest of the M-type. Consist of metals. Just one 1-km M asteroid could supply the world's consumption of industrial metals for decades.

Vesta orbits the sun in 2.4 AU and has a reflectivity of 30%, so it is very bright, and if you know exactly where this minor planet is, it can be observed with the naked eye. Its surface is covered with basalt, which means that the minor planet must have been volcanically active at one time. In any case, this is probably where Differentiation process had taken place. The minor planet had melted, and the heavier elements sank downward as a result of gravity. The group of *Eucrite meteorites* may have originated from Vesta. In 1991 the Spacecraft Galileo radioed close-up images of an asteroid to Earth for the first time: *Gaspra*. It's an S asteroid. The surface is only about 200 million years old, i.e. this object was formed from a collision 200 million years ago from a larger minor planet.

In 1917 *Hirayama* put forward the thesis that there are families of minor planets and each family was formed by explosion or collision of a larger object.

For some asteroids companions have been found (i.e. asteroid *moons*): 243 *Ida* has a small companion *(Dactyl)* (Fig. 6.51), which was found during the *Galileo mission*. The object 4179 *Toutatis* consists of two pieces with diameters of 2.5 and 1.5 km respectively. On September 29, 2004, it came within 1.5 million km of Earth. One of the goals of the Galileo Mission was also to send a probe into Jupiter's atmosphere and an orbiter around Jupiter. The probe was released on the STS-34 mission on 18.10.1989, and a VEGA course was followed. VEGA stands for "Venus-Earth Gravity Assist". So the probe was first sent to the interior of the solar system and then received gravity assists from Venus and passed Earth twice at two-year intervals. On its way to Jupiter, the probe passed through the asteroid belt, where the asteroids Gaspra and Ida were observed, and—as already mentioned—it was able to observe the impact of the fragments of comet Shoemaker

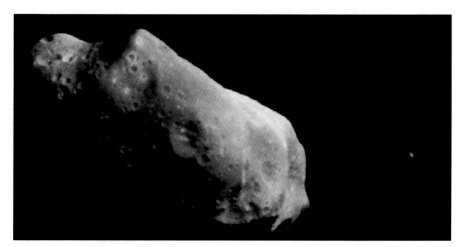

Fig. 6.51 Asteroid Ida with moon Dactyl (right) (NASA)

Levy on Jupiter. Power was supplied by RTGs (Radioisotope Thermal Generators), which provided 570 W.

On 12 February 2001 the mission *NEAR/Shoemaker* landed on the minor planet *Eros*. First the probe approached the small planet up to 26 km, then the descent and a landing with a touchdown speed of 1.5 m/s took place. A gamma ray spectrometer was used to analyse the surface for seven days. The temperatures at the Eros surface at the time of landing were around $-150\,°C$.

In 2005, a Japanese spacecraft took samples from the asteroid *Itokawa* and in 2010 a capsule carrying these samples landed on Earth.

In 2007, the spacecraft *Dawn* was launched. Dawn flew past Mars in 2009 *(Gravity Assist)* and was thereby placed in orbit toward *Vesta*, where it arrived on July 16, 2011 (Fig. 6.52). Images of Vesta's surface show one of the highest mountains in the solar system in the southern hemisphere. Moreover, this area is only $1-2$ billion years old, much younger than the northern hemisphere. This conclusion is arrived at simply by crater counting. Fewer craters means younger surface. The Dawn spacecraft used *ion propulsion*; xenon ions are accelerated by an electric field and the outgoing ion beam accelerates the probe. In this process, the ion propulsion system was in operation for almost 70% of the entire journey time. Per day about 250 g of xenon were consumed, a total of 450 kg Xenon was in the tank.

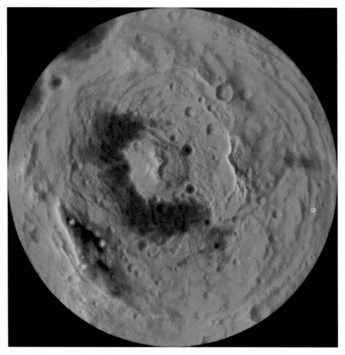

Fig. 6.52 Asteroid Vesta. False-color image shows distinct structures on the surface (NASA/Dawn)

6.8.4 Distribution of Asteroids

The asteroids in the main belt between Mars and Jupiter show peculiarities regarding their distribution, which can be explained by resonances: There are gaps or clusters of minor planets at those distances where the orbital period is in a *commensurable* (integer) ratio to Jupiter's orbit.

- *Gaps* occur at 2:1 (Hecuba gap), 3:1 (Hestia gap), 5:2, 7:3;
- *Accretions* at 1:1 (Trojan), 3:2 (Hilda) 4:3 (Thule) (4:3 means, for example, that the duration of 4 minor planet orbits around the Sun is equal to the duration of 3 Jupiter orbits around the Sun).

The group of *Trojan asteroids* has the same distance as Jupiter from the Sun (5.2 AU). They are located at Lagrange points L4 and L5, i.e. they always form an equilateral triangle with Jupiter and the Sun. In 1990 a similar family of minor planets was found near Mars. Furthermore there are asteroids with special orbits. *Hidalgo* has a large semimajor axis of 5.9 AU and an eccentricity of 0.66. Its farthest point from the Sun (aphelion) lies outside Saturn's orbit. *Chiron* has a large orbital semi-axis of 13.7 AU, and its perihelion lies outside Saturn's orbit, its aphelion outside Uranus' orbit.

Especially interesting for us on earth are *earth crossing asteroids*. About 200 objects are known, with the total number of all objects larger than 1 km estimated at 2000. They can either collide with the terrestrial planets or be accelerated so strongly that they are ejected from the inner solar system. Such an event occurs every 100 million years. One-third of all Earth-orbiting minor planets are likely to crash into Earth, and observational programs are therefore in place to detect all such objects, preferably combined with early warning. The asteroid *Icarus* dips into Mercury's orbit (its perihelion is at 0.19 AU). The objects coming very close to the Earth, *Apollo objects*, are listed in Table 6.11.

Since the minor planets are irregularly shaped, there is a change in brightness as a result of their rotation.

The asteroid 2002 AA discovered in $2002AA_{29}$ has a diameter of about 50−110 m. Its orbit is very similar to Earth's orbit, and it crosses it at aphelion. Also the asteroid discovered in 1986 Cruithne is in a 1:1 resonance with the Earth's orbit. One also speaks of coorbital objects.

Table 6.11 Some Apollo objects

Object	Minimum distance to Earth
Eros	0.15 AU
Apollo	0.07
Adonis	0.03
Hermes	0.004 = 2 times the distance to the moon

Table 6.12 Dates of known NEOs. *Date* means the time of closest approach to Earth, where distance is the distance that will occur at that time

Designation	Date	Distance in AU	diam. m	Circulation a
1998 WT24	16.12.2001	0.0124	1250	0.61
4660 Nereus	22.01.2002	0.0290	950	1.81
1998 FH12	27.06.2003	0.0495	680	1.14
1994 PM	16.08.2003	0.0250	1200	1.80
1998 FG2	21.10.2003	0.00360	220	1.48
1996 GT	12.11.2003	0.0479	860	2.10
1998 SF 36	25.06.2004	0.0137	750	1.53
4179 Toutatis	29.09.2004	0.0104	2400–4600	1.10
1992 UY4	08.08.2005	0.0404	110	4.33
4450 Pan	19.02.2008	0.0408	1570	3.00
1999 AQ10	18.02.2009	0.0118	360	0.91
1994 CC	10.06.2009	0.0169	1100	2.09
1998 FW4	27.09.2013	0.0075	680	3.95
1998 WT24	11.12.2015	0.0277	1250	0.61
4660 Nereus	11.12.2021	0.0263	950	1.82
7482 1994 PC1	18.01.2022	0.0132	1900	1.56
7335 1989 JA	27.05.2022	0.0269	2000	2.35

6.8.5   NEOs

At the acronym NEO (Table 6.12) means *Near Earth Objects,* i.e. objects which are close to the Earth or which could come close to the Earth (associated with this is also the designation *PHAs,* Possible Hazardous Asteroids). With the so-called *Torino Scale* a kind of Richter scale has been established, which is supposed to give an estimate of what an impact of a certain object on Earth could cause. When a new NEO asteroid or comet is discovered, the orbit calculations are still mostly uncertain, but most calculations show that a collision with Earth is unlikely. The levels range from 0 to 10, with 0 meaning that the probability of a collision with Earth is extremely low or that the object will burn up in the atmosphere. 10 means a collision is certain and it will have catastrophic global climate impacts. The color scale means

- White: No practical implications.
- Green: One should monitor the orbit of these objects.
- Yellow: One should be careful; objects should be monitored more closely to better calculate their orbits.
- Orange: Threatening objects; one should immediately calculate their paths more accurately.
- Red: It is best to provide yourself with enough wine or beer. . . .

Fig. 6.53 Image taken when the spacecraft bounced on the asteroid Ryugu (Photo: JAXA)

Important: The assessment of an object according to the Torino scale can change with time, when more precise observations are available.

The object 4660 *Nereus* comes very close to us on 2060–02–14: The closest distance is only 0.008 AU. The minor planet 433 *Eros* (Extension 33 km× 13 km) passed us on 31.01.2012 in only 0.179 AU.

In late September 2018, a Japanese space probe (MASCOT) landed on asteroid *Ryugu*. Due to the asteroid's low gravity the spacecraft explored its surface by hopping—equipped with a special suspension (Fig. 6.53). The asteroid is only about 1 km in size. This asteroid can come as close as 90,000 km to us and is one of the potentially dangerous asteroids, as it could one day collide with Earth.

Table 6.13 Large known Impact crater on earth. Diam denotes the diameter in km

Name	Location	Geograph. latitude	Geograph. longitude	Age 10^6 a	Diam.
Vredefort	South Africa	27.0 S	27.5 E	2023	300
Sudbury	Canada	46.6 N	81.2 W	1850	250
Chicxulub	Mexico	21.3 N	89.5 W	65	170
Manicougan	Canada	51.4 N	68.7 W	214	100
Popigai	Russia	71.7 N	111.7 E	35	100
Chesapeake B.	USA	37.3 N	76.0 W	36	90
Acraman	Australia	32.0 S	135.5 E	590	90
Puchez-Ktunki	Russia	57.1 N	43.6 E	175	80
Morokweng	South Africa	26.5 S	23.5 E	145	70
Kara	Russia	69.2 N	65.0 E	73	65
Beaverhead	USA	44.6 N	113.0 W	600	60
Tookoonooka	Australia	271. S	142.8 E	128	55
Charlevoix	Canada	47.5 N	70.3 W	357	54
Kara-Kul	Tajikistan	39.0 N	73.5 E	5	52
Siljan	Sweden	61.0 N	14.9 E	368	52

In the past, there have been repeated impacts of such objects on Earth (associated with the extinction of many species). Traces of large impact craters are also found on Earth, although here, of course, erosion is rapidly blurring them (Table 6.13).

> The *Chicxulub crater* is the remnant of an asteroid impact 65 million years ago. This event is thought to be responsible for the extinction of the dinosaurs.

6.9 Comets

Comets have always been considered special celestial phenomena, especially since they appear suddenly and do not move along the ecliptic in the sky. They were usually considered harbingers of a near disaster.

6.9.1 Comets: Basic Properties

Comet sightings is already found in records from ancient times. *Halley* recognized that the comet sighted in 1531, 1607, and 1682 was always the same one, reaching its closest point to the sun every 76 years. The first sighting dates back to 239 B.C., and the last sighting was in 1986. In 2061, Halley's comet will again be seen in the sky.

Table 6.14 Data of some known short-period comets. a = large semi-axis

No.	Name	Circulation-period	Perihel-passage	Perihelion distance	a [AU]
1P	Halley	76.1	09.02.1986	0.587	17.94
2P	Encke	3.3	28.12.2003	0.340	2.21
6P	d'Arrest	6.51	01.08.2008	1.346	3.49
9P	Temple 1	5.51	07.07.2005	1.500	3.12
21P	Giacobini-Zinner	6.52	21.11.1998	0.996	3.52
73P	Schwassmann-Wachmann 3	5.36	02.06.2006	0.937	3.06
75P	Kohoutek	6.24	28.12.1973	1.571	3.4
	Hale-Bopp	4000	31.03.1997	0.914	250
	Hyakutake	40,000	01.05.1996	0.230	$\approx$1165

Fig. 6.54 Comet Hyakutake

There are relatively many comets whose aphelia lie near the orbit of Jupiter; they are referred to as the *Jupiter family*. Jupiter's gravitational effects have turned long-period comets into short-period ones (Table 6.14 and Fig. 6.54).

Comets consist of:

- *Nucleus:* 1–50 km in diameter. *Whipple* established the model of a dirty snowball around 1950: Water vapor and other volatiles escape from the nucleus to form the characteristic comet tail. This happens once a comet is within the orbit of Mars. The nucleus of Halley's comet is 8 km by 12 km. The evaporation is not uniform, and it comes to unexpected eruptions.

- *Coma:* extension about 10^5 km.
- *Dust tail* and *ion tail:* extension many million km.

The atmosphere of a comet consists mainly of H_2O and CO_2. UV radiation from the Sun breaks up the water molecules and huge H clouds form around the comet. Comet tails always point away from the Sun. The comet's dust tail is directed away from the Sun by the light pressure of sunlight, but the ion tail (usually bluish) is directed directly by the solar wind. The ion tail is long and narrow, the dust tail is broad and diffuse and often curved; because of its lower velocity, the dust lags behind in the comet's orbital motion (the ejected particles travel in Keplerian orbits around the Sun, particles farther away from the comet slower than particles near the comet, hence the curvature).

> The actual comet nucleus is only a few tens of km in size, but the tail can extend over several million km.

Comets can break up: in 1976, comet *West* broke into four pieces. The comet *Shoemaker Levy* broke up 1993 into 20 pieces, which impacte onto Jupiter in 1994 with an energy release of 100 million megatons of TNT. This event could already be followed on Earth with amateur telescopes. The energy released was equal to that of the impact on Earth 65 million years ago.

The naming of comets is confusing: first with a year with a small letter in the order of discovery, then with a year and Roman numeral in the order of perihelion passage. Short-period comets still get a P: 1810 II-P/Halley. The discoverer is at the very end.

The brightness of a comet depends on the distance comet-sun (r) and the distance comet-Earth (Δ):

$$h = \frac{H}{r^2 \Delta^2} \tag{6.42}$$

h—Intensity of observed brightness, H—absolute brightness at $r = \Delta = 1$.

The *GIOTTO mission* was launched on July 2, 1985, with the goal of studying Comet Halley. The closest approach to Halley (Fig. 6.55) occurred on March 13, 1986, at a distance of only 600 km from the nucleus. The probe was equipped with a dust shield. 14 s before the closest approach, the probe was hit by a large dust particle from the comet, and during 32 min there were disturbances in the data recordings.

Comets Encke and d'Arrest were to be visited as part of the CONTOUR mission, but the probe exploded in 2002 just six weeks after launch.

Fig. 6.55 Halley comet
nucleus (Image ESA/GIOTTO)

Fig. 6.56 Comet Temple 1
with impact site marked by
thick arrow. The white line on
the lower right marks 1 km
(Photo: NASA/UM M. F.
A'Hearn)

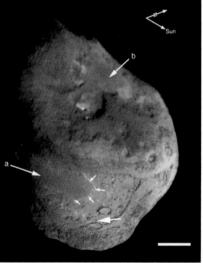

In July 2005, the probe *Deep Impact* bombarded comet *Temple 1* (Fig. 6.56) with a piece weighing 370 kg. The impact created a crater, and the ejected material was analyzed. Organic material was also found inside the comet, and the first definite evidence of water at the comet surface. The target of the 2004 launched ESA's *ROSETTA mission* was comet *67 67P/Churyumov-Gerasimenko*. After flybys of two asteroids (*Steins,* 2008; *Lutetia* 2010),

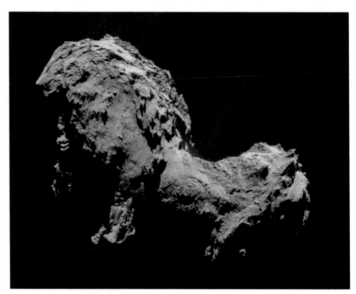

Fig. 6.57 Close-up view of comet 67P/Churyumov-Gerasimenko, September 19, 2014 (Rosetta/ESA)

the Lander *Philae* landed on its surface (Nov.12, 2014). Detailed images of a comet's surface that had never been obtained before (Fig. 6.57) were send to earth.

6.9.2   Kuiper Belt and Oort's Cloud

In 1943. *K. Edgeworth* was the first to suggest the existence of a belt consisting of numerous comets outside Pluto's orbit. *G. Kuiper* then studied it in more detail in 1951, and in 1980, calculations were made to prove that this belt of objects was a source of short-period objects. Then, in 1992, the first member of the Kuiper belt was found: *1992 QB1(D. Jewitt* and *J. Luu).* The object had an apparent magnitude of $22.^m5$ and was discovered with a 2.2-m telescope on Mauna Kea. The dwarf planet Pluto is located within this belt. 1992 QB1 has a diameter of 283 km, and the orbit has a semi-axis of 44.0 AU. In addition, there are also so-called plutinos (this term should be abolished again), which are objects with orbits similar to Pluto's. At present more than 100 objects are known in the Kuiper belt.

One can divide the members of the Kuiper belt into:

- Classical objects: 2/3 of all objects observed so far, $42 < a < 47$ AU, no resonances.
- Resonance objects: Mostly a 3:2 resonance with Neptune; they orbit the Sun twice in three Neptune orbits (they are also called plutinos, because Pluto also has a similar orbit).

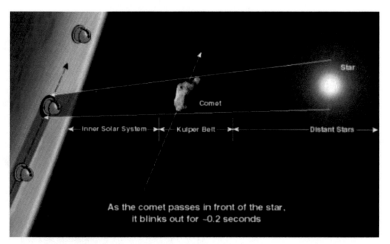

Fig. 6.58 The TAOS project for automatic detection of Kuiper belt objects (Credit: TAOS)

- *Scattered Kuiper Belt Objects (KBO):* These are likely to be on extreme orbits, but only a few objects have been discovered so far, including TL66 in 1996; it has an aphelion distance of 84 AU and a perihelion distance of 30 AU, and its orbital period is 800 years.

The total number of objects larger than 100 km in the Kuiper belt is estimated to be 100,000. The TAOS project (Fig. 6.58) is trying to find these objects. Three small robotic telescopes have been set up on Taiwan along a 7 km line, automatically targeting the same object. If there is an occultation of a star by a KBO, then this is registered with a time shift (which is, however, very small) on all three telescopes.

Between Jupiter and Neptune nine objects are known (among them 2060 *Chiron* and 5145 *Pholus*). Their orbits are unstable, i.e. they are outliers from the Kuiper belt due to gravitational interactions.

In 1950 *J. Oort* postulated the existence of a giant cloud of comets enveloping our solar system. This is based on three observational facts:

- No comet has been observed with an orbit showing that it comes from interstellar space. Thus, there are no hyperbolic velocities (In this context, hyperbolic velocity is the velocity an object would have to reach to leave the solar system).
- The aphelia of many long-period comets appear to be at a distance of 50,000 AU.
- There is no main direction of comets.

From statistics, it follows that in the Oort cloud about 10^{12} Comets could exist. However, because of the great distances involved, these objects cannot be seen directly. The total mass may be more than one Jupiter mass.

Fig. 6.59 Comet McNaught, also brightly visible for a short time in our evening and morning sky (early January 2007). Around the time of the perihelion passage it appeared on images of the solar satellite SOHO. Below the bright comet the planet Mercury is visible. The Sun itself is covered (its size is indicated by a white circular ring) (Photo: ESA/NASA SOHO Mission)

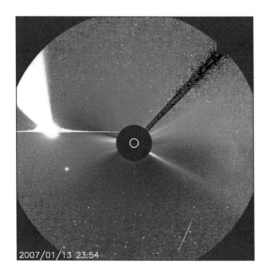

6.9.3   Sungrazer

These are comets that come very close to the Sun (Fig. 6.59). Some move through the corona of the Sun. The SOHO probe (launched in 1995), built to study the Sun, has found about 1000 such objects so far. Their total number is likely to be well over 10^5. They are often torn apart by the strong tidal forces of the Sun. Most of them are only small fragments of a few meters in diameter. The Kreutz Group comes from the disintegration of a larger comet. A large comet passed by the Sun in 1860 at a distance of only 200,000 km. The last bright comet in this group was White-Ortiz-Bolelli in 1970. A very bright comet apparition of objects in this group is thought to occur on average every 20 years.

6.10   Meteoroids

We now come to the smallest objects in the solar system. Their study is important because the matter and structure have changed little since the formation of the solar system.

6.10.1 Nomenclature

We make the following distinctions:

- Meteor: light appearance when a body enters the Earth's atmosphere.
  One can calculate the mass M of a meteor as a function of the geocentric velocity v and the brightness m in the zenith:

$$\log M = 3.6 - 0.4\,m - 2.5\log v \qquad (6.43)$$

- Meteoroid: matter orbiting the sun, in interplanetary space; goes up to micrometeoroids.
- Meteorite: meteoroid that reached the earth's surface.
- Shooting stars: small meteors.
- Fireballs: large meteors.

The parabolic heliocentric velocity of meteoroids at the location of the Earth is 42 km/s. The orbital velocity of the Earth around the Sun is 30 km/s. Therefore, the relative velocity reaches up to 72 km/s. Meteors can therefore be observed in the morning, when the relative velocity is greatest, or in the evening, when it is small. The *meteoroids* are not attracted by the earth, but there is a collision with the earth. Due to the high velocity, melting and burning up occurs as the meteor enters the Earth's atmosphere. In order to see a meteor, it must be within 200 km of the observer. On the whole earth about 25 million meteors fall per day. The typical mass is only 1 g. A *fireball* is already about the size of a golf ball. Per day 100 t of meteor material fall on the earth's surface. The strong deceleration and the flashing occurs at an altitude of 140–100 km, and they extinguish between 90 and 20 km. An ionized air tube is formed which reflects electromagnetic waves, and thus meteors can be observed by radar echoes even during the day and when it is cloudy.

6.10.2 Classification

A subdivision according to origin and orbital shape is made:

- Planetary meteorites: 50%, elliptical orbits of short orbital period, fragments from the asteroid belt.
- Meteorites with near-parabolic orbits: 30%, unknown origin, but certainly from the solar system.
- Cometary meteorites: 20%, from the dissolution of comets. These dissolution products spread along the orbit, and when Earth's orbit and the original comet's orbit intersect, then meteor showers can be observed *meteor showers* (about 50 known). The best known meteor streams are the *Leonids* (maximum 16.11, rest of comet 1866 I), *Perseids* (11.8, remnant of comet Temple-Tuttle) and the *May Aquarids* (5.5, remnant of comet Halley). They are named after the constellation where the point *(radiant)* from which they seem to come from is located. The effect of meteor streams coming from a point in the sky is the same as dense snowflakes hitting the windshield of a moving car.

Larger chunks melt only up to 1 mm and then fall to earth = meteorites. There are 700 finds, whose fall down was observed, and again as many, whose fall down could not be observed. One divides the meteorites into:

- Iron meteorites, where there are the metal meteorites (pure Fe and Ni) and the sulfide meteorites: FeS...

- Stony meteorites: SiO_2, MgO, FeO ....
- Tektite: glass meteorites, mainly SiO_2. Often roundish or circular in shape. They are found only in certain areas (Moldavite in Bohemia).

Other classification:

- Undifferentiated meteorites (chondrites).. They contain about 1 mm large silicate spheres (chondrules); there are also the carbonaceous chondrites.
- Differentiated meteorites: achondrites, metal-rich meteorites.

With masses, which are larger than 10 t and fall to earth, an impact crater is formed: One knows 13 craters with found meteorite. Very well known is the *Arizona crater:* diameter 1 200 m, 170 m deep, impact 60,000 years ago; diameter of Fe-Ni meteorite may have been 50 m, mass 150,000 t and explosion equivalent to 2.5 megatons of TNT (150 Hiroshima bombs); *Nördlinger Ries:* Diameter 20 km, impact 14 million years ago, the meteorite had a diameter of 1 km. In 1908, in Siberia (Tunguska) a bright fireball was observed and within a radius of 70 km the forest was destroyed. However, no meteoritic material was found. Most likely, this was the impact of a comet, which vaporized.

In the case of very small particles (micrometeorites) the air resistance is so high that the particles do not burn up but float to the ground and form deposits.

The iron meteorites have a typical arrangement of their Fe-Ni crystals: If you etch them, you get the *Widmanstetter's etching patterns.*

On February 15, 2013, in *Chelyabinsk*, Russia, more than 1200 km east of Moscow, a spectacular meteorite was seen (Fig. 6.60), which coincidentally occurred a few hours before the approach of an NEO asteroid on the same day. The shock wave generated by

Fig. 6.60 On February 15, 2013, a spectacular meteorite fall occurred in Russia, which was photographed and filmed by many people

the impact of the fragments caused glass fragments from shattering window panes and more than 1000 people were injured. The object, which was about 20 m tall and weighed 11,000 t, exploded in the atmosphere at an altitude of about 23 km, and an energy of 400 kt TNT was released on impact. For comparison, the explosion of the Hiroshima atomic bomb was equivalent to that of 15 kt TNT.

6.10.3 Interplanetary Matter

Still Smaller particles ($<10\,\mu$m, $<10^{-8}$ kg) are summarized as interplanetary dust. In spring, shortly after sunset in the west, or in autumn, shortly before sunrise in the east, because the ecliptic is steeply upward in our latitudes, a faint cone of light can be seen along the ecliptic, the *Zodiacal light*. This is caused by the reflection and scattering of sunlight by 10–80 μm large particles. Opposite the sun one observes the Gegenschein. This is caused by increased backward scattering. The F-Corona is the continuation of the sun's corona (the outermost layer of the sun's atmosphere), where one can see a Fraunhoferspectrum of the sun—it is the link between zodiacal light and Corona. Due to the Poynting-Robertson effect (particles crash into the Sun in spiral orbits), there is a constant loss of mass of interplanetary matter (10^5 g/s). However, this is constantly compensated by comets and asteroids.

The interplanetary gas either comes from interstellar matter itself or is formed by diffusion from planetary atmospheres and comets. The other fraction comes from the *solar wind*. The heating of the corona accelerates it. The magnetic field is carried along, the field lines go radially outwards and are curved due to the rotation of the sun. That is why there are sectors with different field directions (cf. lawn sprinkler). The solar wind consists of currents with high velocity ($v > 650$ km/s) and low velocity ($v < 350$ km/s). Near the earth one measures a proton density of 9 ± 6/cm^3 and a velocity of $v < 470$ km/s. The solar wind changes with solar activity, and the influence on geomagnetic disturbances depends on the velocity. The expansion continues until the kinetic pressure is equal to the interstellar total pressure . This defines the *Heliopause* (up to 80–120 AU).

On board the *Stardust mission* (launched in 1999) there were 130 Aerogel detectors, 2 cm $\times$ 4 cm in size and 3 cm thick, in which decelerated interplanetary particles were trapped. Dust particles from the environment of the small planet *Anne Frank* were captured and brought back to Earth in a capsule (2006).

6.11 Origin of the Solar System

From the study of meteorites, moon rocks (which were brought to Earth by the Apollo astronauts or by unmanned Soviet probes), interplanetary dust (which was studied, for example, by the Giotto mission), the age of the solar system can be estimated: 4.55 billion years.

6.11.1 Extrasolar Planetary Systems

Today, the formation of other planetary systems is directly observed and conclusions can be drawn about the formation of our planetary system. Especially important in this context are the so-called *protoplanetary nebuale* (PPN's). With the Hubble Space Telescope, the VLT of the ESO and IR telescopes in space it is also possible to observe very young stars. Many young stars are surrounded by a gas shell, which is only a few million years old, indicating that planet formation is in progress.

An example of a protoplanetary disk is the star *Beta Pictoris,* whose disk can be observed by new observation techniques (adaptive optics, occultation of the central star) up to a distance of 25 AU from the star. The star itself is 63 light-years away from us and has a diameter of 1.4 solar radii.

In Fig. 6.61, the dimmed star or disk is seen above in visible light. The observer sees almost *edge on* (to the edge), hence the spindle-shaped form. The disk consists of small dust particles (ice, silicate particles) and glows by reflection of the starlight. Since there appears to be no matter in the area around the center, it is assumed that it was captured there by one or more planets. The lower false color image shows more structure. The luminous inner edge is slightly tilted with respect to the plane of the outer disk. This could be from gravitational influences by a planet.

More detailed information on exoplanets can be found in Chap. 16 about astrobiology.

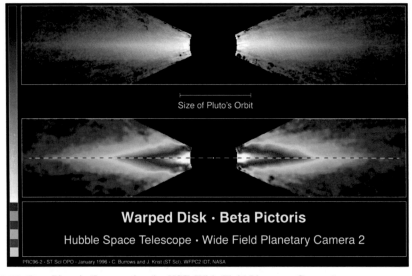

Fig. 6.61 Beta Pictoris (Image taken by HST, Wide Field Planetary Camera)

Table 6.15 Titius-Bode law: comparison between the calculated and the actual distances a in AU from the Sun

Planet	n	$a_{calculated}$	a_{actual}
Mercury	$-\infty$	0.4	0.39
Venus	0	0.7	0.72
Earth	1	1.0	1.0
Mars	2	1.6	1.52
Minor planets	3	2.8	2.9
Jupiter	4	5.2	5.20
Saturn	5	10.0	9.55
Uranus	6	19.6	19.20
Neptune	–	—	30.09
Pluto	7	38.8	39.5

6.11.2 Theories of Formation

Any Theory of the formation of our solar system must following explain special features:

1. All objects in one plane.
2. Orbits nearly circular; orbits and rotation nearly all in the same sense (prograde).
3. Law of distance (Titius-Bode law, n See Table 6.15):

$$a = 0.4 + 0.3 \times 2^n \tag{6.44}$$

4. Sun has 99.87% of the total mass, but only 0.54% of the total angular momentum of the entire solar system.
5. Nature of planets: (a) inner planets: high density, metals, rocks, slow rotation, few (no) moons; (b) outer planets: low density, composition similar to Sun, fast rotation, many moons.
6. Rotational axes of planets (angular momentum vectors) and satellite systems approximately parallel to the total angular momentum vector (perpendicular to the invariant plane).

There are the following theories:

- Old theories: already in the 16./17. century pioneers were *Copernicus* (heliocentric world system), *Galileo* (first telescope observations) and *Kepler* (laws of planetary motion). *René Descartes*In 1644, was the author of the first paper suggesting that the original matter had been in rotation and that vortices had formed as a result; the sun was formed from the large central vortex.
- Collision *(Chamberlain-Jeans):* Close passage of a star past the sun; tidal forces pull out matter that condensed to form planets. Problem: The distances between stars are

extremely large, so there is a very low probability of such a process. If this theory were correct, there would possibly exist only two solar systems in our galaxy! This is refuted by the observation of more than 5000 extrasolar planetary systems.

- Accretion Theory: The sun passes through an interstellar cloud, gathering matter as it goes.
- Primordial Nebulae: *(Kant 1755, Nebular hypothesis):* The protosun is formed as a result of gravitational collapse of an interstellar gas cloud; comes very close to modern ideas (Sect. 11.1 star formation).
- Modern theories: Gravitational collapse of a cloud; gravitational forces outweigh the gas pressure of the cloud's particles, this is called *Jeans instability*; rapid rotation at the beginning (conservation of angular momentum → As the cloud contracts, it rotates faster and faster), causing it to flatten. The dust condenses into what are called Planetesimals. Magnetic fields frozen in the plasma transfer the angular momentum of the protosun to the rotating envelope. The composition of the present-day planets then follows from the condensation sequence of various substances. Methane condenses at 100 K, water at 273 K, silicates at 1000 K, etc. The Sun is hot, and therefore no icy moons could form near the Sun, or large planets massive enough to hold the light elements H and He in their atmospheres.

6.11.3 Protoplanetary Nebula

The composition of the protoplanetary nebula, from which the Sun and planets formed, corresponded to the general cosmic element abundance. 99% consisted of gas (H, He), 1% of solid dust particles, which were about 0.1 μm in size. These were formed in the atmospheres of *Red Giant Stars,* were then physicochemically altered by the bombardment of cosmic rays. Their surface was covered with ice molecules, and due to low temperatures this ice is in a state of crystallization unknown on Earth. Brownian motion causes random collisions in the gas cloud, the particles become larger, the collisions become less frequent, the collision process is thus stopped. Centimetre-sized particles thus decouple from the general gas field of the nebula, fall towards the equatorial plane of the rotating nebula and describe Kepler orbits around the already formed primordial sun. They are decelerated by the surrounding gas and form meter-sized clumps. The orbits of these clumps do not remain stable, and collisions give rise to the *Planetesimals,* clumps several kilometers in size.

The elements hydrogen and helium were formed in the Big Bang and account for almost all observable matter in the universe. It is important to note for further consideration that these elements remain gaseous even at near absolute zero temperatures. 98% of the solar primordial nebula was therefore gaseous. Elements such as carbon, nitrogen, oxygen condense (e.g. H_2O above 110 K). Therefore it is expected that ice was formed in the cooler outer regions of the solar system. All other elements (less than 0.3% of mass of primeval nebula) react with oxygen and form molecules, silicates and so on. These are the Earth-like planets: They consist essentially of a metallic core surrounded by a silicate mantle. The Earth could therefore only form from a protoplanetary nebula of more than 300 present-day Earth masses (0.3% corresponds to about 1/300), as did Venus. For Mercury it was 15 Earth masses, for Mars 30 Earth masses and for the asteroids 0.15 Earth masses. The mass of the nebula to form Jupiter was 1000 Earth masses and for Saturn 500 Earth masses. Altogether then a value for the mass of the nebula, which was necessary for the formation of the planets, results: about 3000 earth masses (=1% of the sun mass). There are also other theories that assume a much more massive primordial nebula.

6.12   Further Literature

We give a small selection of recommended further literature.
The Solar System, Encrenaz, Th., Bibring, J.-P., Blanc, M., Springer, Berlin, 2004
Encyclopedia of the Solar System, McFadden, L.A., Weissman, P., Johnson, T., Academic Press, 2nd ed. 2007
Fundamental Planetary Science, J.J. Lissauer, Cambridge Univ. Press, 2019
Planetary Sciences, I.d.Pater, J.J. Lissauer, Cambridge Univ. Press, 2015
Asteroids, T.H. Burbine, Cambridge Univ. Press, 2016
The Atlas of Mars, K.S. Coles, K.L. Tanaka et al, Cambridge Univ. Press, 2019
Comets and their Origin, U. Meierhenrich, Wiley-VCH, 2014.
The Formation of the Solar System, M. Woolfson, Icp, 2014
Apollo 50, O. Sentinel, Pediment Publ, 2019

Tasks

6.1 Suppose a wooden part of a building still had 80% of the original proportion of ^{14}C. How old is the wood (half-life $t_h = 5730$) years?

Solution
$t = t_h \log_2(0.8) = 5730 \times \log_2(0.8) = -1845. \log_2 a = \log_{10} a / \log_{10} 2$. Therefore, the wood was cut 1845 years ago.

6.2 Why is the interior of the moon cold and the interior of the earth hot?

Solution
Compare (a) the surface area/volume ratio of the two bodies, (b) the masses of the two bodies.

6.3 GPS measurements show that the Atlantic Ocean floor is spreading by about 2.5 cm per year. When did this spreading begin, that is, when was the Atlantic Ocean formed?

Solution

If we take 6000 km as the mean east-west extent of the Atlantic Ocean, then:

$$t = d/v = 6{,}000{,}000/0.025 = 240{,}000{,}000 \text{ Years}$$

6.4 Consider a proton with velocity $v = 10^8$ m/s and a magnetic field with $B = 10^{-4}$ T. What is the radius of gyration?

Solution

Substitution gives $r = 10$ km.

6.5 Compare the solar radiation on Venus with that on Earth!

Solution

$$\frac{F_{\text{Venus}}}{F_{\text{Earth}}} = \frac{E_{\text{Sun}}}{4\pi d^2_{\text{Sun-Venus}}} \Big/ \frac{E_{\text{Sun}}}{4\pi d^2_{\text{Sun-Earth}}} = \frac{1^2}{0.72^2} = 1.9$$

6.6 Why can volcanoes get much taller on Mars than on Earth?

Solution

(a) On Earth plate motions, location of hot spots shifts over time, cf. Hawaii island chain,
(b) On Mars lower gravity, higher structures collapse later under the influence of gravity.

6.7 The escape velocities are approximately the same for our moon and Titan. Why does Titan have an atmosphere unlike our moon?

Solution

Compare surface temperatures and geological activities of both celestial bodies.

6.8 The irradiance from the Sun at the Earth's location is 1 365 W/m². Calculate the corresponding insolation on Pluto.

Solution

Assume by how many times Pluto is farther from the Sun than Earth.

6.9 Calculate the kinetic energy of a 1000 kg meteorite impacting at 30 km/s. Note: 1 kt TNT$= 4\times10^{12}$ J.

Solution

$E_{\text{kin}} = 1/2mv^2 = 1/2(1000)(30{,}000)^2 = 4.5 \times 10^{11}$ J thus: 0.5 kt TNT.

6.10 You observe hydrogen lines in the spectrum of Jupiter. How can you tell if these are from Earth's atmosphere or actually from Jupiter?

Solution
The strength of telluric lines changes with Jupiter's altitude in the sky.

The Sun

7

Our Sun is the only star whose surface we can study in detail due to its proximity. Through space missions are also *in situ* measurements possible. The importance of our sun was already recognized by the ancient civilized peoples. The names of the sun deity (sun cult) are Aton, Ra, Re, Horus (Egyptians), Huitzilopochtli (Aztecs), Sunne, Sol (Germanic peoples), Apollo, Helios (Greeks), Surya (Hindu), Lugh (god of light, Celts), Inti (Inca), Mithra (Persians), Apollo, Sol (Romans), Svarožić (Slavs).

Solar physics has established as an own branch of astrophysics, and here has arisen a line of research topics of great practical relevance: The study of solar-terrestrial relations.[1]

7.1 Basic Data and Coordinates

What are the most important state variables that characterize our Sun, and how can these variables be determined? How can phenomena of the solar atmosphere be defined in a coordinate system?

---

[1] Formerly solar-terrestrial relations, now Space Weather.

© Springer-Verlag GmbH Germany, part of Springer Nature 2023
A. Hanslmeier, *Introduction to Astronomy and Astrophysics*,
https://doi.org/10.1007/978-3-662-64637-3_7

7.1.1 Basic Data

The most important state variables of the Sun are:

Mass	333,000 Earth masses = 1.98×10^{30} kg
Radius	109 Earth radii = 6959×10^8 m
Luminosity	3.826×10^{26} W = 3.826×10^{33} erg/s
Apparent brightness	$m_V = -26\overset{\text{m}}{.}87$
Effective temperature	5777 K
Rotation period	$\approx$25.38 days

In astrophysics, one often gives the data of objects in units of the corresponding value for the Sun, e.g., the mass of a star is 1.4 $M_\odot$.

From its values the Sun is an average star, many stars are even smaller than the Sun and less massive.

7.1.2 Coordinates

A reliable and save method to observe the Sun is the *projection* of its image onto a screen behind the eyepiece of a telescope. To determine the position of phenomena on the surface of the Sun, one uses the coordinates sketched in Fig. 7.1. The *Position angle P* is taken from an astronomical yearbook, in Fig. 7.1 one sees the definition of the sign (+ or −). The *heliographic latitude B_0* of *center of the sun* is also taken from a Yearbook:

- $B_0 > 0$ one looks rather at the northern hemisphere of the sun,
- $B_0 < 0$ one looks rather at the southern hemisphere.

The sun rotates differentially, i.e. faster at the equator than at the poles. Therefore one has introduced the *Carrington rotation* that refers to the solar rotation at 16 degrees of heliographic latitude. . Moreover, solar rotations have been numbered by since November 9, 1853, and the mean rotation period is 27.2753 days (synodic value, sidereal value is 25.38 days). Carrington rotation number 2238 began on November 28, 2020.

Determination of *heliographic coordinates:* One first calculates the latitude B_0 of the sun's center and selects the appropriate grid of degrees onto which the sun's image is projected. Then one sets the east-west direction—one turns the measuring template until the sun's edge or a spot runs exactly on the E-W line. Then the position angle is set. Finally, the coordinates can be read directly.

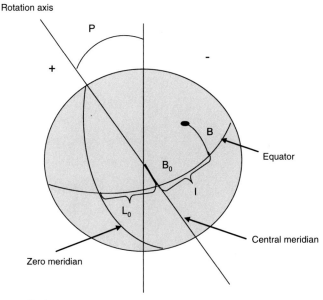

Fig. 7.1 Solar coordinates

7.1.3   Distance

The attempt to determine the distance Earth-Sun is old. *Aristarchus of Samos* (320–250 B.C.) first established the heliocentric world system and tried to determine the distance of the sun (Chap. 3).

Another method is the application of the *third Kepler law* (Sect. 3.3.5). Take a planet which comes closer to the earth than the sun (Venus or Mars), determine its distance in kilometers by means of parallax measurement, and then all distances in the solar system result, since according to Kepler's third law for any two planets 1 and 2 with the orbital semi-axes a_1, a_2 as well as the orbital periods P_1, P_2 are valid:

$$\frac{a_1^3}{P_1^2} = \frac{a_2^3}{P_2^2} \tag{7.1}$$

The distance determination with *Parallax measurement* is simple: sight an object from two locations as far apart as possible whose distance is known. Then this object appears at different angles relative to objects further away, and from the *Parallax* π one has according to

$$\sin \pi = \frac{a}{r} \tag{7.2}$$

Fig. 7.2 Determination of the
Venus parallax during a *Venus
transit (transit)* in front of the
the solar disk. From two distant
observing sites on Earth, the
planet can be seen passing in
front of the solar disk on path 1
and path 2, respectively. The
last observable Venus transit
was 2004, the next will be
2117

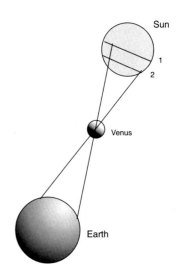

the distance r, where a is the base length. Even simpler is the determination of the distance
from the measurement of *signal propagation times*. One sends a radar beam to Venus and
measures when it arrives again after reflection at the surface of Venus. The method is
however somewhat uncertain, since one does not know the layer, in which the signal is
reflected, sufficiently exactly.

A direct determination of the solar parallax is difficult, because during the day, when the
sun is above the horizon, no reference stars are visible. Figure 7.2 shows the determination
of the Venus parallax.

7.1.4   Solar Mass

As shown in Sect. 3.3.5, the solar mass can be calculated from the exact form of the 3.
Kepler's Law:

$$\frac{a^3}{P^2} = \frac{G}{4\pi^2}\,(M_1 + M_2) \tag{7.3}$$

For example, let's insert: M_1 Mass of the Sun, M_2 Earth mass, P orbital period Earth
around the Sun (1 year), a known Earth-Sun distance. Then we immediately get the mass
of the sun. Earth's mass, M_2 can be neglected here.

7.1.5   Radius

If the solar distance is known, then the solar radius in kilometers follows from the apparent
solar diameter. In the course of a year the apparent Sun radius varies between $32'26''$ (Earth

at perihelion, near the Sun, currently on January 4) and 31′31″ (Earth at aphelion, far from the Sun, currently on July 4).

7.1.6 Luminosity

Principle of determination: one measures the energy received on earth from the sun and then takes into account the distance of the sun. Per square meter we receive on earth a radiant power of $S = 1.37$ kW from the sun, the *Solar constant S*. This value is important for solar pannel operators. But one should not expect 1 m² collector area actually provides this energy, since:

- is the value for outside the earth's atmosphere,
- this is only true for perpendicular incidence of the sun's rays,
- one must take into account the efficiency of the system.

The luminosity of the sun follows from:

$$L = 4\pi r^2 S \tag{7.4}$$

r—Earth-Sun distance.

The solar constant S indicates the amount of energy that can be absorbed at the Earth's location (i.e., in $r = 1.496 \times 10^{11}$ m distance) on an area of 1 m² per second, i.e. $S = 1367 \, \text{Js}^{-1}\text{m}^{-2}$. This energy is distributed over the surface $O = 4\pi R^2 = 4\pi (1.496 \times 10^{11} \, \text{m})^2$, and the total radiant power (luminosity) of the Sun is :

$$L = O \times 1367 \, \frac{\text{J}}{\text{m}^2\text{s}} = 3..845 \times 10^{26} \, \text{W}$$

The solar constant indicates the solar radiant power arriving at the earth.

7.1.7 Effective Temperature

We treat stars as black bodies, since Planck's law of radiation applies to them. In thermodynamic equilibrium (in a cavity of temperature T), the emission and absorption of any volume element must be equal. If the emission is described by an emission coefficient ϵ_ν and the absorption is given by an absorption coefficient κ_ν then in the case of a thermodynamic equilibrium (TE, thermodynamic equilibrium):

$$\epsilon_\nu / \kappa_\nu = B_\nu(T) \tag{7.5}$$

where $B_\nu(T)$ is the Planck function described above (Eq. 5.34); this is known as *Kirchhoff's theorem* .

According to the *Stefan-Boltzmann's law of radiation* the total radiative flux πF of a star (by this is meant the energy flow coming from the interior per square centimetre of the star's surface, for example):

$$\pi F = \sigma T_{\text{eff}}^4 \tag{7.6}$$

and one obtains for the Sun an effective temperature/indexSun!effective temperature eff $=$ 5770 K. The measurement of πF is based on the measurement of the solar constant S.

Thus, we have given the main state variables of the Sun.

In Fig. 7.3 the spectrum of the Sun is shown. Likewise, the radiation curve of a black body with $T = 5800$ K is underlaid.

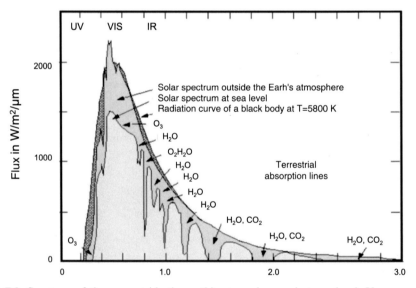

Fig. 7.3 Spectrum of the sun outside the earth's atmosphere and at sea level. You can see the absorption by the Earth's atmosphere. Abscissa: Wavelength in μm

7.1.8 Sun: Observation

Important Never observe the sun with unprotected eyes. The simplest observation is to project the image of the sun on an observation screen (stencil) behind the telescope. Neutral filters in front of the telescope lens can also be used to attenuate the sunlight.

Observation Experiment: Observe the sun with a small telescope in projection. It is very easy to see the center to limb variation, possibly sunspots and flares. By means of sketches one can follow the migration of the spots in a few days which is due to rotation of the Sun.

Special solar telescopes are used in research (see Chap. 5).

- Tower telescopes: At the top of a vertical tower there is a *Coelostat* system that reflects the light into a vertical tower, which is often evacuated to avoid heating up. The tower host the telescope mirrors.
- Coronagraph: The solar disk is darkened by a cone aperture and filters and thus an artificial solar eclipse is produced; one can observe the corona (this outermost part of the solar atmosphere has about the brightness of the full moon). Problem: Stray light in the earth's atmosphere.
- H-Alpha (Hα) telescope: Observation of the sun in the light of the hydrogen line Hα at 656.3 nm; with this one can see sunspots, prominences, filaments, flares. Thus also structures of the chromosphere become observable.
- Spectroheliograph (Hale, 1890): The image of the sun falls on a narrow slit of a monochromator. One observes the sun in a narrow wavelength range, and through a scan one sees the whole sun.
- Lyot Filter: Consists of several birefringent quartz plates. Each plate is half the thickness of the previous one. Due to the property of birefringence, light splits into an ordinary ray I_o and an extraordinary ray I_{eo} each with a different refractive index and phase velocity. Waves in the same polarization state can only occur if the optical path length of I_o, I_{eo} are an integer multiple of the wavelength. By adding a polarizer, to the plates, which is a filter, you get a transmission function with peaks. By using liquid crystals you get a *tunable filter.*
- Radio heliograph: e.g. Nancay, the sun is observed at five frequencies between 150 and 420 MHz. Thereby one detects different heights in the solar atmosphere ($\rightarrow$ the higher the frequency, the deeper one sees into the solar atmosphere, i.e. closer to the solar surface).

An example of a modern solar telescope commissioned in 2012 is GREGOR (Fig. 7.4). It has a diameter of 1.5 m. The telescope is located at an altitude of about 2400 m on the Canary Island of Tenerife.

The *DKIST* solar telescope (see Chap. 5) became operational in 2020 and, with a diameter of 4 m, is the largest solar telescope ever built. In Europe there are plans for an

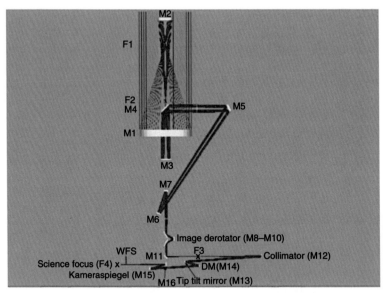

Fig. 7.4 The new GREGOR telescope. Tip Tilt is a tiltable mirror which can partially compensate for air turbulence and allows better solar images. (Source: KIS, Freiburg)

Fig. 7.5 The planned European Solar Telescope, EST [EST consortium]

EST, European Solar Telescope (Fig. 7.5), which could be operational in 2029 and whose diameter would be similar to that of DKIST. Since DKIST is located in Hawaii and EST on the Canary Islands, a round-the-clock solar observation would become possible.

The Sun, our nearest star, is the only star where details can be observed!

7.2 The Structure of the Sun, the Quiet Sun

What layers does the sun consist of and how do these layers come into being? We have to distinguish between the layers of the solar interior and the solar atmosphere.

7.2.1 General Structure of the Sun

The structure of the sun can be roughly subdivided in:

- Sun interior
  - core,
  - radiation zone,
  - convection zone,
- solar atmosphere
  - photosphere,
  - chromosphere,
  - corona.

In general, we can say that our Sun is the only star where surface details can be observed directly. Thereby seen from Earth an angle of $1''$ corresponds to about 720 km on the solar surface. Only recently quasi *in situ* measurements by space probes are possible.

The Sun is a gas sphere in which the density decreases almost monotonically towards the outside. Physical quantities relating to the Sun are denoted by the subscript solar sign $\odot$.

From Table 7.1 one sees that within 0.5 $R_\odot$ 96% of the mass is already concentrated. In the following we discuss the individual layers of the sun and the phenomena of the quiet (i.e. not active) sun occurring in them.

Table 7.1 Variation of the density in the solar interior

$r/R_\odot$	$r/10^{10}$ cm	$T[10^6$ K$]$	$\rho[g/cm^3]$	M_r	$P[N/cm^2]$
0.00	0.00	20	158	0,0	4×10^{12}
0.1	0.7	16	118	0.0089	2.5×10^{12}
0.2	1.39	11	45	0.44	6×10^{11}
0.5	3.48	3.9	0.74	0.96	4×10^9
0.9	6.27	0.44	0.001	1.0	7×10^5

7.2.2   Sun's Interior

Near the center the *Fusion* from Hydrogen, H, to Helium, He, takes place. This *Hydrogen burning* (actually a nuclear fusion) is described in more detail in Chap. 8 on stellar structure. Our Sun has already burned about 50% of the available hydrogen. Fusion produces *neutrinos,* which pass through the sun unhindered due to their small effective cross-section. On Earth, special detectors can be used to detect these neutrinos originating from the interior of the Sun and thus verify the theories of solar structure. Since the measured neutrino fluxes do not correspond to the theory, there exists a *neutrino problem* in solar physics. We have to consider :

- Neutrinos: arrive at Earth with only eight minutes delay, so they reflect the current rate of nuclear fusion in sun;
- Photons: are emitted at the surface of the sun. However, it takes many 10^5 years until a photon reaches the surface from the place of its origin (thereby changing from extremely short-wave to long-wave) and is emitted. This is due to the many scattering processes of the photons. The light of the present sun thus comes quasi from nuclear reactions which took place many 10^5 years ago.

Another way to explore the interior of the sun is through the use of Helioseismology. The sun oscillates, and from the analysis of these oscillations the physical parameters of the sun's interior can be determined. Analogue: From the sound of a bell, the bell material can be inferred.

> The evaluation of the neutrino flux from the Sun and helioseismology make it possible to explore the interior of the Sun directly.

From this, it appears that our ideas about the structure of the Sun agree quite well with observations.

Nuclear fusion in the solar interior can be roughly stated as:

$$4p \rightarrow ^4\text{He} + 2e^+ + 2\nu_e \tag{7.7}$$

So neutrinos are produced, ν_e or, to be more precise, electron neutrinos. The first neutrino experiment goes back to *R. Davis* Some chlorine atoms of the cleaning liquid C_2Cl_4 in a large underground tank are converted to ^{37}Ar. This experiment was performed for the first time in 1964, and only 1/3 of the expected neutrino flux was measured. In the Superkamiokande experiment (1988), a tank filled with 50,000 t of highly pure water is used about 1000 m below the earth's surface. Free electrons are produced when neutrinos react with water molecules in the tank. The neutrinos are detected by the occurring Cerenkov radiation of free electrons.

The measured low neutrino flux is explained by the fact that neutrinos change their properties on their way to Earth. They change into muon and tau neutrinos, but only the electron neutrinos can be measured. These neutrino oscillations would then also require a finite rest mass. Sudbury neutrino experiment (SNO) can detect all three types of neutrino (1999–2006), and neutrino oscillations are considered to be secured to explain the neutrino problem of the Sun.

Energy is released during nuclear fusion in the form of gamma ray quanta. These are absorbed and re-emitted in the radiation zone. Thus, in the radiative zone, energy transport occurs by radiation. In the convection zone, energy transport then starts by convection: Hot plasma flows upward, cools down, sinks downward, heats up again, etc.

The expansion of the three regions of the solar interior corresponds roughly to $(1/3)R_\odot$ in each case , i.e., the convection zone extends, for example, to about 200,000 km below the solar surface.

7.2.3   Photosphere

This layer emits almost the entire Solar radiation, whereby the maximum of the radiant power lies at 500 nm. The thickness of the photosphere is about 400 km and is very thin compared to the solar radius of about 700,000 km. Within the photosphere the temperature decreases from 6000 K to 4000 K, the density from 10^{-7} to 10^{-8} g cm^{-3}. The optical depth is : $\tau = 0.5$ at $T = 5800$ K and $\tau = 0.05$ at $T = 4800$ K.

If you look at a solar image in visible light, you can see *center to limb variation*. The solar disk appears brighter in the center than at the edge. If one looks towards the solar limb, the visual ray must take a longer path through the solar atmosphere, and one therefore sees into less deep and thus cooler layers than in the centre of the solar disc(cf. Fig. 7.6). Important: The center to limb variation is wavelength dependent and becomes more pronounced in the blue than in the red. In the radio wave range at $\lambda > 1$ cm one has *Limb brightening*, since here the radiation comes from higher layers where the temperature increases outward.

> The center to limb variation indicates that the temperature increases inward from the solar surface.

The solar spectrum is shown in Fig. 7.7 shown.

The surface of the Sun in the photospheric region is not homogeneous, but shows a cellular pattern, which, because of its granular appearance, is also known as *granulation* (Fig. 7.8) . These cells have an average extension of about 1000 km and a lifetime of ten

Fig. 7.6 Limb darkening: The
visual ray going to the sun's
limb penetrates less deep

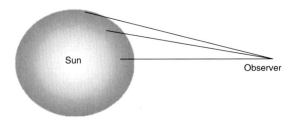

Fig. 7.7 Solar spectrum
(credit: AURA)

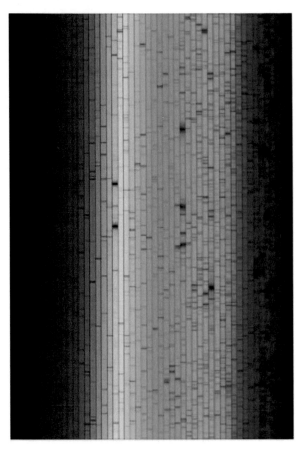

minutes. From a depth of 200,000 km below the surface of the sun the energy transport to
the surface is by *convection*:

Hot plasma flows up, cools, and sinks back down, etc. The bright granules are 200 K
to 300 K hotter than the dark intergranular spaces where the plasma sinks down. If you
bring a spectrograph slit into an image of the Sun near the center of the Sun, the lines
coming from the photosphere appear wiggly (*wiggly lines*, Fig. 7.9). Lines coming from
the granulum are blue shifted, because the matter rises upwards, i.e. towards the observer,
and lines coming from the dark intergranulum are red-shifted, because here the matter

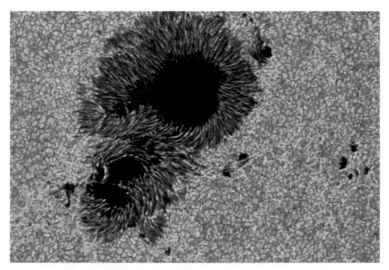

Fig. 7.8 Solar granulation with sunspot; image taken by Hinode satellite

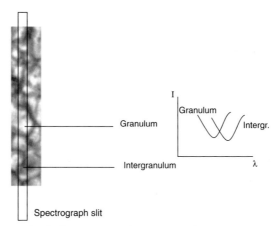

Fig. 7.9 The spectrograph slit lies over granular/intergranular regions. This results in a blue shift for the rising granulum and a red shift for the sinking intergranulum when observed at the center of the solar disk

sinks back into the interior of the sun. If one cannot separate granulum and intergranulum, one gets a superposition of these line profiles, and since the bright and thus ascending elements dominate in size, one gets a mostly C-shaped *Bisector* (which is the line that bisects a line profile at equal intensities). The line profile is therefore asymmetrical. So asymmetric line profiles at stars are always a sign, that there is convection also at their surfaces.

7.2.4   Chromosphere

The chromopshere is extending outward of the photosphere up to about 10^4 km height
(1.00–1.015 $R_\odot$). The density decreases to 10^{-11} g cm^{-3}, the temperature increases again
to 10^5 K. Because of the low density, the radiation of the chromosphere contributes almost
nothing to the total radiation, despite the high temperatures. During a total solar eclipse one
observes the *Spicules,* a bristly structure that aligns with magnetic fields. In the spectrum
of the chromosphere, which is seen to flash only briefly during a total solar eclipse and is
therefore known as the *Flash spectrum* one sees emission lines. These arise as follows:

- Lines that appear in absorption in the photosphere spectrum are visible as emission
  lines in the chromosphere.
- Due to the high temperatures in the chromosphere, lines of highly excited or ionized
  atoms are observed.

Both indicate high temperatures.

The chromosphere can also be observed outside a total eclipse. Spectroheliograms
are used to take monochromatic images. One measures through a *narrowband filter* the
radiation in the centers of strong absorption lines: at $\lambda = 656.3$ nm H$_\alpha$, at $\lambda = 393.3$ nm
and $\lambda = 396.8$ nm the so-called H and K lines of Ca II (singly ionized calcium). In the
latter one sees the *chromospheric network* (supergranulation cells) about 30,000 km in
diameter, at the edges of which are strong magnetic fields. In other lines (e.g. Hα) one
sees prominences protruding beyond the solar limb, which can be seen as dark filaments
in front of the solar disk (Fig. 7.10).

7.2.5   Corona

In the outermost layer of the solar atmosphere, the corona, the density decreases to
10^{-18} g cm^{-3} and the corona merges outwards into the *interplanetary medium.* Its shape
depends on the activity of the Sun: At *maximum* it is symmetrical around the solar disc, in
the *Minimum* it is more concentrated to the equator.

The *spectrum* of the corona consists of the following parts:

- Continuous spectrum, *K corona:* photospheric light scattered by free electrons at
  the corona. But then one should observe absorption lines. But because of the high
  temperatures these are blurred (temperatures of several 10^6, K resp. speed of $v =$
  8000 km/s) and won't come up.
- *L corona:* emission lines; highly ionized elements, e.g. FeXIV (13-fold ionized iron),
  this is the *green corona line* at 530.3 nm. From the ratio FeXIV/FeX one can determine
  the temperature according to the Saha formula.
- *F-Corona:* Normal solar spectrum; originates from scattering by interplanetary dust.

Fig. 7.10 Sun with
prominences projecting
brightly above the solar limb
and appearing dark in front of
the solar disk (source:
SOHO/NASA)

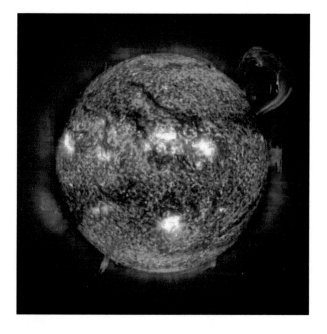

An observation of the corona outside of total solar eclipses is possible in the UV and in the
X-ray range as well as in the radio range (m-waves). In the X-ray light the *coronal holes*
can be seen. Here the magnetic field lines, which determine all structures in the corona,
are open over wide areas. Through these areas the *Solar wind* escapes.

   If one observes the sun in the radio range, the *Plasma frequency* v_0 should be noted;
radio radiation below this frequency v_0 (or above this wavelength) cannot escape because
the refractive index becomes negative, < 0 .

$$v_0 = \sqrt{\frac{e^2 N_e}{\pi m}} = 9 \times 10^3 \sqrt{N_e} \quad \text{[MHz]} \tag{7.8}$$

In this formula, the electron density N_e is given as particles per cm^3. The electron density
in the corona decreases with increasing distance from the solar surface: At a distance of
$1.03\,R_\odot$ it is at most $350 \times 10^6\,cm^{-3}$ at a distance of $2.0\,R_\odot$ it is $3.1 \times 10^6\,cm^{-3}$ and at
$5\,R_\odot$ only more $0.05 \times 10^6\,cm^{-3}$.

   When observing in different frequencies, one can probe different layers of the solar
atmosphere.

If one observes the Sun in the radio range at higher frequencies, then one sees into deeper
layers because the electron density is higher there. So you can do a kind of *tomography* of
the solar corona when observing at different frequencies.

The temperature increases strongly in higher layers of the solar atmosphere. The reason for this is not yet fully understood.

7.3   The Active Sun

Our sun is not a static star—it is an active star (ex. Fig. 7.11), changing occur constantly. We will first discuss the individual phenomena and then their effects on Earth.

7.3.1   Sunspots

The easiest way to determine the solar activity on the surface of the sun is by looking at *Sunspots*. These were observed with the naked eye in ancient times when the sun was low in the sky. For a sunspot to be visible to the naked eye, it must have a total extent of about 40,000 km. Since it has been recognized that phenomena of solar activity affect radio communications, satellite positions, corrosion of pipelines, electric power lines, etc., close attention has been devoted to the study of the *Space Weather.*

The first telescopic observations of the spots were made by *G. Galilei* (1610) *Chr. Scheiner,* and other astronomers. At first sunspots were thought to be inner planets wandering around the Sun. Then it was realized that their migration around the Sun could be explained by the rotation of the Sun.

Spots consist of a dark core part, the *umbra,* which is surrounded by a filamentary brighter *penumbra* (Fig. 7.12). The temperature in the umbra is 4300 K and in the

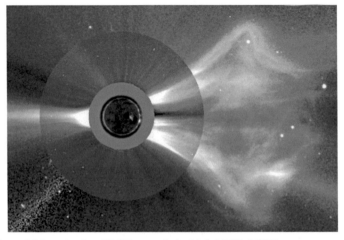

Fig. 7.11 Coronal Mass ejection (CME), recorded with LASCO-SOHO

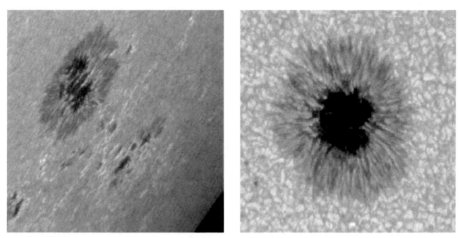

Fig. 7.12 Sunspot with flares at the edge of the sun and in the center of the sun disk; photograph La Palma, Vazquez, Bonet, Sobotka, Hanslmeier

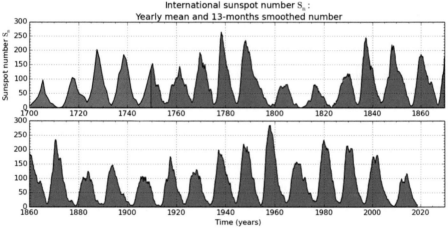

SILSO graphics (http://sidc.be/silso) Royal Observatory of Belgium 2019 April 2

Fig. 7.13   Sunspot relative numbers (source: SIDC)

penumbra 5500 K. The intensity ratio between the spot and the photosphere is 0.13 at a wavelength of 300 nm, and at $\lambda = 1000$ nm 0.46.

The number of sunspots varies with an average period of eleven years, which is the *Sunspot cycle* (Fig. 7.13). The activity cycles have been numbered consecutively since 1760. In 2019, the cycle began with the number 25. Spots usually occur in groups of spots with several individual spots. If g is the number of groups and f is the number of single spots, then we determine the *Relative number R*:

$$R = k(10\,g + f) \tag{7.9}$$

Thereby k is a correction factor that takes into account the influence of the instrument. At the beginning of a new cycle, spots occur at higher heliographic latitudes at $\pm 35°$ on, in the middle of the cycle at $\pm 8°$. Thus one gets a *Butterfly diagram,* if one records the positions of the spots in the course of an eleven-year cycle.

There can be up to two years of overlap between spots of the old cycle and the new cycle. If one observes the spots at the solar limb, the penumbra on the side closer to the center of the sun appears shortened. This is the *Wilson effect.* Lines of equal optical depth are geometrically several 100 km deeper in large spots than in the photosphere; this leads to an asymmetry of the penumbra at the solar limb.

The lifetime of the spots is a few days, for 90% of the spots less than eleven days.

If one examines spectral lines that are created in the spots, one sees a splitting due to the *Zeeman effect.* Therefore spots are related to magnetic fields.

The magnetic flux densities measured in the spots[2] are up to 4000 Gauss[3] (Earth's magnetic field around 0.5 Gauss). The field lines pierce the photosphere vertically in the umbra region and diverge like a razor brush. If H_0 is the strength of the field at the center of the spot and r is the distance from the spot center, then one has:

$$H(r) = H_0(1 - r^2) \tag{7.10}$$

Why are spots cooler than the approximately 6000 K hot solar surface? Outside the field in the sun there is the pressure p_e. In the area of the spot to the pressure p_i there is also amagnetic pressure $B^2/2\mu$ where μ denotes the magnetic permeability. In order to have a stable structure, the following must hold true:

$$p_i + B^2/2\mu = p_e, \tag{7.11}$$

thus $p_i < p_e$ and because of $\rho_i = \rho_e$ it follows that $T_i < T_e$ is. The temperature inside a spot T_i is therefore lower than the surrounding photosphere temperature T_e.

91% of all spots occur in *bipolar groups.* These are magnetic tubes driven by magnetic buoyancy from the solar interior to the surface, and then produce a bipolar group at the two puncture points (Fig. 7.14).

The *Zurich classification* is sketched in Fig. 7.15 (after Bray). Criteria for this are whether a group is unipolar or bipolar, whether a penumbra is pronounced or not, and the heliographic length extent.

The spot preceding in solar rotation is called the p-spot and the one following is called the f-spot. If in a cycle of activity in the northern hemisphere the p-spot has a positive polarity and the f-spot a negative polarity, then in the southern hemisphere it is the other way round: the p-spot then has a negative polarity and the f-spot a positive polarity. On

---

[2] Often one simply speaks of field strengths, although $B = \mu H$, μ =Permeability.

[3] 1 Gauss = 1 G = 10^{-4} Tesla.

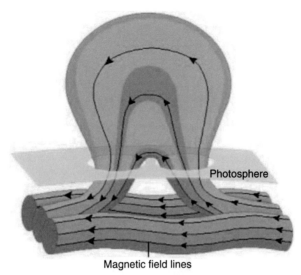

Photosphere

Magnetic field lines

Fig. 7.14 A bipolar spot group is formed when magnetic flux penetrates to the surface of the photosphere

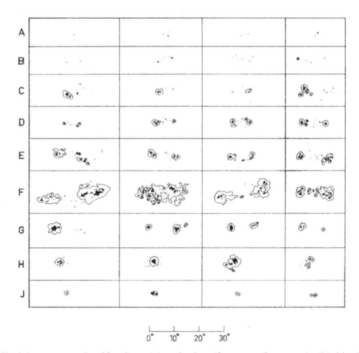

Fig. 7.15 Zurich sunspot classification; (**a**) unipolar (first row from top), (b) bipolar, without penumbra (second row from top), (**c**) bipolar, penumbra at one of the two main spots, (**d**) Penumbra at both main spots, small, etc.

Fig. 7.16 Magnetogram of the
sun. White and black denote
different polarities
(SOHO/MDI)

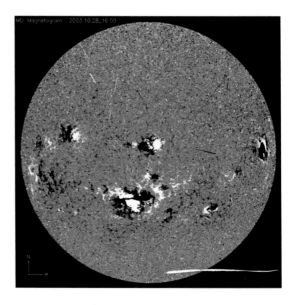

the next cycle, the whole thing is reversed: In the northern hemisphere of the Sun, the
p-spot then has a negative polarity, and so on. This has been found by *Hale* and is known
as *magnetic cycle* (Fig. 7.16).

The spectrum of spots is difficult to observe, it resembles that of a K star. New results
of helioseismology show the dynamics below spots (Fig. 7.17).

> Sunspots are areas of strong magnetic fields where convection is reduced.

7.3.2   Faculae

They are practically the counter part to the spots. They appear as extended areas of
excessive brightness (10% more than in the photosphere). Faculae usually occur in the
vicinity of sunspots; because of the contrast they are especially observable at the solar
limb, i.e. they are an overheating of the higher solar layers. Usually, faculae are found
at the same heliographic latitudes as the spots. The polar faculae occur at unusually high
latitudes in the years before or during a sunspot minimum.

In the range around 430 nm *(G-band)* sees one finds many molecular lines (CH,
vibrational states, rotational states) in the solar spectrum. At very good resolution with
a G-band filter, one finds bright, about 0.1″ extended dots, *G-band bright points,* GBPs,
which are small magnetic elements.

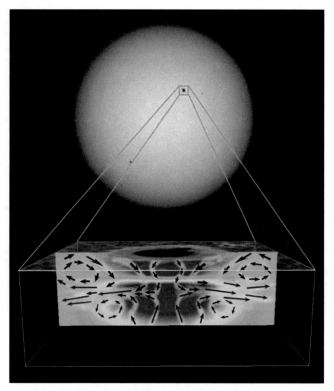

Fig. 7.17 Sunspot with structure below the photosphere (SOHO/MDI)

7.3.3 Prominences

These are clouds of matter in the corona. They can be observed as:

- At the solar limb: bright against the dark sky; either during a total solar eclipse or in the light of a chromospheric line such as Hα;
- on the solar disk: they appear there dark against the bright photosphere, i.e. in absorption, and are called *Filaments*.

Prominences are also subject to the eleven-year cycle. The main zone of their occurrence corresponds to the spot zone. Shortly before the minimum the polar zone appears, which moves poleward and reaches its maximum about two years before the spot maximum at the pole. One distinguishes:

- dormant prominences: long-lived; thickness about 7000 km. Height usually up to 40, 000 km, length 20, 000 km; they are long, thin, and lamellar. They often stand only on single feet in the chromosphere; their life is up to a year. They usually arise in groups

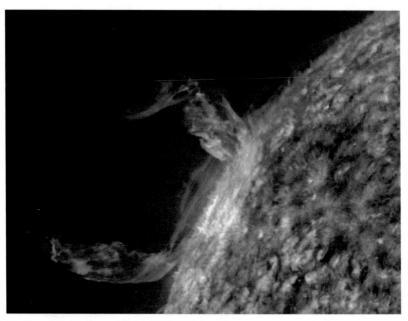

Fig. 7.18 Prominence taken with the SDO/AIA instrument.

of spots or in flares. The sun rotates differentially, i.e. faster at the equator than at the poles, therefore their position becomes more and more parallel to the equator.
- Active or eruptive prominences: usually associated with active spot groups. There are:

  (a) Sprays: explosive rise with 1000 km/s (the maximum observed height was 1.5 million km!).
  (b) Surges (splashes): ascent with 50–200 km/s; occur repeatedly in active spot groups.
  (c) Coronal rain: After eruption, matter flows back like rain.
  (d) Loops: Matter follows magnetic field lines.

The prominences (Figs. 7.10 and 7.18) always appear at the boundary between regions with different magnetic polarities *(neutral line)*. There the field lines run horizontally. The ionized matter is held in place by the magnetic field, and its density is 100 times greater than the ambient density; even in the case of the quiescent prominences, matter is constantly flowing away and being replaced.

7.3.4   Flares and Coronal Mass Ejections

Solar flares can be observed throughout the electromagnetic spectrum. *Carrington* and *Hodgson* observed for the first time in 1859 a *Flare* in visible light (which is very rare).

Hale invented the spectrohelioscope, which could be used to observe the Sun at a particular wavelength, and in 1920 found that flares were visible in the light of the hydrogen line Hα. Then, in 1940, flares were observed in the radio region and, with satellites, later in the UV and in X-rays.

The energy released in flares ranges from 10^{16} J *(nanoflares)* to 10^{25} J (large, so-called *Two Ribbon Flares)*.

A comparison: The *Hiroshima bomb* had an explosive force of 15 kt, with

$$1 \text{kt} (\text{kiloton TNT}) = 4.184 \times 10^{12} \text{J}. \tag{7.12}$$

The largest ever found explosed *Hydrogen bomb* had 50 megatons, and the explosive power of all conventional bombs in World War 2 reached about two megatons.

As an exercise you could estimate how large the energy release in a flare burst is compared to the Hiroshima bomb.

For *classification* of flares one uses two systems:

- Importance: 1, 2, 3, 4 and the additions f for *faint, n* for *normal* and b for *bright.* S indicates a subflare. So the brightest flares are 4b. The higher the value of this optical classification, the longer the apparition usually lasts.
- X-ray classification: Since 1970 there are X-ray observations of flares, and one classifies according to the flux in the range 1–8 Å in units of W/m^2. In powers of ten:

  Class A: $10^{-8} W/m^2$, class B: $10^{-7} W/m^2$, C, M, X.

An M8 flare then has a maximum flux in the 1–8 Å range of. $8 \times 10^{-5} W/m^2$. Note that C1 flares can only be observed at the time of solar activity minimum, when the X-ray background is weak.

If a flare eruption (eruptive flares) occurs, then there is associated:

- shortwave radiation ($<$ 200 nm): of the same order of magnitude as the total solar radiation in this region.
- X-ray radiation: Increased X-ray radiation causes disturbances in the Earth's ionosphere *(Mögel-Dellinger effect)*. Increased ionization of the D layer occurs; radio waves that, in order to be reflected from the F layer, must pass through the D layer twice, are strongly attenuated, thus interfering with radio traffic.
- Corpuscular radiation: The particles with velocities between 1000 km/s and 2000 km/s arrive at Earth one day after the flare outburst, causing magnetic storms, Auroras (green and red oxygen lines, these are forbidden transitions).
- Radio bursts in the m-range.

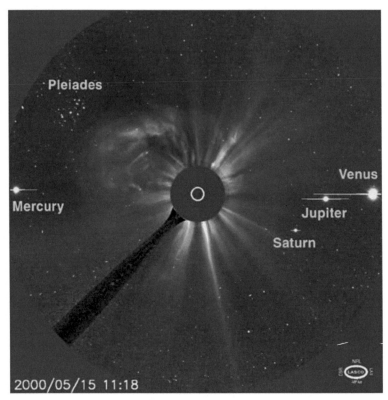

Fig. 7.19 A coronal mass ejection, CME, imaged by the SOHO solar satellite. The Sun itself is obscured and its actual size is indicated by the circular ring (You can also see 4 planets and a few stars in the image) (source: NASA/SOHO)

- Cosmic rays: particles are accelerated almost to the speed of light, in the upper atmosphere of the earth these produce secondary particles and high-altitude showers when they collide with atomic nuclei.

In 1970, *CMEs,* Coronal mass ejections (Figs. 7.11 and 7.19) were observed for the first time. The matter released during a CME leaves the Sun at a speed of up to 2000 km/s. The mass of the ejected matter is $10^{15} \ldots 10^{16}$ g. Eruptive flares are likely to be caused by CMEs. During the eruptive phase of a CME, the field lines are open, but then close again, magnetic reconnection occurs, and an eruptive flare bursts *(gradual flare).* The intersection of these loops with the solar surface appears as two parallel bands in Hα, and it used to be referred to as *two ribbon flares* when the role of CMEs in flare outbursts was not well understood. The same physics seems to be behind eruptive prominences *(disparition brusque).*

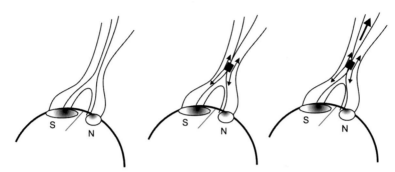

Fig. 7.20 Model for flare formation. At the beginning you have a bipolar group whose field lines are not connected in the corona loop. In the second phase, the flare begins when reconnection of the field lines occurs, and a current flows to the foot points in the photosphere. Finally, the flare moves up into the corona

A *Halo-CME* is a CME that appears to be pointing directly towards the Earth. In this case the CME appears as a halo around the sun. The CME is expanding and appears as an ever-growing halo around the Sun.

Most of the flares are impulsive flares. They can be modeled with a single static magnetic loop (Fig. 7.20). Again, there is a magnetic reconnection with subsequent acceleration of particles, but with no connection to CMEs.

Hale has introduced the following classification for Magnetic structures *(Mt. Wilson classification),* which is also important for the occurrence of flares:

- α: A single dominant spot, usually associated with areas of opposite polarity.
- β: Sunspot pair with opposite polarities.
- γ: Complex group with irregular distribution of polarities.
- δ: Umbrae with opposite polarities within a single penumbra.

In the δ-configuration there occur more flares than in the other groups. Such a configuration can occur in all other groups. Here one has two poles with strong vertical fields close to each other. One can give the following criteria for the occurrence of flares:

- large δ-Spots,
- Umbrae with large elongation,
- high shear in the transverse field or strong gradients in the longitudinal field,
- large spots always have strong flares.

Flares and CMEs arise from reconfigurations of the magnetic field, reconnection.

7.3.5   The Radio Radiation

One must distinguish between solar radio radiation of a slowly variable component and bursts.

The slowly variable component originates from discrete regions of the solar atmosphere, predominantly active regions. The radiation flux is closely correlated with the relative number; it is probably thermal radiation from the corona condensations. Its temperature is about 10^7 K and the wavelength of the radiation is between 1–100 cm, the maximum at 15 cm; the intensity at 10.7 cm is also used as a substitute for spot relative numbers uses.

Radiation bursts produce radiation in the range of 1 cm to 15 m. In the centimeter range, the intensity increases to 20–40 times the normal value, and in the m range, it increases to 10^5-fold of the normal value on.

Let us consider the so-called plasma frequency ω_p. We assume electrons moving with respect to ions at rest. In the case of a gas consisting only of electrons and ionized hydrogen atoms, we find for the charge density:

$$\rho = e n_e \tag{7.13}$$

From $div\mathbf{E} = \rho/\epsilon_0$, $\epsilon_0 = 8.854 \times 10^{-12}$ As/Vm, electric field constant (permittivity of the vacuum) becomes in the one-dimensional case:

$$div(\mathbf{E}) = dE/dx; \rightarrow E = e n_e x/\epsilon_0 \tag{7.14}$$

and the equation of motion is:

$$\ddot{x} = F/m_e = eE/m_e = -\frac{e^2 n_e}{\epsilon_0 m_e}x \tag{7.15}$$

this is the equation of a harmonic oscillator, and one obtains the Plasma frequency To

$$\omega_p = \sqrt{\frac{n_e e^2}{\epsilon_0 \epsilon m_e}} \tag{7.16}$$

when the propagation in a medium of permittivity $\epsilon > 1$ is considered.

The higher the electron density, the higher the plasma frequency. The plasma frequency is also responsible for the reflection of radio waves in the Earth's atmosphere, or how deep into the corona one can look in the radio region.

Fig. 7.21 Course of electron density and temperature in the different layers of the solar atmosphere; the transition zone from chromosphere to corona is also called the *transition region*.

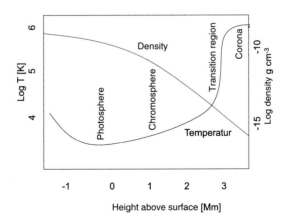

> Corona: Observations at high high frequencies you see regions of high electron density and therefore you look into deeper layers.

If you plot the *bursts* in the time-frequency diagram, one can see how the regions of stronger radio radiation move through the solar atmosphere. The electron density decreases outward and so does the plasma frequency (Fig. 7.21).

The *Bursts* are classified into:

1. Type I: m-range; short, steeply rising bursts; total duration less than 1 s; main component of noise storms.
2. Type II: m range; emission in a narrow frequency band, shifting from high to low frequencies with time; drift velocity in the frequency band is between 0.5 and 1 MHz/s. Duration about ten minutes; matter thus passes through the corona, i.e. from the frequency drift one can read the height and the velocity from 400–1000 km/s. Occurrence: relatively rare, in activity maximum one outburst every 50 h; correlation with geomagnetic storms (these start 2–3 days after the outburst).
3. Type III: m range; narrow-band, but faster frequency drift than type II (20 MHz/s). Duration 10 s, ascent rate 0.4 c. Height about 1 $R_{\odot}$; during the sunspot maximum three type III bursts per hour are observed. Occurrence mostly at the beginning of a flare (Fig. 7.22).
4. Type IV: throughout the radio range; fast electrons; height above photosphere 0.3–0.4 $R_{\odot}$; no more plasma oscillations, because matter density is too low.
5. Noise Storms: m range; large number of individual radiation bursts; duration hours to days; frequent at spot maximum. Type I bursts occur repeatedly; synchrotron radiation of fast electrons; 0.3–1 $R_{\odot}$ over the photosphere; preferably over spot regions.

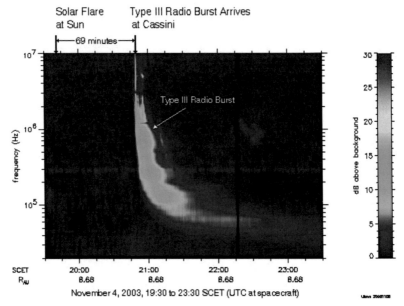

Fig. 7.22 Type III burst imaged by the Cassini spacecraft studying Saturn. More than an hour after the event on the Sun, Cassini shows the radio emission and, in the frequency-time diagram, the emergence of the radio emission at different altitudes in the solar atmosphere (lower frequencies at higher altitudes) (source: Cassini/NASA)

7.3.6 X-rays of the Corona

In visible light, the corona can be seen during a total solar eclipse (Fig. 7.23). The missions *SOHO, TRACE, YOHKOH* an d *RHESSI* enable also observations of the sun in the extreme UV- and X-ray range where the radiation of the Sun varies by the following factors during the eleven-year cycle:

- Variation in the UV range: ≈ 2,
- Variation in the X-ray range ≈ 100.

In Fig. 7.24 the sun is shown in the X-ray region shown. This is emitted due to the high temperatures in the corona. One can clearly see the difference between the emissions near the maximum and near the minimum of solar activity.

The energy of X-rays is a measure of the energy of the electrons that produce them. It is useful to remember:[4]

$$1 \text{keV} = 1.6 \times 10^{-16} \text{J} \tag{7.17}$$

---

[4] In visible light at a few eV.

Fig. 7.23 Total solar eclipse of 29.03.2006. Shortly before totality (diamond ring phase) some red glowing prominences are visible, lower right: totality. Corona well visible (image A. Hanslmeier)

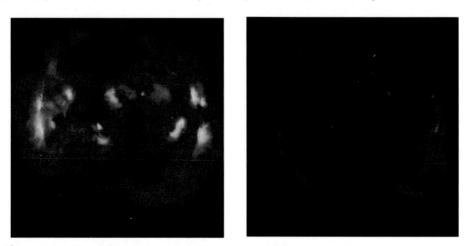

Fig. 7.24 The Sun in the X-ray region — left March 1993 (Sun very active), right near minimum March 1995 (photo: YOHKOH)

According to this, X-rays can be divided into: (a) hard X-rays 10–100 keV, (b) soft X-rays 1–10 keV.

Consider the time evolution of *flares:* Hard X-ray radiation is mostly released during the impulsive initial phase, soft X-ray radiation reaches the maximum with a time delay (some minutes). For the first time, hard X-ray radiation was observed by the *Solar Maximum*

Mission (SMM) around 1980. It was found that there are two sources of this X-ray radiation localized at the foot points. The suggests non-thermal electrons moving down to the foot points from the corona into the denser chromosphere. The microwave emission is also related to this. The slow increase in the soft X-ray region corresponds to the time integral of the hard X-ray emission radiation→*Neupert effect.*

The high temperatures in the course of flares lead to different radiation mechanisms:

- *Thermal bremsstrahlung:* protons attract free electrons; electrons change their velocity, bremsstrahlung occurs, observed as soft X-rays. In thermal bremsstrahlung the particles of the emitting plasma have a defined velocity v, and the distribution function corresponds to of a *Maxwell distribution:*

$$f(v) = 4\pi \left(\frac{m}{2\pi kT}\right)^{3/2} v^2 \exp\left[\frac{-mv^2}{2kT}\right] \tag{7.18}$$

  The emission from this plasma is then called thermal bremsstrahlung, and the radiation power (in [W m^{-3}]) is

$$P_{BR} = \frac{Z_i^2 n_i n_e}{(7.69 \times 10^{18})^2} T_e[\text{eV}]^{1/2} \tag{7.19}$$

  where n_i, n_e ion or electron density, Z_i the charge number of the deflecting charge and T is the temperature below which the energy is insufficient to emit a photon (frequency v) in the X-ray range ($kT = \hbar v$).
- Synchrotron radiation: electrons with very high velocities are accelerated along magnetic field lines. High energy electromagnetic waves are emitted tangential to the direction of motion.
- Non-thermal bremsstrahlung: electrons with very high energy, no Maxwell distribution.

Coronal Holes Almost always to be observed at the poles of the sun ; they appear dark in X-ray light, the gas density (4×10^{14} electrons m^{-3}) as well as the temperature ($\approx 1 \times 10^6$ K) are lower than in the surrounding area. The expansion is about 700 to 900×10^6 m. The magnetic fields in the coronal holes are open and thus charged particles can escape at high speed; this is the high speed solar wind. Before the minimum of the activity cycle they cover the poles, near the maximum the polar holes shrink, and near the solar equator smaller holes appear.

Other phenomena of the corona are arcs of active regions ("active region loops"), which have an extent of about 10^7 m, temperatures of $2 - 4 \times 10^6$ K and an electron density of $1 - 7 \times 10^{15}$ m^{-3} as well as X-ray bright points whose extent is $5 - 20 \times 10^6$ m, and $T = 2.5 \times 10^6$ K and $\rho_e = 1.4 \times 10^{16}$ m^{-3}.

At the time of the sunspot maximum, there are many bright coronal arcs *(loops),* at the time of the minimum the Sun appears very faint in X-ray light.

In the chromosphere and the corona the plasma follows the magnetic field lines because of the low density.

7.4   The Space-Weather-Solar-Terrestrial Relations

We have discussed numerous phenomena on the Sun or in its atmosphere, where energetic short-wave radiation or particles are released.

Under the term space weather we summarize their effects on the physics of the Earth and near-Earth space, or in general their influence on the planets.

In the solar system these processes can be studied in detail and they are also of great importance in clarifying the question of habitability of Exoplanets.

7.4.1   The Solar Activity Cycle

Sunspot counts are among the oldest scientific records ever. After the first telescopic observations in 1610 by Galileo and others, interest in sunspots waned. The cause, as we now know, was the fact that there were almost no sunspots, especially between 1645 and 1715 → Maunder Minimum.

The German amateur astronomer *Samuel Heinrich Schwabe* (1789–1875) wanted to find planets within Mercury's orbit, so he observed sunspots very carefully. He did not discover an intramercurial planet, but he did discover the eleven-year sunspot cycle. The Swiss astronomer *Rudolf Wolf* (1816−1893) then introduced the *Sunspot relative number R* ("Zurich number"). Today the relative number R is defined as the mean of observations from different observatories,[5] and one forms monthly averages. The Greenwich Heliophysical Observatory made records from 1874 to 1976, then the Debrecen Heliophysical Observatory.

With the *Sunspot cycle* other solar activity phenomena also change, such as the frequency of flares and CMEs.

---

[5] SIDC Sunspot number, Solar Influences Data Center, Brussels.

- Variation of CMEs: 1/day at minimum–6/day at maximum
- Solar wind

  (a) Minimum: fast component (800 km/s) emitted almost all over the Sun, slow component (400 km/s) in low latitude regions.
  (b) Maximum: slow component becomes dominant; greater symmetry.

- Flares: the number of M and X flares, N_M resp. N_X, is correlated with the *sunspot relative number R*, and one finds the following approximate formula:

$$N_M = 2.86R \qquad N_X = 0.23R \qquad (7.20)$$

Besides the variation with the activity cycle, there appear to be other cycles of flare activity with periods of multiples of 24 days.

Other longe periods of solar activity are the *Gleissberg cycle* (about 90 years) as well as the 22-year magnetic Cycle *(Hale cycle)*.

7.4.2   Time Series, Period Analysis

Such periods can be analyzed by means of a *power spectrum* of relative numbers. The power spectrum of a time series is essentially the square of its Fourier transform (Chap. 17). Let us take a sine-function. This has a period, and the associated power spectrum therefore has one *peak*. For a function with two periods, two peaks are found, and so on. From the individual peaks in the power spectrum one can therefore conclude on periods of a signal.

There are numerous programs for data analysis in the field of astronomy, which can be downloaded from the internet. Especially widespread in the field of professional astrophysics are IRAF, MIDAS, ANA and of course the programming language Python, in which there are numerous ready-made astronomy packages. These are freely downloadable without license fees. IDL has a fee, but is very often used in the field of solar physics; a demo version, which runs for about eight minutes without restrictions, can also be obtained from the Internet. All mentioned programs work with a PC under Windows, Linux or MacOS. As a good exercise try to download such a program and perform experiments with power spectra.

To clarify the question whether the parts of the power spectrum at high frequencies are periodic or not, one examines the underlying attractor. An attractor in the *phase space* is a subspace of the phase space when, if $t \rightarrow \infty$, is taken. One uses the correlation integral to determine the *fractal dimension* of this attractor. A high fractal dimension means *Turbulence*, Chaos. Periodic or multiply periodic systems have an integer dimension, the value of which depends on the number of modes. If one has a low fractal dimension, then the system can be described with a few ordinary differential equations. Suppose the data are given as time series:

$$x(t_1), \ldots, x(t_n)$$

$$x(t_1 + \Delta t), \ldots, x(t_N + \Delta t)$$

$$x(t_1 + (d-1)\Delta t), \ldots, x(t_N + (d-1)\Delta t)$$

Let $x_1 = [x(t_i), x(t_i + \Delta t), \ldots, x(t_i + (d-1)\Delta t)]$ a be a d-dimensional vector describing a point in the d-dimensional space. So we can reconstruct an attractor from a scalar time series in this way, from a single variable. The real attractor, which has generated this data sequence, is of course embedding this d-dimensional attractor, if $d \to \infty$. Therefore one determines $|x_i - x_j|$ for $d = 1$ and calculate the correlation integral of q-th order $C_{d=1}^{(q)}(r)$. Then one increases the dimension of the artificial phase space and calculates $C_{d=2}^{(q)}(r)$. This is repeated until of increase of the correlation integral does not change any more. The Embedding theorem from *Takens* says that $d \geq 2n + 1$ is sufficient, where n is the actual but unknown dimension of the attractor.

This shows that one can arrive at a complete description of a physical system using a time series.

7.4.3   The Solar Irradiance

Is the brightness of the Sun changing, and if so, what effect might this have on the Earth's climate?

Since 1979 the *solar irradiance* has been measured by satellites (ACRIM, ERBE, SOHO, ...). In principle, there are several components of its change, probably on different time scales:

$$\frac{\Delta S_\odot}{S_\odot} = \frac{\Delta S_S}{S_\odot} + \frac{\Delta S_F}{S_\odot} + \frac{\Delta S_N}{S_\odot} + \frac{\Delta S_{DO}}{S_\odot} + \frac{\Delta S_{NM}}{S_\odot} \tag{7.21}$$

Thus, the solar irradiance changes go back to:

- ΔS_S Spots;
- ΔS_F flares;
- ΔS_N network, this is seen well in Ca light; it coincides with the *Supergranulation cells* (about 30,000 km diameter of each cell);
- ΔS_{DO} caused by deep magnetic fields; these cause disturbances in the heat transport at the base of the *Convection zone* and could show up as brightenings or darkenings at the surface;
- ΔS_{NM} non-magnetic origin.

The Maunder Minimum

The American astronomer *Douglass* studied tree rings on felled trees around 1900 and found that they had a specific pattern that was repeated at eleven-year intervals. Between 1645 and 1715 this periodicity disappeared. The English astronomer *Maunder* sent in 1922 Douglass a letter stating that virtually no sunspots were seen at that very period. Climate records from this period revealed what is known as the Little Ice Age in Europe. That gives you the context:

> High solar activity → global temperature increase on Earth.

Since sunspots are more than 1000 K cooler than the surrounding photosphere, this at first seems like a contradiction. Less energy is emitted in sunspots, but this radiation deficit is overcompensated by enhanced radiation in the bright and hot flare regions.

A few years later *Eddy* found the connection between solar activity and the ^{14}C-Concentrations in tree rings. The sun emits electrically charged particles. But energetic particles also reach us from sources outside the solar system (supernova explosions, the nucleus of the galaxy, etc.). If the earth had no magnetic field, these particles would hit the earth's surface unhindered.

We are doubly protected from cosmic ray particles:

- *Earth's magnetosphere* mainly protects us from cosmic ray particles, rays coming from the sun itself.
- The *magnetic field of the Sun* extends into interplanetary space, as do solar wind particles in this space, which is called the *heliosphere*. The *Heliopause* is located at about 100 AU. In the area of the *Termination Shock* the solar wind particles are decelerated from supersonic velocity of the particles to a velocity below the local sound velocity of the plasma, i.e. from about 350 km/s to 150 km/s. The medium is compressed, heating occurs. The medium compresses, heating occurs. Behind the termination shock is the *heliosheath,* which has an extension of several 10 AU. At the heliopause the influence of the solar wind ends. On 30 August 2007 Voyager 2 reached[6] the Termination Shock, which was located at a distance of 84 AU. The distance of the shock is determined by the solar activity.

The *Solar Activity* is linked to the Sun's magnetic field: When activity is high, the density of the interplanetary magnetic field is higher than when solar activity is low. Therefore, when solar activity is low, more energetic fractions of cosmic rays can reach the Earth's atmosphere, where they then produce secondary particles and neutrons. When a neutron

---

[6] Launched on 20 August 1977.

collides with a nitrogen nucleus ^{14}N the radioactive ^{14}C isotope is produced. The plants then take this up, and you can therefore measure from the ^{14}C how much solar activity there has been. However, the whole thing only occurs with a time delay (about 20 years). Since year 1, the following minima or maxima have been found:

- Maunder minimum: 1645–1715,
- Spörer minimum: 1460–1550,
- Medieval maximum: 1100–1250 (unusually warm period, Greenland means green land),
- Medieval minimum: 640–710,
- Roman Maximum: 20–80.

In addition to the eleven-year cycle, there is probably the approximately 90-year Gleissberg cycle.

So we can use the following indicators *(proxies)* back into the past. :

- Relative numbers, spots;
- Cosmogenic isotopes: ^{14}C; Be-isotopes .
- Polar Ice Drilling : The particles cannot move crosswise to the magnetic field lines and preferentially arrive at the poles; thus the effects are greatest here.

So does the sun change our climate? Generally speaking *Climate changes* occur due to the following causes:

- *Circulation* in the Earth's atmosphere or in the oceans (cf. Gulf Stream); this is related to the movement of continental plates over the course of Earth's history (plate tectonics).
- *Mountain formation:* Mountains and plateaus have great influence on climate; on the side where winds rise you have precipitation, on the other side you have drought. Ex: Andes: On the eastern slopes you have dense forests, on the western slopes deserts.
- Changes of the *Earth's orbit:* Due to the interference of the other planets, the eccentricity of the Earth's orbit changes by 6% in a cycle of 100,000 years. The tilt of the Earth's axis changes by about 3° with a period of 41,000 Years.
- *Greenhouse Effect:* Since the beginning of industrialization around 1750, the CO_2-content in the earth's atmosphere increased from 280 ppm to 360 ppm. But not only the CO_2 is responsible for the greenhouse effect, but also the methane content, which has more than doubled. This greenhouse effect could raise the global temperature between 1 °C and 3.5 °C (associated with a rise in sea level of 50 cm, which would currently affect 100 million people).

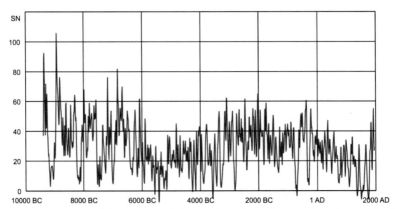

Fig. 7.25 Reconstruction of the sunspot relative numbers. It can be seen that solar activity has been increasing again since 1900 and that there are long periods of varying solar activity (according to Solanki)

- Solar activity: at the time of maximum, the Sun is about 0.1% brighter, corresponding to 1.3 W/m^2. Studies have shown that a 0.1% change in solar radiation leads to a global temperature change of 0.2 °C.

Since solar activity has been increasing overall since 1900, some of the temperature increase could be caused by the sun. Note here the increase since 1900 in Fig. 7.25. However, the observed increase in solar activity since 1900 is only sufficient to explain the Earth's temperature increase up to about 1970, i.e., the presently observed rapid temperature increase since 1970 is certainly due to the *anthropogenic greenhouse effect* (CO_2-increase).

So in the long run, the sun certainly determines our climate.

Space Weather

The Earth's magnetic field and the effects of the solar wind have already been discussed in Sect. 6.3.3 also the effects of incoming particles (Van Allen belt, auroras).

Changes in solar activity can influence currents in the Earth's atmospherere *(GIC, geomagnetically induced currents)*. These can cause surges in overhead power lines and thus destroy transformers, paralyzing the power supply to large areas (this occurred in Quebec in 1989). Geomagnetic activity is described by the *K-*, *Ap* and *Kp index*.

X-rays are produced during the eruption of a solar flare. This increases the ionization in the Earth's atmosphere, the ion density as well as the electron density increase. A sudden ionospheric perturbation occurs, *sudden ionospheric disturbance, SID*). Radio signals are already absorbed in the D layer of the atmosphere and radio communication is interrupted. On the other hand, at the maximum of solar activity, higher frequencies are reflected in the Earth's atmosphere, increasing the range for RF transmissions. High-energy particles from the solar wind cause the effect of bit reversal in computers, resulting in incorrect commands; satellites can get out of control. The Earth's upper atmosphere is heated by increased shortwave radiation during a flare, it expands, and near-Earth satellites can thus be severely slowed and crash to the surface. Solar flares also pose a significant radiation threat to astronauts.

That's why people all over the world are now trying to predict what is known as Space Weather. When and with which intensity is the outbreak of a solar flare or CME to be expected? What precautionary measures can be taken?

7.5   Helioseismology

In the year 1962 it was discovered that the upper photosphere oscillates up and down with a period of 5 min *(5 minute oscillation.)* In reality there is a superposition of many oscillations. Waves wander into sun-inside, where temperature is much higher, and there waves are reflected upward again (Fig. 7.26). The reflection at the solar surface occurs due to the extreme decrease in density and temperature.

> The sun therefore oscillates like a *(resonant cavity)*. By studying different frequencies, one can therefore detect different layers.

The propagation speed of sound waves is

$$v = \sqrt{\kappa \Re T / M} \tag{7.22}$$

$\kappa = c_p/c_v$, $\Re = 8.31\,\mathrm{J/(mol K)}$.

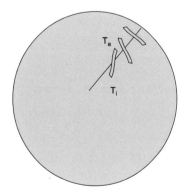

Fig. 7.26 Propagation of a wave front into the solar interior. Because of $T_i > T_a$ the speed of sound is also greater there, and the wave front is bent upwards

Table 7.2 Absorption lines
and oscillations observed in the
centres of the lines

Line [nm]	Element [nm]	Period [s]	v [km/s]
39.37	Ca II	150	2.00
656.28	Hα	180	1.34
516.87	Ni I	300	0.31

The most pronounced is the 5-minute oscillation. The higher the spectral lines originate in the solar atmosphere (photosphere), the greater the amplitude of the oscillation. In addition, the period (frequency) changes with altitude (cf. Table 7.2).

7.5.1   Mathematical Description

As will be shown in more detail in the chapter on stellar structure, one can start from the equations of an equilibrium state of a star: In such a state of equilibrium one has $\rho = \rho_0(r)$, $P = P_0(r)$, $\Phi = \Phi_0(r)$, $\mathbf{v} = 0$ i.e. density, pressure and gravitational potential depend only on the distance from the center r . If Φ the gravitational potential, $\mathbf{v}$ the velocity of a fluid, Γ the ratio of the specific heats, then holds:

$$\rho \frac{d\mathbf{v}}{dt} = - \operatorname{grad} P + \rho \operatorname{grad} \Phi \qquad (7.23)$$

$$\frac{d\rho}{dt} + \rho \operatorname{div} \mathbf{v} = 0 \qquad (7.24)$$

$$\frac{1}{P}\frac{dP}{dt} = \frac{\Gamma}{\rho}\frac{d\rho}{dt} \qquad (7.25)$$

$$\nabla^2 \Phi = - 4\pi G\rho \qquad (7.26)$$

The first equation is the equation of motion, the second the equation of continuity, the third the adiabatic equation, and the fourth the Poisson's Equation. Now suppose one can take any variable f in the form: $f = f_0 + f_1$ with

$$f_1 = \operatorname{Re}\left[\exp(i\omega_{nl}t)\bar{f}_1(r)Y_l^m(\Theta, \phi)\right] \qquad (7.27)$$

where Re is the real part and $Y_l^m(\Theta, \phi)$ the spherical function.

$$Y_l^m(\Theta, \phi) = P_l^m(\Theta)\exp(im\phi), \qquad (7.28)$$

where $P_l^m(\Theta)$ is the associated Legendre function. If a star is spherical, then its oscillation frequency does not depend on m. Only when the deviations from the spherical shape are small do the above formulas hold. In the case of the sun one has different m-modes with different frequencies. The oscillation frequency depends on l and n. The numbers n, l,

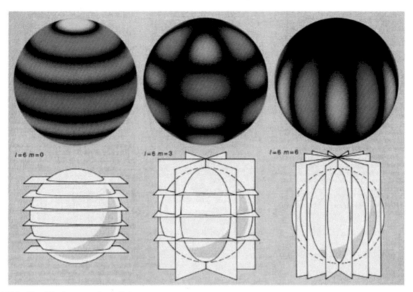

Fig. 7.27 Oscillation modes of the sun

m indicate how often the function f_1 in the radial as well as in the θ- and ϕ-direction vanishes. Furthermore $|m| \leq l$ (Fig. 7.27).

In solving the system of equations one has the problem of boundary conditions. At the center of a star, everything is clear. But stars do not have a defined surface. One could simply assume that all waves are reflected back at the surface of the Sun (since there, by definition P, ρ disappear). In reality, however, some energy is transferred outward into the solar atmosphere.

If the changes in gravitational potential due to the oscillations are negligible, we have a 2nd order DE (differential equation). This is sufficient for most perturbations: certain parts of the star move outward, other parts move inward. In the case of radial oscillations only, a perturbation vector ξ may be introduced as:

$$\mathbf{v} = \frac{d\xi}{dt} \tag{7.29}$$

and further:

$$\Psi = c_s^2 \rho_0^{1/2} \operatorname{div} \xi \tag{7.30}$$

Thereby c_s is the speed of sound in the undisturbed star:

$$c_s = (\Gamma P_0 / \rho_0)^{1/2} \tag{7.31}$$

The equation for the radial part is then simply:

$$\frac{d^2\Psi}{dr^2} = -\frac{1}{c_s^2}\left[\omega^2 - \omega_c^2 - S_l^2\left[1 - \frac{N^2}{\omega^2}\right]\right]\Psi \tag{7.32}$$

We therefore have, in addition to ω the three frequencies:

- ω_c *acoustic cutoff frequency*

$$\omega_c^2 = \left(c_s^2/4H_p^2\right)(1 - 2dH_p/dr) \tag{7.33}$$

- S_l *Lamb frequency*

$$S_l = c_s[l(l+1)]^{1/2}/r \tag{7.34}$$

- N *Brunt-Väissälä frequency:*

$$N^2 = g\left[\frac{1}{\Gamma P}\frac{dP}{dr} - \frac{1}{\rho}\frac{d\rho}{dr}\right] \tag{7.35}$$

where the pressure scale height is given by:

$$H_p = |\rho/(d\rho/dr)| \tag{7.36}$$

and

$$g = GM/r^2 \tag{7.37}$$

It can be seen that: S_l is always real, ω_c, N can become imaginary. If N^2 becomes imaginary, one has convection (chapter about star-construction!).

Equation 7.32 can be written as:

$$\frac{d^2\Psi}{dr^2} + K_r^2\Psi = 0 \tag{7.38}$$

- $K_r^2 > 0$: behavior of solution depends on radius.
- $K_r^2 < 0$: exponential behaviour, the modes fall exponentially (*evanescent modes*).

There are two domains where K_r^2 is positive:

- High frequencies $\omega > S_l, \omega_c$: Then the pressure fluctuations become important, one speaks of p-modes.
- Low frequencies $\omega < N$: One speaks of *Gravity mdoes* (engl. *gravity modes*), *g-modes*.

S_l, ω_c, N depend on location in the sun. S_l decreases monotonously from center to surface, N becomes imaginary in the *Convection zone*. The p-modes can propagate into the solar interior up to the Lamb frequency, and at the surface they terminate at the acoustic cut off frequency. The g-modes are absorbed or reflected in the convection Zone. Therefore the p-modes can be observed more easily than the g-modes, because g-modes are exponentially attenuated in the convection zone. The p-modes of smallest order (l) penetrate deepest into the Sun. Together with the g-modes, they thus provide clues to the deep interior of the Sun.

The Sun is not strictly spherical: one must consider rotation and magnetic field. For a rotating star, the oscillation frequency depends on m, each frequency ω_{nl} thus splits into the $2l + 1$ frequencies ω_{nml}. From this rotational splitting, one can study how the Sun's rotational velocity changes in the interior.

7.5.2   Observational Results

The SOHO-MDI- the VIRGO instruments have been used to measure vertical velocities propagating through the Sun. With VIRGO oscillations of the solar brightness were measured. Figure 7.28 shows deviations from the mean rotation of the Sun (left) and temperature deviations in the interior of the Sun (right). A temperature increase has been detected at the transition zone between the convection zone and the radiation zone. At this zone, called *tachoclyne,* also occurs a jump in the rotation speed of the sun:

- The interior rotates slower and like a rigid body,
- the outer layers rotate faster and differential.
- Therefore, shearings occur; these shearings cause the solar dynamo.

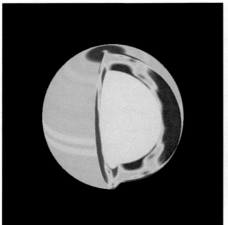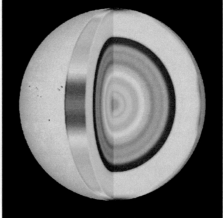

Fig. 7.28 Temperature deviations in the Sun (right), deviations from the mean solar rotation (left) (NASA/ESA SOHO)

Fig. 7.29 Course of solar
rotation at the surface
(differential, at $r/R = 1$); from
the so-called tachoclyne the
Sun rotates like a rigid body
(SOHO-MDI, ESA/NASA)

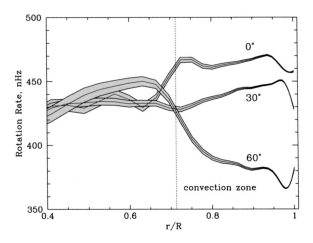

The zone in the centre of the sun, where the nuclear reactions take place, indicates that this is 0.1% cooler than the assumed 15 million K. So possibly the sun is producing less energy today.

With the *Michelson Doppler Imager* (MDI) on board SOHO, the rotation of the Sun has also been measured. In Fig. 7.28, left, dark means faster rotation than average, light means smaller. A band-like structure is found that rotates slightly faster than the surrounding area, and this structure extends about 20 000 km into the depths. Sunspots form at the edges of these bands.

In Fig. 7.29 results of SOHO-MDI observations are reproduced, showing the pattern of rotation as a function of distance from the center of the Sun. At the solar surface ($r/R = 1$) one can clearly see the differential rotation, i.e. higher frequency at the equator than e.g. at 60° heliographic latitude. At the transition layer between the convection zone and the radiation zone (tachoclyne) the solar rotation changes into a rigid rotation. This results in strong shear in the tachoclyne region.

SOHO/VIRGO was used to measure the brightnesses of the Sun at various wavelengths. The data can then be transformed into Fourier space and the *power spectrum* calculate. This shows over which frequencies a signal is distributed. In the data one can see the p-modes as well as the granulation and supergranulation.

Since February 2010 the NASA mission SDO (Solar Dynamics Observatory) has been sending data. The EVE (Extreme UV Variability Experiment) can be used to obtain images of the Sun every 10 s in the range 0.1 to 105 nm. With HMI (Helioseismic and Magnetic Imager) one can determine the variation of the magnetic field, and with AIA (Atmospheric Imaging Assembly) one gets an image every 10 s in nine different UV-ranges and at visible range.

Objective of SDO: Evolution and formation of activity regions on the Sun.

7.6   Magnetohydrodynamics of the Sun

The *magnetohydrodynamics (MHD)* studies the connection between magnetic fields and plasma motions.

We give a brief overview with applications to solar physics.

7.6.1   Maxwell's Equations

We want to give here briefly the basic equations. Starting point are the Maxwell's equations:

$$\nabla \times \mathbf{B} = \mu \mathbf{j} + \frac{1}{c^2} \frac{\partial \mathbf{E}}{\partial t} \tag{7.39}$$

$$\nabla.\mathbf{B} = 0 \tag{7.40}$$

$$\nabla \times \mathbf{E} = -\frac{\partial \mathbf{B}}{\partial t} \tag{7.41}$$

$$\nabla.\mathbf{E} = \frac{\rho_e}{\epsilon} \tag{7.42}$$

B—magnetic induction, **E**—electric field strength, **j**—electric current density, ρ_e—electric charge density, ϵ—dielectric constant, μ—magnetic permeability.

In astrophysics we assume ϵ, μ as constant and $\epsilon \approx \epsilon_0 = 8.854 \times 10^{-12}\,\mathrm{F\,m^{-1}}$, $\mu \approx \mu_0 = 4\pi \times 10^{-7}\,\mathrm{H\,m^{-1}}$. Further it is to be noted that the properties of the plasma are isotropic except for one exception: The *thermal heat conduction* κ is preferentially along the magnetic field lines!

The simplified *Ohm's law* is:

$$\mathbf{j} = \sigma\,(\mathbf{E} + \mathbf{v} \times \mathbf{B}) \tag{7.43}$$

In the case one has a plasma consisting of electrons and only one type of ions, then:

$$\mathbf{j} = n_i\,Z_i e \mathbf{v_i} - n_e e \mathbf{v_e} \tag{7.44}$$

$$\rho_E = n_i\,Z_i e - n_e e \tag{7.45}$$

n_i—Number of ions, n_e—number of electrons, $\mathbf{v}_e$—velocity of the electrons. $Z_i e$—charge of the ion, $-e$—charge of the electron.

In general, in almost all astrophysical cases one assumes that magnetic fields are permanent and electric fields can be neglected.

There are no magnetic monopoles ($\nabla \mathbf{B} = 0$). Electric fields are produced by changing magnetic fields. Magnetic fields, which are generated by the displacement current $1/c^2 \partial \mathbf{E}/\partial t$ are negligible. If one has a high current density $\mathbf{j}$ then magnetic fields are generated (cf. 1st Maxwell equation). For a stationary medium holds:

$$\nabla \times B = \mu \mathbf{j} \tag{7.46}$$

$$\nabla \times \mathbf{E} = - \partial \mathbf{B}/\partial t \tag{7.47}$$

$$\mathbf{j} = \sigma \mathbf{E} \tag{7.48}$$

This gives the following equation insert (7.48 into 7.46 and then form the vector product):

$$\frac{\partial \mathbf{B}}{\partial t} + \frac{1}{\mu_0 \sigma} \nabla \times \nabla \times \mathbf{B} = 0 \tag{7.49}$$

and because of $\nabla \times \nabla \times \mathbf{B} = \mathrm{grad\,div}\mathbf{B} - \nabla^2 \mathbf{B}$ and $\mathrm{div}\mathbf{B} = 0$:

$$\frac{\partial \mathbf{B}}{\partial t} = \frac{1}{\mu_0 \sigma} \nabla^2 \mathbf{B} \tag{7.50}$$

In Cartesian coordinates, one then obtains, e.g.:

$$\frac{\partial \mathbf{B_x}}{\partial t} = \frac{1}{\mu_0 \sigma} \left[\frac{\partial^2 \mathbf{B_x}}{\partial x^2} + \frac{\partial^2 \mathbf{B_x}}{\partial y^2} + \frac{\partial^2 \mathbf{B_x}}{\partial z^2} \right] \tag{7.51}$$

The solution of this equation indicates how magnetic fields decay, along with the currents they produce. One can estimate the approximate decay time:

$$\tau_D = \mu_0 \sigma L^2 \tag{7.52}$$

$L \ldots$ within this distance the currents should change. From the DG 7.51 it also follows that if at time $t = 0$ a field of the form $B_x = B_0 \exp(iky)$ is given, the equation is:

$$B_x = B_0 \exp(iky) \exp(-k^2 t/\mu_0 \sigma) \tag{7.53}$$

$\lambda = 2\pi k$ Is the wavelength of the spatial change of the field.

In the laboratory, the currents decay very quickly because the material has a low expansion. In stars, one has a high conductivity as well as a large L. That is why there is a fossil field here, which comes from the time of star formation.

7.6.2 Induction Equation

With the *induction equation* we describe how a magnetic field develops with plasma movements.

From 7.43 one eliminates $\mathbf{E} = \mathbf{j}/\sigma - (\mathbf{v} \times \mathbf{B})/\sigma$. Thus, Eq. 7.41 to:

$$\nabla \times \frac{1}{\sigma}[\mathbf{j} - \mathbf{v} \times \mathbf{B}] = -\frac{\partial \mathbf{B}}{\partial t}$$

Since the velocities under consideration are small compared to c, one can use in 7.39 the term $1/c^2(\partial \mathbf{E}/\partial t)$ omit and $\nabla \times \mathbf{B} = \mu_0 \mathbf{j}$ (*induction equation*):

$$\frac{\partial \mathbf{B}}{\partial t} = \nabla \times (\mathbf{v} \times \mathbf{B}) + \eta_0 \nabla^2 \mathbf{B} \tag{7.54}$$

where $\eta_0 = (\mu_0 \sigma)^{-1}$ is the magnetic diffusivity, σ the electrical conductivity.

- The ratio $\nabla \times (\mathbf{v} \times \mathbf{B})/(\eta_0 \nabla^2 \mathbf{B})$ is known as *magnetic Reynolds number* $R_m = l_0 v_0 / \eta_0$.
- Diffusion $\sim \eta_0 \nabla^2 \mathbf{B}$.
- $R_m \gg 1$ then one can neglect the diffusion $\to$ magnetic field is frozen in the plasma and one can also write $\mathbf{E} = -\mathbf{v} \times \mathbf{B}$.

7.6.3 Plasma Equations

Now we can write down how plasma moves. Here the total derivative with respect to time is composed of

- Space point fixed, time derivative, thus $\partial/\partial t$
- Time point fixed, difference of velocity at different space points: $\mathbf{v}\nabla$.
    Therefore is the *substantial derivative:*

$$\frac{D}{Dt} = \frac{\partial}{\partial t} + \mathbf{v}\nabla \tag{7.55}$$

One has the *Mass continuity:*

$$\frac{D\rho}{Dt} + \rho\nabla\mathbf{v} = 0 \tag{7.56}$$

The *Lorentz force* is given by :

$$F_L \approx \mathbf{j} \times \mathbf{B} \tag{7.57}$$

The *Equation of motion* of plasma becomes:

$$\rho\frac{D\mathbf{v}}{Dt} = -\nabla p + \mathbf{j} \times \mathbf{B} + \rho\mathbf{g} + \rho\nu\nabla^2\mathbf{v} \tag{7.58}$$

and thereby is: ∇p a pressure gradient, $\mathbf{j} \times \mathbf{B}$ the Lorentz force, $\rho\mathbf{g}$ the force of gravity (gravitational acceleration $g_\odot =264$ m s^{-2}), $\rho\nu\nabla^2\mathbf{v}$ viscosity.

Furthermore, one still needs an energy equation. Into this goes the thermal conductivity. For the corona, note:

$$\frac{\kappa_\perp}{\kappa_\parallel} \approx 10^{-12} \tag{7.59}$$

I.e. heat conduction occurs mainly along the field lines.

The magnetic force on a moving charge ($\mathbf{v}$) is

$$\mathbf{F} = q\mathbf{v} \times \mathbf{B} \tag{7.60}$$

If one has a magnetic field in a conducting fluid, then the force exerted by it is:

$$\mathbf{F_{mag}} = \mathbf{j} \times \mathbf{B} = \frac{1}{\mu_0}\nabla \times (\mathbf{B} \times \mathbf{B}) \tag{7.61}$$

(here we have Eq. 7.46 used). This can be written as:

$$\mathbf{F_{mag}} = -\mathrm{grad}\,(B^2/2\mu_0) + \mathbf{B}\nabla\mathbf{B}/\mu_0 \tag{7.62}$$

The two right-hand terms mean:

1. Gradient of an isotropic pressure,
2. Stress along the field lines.

Therefore, if one has a *magnetic flux tube* with the pressure P_i and if the pressure in the surrounding medium is P_o, then the equilibrium is:

$$P_o = P_i + \frac{B^2}{2\mu_0} \tag{7.63}$$

$P = \Re\rho T/\mu$... gas pressure. If holds $T_i = T_o$ then it follows:

$$\rho_i < \rho_a \tag{7.64}$$

So our flux tube is lighter than the surroundings and rises upwards, *magnetic buoyancy.*

There is a connection between the plasma and the fields:

• The plasma determines the motion of the magnetic fields (photosphere of the Sun); we speak of frozen field lines.

- The magnetic field determines the motion of the plasma (when the density of the plasma is low, e.g. corona).

The speed of sound in a gas is:

$$c_s = \sqrt{\gamma P/\rho} \tag{7.65}$$

Hydromagnetic waves propagate with the *Alfvén velocity*:

$$c_H = \sqrt{B^2/\mu_0 \rho} \tag{7.66}$$

7.6.4   Motion of a Particle in a Magnetic Field

Now let's consider the motion of a single charged particle in an electromagnetic field; q...particle charge; the equation of motion (Lorentz force) reads:

$$m\frac{d\mathbf{v}}{dt} = q(\mathbf{E} + \mathbf{v} \times \mathbf{B}) \tag{7.67}$$

If **b** is a unit vector, then we decompose the fields into two components:

$$\mathbf{B_0} = B_0 \mathbf{b} \tag{7.68}$$

$$\mathbf{E} = E_\| \mathbf{b} + \mathbf{E}_\perp \tag{7.69}$$

$$\mathbf{v} = v_\| \mathbf{b} + \mathbf{v}_\perp \tag{7.70}$$

Since $\mathbf{v} \times \mathbf{B_0} = B_0(\mathbf{v}_\perp \times \mathbf{b})$ perpendicular to **b**, we get from the equation of motion above:

$$m\frac{dv_\|}{dt} = q E_\| \tag{7.71}$$

$$m\frac{dv_\perp}{dt} = q[\mathbf{E}_\perp + B_0(\mathbf{v}_\perp \times \mathbf{b})] \tag{7.72}$$

From the first equation, we immediately have:

$$v_\| = (q E_\|/m)t + v_{\|0} \tag{7.73}$$

So: the motion of the particles is parallel to the field lines. Particles of different charges q move in opposite directions! The solution of the second equation results in a circular motion around **b** with the Frequency $|q|B_0/m$ and the *Gyration radius* $r_g = m v_{\perp 0}/|q|B$.

Example

An electron has the Gyration frequency $1.8 \times 10^{11}(B/\text{T})$ Hz and a gyration radius of $6 \times 10^{-9}(v_{\perp 0}/\text{km s}^{-1})/(B/T)$ m.

If there is an additional non-magnetic force $\mathbf{F}$ normal to $\mathbf{B}$ then the result is a Drift velocity of:

$$\mathbf{v}_{\text{DF}} = \mathbf{F} \times \mathbf{B}/q\,B^2 \tag{7.74}$$

If $\mathbf{F} = m\mathbf{g}$, i.e. gravitation, then one sees that the *Drift velocity* is proportional to the ratio m/q, i.e. the *Ion drift* is much larger than *electron drift*. This drift of particles in different directions produces currents.

Now to understand understand Flares, one needs the concept of *magnetic reconnection.* of Suppose two oppositely directed magnetic field lines are brought closer together by plasma motion. When the field lines approach each other, the gradient becomes large, and between them there is a neutral point where the field disappears. Dissipation now causes reconnection of the field lines, and the plasma then moves in the opposite direction to the original motion. This causes a release of magnetic energy in the vicinity of the neutral point.

Solar Dynamo

The activity cycle of the sun can be explained with the *dynamo theory*. The starting point is a poloidal field extending from pole to pole along the z-axis. Due to the differential rotation of the sun (rotation is faster at the equator than at the poles, ω-Effect), the magnetic field lines are wound up in the toroidal direction, and a toroidal field is formed (Fig. 7.30). From the toroidal field, Coriolis force and convection again produce a poloidal component (α-effect).

For the mathematical description the principle of *mean field electrodynamics* is used: The magnetic induction and the velocity are divided into a mean and a fluctuating part written:

$$\mathbf{B} = \mathbf{B_0} + \mathbf{b}, \qquad \mathbf{v} = \mathbf{v_0} + \mathbf{u} \tag{7.75}$$

The average magnitudes $\mathbf{B_0}$, $\mathbf{v_0}$ change only slowly with time. $\mathbf{u}$ let be given, $\mathbf{b}$ shall be found: Put above approach and get for the mean and fluctuating part:

$$\frac{\partial \mathbf{B_0}}{\partial t} = \nabla \times (\mathbf{v_0} \times \mathbf{B_0}) + \nabla \times <\mathbf{u} \times \mathbf{b}> +\eta \nabla^2 \mathbf{B_0} \tag{7.76}$$

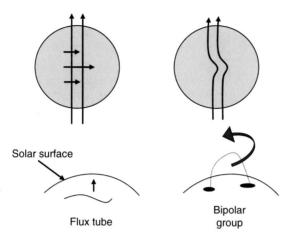

Fig. 7.30 Principle of the solar dynamo. The ω-effect is the winding up of the field lines at the equator due to differential rotation. A poloidal field is transformed into a toroidal one. Flux tubes rise from the interior of the sun to the surface due to magnetic buoyancy and pierce the photosphere forming a bipolar group. Due to the α effect (twisting of the field lines by convection and Coriolis force) the toroidal field again becomes an inverted poloidal field

and:

$$\frac{\partial \mathbf{b}}{\partial t} = \nabla \times (\mathbf{b_0} \times \mathbf{b}) + \nabla \times (\mathbf{u} \times \mathbf{B_0}) + \nabla \times (\mathbf{u} \times \mathbf{b} - <\mathbf{u} \times \mathbf{b}>) + \eta \nabla^2 \mathbf{B} \qquad (7.77)$$

$\eta = 1/\mu_0\sigma$—magnetic resistance, $<>$—mean value.

Corona Heating
The following energy densities can be given for the corona:

- Kinetic energy

$$E_{\text{kin}} = \frac{1}{2} m_p n v^2 \qquad (7.78)$$

with $n = 10^{15}\,\text{m}^{-3}$ as particle density and $v = 1000\,\text{km/s}$ one obtains for the energy density $8 \times 10^{-4}\,\text{J\,m}^{-3}$.
- Thermal energy

$$E_{\text{therm}} = nkT \qquad (7.79)$$

with $T = 10^6$ K one obtains for the energy density $1 \times 10^{-2}\,\text{J\,m}^{-3}$

- Potential energy

$$E_{\text{pot}} = nm_p gh \tag{7.80}$$

m_p mass of the proton, $h = 10^5$ km The energy density is $5 \times 10^{-2}\,\mathrm{J\,m^{-3}}$
- Magnetic energy:

$$E_{\text{magn}} = \frac{B^2}{2\mu_0} \tag{7.81}$$

with 10^{-2} T the energy density is $40\,\mathrm{J\,m^{-3}}$. The magnetic energy dominates the corona.

Finally, we address the problem of *Heating* of the corona.

- Earlier idea: the solar surface (lower photosphere) is convective, sound waves are generated there; these have an energy density of:

$$\frac{1}{2}\rho v^2 \tag{7.82}$$

Thereby ρ the density of matter and v is the velocity of the particles carrying the waves. This energy is conserved as the waves propagate upward. As the wave moves into a region of lower density (the density naturally decreases from the photosphere into the corona), the wave will increase in amplitude. Eventually it will become a shock wave, releasing its energy into the surrounding medium.
- MHD waves: Energy dissipation as soon as the *Alfvén velocity* c_H is greater than the local speed of sound c_s is. Another possibility would be magnetic reconnection. Magnetic field lines often enter the corona in large arcs, their foot points migrating by convection. From the corona holes, which correspond to open magnetic field configurations, the solar wind flows away. In the ecliptic plane, the typical velocity is 400 to 500 km/s. Ulysses was a space mission in which a probe was to study primarily the poles of the Sun. To do this, the probe was first sent to Jupiter, where it received the necessary *gravity assist* Got to get out of the ecliptic plane. To the great surprise of solar physicists, it was found in 1995 (minimum solar activity) that there are also coronal holes at the poles from which the solar wind flows off at up to 700 km/s.
- Spicules:

  (a) Type I; amounts to $T < 100{,}000$ K, then cool plasma sinks back down,
  (b) type II; amounts to $T > 1{,}000{,}000$ K, then hot plasma rises upward into the higher corona. This has been established with HINODE satellite observations. In most spicules, plasma heating occurs only up to about 100,000 K, but there are some where the temperature rises to much higher values.

Heating causes the corona to expand → Solar wind.

7.7 Further Literature

We give a small selection of recommended further reading.
An Introduction to Waves and Oscillations in the Sun, S. Narayanan, Springer, 2012
Magnetohydrodynamics of the Sun, E. Priest, Cambridge Univ. Press, 2017
The Sun from Space, K. Lang, Springer, 2008
The Sun, M. Stix, Springer, 2004
The Sun and Space Weather, A. Hanslmeier, 2006

Tasks

7.1 Compare the solar radiation on Venus with that on Earth!

Solution

$$\frac{F_{\text{Venus}}}{F_{\text{Earth}}} = \frac{E_{\text{Sun}}}{4\pi d_{\text{Sun-Venus}}^2} \Big/ \frac{E_{\text{Sun}}}{4\pi d_{\text{Sun-Earth}}^2} = \frac{1^2}{0.72^2} = 1.9$$

7.2 The sun loses about $3 \times 10^{-14}\,M_\odot/yr$. How much of this does the earth absorb?

Solution

The earth takes the fraction
$$\frac{A_{\text{Earth}}}{A_{1\,AU}} = \frac{\pi R_{\text{Earth}}^2}{4\pi R_{1\,AU}^2} = \frac{(6\times10^6)^2}{4(1.5\times10^{11})^2} = 4 \times 10^{-9}$$
and thus $M = 8.8 \times 10^9\,\text{kg/day}$. So the captured solar wind makes the Earth heavier by almost nine billion kg per day. What simplifications have been made here?

7.3 Jupiter is about 5 times as far from the Sun as the Earth. How large does the sun appear in Jupiter's sky?

Solution

1/5 of the diameter in Earth's sky.

7.4 Determine the kinetic energy of a proton in the solar wind and compare it with the energy of a X-ray photon.

Solution

$E_{\text{kin}} = 1/2mv^2 = \frac{1}{2}(1.7 \times 10^{-27}) \times 450,000^2\,\text{J} = 1.7 \times 10^{-16}\,\text{J}$ (Solar wind: $v = 450$ km/s). An X-ray photon at a frequency of 10^{18} Hz has an energy of $E = h\nu = 6.626 \times 10^{-34} \times 10^{18}\,\text{J} = 6.626 \times 10^{-16}\,\text{J}$ Thus protons have an energy of the same order of magnitude as X-ray photons and can destroy cells, for example.

7.5 Calculate how long it takes the solar wind to reach the system Alpha Centauri ($d =$ 1.33 pc) to reach.

Solution

1.33 pc $= 3.9 \times 10^{13}$ km and $t = d/v = (3.9 \times 10^{13})$ s$/450 = 8.7 \times 10^{10}$ s $= 2700$ years

7.6 Estimate the magnetic Reynolds number in an active region of the solar surface! Typical values are: $l_0 \approx 700$ km, $\eta_0 = 1 \, \text{m}^{-2}\,\text{s}^{-1}$, $v_0 \approx 10^4$ m/s.

Solution

$R_m = 7 \times 10^9 \gg 1$, therefore magnetic field is frozen in the plasma.

State Variables of Stars

8

In this section we deal with the determination of the most important properties or state variables of a star: radius, temperature, mass, density, gravitational acceleration, chemical composition, magnetic field and rotation. First, we briefly review the trigonometric distance determination. Although the distance of a star is not a state variable characterizing the star itself, its knowledge is necessary for the determination of other physical parameters of stars.

8.1 Distance, Magnitudes

The apparent brightness of a star depends on its true luminosity and its distance. Thus, although the distance of a star is not a quantity characterizing the star itself, it is important for deriving other essential physical state quantities.

8.1.1 Apparent Brightness

How bright is a star? The brightness depends on:

- the true luminosity of the star,
- the distance of the star.

In antiquity the concept of *magnitude classes,* lat. *magnitudo,* was introduced. The brightest stars called 1st magnitude stars, the faintest stars just visible to the naked eye are then 6th magnitude stars.[1]

---

[1] Because of light pollution in big cities you can only see stars up to about 3rd magnitude.

© Springer-Verlag GmbH Germany, part of Springer Nature 2023
A. Hanslmeier, *Introduction to Astronomy and Astrophysics,*
https://doi.org/10.1007/978-3-662-64637-3_8

Table 8.1 Apparent
brightnesses Stars

Sun	$-26\overset{m}{.}8$
Full Moon	-12^{m}
Venus	$-4\overset{m}{.}5$
Sirius	$-1\overset{m}{.}6$
Polaris	$+2\overset{m}{.}12$
Faintest stars visible to the naked eye	$+6\overset{m}{.}0$

Note Magnitude classes in astrophysics refer to stellar brightnesses and have nothing to do with the size (diameter) of a star!

Sensory perceptions (eye, ear) are always proportional to the logarithm of the stimulus; this is known as the *Weber-Fechner law*. To accommodate both the brightness scale of ancient astronomers and Weber-Fechner law it was defined: Given are two stars with the apparent magnitudes m_1, m_2. Let the intensity of their radiation be I_1 respectively. I_2. Then:

$$m_1 - m_2 = -2{,}5 \log(I_1/I_2) \tag{8.1}$$

$$I_1/I_2 = 10^{-0{,}4(m_1-m_2)} \tag{8.2}$$

m comes from the Latin term magnitudo. Furthermore, it was determined that for a difference of $\Delta m = 1$ the intensity ratio is 2.512. If $\Delta m = 2$ then the intensity ratio is 2.512×2.512 etc. To a $\Delta m = 5$ corresponds to an intensity ratio of 100 (2.512^5). This scale can also be used to indicate very bright objects as the sun and the moon (Table 8.1).

One can also convert the astronomical magnitudes into the usual physical values: The magnitudes depend on the spectral range—take, for example, the visual range, V, which is defined around a central wavelength of 550 nm, then the following holds: The flux at $V = 0$ is 3640 Jy, the $d\lambda/\lambda = 0.16$, where
   1 Jy = 1.51×10^7 Photons s^{-1} m^{-2} $(d\lambda/\lambda)$.

Let's calculate how many photons arrive outside the Earth's atmosphere (a) for a star $V = 0$ and (b) $V = 20.0$.

Solution
(a) $1.51 \times 10^7 \times 0.16 \times 3640 = 8794 \times 10^6$ photons m^{-2} $\approx 10^6$ photons cm^{-2}.
(b) At a brightness of $V = 20$ holds $10^{-0.4V} = 10^{-8}$ and thus $10^{-8} \times 1.51 \times 10^7 \times 0.16 \times 3640 = 8794 \times 10^{-2}$ photons m^{-2} $\approx 10^{-2}$ photons cm^{-2}.

How do you set the zero point of the scale? This is done using the international *pole sequence,* a series of Stars around the celestial north pole.

Star magnitudes are given in magnitude classes. The larger this value, the fainter the star.

8.1.2 Distance

> The annual movement of the earth around the sun leads to the *stellar parallaxes.*

A nearby star is seen at different angles against more distant background stars, so the position of a nearby star in the sky shifts slightly with a period of one year.

The main objection to the *heliocentric system* was always that such parallaxes could not be detected. Only in 1838 *Bessel* determined the parallax of the star 61 Cygni and *Struve* those from Wega. The problem lay in the accuracy of the measurement, since all stellar parallaxes are below $1''$. If a is the distance sun-earth and r the distance earth-star, then for the parallax of a star is valid

$$\pi [\text{rad}] = a/r \tag{8.3}$$

A star is located at a distance of 1 pc (1 pc = 206,265 AU), if its parallax is $1''$ is 1.

$$r[\text{pc}] = 1/\pi[''] \tag{8.4}$$

8.1.3 Absolute Brightness, Distance Modulus

As already mentioned, the *apparent magnitude m* of a star depends on the luminosity and the distance:

$$m = m(L, r) \tag{8.5}$$

The apparent brightness says nothing about the true brightness (luminosity).

Therefore one still has the term *absolute brightness, M.* The absolute brightness is understood to be the apparent magnitude that a star would possess at a unit distance of 10 pc = 32.6 light-years.

Relationship between apparent and absolute brightness:

$$m - M = 5 \log r - 5 \tag{8.6}$$

The expression

$$m - M \tag{8.7}$$

is called *Distance modulus.*

Table 8.2 Absolute
magnitudes of various objects

Object	Absolute brightness
Brightest galaxies	−23
Supernova 1987 A	−15.5
Globular cluster	−10...−6
Brightest stars	−9
Sun	+4.79
Weakest stars	∼ 20

Table 8.3 HIPPARCHOS
parallax measurements of some
bright stars

Star	HIP no.	Brightness in V	Parallax in ″
Sirius	32349	−1.44	0.37931
Canopus	30438	−0.62	0.01043
Rigil Kent	71683	0.01	0.74212
Arcturus	69673	−0.05	0.08885
Vega	91262	0.03	0.12893

The absolute brightness of the sun is about $4.^m8$. At a distance of 10 pc it would therefore be an inconspicuous star, just visible to the naked eye.

In Table 8.2 the absolute luminosities of some objects are given. Thus, a supernova at a distance of 10 pc would shine much brighter than the full Moon.

Using the HIPPARCHOS satellite one could measure the parallaxes of 18,000 stars with an accuracy of 10^{-3}. The mission (carried out by the European Space Agency ESA) took place between 1989 and 1993. A total of 120,000 astrometric and photometric data were recorded (Table 8.3).

ESA's *GAIA* (Global Astrometric Interferometer for Astrophysics, launch Dec. 2013) mission measures parallaxes of more than 10^9 Stars in our Milky Way. Similar to the solar satellite SOHO, GAIA was positioned at one of the Lagrange points in the Sun-Earth system (in this case L_2). The satellite measured each star about 70 times during the five-year mission. The measurement accuracy for stars up to the 15th magnitude class is 20 μas where 1 μas $= 10^{-6}$ arcseconds. In April 2018, the DR2 catalog was released, covering more than 1.7 billion stars.

The main objectives of the GAIA mission (see Fig. 8.1) were:

- Astrometry: accurate position determination, parallax determination, determination of proper motion of objects.
- Radial velocity measurements of more than 100 million stars (brighter than 17^m).
- Determination of important physical parameters such as mass, temperature.
- Accurate tests of general relativity (by measuring curvature of space).
- Discovery of exoplanets.

Other methods of determining the distance of stars are discussed in the following chapters.

Fig. 8.1 The GAIA satellite. ESA

8.1.4 Bolometric Brightness

Stars do not only radiate in the visible range, but also e.g. in the UV, IR, X-ray range. The measured luminosities in the visible spectral range are therefore too low. In order to get a measure for the total radiation of a star, one introduced the term *bolometric brightness*.

The *bolometric correction B.C.* is used to correct for these missing amounts of energy. The bolometric brightness refers to the entire spectrum.

Example The absolute brightness bolometric brightness of the Sun is $4.^M87$. The absolute bolometric magnitude $4.^M74$.

The brightest stars reach -9^M, the faintest $+17^M$.

All magnitudes must still,be corrected for the amount of interstellar extinction.

The bolometric magnitude measures the total radiation of a star.

8.2   Stellar Radii

In this section we address the problem of determining the true stellar diameters. Since the stars are far away, it is very difficult to determine the angular diameters.

8.2.1   Basic Principle

An object with the true diameter D appears at a distance r at an angle Θ:

$$\Theta\,[''] = 206,\,265\,\frac{D}{r} \tag{8.8}$$

This formula is valid only for small angles.

For the *sun* one determines the radius trigonometrically. The apparent solar diameter is $32'$ and with the solar distance of $r = 149 \times 10^6$ km one obtains the Solar radius to

$$R_\odot = 6.96 \times 10^5 \text{ km} \tag{8.9}$$

For stars, the problem is to measure the extremely small apparent diameter as a consequence of their large distance.

8.2.2   Stellar Interferometer

The starlight is directed into a normal telescope by two mirrors placed at a distance D (several meters) apart (Fig. 8.2). Since stars are extended objects, albeit of extremely small angular extent. For simplicity, let us imagine a stellar disk consisting of two halves separated by a distance of $\alpha/2$ from each other. The light passes through both mirrors. As a result of the angular extent of the star, the wavefronts are inclined with respect to each other by the angle Φ. Therefore, the following superpositions occur:

- First wave: comes, for example, from the left edge of the star. After reflection from the two mirrors, the rays are combined, interference occurs, and amplification occurs when the path difference is a multiple of the wavelength: $\Phi = n\frac{\lambda}{D}$.
- Second wave: comes from the right edge of the star. It forms with the first one an angle of $\alpha/2$. Thus again a *Interference occurs*.

Both systems amplify when

$$\frac{\alpha}{2} = n\frac{\lambda}{D}. \tag{8.10}$$

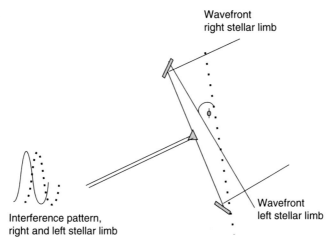

Fig. 8.2 Schematic of the Michelson interferometer Michelson interferometer

As long as $\alpha \ll \Phi$ the star is also point-like for the interferometer. But as soon as

$$\Phi = \alpha = \frac{\lambda}{D} \tag{8.11}$$

the interference system disappears. D, the distance between the two mirrors, can therefore be varied until the interference system disappears. But from this then follows the angular diameter α of the star. If one still knows the distance, then one has the linear diameter.

This type of interferometer is also called *Michelson interferometer* (*Michelson*, around 1920). There are also intensity interferometers.

8.2.3 Stellar Occultations by the Moon

The Moon runs in the course of time on an orbit slightly inclined to the ecliptic over the sky. If the moon occults a star, a diffraction figure is formed by the diffraction of the light at the moon's edge, and due to the small but finite diameter of the star, it does not disappear immediately behind the moon's limb. Thus one can determine the diameter of a star, but this is only possible for stars which can be occulted by the moon, i.e. which are located along a narrow band around the ecliptic (Fig. 8.3).

Theoretically one could argue that such star occultations can also be produced artificially at the telescope and so the diameter can be determined. The problem here, however, is that as a result of the air turbulence in the Earth's atmosphere, there are constant changes in the position of the stars's position, making measurements impossible. In the moon-star system, on the other hand, the atmospheric conditions are the same for both the lunar limb profile and the star at the moment of occultation.

Fig. 8.3 Shortly before the
moon occults a star, there are
diffraction phenomena because
of the lunar limb and the
unevenness at the lunar limb

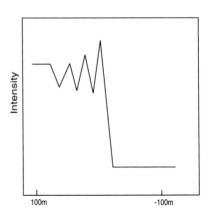

8.2.4   Eclipsing Variable Stars

At least half of all eclipsing variable stars are double or multiple systems. If we lie in the
line of sight of a double star system, then it comes to *occultations*. A star 1 has a diameter
D and is eclipsed by a star 2 with diameter d orbiting with the velocity v. The whole system
has a relative velocity of V. If star 2 approaches the observer as a result of its orbit, then
one measures the following *Doppler shift:*

$$\frac{\Delta \lambda_1}{\lambda_0} = \frac{V + v}{c} \tag{8.12}$$

If star 2 moves away because of its orbit, then:

$$\frac{\Delta \lambda_2}{\lambda_0} = \frac{V - v}{c} \tag{8.13}$$

and thus:

$$\frac{\Delta \lambda_1 - \Delta \lambda_2}{\lambda_0} = 2\frac{v}{c} \tag{8.14}$$

Consider, a binary star system as in Fig. 8.4, consisting of a large star 1 and a smaller
companion star 2.

If star 2 disappears behind 1, an occultation occurs (Fig. 8.4, 1'–4'), the brightness of
the whole system decreases. Similarly, if star 2 passes in front of star 1 as seen from us
(transit), the brightness of the system decreases. The time t_1 is the time when star 2 starts
to cover star 1, at t_2 it is completely in front of star 1 and so on. We therefore obtain for

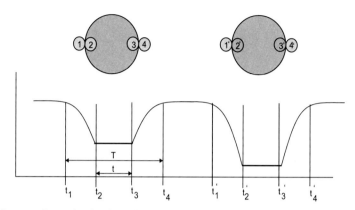

Fig. 8.4 Diameter determination in eclipsing variable stars. For 1 to 4, a transit occurs; for 1'–4', the smaller star is eclipsed by the larger one

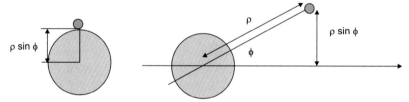

Fig. 8.5 Geometry of eclipsing stars

the diameters of the two components:

$$D + d = v(t_4 - t_1) \tag{8.15}$$

$$D - d = v(t_3 - t_2) \tag{8.16}$$

However, there are several uncertainties in this method:

- instead of circular, orbit is elliptical;
- star is not exactly spherical;
- surface brightness of stars is not uniform (cf. center to limb variation of the Sun);
- in reality both stars move around the common center of gravity.

In Fig. 8.5 it is shown that the inclination of the orbit must be very small in relation to the line of sight of the observer, in order to observe occultation.

Interesting is the star Betelgeuse (α Ori), which is about 200 pc away and is a triple system. The closest component orbits the main star in only 2.1 years, passing through the widely extended Chromosphere of Betelgeuse (Fig. 8.6). On December 12, 2023,

Fig. 8.6 α Ori with its inner
component

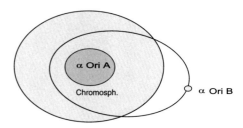

the asteroid Leona will occult Betelgeuse. The extent of Betelgeuse varies between 290
million and 480 million km, so Earth's orbit would fit comfortably within the star.

8.2.5 Speckle Interferometry

The Speckle interferometry technique uses short exposures of stars to eliminate the *Seeing*.
The turbulence of the Earth's atmosphere has longer periods, so exposures are made around
1 ms and therefore images are obtained that are close to the resolution limit of a telescope.
If one has an unresolved point light source, then its image brightness depends on the
exposure time. In *speckle interferometry* the optimal integration time is $\approx \lambda^{1.2}$. If one
has longer exposure times, the image becomes blurred to a seeing disc *(blurring)* and is
about 1″ large. If the exposure times are smaller than the coherence time $t_c \approx 10$ ms (in the
optical range, in the IR 100 ms), then we obtain a group of bright *Speckles* whose size is
approximately that of the Airy disk (r_0).[2] Thus, the influence of the atmosphere is frozen.
The speckles are distributed over an area of diameter λ/r_0, the number of speckles is of
the order of the sub-apertures D^2/r_0^2. The observed image I' is a convolution between the
true image and the so-called *point spread function* (PSF) of the telescope. This indicates
how an ideal point source of light behaves after passing through the telescope. In classical
spectral analysis one tries to restore the original amplitude, and in bispectral analysis one
also tries to restore the phases. One has thus (F stands for the Fourier transform):

$$F(I') = F(I)F(b) \tag{8.17}$$

I—Original intensity distribution of the object, *b*—PSF of the telescope. High spatial
frequencies are not affected by the Earth's atmosphere, but low ones are, and therefore
low spatial frequencies are disturbed by seeing.

---

[2] Corresponds to the resolving power of a telescope.

8.2.6 Microlensing

With the methods described above, radii can only be determined for relatively large evolved stars (cf. HRD, Sect. 8.5). For small main sequence stars, the determination of radii is difficult because their angular diameter is too small. Under the term *Microlensing* one understands the deflection of light by an object according to the general theory of relativity, which, however, does not lead to separate images as in the case of galaxies (multiple images of a galaxy or quasar—due to lens effect of a n unseen galaxy lying between the observer and the galaxy), but only to an increase in brightness. The lens moves through the connecting line earth-star, and the latter changes its brightness analogously. The change in brightness is symmetrical, and the event lasts a few weeks or months. By evaluating the photometric light curve one can determine the diameters of the objects. Such events are more frequent in double stars.

8.3 Stellar Masses

The Mass of stars is the most difficult to determine; however, it is a fundamental state variable on which many other parameters depend: Stellar evolution, age of stars, nuclear fusion, etc.

8.3.1 Kepler's Third Law

Directly derivable is the mass only, if a star with a mass M_1 has a companion with a mass M_2 so it is a double star or multiple system. One can also use exoplanets to determine the mass of a star. The third Kepler's Law reads:

$$\frac{(M_1 + M_2)}{a^3} U^2 = \text{const} = 1 \tag{8.18}$$

The units here are: mass M in solar masses, a the distance between star and companion in AU and the orbital period U in years. We therefore get the mass sum:

$$M_1 + M_2 = \frac{a^3}{U^2} \tag{8.19}$$

How can one now determine the necessary quantities a, U?

- Visual binary stars: both components observable, $\rightarrow a$ in AU once their distance is known. Then follows the mass sum. The mass ratio follows from the analysis of the

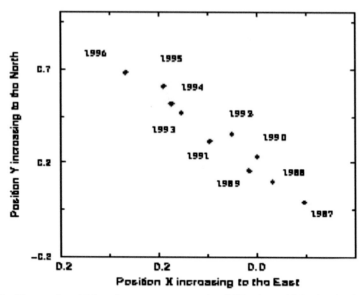

Fig. 8.7 Double star in which only one component is visible, but which clearly shows regular variations (HIPPARCHOS)

absolute orbit, i.e., one must know the motion of the two components about their common center of mass.

$$M_1/M_2 = a_2''/a_1'' \tag{8.20}$$

- Only bright component visible: in most cases, however, the fainter component is not directly observable and only the absolute orbit of the brighter component is known (Fig. 8.7).

$$a_1/a = M_2/(M_1 + M_2) \tag{8.21}$$

This is estimated according to a resolved and inserted:

$$(M_1 + M_2)\left(\frac{M_2}{M_1 + M_2}\right) = \frac{a^3}{U^2} \tag{8.22}$$

- Exoplanets: This can be simplified for extra solar planets where: $M_2 \ll M_1$ and therefore $M_2 + M_1 \approx M_1$.
- Spectroscopic binary stars: One does not see the double stars separately, but due to the motion of the two components there is a periodic shift of the spectral lines due to the

Fig. 8.8 Spectroscopic double star (α Aur). The masses of the two components are about 2.5 solar masses [HIPPARCHOS]

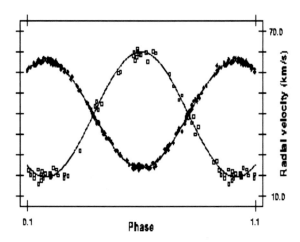

Doppler effect. Problem: Orbital inclination i is unknown. Let v is the orbital velocity in the direction of the observer, then we get:

– $a_1 \sin i$, when only one spectrum is known;
– $(a_1 + a_2) \sin i = a \sin i$ when both spectra are visible and only relative line shifts can be measured;
– $a_1 \sin i$ and $a_2 \sin i$ when both spectra are visible and absolute shifts have been measured (Fig. 8.8).

   The problem is the orbital inclination i. From mathematics one has for the averaging over the orbital inclinations i:

$$\overline{\sin^3 i} = 0.59 \tag{8.23}$$

– Spectroscopic eclipsing variability: Here is $i = 90°$.

Overall, there are good mass determinations for only a few 100 stars.

8.3.2 Gravitational Red Shift

For stars with large gravitational acceleration, the *relativistic red shift* apply: Light consists of photons of energy $E = h\nu$. One can attribute the photons a mass m (rest mass = 0):

$$E = mc^2 = h\nu \rightarrow m = h\nu/c^2 \tag{8.24}$$

The photons did work against gravity, lose energy, which $\propto \lambda^{-1}$ and the light appears reddened according to the classical derivation:
Gravitational red shift:

$$\frac{\Delta\lambda}{\lambda} = \frac{GM}{Rc^2} \tag{8.25}$$

Where R is the radius of the star. From this formula, we see that the relativistic red shift is large for stars of large mass that are very compact (small R).

As an example, consider white dwarfs. These final stages of the evolution of stars below 1.4 solar masses are very compact; $M \approx 1\,M_\odot$, $R \approx 0.01\,R_\odot$; white dwarfs have a relativistic red shift of is 10^{-4}.

8.3.3   Microlensing

Microlensing can also be used for mass determination. There are several projects. One of them is to find neutron stars and black holes using the Hubble Space Telescope. So far, the masses of neutron stars and black holes are only known if they are companions in a binary system. Masses of isolated neutron stars and black holes can be determined by microlensing. For this purpose one regularly records star fields and

- examines them for symmetrical variations in brightness (in the case there occurs an alignment of earth-neutron star or black hole—another star in thee field)
- make precise position measurements of the stars, astrometry.

To increase the probability of a microlensing event, one chooses dense stellar fields in the Galactic plane, since the stellar density is high there.

Important: The mass of a star determines its lifetime and evolution.

8.3.4   Derived Quantities

Density is directly obtained from the mass and radius

$$\rho = \frac{M}{\frac{4\pi}{3}R^3} \tag{8.26}$$

Also one immediately gets the *Gravitational acceleration* of a star:

$$g = GM/R^2 \tag{8.27}$$

A first evaluation for many stars results in the following value ranges for:

- Mass: 0.2–60 $M_\odot$,
- Radius: 0.1–500 $R_\odot$.

Although there are several methods for determining stellar masses, their determination is difficult but extremely important because it determines the further evolution of a star.

8.4 Stellar Temperatures

There are many different temperature terms. The values derived from the different methods do not match exactly because stellar radiation does not behave exactly like that of a black body.

8.4.1 Stars as Black Bodies

Here there are different definitions. Basically one assumes that the stars radiate like a *Black body*, and then defines:

- *Effective temperature:* The temperature of a star that corresponds to the radiation from a blackbody emitting the same energy per unit area as the star.
- *Radiative Temperature:* The temperature which corresponds to the radiation of a black body in a narrow wavelength range.
- *Color temperature:* The Temperature corresponding to the radiation of a black body in a spectral range (= color).
- *Gradation temperature:* The temperature whose intensity-wavelength curve at a given wavelength has the same slope as that of the intensity curve of a black body.
- *Wien's temperature:* Follows from the maximum of the intensity distribution.

For an A0-V star, one has (e.g. $T_F(500)$ is the temperature at a wavelength of 500 nm):
$$T_{\text{eff}} = 9500 \,\text{K} \qquad T_F(425 \,\text{nm}) = 16{,}700 \,\text{K} \qquad T_F(500 \,\text{nm}) = 15{,}300 \,\text{K}.$$
The differences between these temperature terms indicate that the radiation of stars is only to a first (but mostly good) approximation that of a black body.

8.4.2 Other Temperature Terms

Temperature can also be defined in terms of kinetic temperature, or in terms of atomic states of excitation and ionization.

- *Kinetic temperature:* From of the kinetic theory of gases we know that particles move with the most probable velocity

$$v_{\text{th}} = \sqrt{\frac{2\Re T}{\mu}} \tag{8.28}$$

($\Re$ is the gas constant). This thermal velocity is also partly responsible for the *width* of the spectral lines.

The thermal velocity of particles depends on the temperature and the particle mass.

- *Electron temperature T_e:* Defined through the kinetic temperature of the electrons (note: $m_e = (1/1800)m_p$, electrons have only 1/1800 of the proton mass).
- *Ionization temperature T_{ion}:* Results from the ratio of the number of atoms in different ionization states (Saha formula, is a function of temperature and pressure).
- *Excitation temperature T_{exc}:* The Boltzmann formula is used to obtain the relative atomic numbers in different excited states (which depends on temperature).
- *Band temperature:* molecules generate dark bands in the spectrum due to their rotational or vibrational transitions.

8.5 Classification of Stars, HRD

We now come to one of the most important diagrams in astrophysics. If we plot the temperatures of the stars against their luminosities, we obtain the Hertzsprung-Russell diagram, HRD.

Stars can be classified according to their line spectrum.

8.5.1 Spectral Classification

A single star spectrum is obtained when starlight is focused through a telescope and falls on a spectrograph where it is split. The spectrum itself is recorded either photographically or by CCD.

Angelo Secchi has introduced *Spectral types* around 1863. The classification still in use today was introduced in 1910 by *Annie Cannon*. This is also known as the *Harvard Classification.* Cannon herself classified 400,000 stars *(Henry Draper catalogue).*

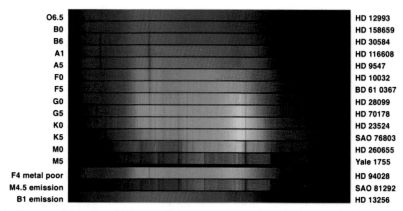

O6.5		HD 12993
B0		HD 158659
B6		HD 30584
A1		HD 116608
A5		HD 9547
F0		HD 10032
F5		BD 61 0367
G0		HD 28099
G5		HD 70178
K0		HD 23524
K5		SAO 76803
M0		HD 260655
M5		Yale 1755
F4 metal poor		HD 94028
M4.5 emission		SAO 81292
B1 emission		HD 13256

Fig. 8.9 Spectra of the stars. Hydrogen lines are most prominent in A stars, the later the spectral type, the more lines are visible in the spectrum (Copyright: KPNO 0.9-m Telescope, AURA, NOAO, NSF, Princeton)

Originally stars were classified according to the strength of their hydrogen lines. The stars with the strongest hydrogen lines were given the designation A and so on. Today one uses a sequence of descending temperature and therefore the strange order:
O – B – A – F – G – K – M
These are simply remembered: **o**h **b**e **a** **f**ine **g**irl **(g**uy) **k**iss **m**e. Further, one still subdivides decimally. We briefly discuss the individual types (Fig. 8.9):

- O: The hottest stars, bluish, He-II lines; He-I lines increase from type O5; there are also lines of Si IV, O III, N III, and C III. *Balmer lines* of hydrogen are present, but relatively weak compared to the other lines.
- B: White-blue stars; He I dominates, no He II; the hydrogen lines become stronger; Mg II and Si II. Very often they are surrounded by envelopes, which are noticeable by characteristic shapes of the spectral lines (Fig. 8.10).
- A: White stars; H lines dominate; at A0 the strength of H lines is greatest; lines of ionized metals (Fe II, Si II, Mg II). Ca II lines become stronger (Fig. 8.11).
- F: White to slightly yellowish stars, strength of hydrogen lines decreases, neutral *metal lines* appear. Ca II H and K lines become stronger; neutral metals (Fe I, Cr I).
- G: Yellowish stars; Ca II dominates; at G2 the H and K lines of ionized calcium (Ca II) are strongest; neutral metals become stronger, no more ionized metals (Fig. 8.12).
- K: Reddish stars; first appearance of *molecular bands;* neutral metals. Bands of TiO.
- M: Coolest, red stars; neutral metal lines strong, molecular bands dominant.

Since the radiation of stars conforms to a Planck curve to a first approximation, one has problems at both ends of the spectral sequence:
For very hot stars the maximum of the intensity distribution is in the UV region, for very cool stars it is in the IR region.

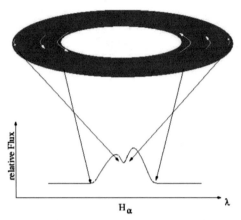

Fig. 8.10 Stars with an envelope: formation of the characteristic line profile of the hydrogen line H_α. The emission comes from the envelope, the displacement from its rotation

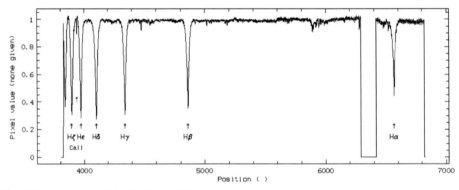

Fig. 8.11 Spectrum of the star Vega (A0)

Fig. 8.12 The line-rich spectrum of a G star

8.5.2   The Hertzsprung-Russell Diagram

In 1911 *Hertzsprung* established the most important diagram for stellar astrophysics. The absolute brightness of the stars (corresponding to the true luminosity of the stars) is plotted against their spectral type (Fig. 8.13). However, one must consider selection effects:

- If one creates a *Hertzsprung-Russell diagram* for stars in the solar neighborhood, mostly faint stars are found there.

Fig. 8.13 The
Hertzsprung-Russell diagram

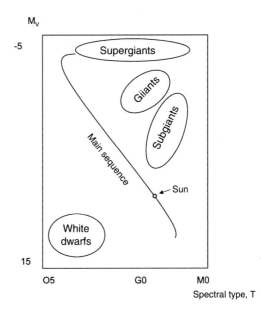

- If one generates such a diagram for bright stars, then obviously stars of large luminosity are preferred.

→ Most important result: the stars are not randomly arranged in an HRD:

- Diagonally runs in the HRD the main sequence from top left to bottom right.
- On the upper left there are hot luminous stars,
- and on the lower right, cool stars of low luminosity.
- More than 90% of all stars fit into this *main* sequence. *main sequence*).
- Furthermore, the HRD shows that there are stars which are clearly above the main sequence at a given temperature, but whose luminosity is higher.

So in the HRD, we plot: spectral type and luminosity of the stars. Since stars radiate like black bodies in a good approximation, one can use the temperature instead of the spectral type or even more simply the color or the color index of the stars. Color index *(color index, CI)* is the difference of brightnesses:

$$\text{color index} = m_{\text{short wave}} - m_{\text{long wave}} \qquad (8.29)$$

The Hertzsprung-Russell diagram is fundamental in astrophysics. One plots the temperature (spectral type, color) against the luminosity (absolute brightness) of the stars.

Standardized *color systems* have been introduced. Very often the *UBV system* is used:

- U stands for brightness in the near UV,
- B for brightness in the blue range,
- V for brightness in the visual range.

So one measures brightnesses with these filters and then simply speaks of a brightness in U, B, V and often writes instead of m_U, m_B, ... simply $U, B, \ldots$ For the spectral type A0 is defined : $U - B = B - V = 0$.

The *color index* C.I. is given by the difference of brightnesses in different wavelengths:

$$\text{C.I.} = m_{\text{short wave}} - m_{\text{long wave}} \tag{8.30}$$

From the color index, one can immediately determine *temperature*; e.g., for $B - V$:

$$B - V = 7090\frac{1}{T_{\text{eff}}} - 0.71 \tag{8.31}$$

In Fig. 8.14 a *Color-brightness diagram*, CBD, is shown.

In Table 8.4 shows some photometric systems.

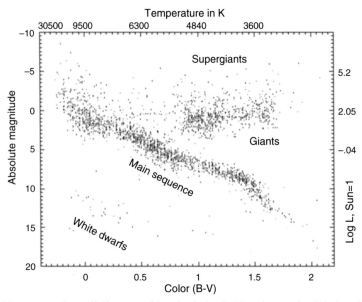

Fig. 8.14 Hertzsprung-Russell diagram with color index B-V as the x-axis. This is called a color-brightness diagram (CBD)

Table 8.4 Photometric systems

Designation	Central wavelength [nm]	Filter width [nm]
U	365	70
B	440	100
V	550	90
R	720	138
I	806	149
Z	900	–
Y	1020	120
J	1220	213
H	1630	307
K	2190	390
L	3450	472
M	4750	460
N	10,500	2500
Q	21,000	5800

What is the sign of $U - B$ for our sun? Our Sun is a cool star with maximum intensity at 550 nm. Therefore its brightness in U is lower than in B, lower brightness means higher value for the magnitude class, therefore $m_U > m_B \rightarrow m_U - m_B > 0 \rightarrow U - B > 0$

An important parameter is the metallicity (*abundance of metal*), defined as

$$\text{Fe/H} = \log \frac{(\text{Fe/H})_{\text{Star}}}{(\text{Fe/H})_{\text{Sun}}} \tag{8.32}$$

Metal Poor Stars have a negative metallicity. A metallicity of -1 means that the object shows only about 1/10 of the metal abundance in the spectrum as our Sun.

8.5.3 Luminosity Classes

The *luminosity* of a star is given by its surface $4\pi r^2$ and the *Stefan-Boltzmann law* σT_{eff}^4:

$$L = 4\pi r^2 \sigma T_{\text{eff}}^4 \tag{8.33}$$

> If a star has a much higher luminosity at a given temperature, then its surface area must be larger, therefore it is called a *giant star.*

Temperature alone is not sufficient to define the location of a star in the HRD. Therefore, a star of given temperature may be a normal main sequence star or may be above it; then it is a giant star. For this reason in 1937 the *Morgan-Keenan luminosity classes* have been introduced:

- I: supergiants (Ia, Ib, Ic);
- II: bright giants (IIa, IIb, IIc),
- III: giants (IIIa, IIIb, IIIc),
- IV: subgiants, (IVa, IVb, IVc),
- V: dwarf stars (Va, Vb, Vc), main sequence stars.
- VI: subdwarfs.

In the HRD, therefore, above the main sequence on the right are the supergiants, below them the bright giants, and so on. A G2 supergiant is 12.5 magnitudes brighter than our Sun.

The absolute brightness of the Sun is $4^m\!.6$; a supergiant of similar spectral type (G2) at this distance would have a brightness of $-7^m\!.9$; for comparison, Venus reaches about $-4^m\!.5!$

The next star is α Cen and has the same spectral type as our Sun (G2V). Its brightness is $V = -0.33$ the distance is 1.3 pc, and the absolute magnitude is $4^m\!.5$. The star Capella (α Aur) is of the type a G8 III, has $V = 0.09$ and a distance of 14 pc. Its absolute brightness is therefore -0.59. Here,V means the brightness measured in the visual band.

In Table 8.5 the characteristic state variables of the main sequence stars are listed.

> The position of stars in the HRTD: main sequence stars (about 80% of all stars) in the diagonal; hot, massive stars are on the left; cool, low-mass stars are on the lower right.

Table 8.5 Characteristic state variables from main sequence stars

Spectral type	Mass[$M_\odot$]	Luminosity[$L_\odot$]	Temperature[K]	Radius Sun = 1
O5	40	7×10^5	40.000	18
B0	16	27×10^4	28.000	7
A0	3.3	55	10.000	2.5
F0	1.7	5	7500	1.4
G0	1.1	1.4	6000	1.1
K0	0.8	0.5	5000	0.8
M0	0.4	0.05	3500	0.6

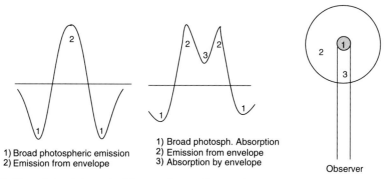

1) Broad photospheric emission
2) Emission from envelope

1) Broad photosph. Absorption
2) Emission from envelope
3) Absorption by envelope

Observer

Fig. 8.15 The origin of the typical line profile a stellar envelope

Special features in the spectrum are indicated by:

- n: diffuse blurred lines; nn very diffuse; indicates rapid rotation of the star;
- s (sharp) sharp lines;
- e: emission lines (suggesting an extended envelope around the star from which these lines originate) (Fig. 8.15);
- v (variable); spectrum variable;
- k (K line); strong interstellar Ca-II line;
- p (peculiar); any peculiarities in the spectrum. Here Metallic stars there are the Ap stars or the Am stars *(metallic line stars)*.

8.5.4 Balmer Discontinuity

In the so-called *Paris classification* one determines:

- D: Size of the Balmer*discontinuity* at $\lambda = 370$ nm. This results from the extended continua in the short-wavelength and in the long-wavelength region of 370 nm.
- λ_1: Location of the Balmer discontinuity = intersection of the registered continuum with the parallel to the long-wavelength continuum through the bisector of the Balmer discontinuity.

The Balmer discontinuity D does not occur at the theoretical series limit, but occurs earlier because the higher order transitions converge. This happens earlier the higher the electron pressure. Therefore D is determined by the temperature and thus defines the spectral type; λ_D Is defined by pressure and is therefore characteristic of luminosity. Giant stars have a very extended atmosphere and therefore low pressure, and their spectral lines appear sharp.

Another possibility of spectral classification is the *narrow band classification,*where one measures the intensity within a narrow bandwidth (3–15 nm) at defined points in the spectrum.

Objective prism spectra have only a low *dispersion* ($\approx$ 50 nm/mm), and therefore one uses as a criterion:

- Size and sharpness of the Balmer discontinuity,
- intensity of the Balmer lines,
- Ca-II line intensity.

8.5.5   Star Population and FHD

> In astrophysics, all elements heavier than He are called *Metals* .

According to the metal content one distinguishes between two *stellar populations:*

- Population I: young Stars, 2−4 % heavy elements,
- Population II: old stars, low content of heavy elements (less than 1%).

We will talk about this later in the stellar evolution.

From the observational side, a *color-brightness diagram, CBD,* is still easier to obtain. One plots color (a measure of temperature and therefore spectral type) against luminosity (absolute brightness). To determine the color, one simply measures the brightness with two different color filters (e.g., in the B and V filters, Fig. 8.16). In a *two-color diagram* for example, one plots B-V against U-B. The result is a wave in the diagram. At A0 the Balmer depression decreases, U becomes brighter, opposite to the temperature response; at F5 the Balmer depression decrease predominates again. The influence of a reddening

Fig. 8.16 Determination of the B-V brightness. Let a star **a** have a temperature of 10,000 K, and the maximum of its radiation, given by a Planck curve, lies in the B region. A second star **b** has the maximum of its radiation in the V-range

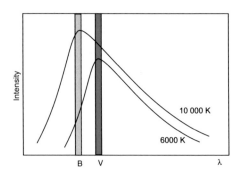

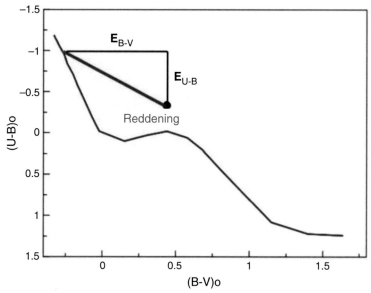

Fig. 8.17 In the two-color diagram, the color index U-V is plotted against B-V. A possible reddeing of the colors of the stars of a star cluster, caused by interstellar matter, is indicated

by interstellar matter has the effect of a shift in the diagram. Let us assume that we are looking at a two-color diagram of the stars of a star cluster, which are practically all at the same distance from us. Then the reddening due to interstellar matter affects all stars equally and is shown in Fig. 8.17.

8.5.6 The Mass-Luminosity Relation

For main sequence stars there is a *mass-luminosity relationship*:

$$L \approx M^{3.5} \tag{8.34}$$

Thus, if we know the luminosity of a main sequence star, it follows its a mass. The masses decrease along the main sequence from the top left to the bottom. Right downwards.

From the position of a star in the HRD follows its temperature, luminosity and mass.

8.6   Rotation and Magnetic Fields

Both effects are important for the activity of a star. In the spectrum, rotation is noticeable by an elliptical line profile, magnetic fields also lead to a broadening *(Zeeman effect)*.

8.6.1   Rotation

Solar rotation can be determined directly:

- by so-called tracers, e.g. migration of sunspots due to solar rotation, the equatorial solar rotation velocity is $v_{\text{rot}\odot} = 2\,\text{km/s}$,
- spectroscopically.

For some eclipsing binaries, one also proceeds by the spectroscopic method. If the right or the left edge of the star becomes visible for a short time, there is a bump in the radial velocity curve (Fig. 8.18).

Indirectly, the rotation of a star can only be determined from the width of its spectral lines. This *line broadening* occurs because, in effect, one half of the star is moving toward the observer (hence blue shift) and the other away from it (hence red shift). Rotating stars provide an elliptical line profile. Again, of course, the inclination of the rotation axis is

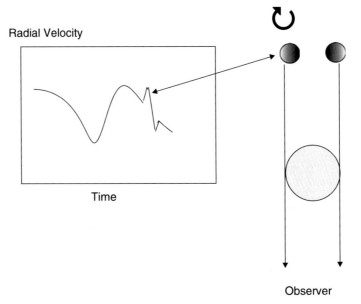

Fig. 8.18 So-called bump in the radial velocity curve of a star, when a component is close to before or after total occultation

Fig. 8.19 Line broadening; thermal (left) and due to rotation of a star (right)

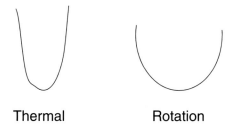

Thermal   Rotation

unknown, and we obtain (v_{eq}—rotational velocity at the equator):

$$v_{eq} \sin i. \tag{8.35}$$

If $i = 0$ then one looks at the pole of the star, and one observes no rotation broadening. The limit for rotation follows from the condition:
Gravitational acceleration = Centrifugal acceleration.

$$\frac{GM}{R^2} = \frac{v_{rot}^2}{R} \qquad v_{rot} = 440\sqrt{M/R} \text{ km/s} \tag{8.36}$$

Rotation changes as stars evolve.

Protoplanetary disks slow down the rotation. The Spitzer-telescope;[3] is equipped by a 0.85-m mirror, and one observes in the IR between 3 and 180 μm. It has been found that young stars in the Orion Nebula region (a so-called *star forming region*) rotate more slowly when there are protoplanetary disks around them that show up in IR radiation. Stars without detectable protoplanetary disks on the other hand rotate more rapidly.

Stellar rotation (Fig. 8.19) influences stellar evolution and star formation.

There are already some 10^5 Stars whose $v \sin i$ is know:

- Early types, i.e. spectral type O, B, A, and early F: 50−400 km/s, i.e. high rotation rate.
- Late types, i.e. spectral type G, K, M: rotate slower, $v_{rot} \sin i < 50$ km/s
- Trend in HRD: rotation rate decreases with stellar mass; the lower the mass, the slower stars rotate.

Rotation mixes elements or can explain anomalies; rotation affects stellar activity (starspots, winds, ...). High rotation rates cause an outflow of stellar matter at the equator, which accumulates in a shell *(shell stars)*.

---

[3] Launched 2003.

Fig. 8.20 Principle of Doppler imaging for the detection of starspots. The starspot makes the line profile appear less deep (sketched below); rotation of the star changes the position of this bump

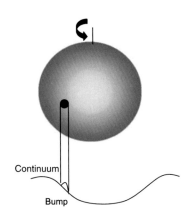

Continuum

Bump

8.6.2   Magnetic Fields

The direction of the angular momentum axis of the atoms is quantized; in the presence of a magnetic field, the terms split → Line splitting, Zeeman effect.

By means of "Doppler Imaging" (*Vogt, Penrod,* 1983) we can detect *stellar spots* as magnetically active regions. Let us take a stellar photospheric line which is Doppler broadened by the rotation of the star. If a large starspot now moves across the stellar disk as a result of the rotation, then this can be seen in the spectrum. Due to he spot

1. Continuum shifted down to entire line profile;
2. Component of the spot Doppler-shifted due to rotation.
3. Measured line: subtract 1−2; less light is subtracted at the point of Doppler shift.
4. → Bright "bump" due to spot in line profile (Fig. 8.20).

Another way to detect starspots is by using a technique called *Zeeman Doppler imaging* or by observing eclipses of stellar spots in active binary stars. Many magnetically active stars are close binary systems *(close binaries),* and one or both components rotate extremely rapidly due to tidal interaction.

One can also see fine structures in the *Balmer lines of* hydrogen, which can be traced back to matter motions in the stellar chromospheres or coronae → stellar prominences.

Here again the hydrogen line Hα is a useful proxy. This line originates in stellar atmospheres usually in the chromosphere. The calcium lines (H and K lines) are also used to determine the magnetic activity of stars: One measures their *equivalent width.*

Starspots, prominences etc. can only be detected by special spectroscopic techniques.

8.7 Peculiar Stars

We discuss here some special stars. First the particularly bright stars (bright therefore, because they are either very close to or actually shine very brightly); then some special types of stars, such as eclipsing variable stars, where there a *Mass Exchange* between the components occurs.

8.7.1 Bright Stars

We give in Table 8.6 the brightest stars visible in the northern sky.

Compare in Table 8.6 the actual luminosities of the stars Deneb and Sirius! As a measure of Luminosity we use the absolute Brightness.

$m - M = 5 \log r - 5$, where r in [pc]. Deneb: $m = 1.25$ and $r = 3227.7/326 = 990$ pc $\rightarrow M = -8.73$. Sirius: $M = 1.43$. Deneb is therefore 10.15 magnitudes brighter than Sirius.

Table 8.6 Brightest Stars at north sky

Name	Designation	Brightness (V)	Distance [Ly]	Spectral type
Sirius	α CMa	−1.46	8.6	A1Vm
Arcturus	α Boo	−0.04	312.6	K1 III
Vega	α Lyr	0.03	25.3	A0V
Capella	α Aur	0.08	42.2	G5IIIe
Rigel	β Ori	0.12	772.5	B8a
Procyon	α CMi	0.38	11.4	F5IV-V
Betelgeuse	β Ori	0.50	427.3	M1Ia
Altair	α Aql	0.77	16.8	A7V
Aldebaran	α Tau	0.85	65.1	K5III
Antares	α Sco	0.96	603.7	M1I
Spica	α Vir	0.98	262.1	B1III
Pollux	β Gem	1.14	33.7	K0III
Fomalhaut	α PsA	1.16	25.1	A3V
Deneb	α Cyg	1.25	3227.7	A2Ia
Regulus	α Leo	1.35	77.5	B7V
Bellatrix	γ Ori	1.64	242.9	B2III

Table 8.7 Some known eclipsing binaries

Name, designation	Distance [Lj]	Period	Variation
Algol, β Per	93	2d 20h 48m	$2^{m}3$–$3^{m}5$
Sheliak, β Lyr	1000	12.9 d	$3^{m}4$–$4^{m}6$

8.7.2   Algol and Eclipsing Binaries

Eclipsing variables like Algol Stars *(eclipsing binaries)* have already been mentioned in the discussion of the methods for the mass- and diameter determination of stars. *Algol*[4] is the second brightest star in the constellation Perseus (β Pers). Its name (Devil's Star) comes from the fact that it is variable (Table 8.7). It is a triple system; two components, closely adjacent and occulting each other as seen from us, are orbited by a third far away. In the *Algol stars*, the occulting components are still round, i.e., not deformed by gravitational interaction; for the *Beta Lyrae stars* (named after the prototype β Lyr), the two components are elliptically deformed. The data of β Lyr can also be found in Table 8.7.

In the spectrum of β Lyr one sees the continuum with emission lines of the B8 giant, an absorption spectrum of the B5 star and an emission spectrum. In the B8 spectrum one sees periodic Doppler shifts of ± 180 km/s while the lines of the B5 spectrum show no shifts and originate from the gas envelope.

The Roche sphere *(Roche lobe)* is the region around a star within which matter is gravitationally bound to the star. In a binary star, one of the components may expand over the Roche sphere $\rightarrow$ Matter outside the Roche sphere falls on the other star. The point at which the Roche spheres of the two stars M_1, M_2 touch, corresponds to thereby the Lagrange point L1 (Fig. 8.21).

For the β-Lyr stars, the B8 component fills the Roche sphere; the second component is smaller and less luminous, but has the larger mass. Just before the occultation of the B8 component, so-called satellite lines are produced, which are red shifted. These are produced by the rotation of the gas disk (accretion).

For Algol stars, one component fills the Roche limit.

Narrow binary systems are classified as follows (Fig. 8.22):

- D, *detached*, separated systems; the two masses M_1, M_2 are much smaller than the Roche interface.
- SD, *semi detached*, semi separated systems. One component extends to the interface.
- C, *contact*, contact systems, both components reach the interface.

The *W-Ursae-Majoris stars* are short-period contact systems.

One can estimate how large the mass transfer is by measuring the change in period.

---

[4] al-gu-l, the demon.

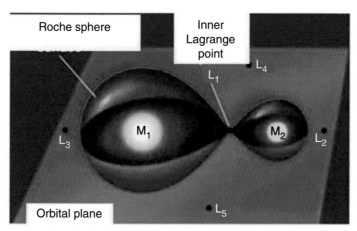

Fig. 8.21 The Roche boundaries around two stars M_1, M_2 where $m_1 > m_2$. As soon as one star fills its Roche sphere (up to the point L1), mass transfer occurs to the other star

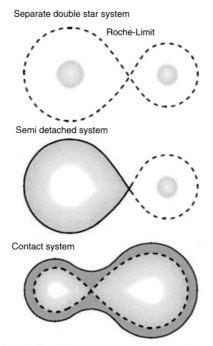

Fig. 8.22 Kopal classification of close binary star systems. From Stars and Space

8.8   Further Literature

We give a small selection of recommended further reading.
Astrophysics for Physicists, A.R. Chourdhuri, Cambridge Univ. Press 2010
An Introduction to Modern Astrophysics, B.W. Caroll, D.A. Ostlie, Cambridge Univ. Press, 2017
Astrophysics—a very short introduction, J. Binney, Oxford Univ. Press, 2016
An Introduction to Stellar Astrophysics, F. LeBlanc, Wiley, 2010

Tasks

8.1 Calculate the absolute brightness of the sun: $m = -26.7$, distance of the Sun in pc = 1/206,265.

Solution
$$M = -5 \log r + 5 + m = -5 \log(1/206.265) + 5 - 26.7 = 4.87$$

8.2 Take a cool supergiant with $T = 3000$ K with a luminosity of $10^4 L_\odot$. What is its magnitude?

Solution
$$R_* = 400 \, R_\odot.$$

8.3 A galaxy has an absolute luminosity of -20^M. How bright does it appear to us at a distance of (a) 1 Mpc, (b) 1000 Mpc?

Solution
(a) 5^m, (b) 20^m.

8.4 Discuss which sign(s) the bolometric correction has.

Solution
Always negative.

8.5 Aldebaran (α Tau) has $\alpha = 0.023''$. The Distance d is 20.8 pc. Calculate the true diameter.

Solution

$D = (1.1 \times 10^{-7})(20.8\,\text{pc}) = 2.3 \times 10^{-6}\,\text{pc} = 7.2 \times 10^7\,\text{km}$. Thus Aldebaran has 50 times the diameter of the sun.

8.6 Calculate the relativistic red shift for our sun.

Solution

$\Delta\lambda/\lambda = 10^{-6}$.

Stellar Atmospheres

9

The term stellar atmosphere is very broad: In principle, it is understood to mean those layers of a star in which the spectral lines are formed. In our *sun* this essentially occurs in the *photosphere,* (only 400 km thick) but also in Chromosphere and even in the Corona.

The physics of stellar atmospheres deals with the formation and interpretation of spectra $\rightarrow$ qualitative and quantitative spectral analysis. First we want to sketch a quantum mechanical description of the emission-absorption processes.

Important input parameters for the physics of stellar atmospheres are temperature and gravitational acceleration.

The homogeneity of stellar atmospheres is no longer given as soon as one considers e.g. center to limb variation or large spots. Magnetic fields or stellar winds also cause anisotropy of the plasma. Let us consider two extreme cases:

Sun: Atmosphere (photosphere) very thin, plane-parallel approximation possible.
Super giants: Atmosphere extremely extended, so spherical models better.

9.1 Quantum Mechanical Description

In this section, we will give a brief overview of the quantum mechanical description of a particle to understand the origin of the spectral lines, as well as the quantum mechanical parameters to describe the electron configuration.

9.1.1 Description of a Particle

$\psi(x, y, z, t)$ let be a complex function with which the state of a particle can be completely described quantum mechanically.

© Springer-Verlag GmbH Germany, part of Springer Nature 2023
A. Hanslmeier, *Introduction to Astronomy and Astrophysics*,
https://doi.org/10.1007/978-3-662-64637-3_9

The probability $dw(x, y, z, t)$ of finding a particle at time t at the location $\mathbf{r} = (x, y, z)$ is given in quantum mechanics by the absolute square of the *Wave function*:

$$dw(x, y, z, t) = |\psi(x, y, z, t)|^2 dV \tag{9.1}$$

The normalization simply states that the probability of finding the particle anywhere must equal 1:

$$\int |\psi^2| dV = 1 \tag{9.2}$$

9.1.2   Schrödinger Equation

The classical *energy* is given by the sum of kinetic and potential energy:

$$E = \frac{\mathbf{p}^2}{2\,m} + V(\mathbf{r}, t) \tag{9.3}$$

In quantum mechanical description, one replaces energy and momentum by the *operators:*

$$E \rightarrow i\hbar \frac{\partial}{\partial t} \tag{9.4}$$

$$p \rightarrow -i\hbar\nabla \tag{9.5}$$

The operator on the right-hand side of the Schrödinger equation is also known as the *Hamilton operator* and denoted by H. With this the *Schrödinger equation* becomes

$$i\hbar \frac{\partial}{\partial t}\psi(\mathbf{r}, t) = H\psi(\mathbf{r}, t) \tag{9.6}$$

respectively:

$$i\hbar \frac{\partial}{\partial t}\psi(\mathbf{r}, t) = \frac{\hbar^2}{2\,m}\nabla^2\psi(\mathbf{r}, t) + V(\mathbf{r}, t)\psi(\mathbf{r}, t) \tag{9.7}$$

The Schrödinger equation is a non-relativistic equation. If one wants to understand spin, for example, one needs the relativistic Dirac equation (it predicts, among other things, the

Fig. 9.1 Hydrogen atom, in the center is the proton, p^+, mass m_p, surrounded by an electron e^- mass m_e

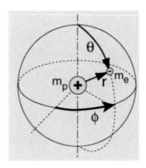

spin of the electron, and the existence of antiparticles[1]).

$$i\hbar\frac{\partial\psi}{\partial t} = \left(c\alpha\mathbf{p} + \beta mc^2\right)\psi \tag{9.8}$$

And $\alpha = \left(\alpha_x, \alpha_y, \alpha_z\right)$ consists of 4×4-matrices and $\mathbf{p} = \frac{\hbar}{i}\nabla$.

$$\beta = \begin{pmatrix} 1 & 0 & 0 & 0 \\ 0 & 1 & 0 & 0 \\ 0 & 0 & -1 & 0 \\ 0 & 0 & 0 & -1 \end{pmatrix} \qquad \alpha_x = \begin{pmatrix} 0 & 0 & 0 & 1 \\ 0 & 0 & 1 & 0 \\ 0 & 1 & 0 & 0 \\ 1 & 0 & 0 & 0 \end{pmatrix} \tag{9.9}$$

$$\alpha_y = \begin{pmatrix} 0 & 0 & 0 & -i \\ 0 & 0 & i & 0 \\ 0 & -i & 0 & 0 \\ i & 0 & 0 & 0 \end{pmatrix} \qquad \alpha_z = \begin{pmatrix} 0 & 0 & 1 & 0 \\ 0 & 0 & 0 & -1 \\ 1 & 0 & 0 & 0 \\ 0 & -1 & 0 & 0 \end{pmatrix} \tag{9.10}$$

9.1.3 Wave Functions for Hydrogen

Only for the hydrogen atom, the Schrödinger equation can be solved analytically. Let us use *Polar coordinations* r, Θ, Φ (Fig. 9.1). One separates the wave function in

$$\psi(r, \Theta, \Phi) = R_{nl}(r)Y_{lm}(\Theta, \Phi) \tag{9.11}$$

- radial solutions: $R_{nl} \to$ *Laguerre polynomials*. The square again gives the probabilities at what distance one can find the particle.
- Y_{lm} *Spherical harmonics.*

---

[1] Discovery of the positron in 1932.

The Laguerre polynomials are solutions of the Laguerre differential equation

$$x\, y''(x) + (1 - x)\, y'(x) + ny(x) = 0 \qquad n = 0, 1, \ldots \tag{9.12}$$

And one finds :

$$L_0(x) = 1$$

$$L_1(x) = -x + 1$$

$$L_2(x) = \frac{1}{2}\, (x^2 - 4x + 2)$$

$$L_3(x) = \frac{1}{6}\, (-x^3 + 9x^2 - 18x + 6)$$

resp. the formula of *Rodriguez:*

$$L_n(x) := \frac{e^x}{n!}\, \frac{d^n}{dx^n}\, (x^n e^{-x}) \tag{9.13}$$

For the associated Laguerre polynomials the DE is:

$$z\, y''(x) + (k + 1 - x)\, y'(x) + (p - k)\, y(x) = 0 \tag{9.14}$$

where $n = 0, 1, \ldots k \le n$, and one finds:

$$L_0^k(x) = 1$$

$$L_1^k(x) = -x + k + 1$$

$$L_2^k(x) = \frac{1}{2}\, \left[x^2 - 2\,(k + 2)\, x + (k + 1)(k + 2) \right]$$

The formula of *Rodriguez* reads:

$$L_n^k(x) = \frac{e^x\, x^{-k}}{n!}\, \frac{d^k}{dx^k}\, (e^{-x}\, x^{n+k}). \tag{9.15}$$

In the radial part of the wave function one has then:

$$R_{nl}(r) = D_{nl}\, e^{-\kappa r}\, (2\,\kappa\, r)^l\, L_{n+l}^{2l+1}(2\,\kappa\, r) \tag{9.16}$$

Where D_{nl} is a normalization constant, κ a characteristic length and n is the principal quantum number, l the orbital angular momentum quantum number.

The spherical harmonic functions are defined as follows:

$$Y_{lm}(\theta, \phi) = \frac{1}{\sqrt{2\pi}} N_{lm} P_{lm}(\cos\theta) e^{im\phi} \qquad (9.17)$$

where $P_{lm}(x)$ are the assigned Legendre polynomials

$$P_{lm}(x) := \frac{(-1)^m}{2^l l!} (1-x^2)^{\frac{m}{2}} \left(\frac{\partial}{\partial x}\right)^{l+m} (x^2-1)^l \qquad (9.18)$$

and N_{lm} are the normalization factors:

$$N_{lm} = \sqrt{\frac{2l+1}{2} \cdot \frac{(l-m)!}{(l+m)!}} \qquad (9.19)$$

For bound states, the solution to the Schrödinger equation is:

$$E_n = -\frac{\mu Z^2 e^4}{8 h^2 \epsilon_0^2} \frac{1}{n^2} = -R \frac{Z^2}{n^2} \qquad (9.20)$$

For the transitions of an electron between two states n_1, n_2 (see Table 9.1), we find the following wavelength of emission or absorption of a photon:

$$\frac{1}{\lambda} = \frac{1}{hc}\left(E_{n_1} - E_{n_2}\right) = R_H\left(\frac{1}{n_1^2} - \frac{1}{n_2^2}\right) \qquad (9.21)$$

Table 9.1 Hydrogen atom: important transitions

Series		n_i	n_j	λ
Lyman	$L\alpha$	1	2	121.6 nm
	$L\beta$	1	3	102.6
	$L\gamma$	1	4	97.3
	$L\delta$	1	5	94.9
Balmer	$H\alpha$	2	3	656.3
	$H\beta$	2	4	486.1
	$H\gamma$	2	5	434.0
	$H\delta$	2	6	410.2
Paschen	$P\alpha$	3	4	1875.1
	$P\beta$	3	5	1281.8
	$P\gamma$	3	6	1093.8
	$P\delta$	3	7	1005.0

R_H the *Rydberg constant.* It holds

$$R_H = \frac{\mu}{m_e} R_\infty = \left(\frac{M_H}{M_{H+m_e}} \right) R_\infty \tag{9.22}$$

and $R_\infty = 109{,}737.31 \text{ cm}^{-1}$.

In addition to hydrogen, the universe also contains Deuterium, but only at about 2×10^{-4} of the hydrogen abundance. The Rydberg constant for deuterium is:

$$R_D = \frac{\mu_D}{H} R_H \tag{9.23}$$

and the reduced masses are:

$$\mu_H = \frac{M_H + m_e}{m_H m_e} \qquad \mu_D = \frac{M_D + m_e}{M_D + m_e} \tag{9.24}$$

and $\mu_H / \mu_D = 1.00027$. From this follows for the wave numbers of the Hα-line:

H: 15238.7 cm^{-1} D: 15233 cm^{-1}.

9.1.4   Quantum Numbers

One has four quantum numbers:

- three quantum numbers give space geometry,
- the 4th quantum number the spin.

Thus:

- $R(r)$ *principal quantum number,* $n = 1, 2, 3, \ldots$
- $\Theta(\theta)$: orbital number, $l = 0, 1, 2, \ldots n - 1$; identifies the *Orbital angular momentum* of the electron, gives the shape of the orbital.
- $\Phi(\phi)$: *magnetic quantum number,* $-l, -l + 1, \ldots, l - 1, l$; $2l + 1$ Values
- $m_s = \pm\frac{1}{2}$ *Spin quantum number.*

In the field of IR spectroscopy the *rotational quantum number J* and the *vibrational quantum number v* are important, and k describes the *precession motion* of a molecule around its rotation axis. Furthermore, there is also the *nuclear spin quantum number.*

In IR spectroscopy, the excitation of the molecules plays an important role: they can oscillate or rotate, but these states are quantized.

Table 9.2 Quantum numbers and orbitals

Quantum number	Character	Range	Designation
Main quantum number	n	1,2,3,...	K,L, M,...
Secondary quantum number	l	0,..., $n-1$	s,p,d, f,...
Magnetic quantum number	m	$-1,...,1$	
Spin quantum number	s	$-1/2,+1/2$	

9.1.5 Electron Configurations

Pauli Principle In a *Quantum Cell* electrons must differ in at least one quantum number. Therefore, all electrons in an atom distribute themselves to different states.

A summary is given in Table 9.2.

The principal quantum numbers define the *shells;* the minor quantum numbers define the subshells.

Each shell can be occupied at most by $2n^2$ electrons, and one denotes the shells by K (for $n = 1$), L (for $n = 2$), M (for $n = 3$), etc. The outermost shell determines the chemical behavior *(valence shell)*. The occupation is in this order:

1. 1 s
2. 2 s 2p
3. 3 s 3p
4. 4 s 3d 4p (Hund's rule)...

Electron configuration is important for labeling radiative transitions. Suppose there are five electrons in the 2nd sub shell of the 2nd shell, then they are given as follows:

$2p^5$

Where p stands for the 2nd subshell.

9.1.6 Hydrogen Fine Structure

The *electron spin* is a consequence of the relativistic treatment of quantum mechanics → Coupling between Orbital angular momentum l and spin s ⟹ total angular momentum j of the electron.

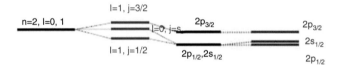

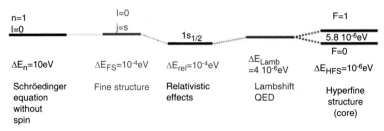

Fig. 9.2 Fine structure or hyperfine structure in the hydrogen atom.

Hydrogen: $s = 1/2$, therefore for $l \neq 0 : j = l \pm 1/2$.

Besides there is the *hyperfine structure*. Here the *nuclear spin, i*, must be taken into account. Thus one has

$$f = j \pm \frac{1}{2} \tag{9.25}$$

Because of the nuclear spin, the ground state can be divided into two levels $f = 0$ resp. $f = 1$ can be split. The energy difference is only 6×10^{-6} eV and the transition corresponds to a frequency of 1420.4 MHz or a wavelength of 21 cm (chapter on galaxies) (Fig. 9.2).

9.1.7   Complex Atoms

For an atom with N electrons and charge number Z is the Schrödinger equation:

$$\left[\sum_{i=1}^{N} \left(-\frac{\hbar^2}{2m_e} \nabla_i^2 - \frac{Ze^2}{4\pi\epsilon_0 r_i} \right) + \sum_{i=1}^{N-1} \sum_{j=i+1}^{N} \frac{e^2}{4\pi\epsilon_0 |\mathbf{r}_i - \mathbf{r}_j|} - E \right] \psi(\mathbf{r}_1, \dots \mathbf{r}_N) = 0 \tag{9.26}$$

In the first sum stands the kinetic energy as well as the Coulomb attraction between electron and atomic nucleus. In the second sum stands the expression for the repulsion between electron-electron. This makes the equation no longer analytically solvable.

Atomic ions containing the same number of electrons belong to the so-called isoelectronic series.

A splitting of the spectral lines occurs trough[2] interaction of the orbital angular momentum of an electron with the spin. Because of the electron spin, the electron has a magnetic moment. This is coupled to the magnetic moment of the nucleus → for one spin direction the energy increases, for the other it decreases. → Increase of the number of the Energy levels, hence more lines.

One distinguishes:

- light atoms (e.g. C): spin-orbit coupling, LS coupling (Russel-Saunders coupling). Here the electrostatic interaction of all electrons is large compared to the spin-orbit interaction of single electrons. Thus the spin-orbit coupling of an electron is broken, and the total momentum is:

$$\mathbf{J} = \sum_i \mathbf{l}_i + \sum_i \mathbf{s}_i \tag{9.27}$$

- jj-coupling applies to heavy atoms. The electrostatic interaction of all electrons is small compared to the sum of all spin-orbit interactions of individual electrons:

$$\mathbf{J} = \sum_i \mathbf{j}_i \tag{9.28}$$

9.2   Excitation and Ionization

In this section we discuss the distribution of electrons/atoms among the different

- Excitation states: this leads to the Boltzmann formula,
- ionization states: this leads to the Saha formula.

This can then be used to explain why hydrogen lines are only faintly visible in very hot stellar atmospheres.

---

[2] Very well known: Splitting of the D line of Na.

9.2.1   Thermodynamic Equilibrium

In thermodynamic equilibrium, TE, the internal energy is distributed uniformly by collisions among all particles.

Example of a non-equilibrium: an extremely thinly distributed gas (e.g. interstellar matter); collisions hardly occur there because of the low density; likewise in thin atmospheres (corona of the sun, etc.).

Within TE one can calculate particle-speeds by the *Maxwell distribution:*[3]

$$F(v) = \sqrt{\frac{2}{\pi}} \left(\frac{m_M}{kT}\right)^{3/2} v^2 \exp\left(-\frac{m_M v^2}{2kT}\right) \tag{9.29}$$

Where $F(v)$ is the distribution function of the velocities, m_M is the particle mass, $k = 1.38 \times 10^{-23}\,\mathrm{JK}^{-1}$ Boltzmann constant. The probability that a particle has a velocity in the interval v_1, v_2 is calculated from

$$w = \int_{v_1}^{v_2} F(v)\,dv \tag{9.30}$$

and the fraction of particles in a small velocity interval Δv is approximately

$$f = F(v)\Delta v \tag{9.31}$$

The most probable velocity is obtained from the maximum of the distribution function, i.e., one sets the derivative of Eq. 9.29 to zero.

$$v_{\max} = \sqrt{\frac{2kT}{m_M}} \tag{9.32}$$

The root mean square velocity is found from the kinetic theory of gases:

$$pV = \frac{1}{3}n M \bar{v}^2 \qquad pV = n\Re T \tag{9.33}$$

---

[3] 1860 Maxwell, Boltzmann.

Hence

$$\sqrt{\overline{v^2}} = \sqrt{\frac{3kT}{m}} = \sqrt{\frac{3\Re T}{M}} \tag{9.34}$$

A doubling of the temperature increases the root mean square velocity by a factor of $\sqrt{2}$. For the mean kinetic energy per particle we find:

$$E_{\text{kin}} = \frac{3}{2}kT \tag{9.35}$$

Maxwell distribution is valid for thermodynamic equilibrium.

9.2.2 Boltzmann Formula

We assume thermal equilibrium. The average state of the atoms is not supposed to change. Any excitation in which an electron jumps from a level A to a level B is compensated by a transition from B to A, one therefore has equilibrium:

$$A \rightarrow B = B \rightarrow A \tag{9.36}$$

N_A, N_B let be the number of atoms in state A and B respectively, where state B is said to have a higher energy than A , $B > A$.

The Boltzmann formula gives the distribution over the different excited states.[4]

$$\frac{N_B}{N_A} = \frac{g_B}{g_A} \exp[(E_A - E_B)/kT] \tag{9.37}$$

g is the multiplicity of the level *(statistical weight)* and E is the energy. Please note:

1. N_B/N_A increases with increasing temperature
2. If the temperature is given, then the ratio N_B/N_A increases when $E_B - E_A$ decreases between two energy levels.
3. In plasma physics, one often computes with:

$$\frac{1\text{eV}}{k} = \frac{1.60 \times 10^{-19}\,\text{J}}{1.38 \times 10^{-23}\,\text{J/K}} = 1604\,\text{K} \tag{9.38}$$

---

[4] *Ludwig Boltzmann, 1844–1906.*

4. Consider a volume of gas containing H and He atoms. Both atoms produce spectral lines, but which lines are stronger? Let the number of H atoms in state 2 compared to state 1 be $N_2/N_1 = 1/10$. The ratio N_2/N_1 for He, however, will be much lower, since other (higher) excitation energies are required.

> The strength of a line therefore depends on the element and the temperature T.

Let us examine the effect of temperature on the distribution of hydrogen atoms in a stellar atmosphere in the ground state and in the first excited state.

1. Number of hydrogen atoms in the ground state N_0, statistical weight $g_0 = 2$.
2. Number of hydrogen atoms in the first excited state N_1, statistical weight $g_1 = 8$.

Further $E_1 - E_0 = 10.2\,\text{eV}$, and from the Boltzmann formula

$$\frac{N_1}{N_0} = \frac{g_1}{g_0}\exp(-(E_1 - E_0)/(kT)) \tag{9.39}$$

we get $T = 3000\,\text{K}$:

$$\frac{N_1}{N_0} = \frac{8}{2}\exp\left(-\frac{10.2 \times 1.6 \times 10^{-19}}{1.38 \times 10^{-23} \times 3000}\right)$$

$$= 3 \times 10^{-17}$$

$$T = 6000\,\text{K} : 10^{-8}$$

$$T = 12,000\,\text{K} : 10^{-4}$$

Up to a temperature of about 10,000 K the intensity of the spectral line increases according to this formula, which agrees with observations. However, at higher temperatures, it must be taken into account that the ionization also increases!

9.2.3   Saha Equation

At higher temperatures, the atoms become ionized. Therefore, a hot gas consists of neutral atoms, ions, and free electrons. The higher the electron density N_e the higher the probability that an ion captures an electron and becomes neutral. Therefore, two processes are relevant to the distribution among the different ionization states:

- Ionization
- Recombination

If the rate of ionization equals the rate of recombination, then again we have equilibrium:

$$X \rightleftharpoons X^+ + e^- \tag{9.40}$$

The Indian physicist *Saha* has established the equation named after him: N_{i+1} is the number of ions in the $(i+1)$-th ionization state, N_i the number of ions in the i-th ionization state, χ_i is the ionization potential.

$$\frac{N_{i+1}}{N_i} = \frac{A(kT)^{3/2}}{N_e} \exp\left(-\chi_i/kT\right) \tag{9.41}$$

In A stand atomic constants as well as the partition function (degeneracy) resulting from the statistical weights.

We consider N_+/N_0 for hydrogen as a function of temperature. It is found that below 7000 K most H is neutral. The ratio N_2/N_1 increases with T, but at high temperature T there are no more neutral H atoms. The curve N_2/N reaches a maximum at 10,000 K. The transitions from $n > 2$ to $n = 2$ is called *Balmer series.* The strength of the Balmer lines is greatest at $T = 10,000$ K.

The general form of the Saha equation is (u_i partition function or degeneracy, P_e electron pressure):

$$\frac{N_{i+1}}{N_i} P_e = 2 \frac{u_{i+1}}{u_i} \frac{(2\pi m_e)^{3/2}(kT)^{5/2}}{h^3} \exp\left(-\frac{\chi_i}{kT}\right) \tag{9.42}$$

Written logarithmically:

$$\log\left(\frac{N_{i+1}}{N_i} P_e\right) = -\chi_i \Theta + \frac{5}{2}\log T - 0.48 + \log\frac{2u_{i+1}}{u_i} \tag{9.43}$$

Where P_e in dyn (1 dyn = 1 cm g s^{-2} = 10^{-5} N), χ in eV and $\Theta = 5040/T$.

One can make a very simple estimate of the temperature of a stellar atmosphere. Suppose a particular ion is very abundant, i.e., its associated spectral line is very strong. Then its ionization potential becomes $\approx kT$.

9.3 Radiation Transport

The transport of radiation is described by a transfer equation. This involves (i) emission, (ii) absorption; both processes are described by macroscopic coefficients.

9.3.1   Transfer Equation

Inside of a stellar atmosphere there is a flow of energy outwards. Energy transport is possible in principle by:

- Heating, conduction,
- Radiation,
- Convection.

In radiative transfer, which is most important to the physics of stellar atmospheres, photons are continuously absorbed and then re-emitted in all directions. As a result, less intensity reaches the observer, and dark spectral lines are seen. What is important here is the temperature gradient, the amount by which the temperature decreases with altitude. You have lower and hotter layers radiating energy to the higher lying, cooler layers.

We now examine the main laws of this radiative transfer. The radiation field can be described by the intensity I_ν. In general, this also depends on the direction of the radiation. The frequency dependence is described by the index ν. The radiation energy dE_ν that is emitted in the frequency interval $[\nu, \nu + d\nu]$ during the time dt through a perpendicular surface dF into the solid angle $d\omega$ is:

$$dE_\nu = I_\nu dt \, dF \, d\omega \, d\nu \qquad (9.44)$$

Note: In exact thermal equilibrium, the radiation is independent of the direction of radiation, hence isotropic. The intensity distribution is then given by the Planck function:

$$I_\nu = B(\nu, T) \qquad (9.45)$$

Now we make a balance: what happens when the radiation passes through a thin layer of matter of thickness ds (Fig. 9.3)?

$\rightarrow$ **Absorption:** the change in intensity when passing through ds is:

$$dI_\nu = -I_\nu \kappa_\nu ds \qquad (9.46)$$

For this an *Absorption coefficient* κ_ν is introduced. This has the dimension $1/L$, L—length, and is a function of chemical composition of the matter and of the degree of ionization and excitation, i.e. of T and P (cf. Saha equation).

$\rightarrow$ **Emission:** this causes an increase in intensity, and we introduce the *Emission coefficient* ϵ_ν:

$$dI_\nu = \epsilon_\nu ds \qquad (9.47)$$

Again, this coefficient depends on temperature and pressure.

Fig. 9.3 Change in the instensity I coming from the stellar interior as it passes through a layer ds. Absorption changes the intensity to $I - dI$

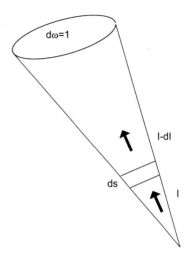

If thermal equilibrium holds, then:

$$\frac{\epsilon_\nu}{\kappa_\nu} = B(\nu, T) \tag{9.48}$$

which is known as *Kirchhoff's theorem*. Note:

In general, a stellar atmosphere is not in thermal equilibrium, because this would mean the same temperature for every depth T and isotropic radiation.

In reality, we observe a net outward radiation flux, and the temperature drops. If we use the approximation for thermal equilibrium locally, then we speak of a *LTE (local thermodynamic* equilibrium).

Now we complete the balance: The change of intensity dI_ν when passing through a layer ds is equal to emission minus absorption:

$$dI_\nu = -I_\nu \kappa_\nu ds + \epsilon_\nu ds \tag{9.49}$$

We divide this by ds: $\frac{dI_\nu}{ds} = -I_\nu \kappa_\nu + \epsilon_\nu$.

One introduces the *optical depth* τ_ν:

$$\tau_\nu = \int \kappa_\nu ds \tag{9.50}$$

Fig. 9.4 Relationship between
optical depth τ, geometric
depth t resp. geometricdepth s,
seen from an angle Θ

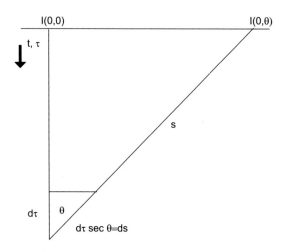

If one has only one absorption, it follows from 9.49

$$dI_\nu = - I_\nu \kappa_\nu ds$$

$$-\int dI_\nu / I_\nu = \int \kappa_\nu ds = \tau_\nu$$

$$I_\nu = I_{\nu,0} e^{-\tau_\nu}$$

Extreme cases:

- $\tau_\nu \ll 1$ *optically thin,*
- $\tau_\nu \gg 1$ *optically thick.*

Let's estimate the thickness of the solar photosphere. The mean value for the absorption coefficient is:[5] $\bar{\kappa} = 3 \times 10^{-8}\,\mathrm{cm}^{-1}$. For an optical depth of 1, according to the above formula, the intensity falls to the $(1/(e)$-th part. Therefore, we find the thickness of the photosphere Δs from the relation

$$\bar{\kappa} \Delta s = \tau \sim 1$$

and thus $\Delta s = 300\,\mathrm{km}$.

Let us go back again to our considerations about the radiation intensity. Let us assume that our surface element dF forms an angle with the normal θ. The depth t and the path distance s are related via (Fig. 9.4):

$$\cos \theta = -d\tau/ds \qquad (9.51)$$

---

[5] Averaging over all frequencies ν.

Applying Kirchhoff's theorem, we find:

$$\cos\theta\frac{dI_\nu(t,\theta)}{d\tau_\nu} = I_\nu(t,\theta) - B_\nu(T(t)) \tag{9.52}$$

At radiative equilibrium, the radiative flux is independent of the depth t:

$$F = \int_{\theta=0}^{\pi}\int_{\nu=0}^{\infty} I_\nu(t,\theta)2\pi\cos\theta\sin\theta d\theta d\nu = \sigma T_{\text{eff}}^4 \tag{9.53}$$

At the stellar surface the incident radiation vanishes, i.e. $I_\nu(0,\Theta)$ for $0 < \theta < \pi/2$. The solution becomes simple if one assumes that the absorption coefficient is independent of the frequency: this is called as *grey stellar atmosphere*. One can also assume an averaged absorption coefficient for example the *Rosseland's opacity coefficient*:

$$\bar{\tau} = \int_{-\infty}^{t} \bar{\kappa}dt \tag{9.54}$$

9.3.2 Solutions of the Transfer Equation

The *source function* is:

$$S_\nu = \epsilon_\nu/\kappa_\nu, \tag{9.55}$$

With this the transport equation becomes:

$$\cos\theta\frac{dI_\nu}{d\tau_\nu} = I_\nu(\tau,\theta) - S_\nu(\tau) \tag{9.56}$$

In American literature one often finds : $\mu = \cos\theta$. To solve, substitute:

$$\sec\theta = 1/\cos\theta \qquad \tau/\cos\theta = \tau\sec\theta = \xi$$

and substitute this into the transgfer equation:

$$\frac{dI}{d\xi} - I = -S$$

Now we multiply this by the integrating factor $e^{-\xi}$:

$$\frac{d(Ie^{-\xi})}{d\xi} = -Se^{-\xi} \qquad \frac{dIe^{-\xi}}{d\xi} = \frac{dI}{d\xi}e^{-\xi} - e^{-\xi}I$$

and we get

$$Ie^{-\xi} = -\int Se^{-\xi'}d\xi'$$

and therefore:

$$Ie^{-\tau\sec\theta} = -\int Se^{-\tau'\sec\theta}d\tau'\sec\theta$$

and finally

$$I = -\int_{\infty}^{\tau} S(\tau')e^{-(\tau'-\tau)\sec\theta}d\tau'\sec\theta$$

Let us now consider the most important case for us: radiation at the surface: Then $\tau = 0$, and we get (the $'$ omitted):

$$I_\nu(0,\theta) = \int_0^\infty S_\nu(\tau)e^{-\tau_\nu\sec\theta}d\tau_\nu\sec\theta \tag{9.57}$$

So this is the intensity reaching the observer. If one looks vertically into a stellar atmosphere ($\theta = 0$), and the same temperature prevails (S_ν independent of τ_ν):

$$I_\nu = S_\nu \int e^{-\tau_\nu}d\tau_\nu = S_\nu(1 - e^{-\tau_\nu})$$

Now we can apply to this the two cases optically thick and optically thin:

- optically thick: $\tau \gg 1$, then $e^{-\tau} = 1/e^\tau \to 0$ and you have $I_\nu = S_\nu$;
- optically thin: $\tau \ll 1$, from $e^{-\tau} = 1 - \tau$ follows: $I_\nu = S_\tau\tau_\nu$.

The general solution of 9.57 can be integrated numerically. However, we study one approach.

Eddington-Barbier Approximation
One expands the soruce function:

$$S(\tau) = S(\tau') + (\tau - \tau')\frac{dS}{d\tau}\bigg|_{\tau'} + \frac{(\tau - \tau')^2}{2}\frac{d^2S}{d\tau^2}\bigg|_{\tau'} + \dots$$

This is put into the formula 9.57, and obtains for the first two terms:

$$I_v(0, \theta) = \int_0^\infty \left(S(\tau') + (\tau - \tau')\frac{dS}{d\tau} \right) \exp^{-\tau \sec \theta} d\tau \sec \theta$$

$$= - S(\tau')\exp^{-\tau \sec \theta}\Big|_0^\infty + (\cos \theta - \tau')\frac{dS}{d\tau}\Big|_{\tau'}$$

and integrating term by term.

If $\tau' = \cos \theta$ is, then the second term disappears, and it follows:

$$I(0, \theta) = S_{\tau=\cos \theta} \tag{9.58}$$

One can easily show that then the next term of the series expansion becomes a minimum. We have therefore, to a good approximation:

The outgoing intensity is equal to the source function at the optical depth $\tau = 2/3$.

This also provides a simple relation for the *Center to limb variation* .

Gray Atmosphere, Milne Solution
One replaces the frequency-dependent absorption coefficient with an appropriate mean value, $\kappa_v \rightarrow \kappa$, and then obtains:

$$\cos \theta \frac{dI(\tau, \theta)}{d\tau} = I(\tau, \theta) - S(\tau)$$

$$I = \int_0^\infty I_v dv \qquad S = \int_0^\infty S_v dv$$

The source function is then equal to $S(\tau) = J(\tau) = \int \int I(\tau, \theta)d\omega/4\pi$, i.e. the intensity integrated over the sphere. With the conditions $F^+ = F$, $F^- = 0$ one obtains

$$S(\tau) = \frac{3}{4\pi}F[\tau + q(\tau)] \tag{9.59}$$

Where $1/2 \leq q(\tau) \leq 1$; $q(0) = 1/\sqrt{3} = 0.5774$. In practice one calculates with:

$$q(\tau) = 0.7104 - 0.1331\exp^{-3.4488\tau} \tag{9.60}$$

Eddington's Solution

Multiply the transport equation 9.59 with $\cos\theta$ and integrate over the solid angle elements, then arrives at:

$$c\frac{dP_s}{d\tau} = \Phi \tag{9.61}$$

Here is P_s the radiation pressure:

$$P_s = \frac{1}{c}\int\int I\cos^2\theta\, d\omega \tag{9.62}$$

and Φ:

$$\Phi_\nu = \int_{\phi=0}^{2\pi}\int_{\theta=0}^{\pi} I_\nu(\theta,\phi)\cos\theta\sin\theta\, d\theta\, d\phi \tag{9.63}$$

Now one approximates:

$$cP_s = \int\int I\cos^2\theta\, d\omega \sim \overline{\cos^2\theta}\int\int I\, d\omega = \frac{4\pi}{3}J = \frac{4\pi}{3}S$$

Since $\Phi = \text{const}$, one obtains:

$$S(\tau) = \frac{3}{4\pi}\Phi\tau + \alpha \tag{9.64}$$

and the integration constant can be determined, leading to the following final result:

$$S(\tau) = \frac{3}{4\pi}\Phi\left(\tau + \frac{2}{3}\right) \tag{9.65}$$

In high atmospheric layers (close to $\tau = 0$) this solution is no longer valid, because there the radiation becomes extremely anisotropic. For the outgoing intensity we find:

$$I(\tau,\theta) = \frac{3}{4\pi}\Phi\left(\cos\theta + \tau + \frac{2}{3}\right) \tag{9.66}$$

Thus we get an expression for the center to limb variation in a stellar atmosphere ($I(0,\Theta)$—intensity at optical depth $\tau = 0$ under the angle Θ):

$$\frac{I(0,\theta)}{I(0,0)} = \frac{2}{5}\left(1 + \frac{3}{2}\cos\theta\right) \tag{9.67}$$

If we look at the center of the stellar disk (solar disk), we have $\theta = 0$, and we have the value $I(0, \theta)/I(0, 0) = 1$. For $\theta = 90°$ the value is 0.40.

9.4   Absorption Coefficients

The absorption coefficient determines how deep we can see into a star's atmosphere from the outside. What is the absorption coefficient made up of? There is a basic distinction between continuous and discrete absorption. Discrete absorption becomes important in spectral lines.

9.4.1   Continuous Absorption

First we examine the *continuous absorption coefficient*, which arises by:

- Bound-free transitions,
- Free-free transitions.

Here, bound-free means that an electron jumps from a bound energy state to a free energy state by absorbing a photon (i.e., the atom is ionized). In the case of *neutral hydrogen* one has:

- Lyman series at 91.2 nm, the transitions occur from the ground state (1st level); the bounded-free transition corresponds to ionization from the ground state;
- Balmer series at 364.7 nm, the transitions occur from the 2nd level; the bound-free transition corresponds to ionization from the 2nd level;
- Paschen series at 820.6 nm, the transitions occur from the 3rd level; the bound-free transition corresponds to ionization from the 3rd level ...

On the short wavelength limit of a series there is the corresponding continuum, absorption decreases with ν^{-3}. At long waves, the boundary continua cluster and merge into the free-free continuum of the H I . In cooler stellar atmospheres (including the Sun), the bound-free and free-free transitions of the *negative hydrogen ion H^-* are essential. This is formed by the addition of a second electron to a neutral H atom. The ionization energy of H^- is only 0.75 eV, and therefore it plays no role in stellar atmospheres of higher temperature. The long-wavelength limit of the bound-free continuum is in the IR at 1.655 μm. The free-free absorption increases towards long waves.

The *atomic coefficients κ_ν* follow from quantum mechanics. For this purpose, one uses the formulas of Boltzmann and Saha.

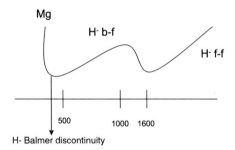

Fig. 9.5 Course of the continuous absorption coefficient for sunlike stars. Between the Balmer limit (364.7 nm) and 1655 nm (1.655 μm) the H^--bounded-free absorption (b-f, bounded-free) is predominant. At longer wavelengths the H^--free absorption

Let us summarize what can be noted in terms of the absorption coefficient for hot stars and for cool stars:

- Hot stars: $T \geq 7000\,K$, continuous absorption of H atoms; also, free-bound and free-free absorption of He I and He II (in very hot stars).
- Cooler stars: H^--Absorption (Fig. 9.5). Furthermore, C I, Si I.

9.4.2   Scattering

In addition to these absorption processes, the *Scattering* of light is of importance:

- Hot stars: *Thomson Scattering* on free electrons. The scattering coefficient is

$$\sigma_{el} = 0.66 \times 10^{-24}\,cm^2 \tag{9.68}$$

  and is therefore independent of the wavelength λ.
- Cool stars: *Rayleigh scattering* from neutral hydrogen atoms.

$$\sigma_{at} \propto \lambda^{-4} \tag{9.69}$$

It depends strongly on the wavelength!

9.4.3   Theory of Absorption Lines

So far, we have the continuous absorption coefficient κ which changes only slowly with frequency. It determines how far you can see into a stellar atmosphere. Now we treat the

Fig. 9.6 Transitions in the atom

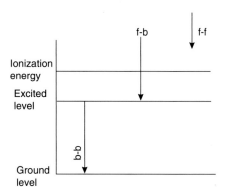

Line absorption coefficient κ_{vl} (Fig. 9.6), which is a function of the distance Δv from the line center and decreases relatively quickly from a high maximum to zero.

Adding the two, we get:

$$x_v = \int_{-\infty}^{t} (\kappa_v + \kappa_l) dt' \qquad (9.70)$$

Therefore, let us consider the intensity of the radiation emerging from the surface of the atmosphere at the angle θ:

- For the line:

$$I_v(0, \theta) = \int_{0}^{\infty} S(x_v) \exp^{-x_v/\cos\theta} dx_v / \cos\theta \qquad (9.71)$$

- Adjoint continuum:

$$I(0, \theta) = \int_{0}^{\infty} S(\tau) e^{-\tau/\cos\theta} d\tau / \cos\theta \qquad (9.72)$$

And further:

$$\frac{d\kappa_v}{d\tau} = \frac{x_v + \kappa}{\kappa} \qquad \kappa_v = \int \frac{x_v + \kappa}{\kappa} d\tau \qquad (9.73)$$

Thus, we obtain for the dip of the line:

$$r_v(0, \theta) = \frac{I(0, \theta) - I_v(0, \theta)}{I(0, \theta)} \qquad (9.74)$$

An important measure of line absorption is its *equivalent width* (Fig. 9.7). To do this, we calculate the area of the line dip under the continuum (which we normalize to 1). Then the equivalent width is equal to the side length of the equal area rectangle (where two sides

Fig. 9.7 Definition of the equivalent width as the width of the shaded rectangle having the same area as the area of the line depression under the continuum (normalized to 1).

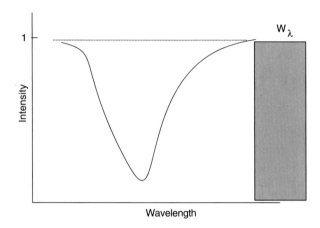

are equal to 1 because of normalization); therefore, the equivalent width is:

$$W_\lambda = \int r_\lambda(0, \theta) d\lambda \tag{9.75}$$

If one uses the *Eddington-Barbier approximation*, then the dip is:

$$r_\nu(0, \theta) = \frac{S_\nu(\tau = \cos\theta) - S_\nu(x_\nu = \cos\theta)}{S_\nu(\tau = \cos\theta)} \tag{9.76}$$

Assuming a local thermodynamic equilibrium (LTE): $S_\nu(\tau) = B_\nu(\tau_\nu)$. We can neglect the small frequency differences in the area of the line. We see:

The radiation in the line originates from a higher layer $x_\nu = \cos\theta$ and thus from a layer of lower temperature than the neighbouring continuum, which comes from $\tau = \cos\theta$.

For the stars one uses the *radiative flux F*, because here one cannot observe a center-limb variation. In the LTE case:

$$R_\nu(0) = \frac{F(0) - F_\nu(0)}{F(0)} \approx \frac{B_\nu(T(\tau = 2/3)) - B_\nu(T(x_\nu = 2/3))}{B_\nu(T(\tau = 2/3))} \tag{9.77}$$

resp. for the equivalent width in the spectrum:

$$W_\lambda = \int R_\nu(0) d\lambda \tag{9.78}$$

For weak lines or in the wings of strong lines, the following approximation holds: $\kappa_\nu \ll \kappa$ and

$$\kappa_\nu \approx \frac{2}{3}\left(1 + \frac{\kappa_\nu}{\kappa}\right) \tag{9.79}$$

resp. if we expand $B_\nu(x_\nu = 2/3)$ into a series, the following results for the dip:

$$R_\nu(0) \approx \frac{2}{3}\frac{\kappa_\nu}{\kappa}\left(\frac{d \ln B_\nu}{d\tau}\right)_{\tau=2/3} \tag{9.80}$$

Thus, in a stellar spectrum, the continuum and the wings of the lines arise mainly in the layers of the atmosphere whose optical depth is 2/3. Furthermore, one sees that the line strength depends strongly on the temperature gradient. If the temperature rises inward one observes absorption lines.

9.5 Line Profiles

A close analysis of stellar spectra reveals line profiles. What can we learn about the physical state of a stellar atmosphere from the shape of such line profiles?

9.5.1 Damping

In this line profile section we study the profile function $\phi(\nu)$ as a function of temperature T, electron pressure P_e or gas pressure P_g at the distance $\Delta\nu$ from the line center.

Let us first consider a time-limited wave packet in classical optics with duration τ. This corresponds to a spectral line whose absorption coefficient has a typical *Lorentz distribution*

$$L(\nu) = \frac{\gamma}{(2\pi\Delta\nu)^2 + (\gamma/2)^2} \tag{9.81}$$

If one integrates over the line, it should be normalized to 1:

$$\int I(\nu)d\nu = 1. \tag{9.82}$$

Important is the *Damping constant* $\gamma = 1/\tau$. This is at the same time the half-width of the absorption coefficient.

A distinction is made between:

- Radiation damping: Limitation of the radiation process by the radiation of the atom itself;
- Collisional damping: Limitation of the radiation process by collisions with other particles.

According to quantum theory, one has for the line profile:

$$\gamma = \gamma_{\text{rad}} + \gamma_{\text{coll}} \tag{9.83}$$

The *Radiative damping* γ_{rad} is equal to the sum of the two reciprocal lifetimes of the energy levels involved in the absorption process n, m. Neglecting the stimulated emission processes, holds:

$$\gamma_{\text{rad}} = \sum_{l<n} A_{nl} + \sum_{l<m} A_{ml} \tag{9.84}$$

For allowed transitions we have $\gamma_{\text{rad}} = 10^7 \ldots 10^9$ therefore we get half-value widths at e.g. $\lambda = 400$ nm of 10^{-6}–10^{-4} nm.

The *Collisional Absorption Constant* $\gamma_{\text{coll}} = 2 \times$ Number of effective collisions per s. To this one counts those close encounters between an interfering particle and a radiating particle, at which the phase of the oscillation is shifted by more than 1/10 of the oscillation wavelength.

In cooler stars (Sun), hydrogen is largely neutral. Therefore, collisiional damping by neutral H atoms predominates. These influence the radiating atom by *Van der Waals forces* (interaction energy proportional to d^{-6}, d—distance). At smaller distances, repulsive forces become effective. γ_{coll} is proportional to the gas pressure P_g.

Collisional damping by free electrons predominates for spectral lines that have a large quadratic *Stark effect*[6] and in Ionized atmospheres. Here the interaction goes with the square of the field strength produced by the electron at the location of the radiating particle (proportional to d^{-4}). The damping constant is proportional to the electron pressure P_e. At 10^9 effective collisions per s one has a half width of 10^{-4} nm. In solar-like stars the collisions come from H atoms, in hot stars from electrons.

The quadratic Stark effect leads to a shift of the energy levels proportional to the square of the field strength:

$$E_{\text{el}} \propto \alpha \mathbf{E}^2 \tag{9.85}$$

---

[6] *Johannes Stark*, 1913.

This is a 2nd order perturbation term, and it only occurs for atoms that do not have a permanent dipole moment. The linear Stark effect occurs at degenerate energy levels.

The lines of H and He II show a large linear Stark effect splitting in electric fields. Their broadening in partially ionized gases is primarily due to the quasi-static Stark effect of the randomly distributed fields generated by the much slower moving ions. It is shown that the absorption coefficient in the line wings is approximately proportional to $\lambda^{-5/2}$.

9.5.2 Doppler Broadening

In addition to the broadening due to radiation and collisional damping, the broadening due to the Doppler effect due to the motion of radiating atoms because of the high temperatures is important; the Doppler formula is:

$$\frac{\Delta \nu_D}{\nu_0} = \frac{\Delta \lambda_D}{\lambda_0} = \frac{V_0}{c} \tag{9.86}$$

There $\Delta \nu_D$ resp. $\Delta \lambda_D$ is the *Doppler width*, V_0 The most probable speed, given by

$$V_0 = \sqrt{2kT/m} \tag{9.87}$$

and m the mass of the absorbing atom.

Assuming the *Maxwell-Boltzmann distribution* one finds the Doppler profile to:

$$D(\nu) = \frac{1}{\sqrt{\pi} \Delta \nu_D} \exp[-(\Delta \nu/\Delta \nu_D)^2] \tag{9.88}$$

ν_0, λ_0 stand for the centre of the line pofile.

Let us consider Fe atoms in the solar atmosphere at $T = 5700$ K. One obtains from (Eq. 9.87) $V_0 = 1.3$ km/s and, given an Fe-I line at $\lambda = 386$ nm a Doppler width of $\Delta \lambda = 1 \times 10^{-3}$ nm results.

In addition to these thermal motions, in a stellar atmosphere also *Turbulence* will occur. This then has an additional additive effect. In this context one speaks of a *Microturbulence*. On the line of sight through the atmosphere there are several elements shifted against each other. In the case of *Macroturbulence* the line as a whole is shifted, on the line of sight there is only one element.

The effect of the micro turbulence, which causes an additional line broadening, can be calculated easily. If the temperature of a star was determined with other methods and one puts the value found for it into the Doppler broadening, then the line profile should agree with the measured line width. However it is to be noted that also e.g. stellar rotation can broaden lines.

9.5.3   Voigt Profile

The *Voigt profile* is the convolution of Doppler effect and damping:

$$\phi(v) = \int_{-\infty}^{\infty} L(v - v')D(v')dv' \tag{9.89}$$

Here one normalizes to the value at the line center equal to 1. Consider the ratio of half the damping constant $\gamma/2$ to the Doppler width $\Delta\omega_D = 2\pi\,\Delta\nu_D$:

$$\alpha = \gamma/(2\Delta\omega_D) \tag{9.90}$$

This is in almost all stellar atmospheres <0.1. Nevertheless:

- Doppler broadening proportional to exponential decay,
- damping proportional to $1/\Delta\lambda^2$.

For this reason, the result for a line profile is:

- sharp Doppler nucleus at Line center,
- damping wing further away from line center.

The exact calculation gives for the *Doppler core:*

$$\kappa_v = \sqrt{\pi}\,\frac{1}{4\pi\,\epsilon_0}\frac{e^2}{mc^2}Nf\,\frac{\lambda_0^2}{\Delta\lambda_D}\exp\left[-\left(\frac{\Delta\lambda}{\Delta\lambda_D}\right)^2\right] \tag{9.91}$$

And for the damping wings:

$$\kappa_v = \frac{1}{4\pi}\frac{1}{4\pi\,\epsilon_0}\frac{e^2}{mc^2}Nf\gamma\,\frac{\lambda_0^4}{c\Delta\lambda^2} \tag{9.92}$$

(This formula does not apply to the lines of H I and He II broadened by the linear Stark effect).

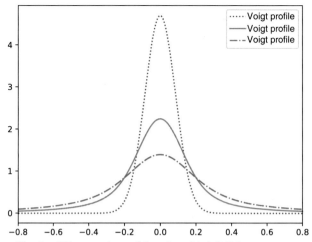

Fig. 9.8 Voigt profiles for different values of damping. (c) A.J. Friss

In Fig. 9.8 three Voigt profiles are given for different damping values.

When examining a line profile, one can roughly speak of a Doppler core and of damping wings.

9.6 Analysis of Stellar Spectra

In this section we deal with the analysis of stellar spectra, first discussing the concept of curves of growth and then quantitative spectral analysis.

9.6.1 Curves of Growth

Here we deal with the question: how does the line profile and the equivalent width W_λ change if one varies the product of the number of absorbing atoms N and the oscillator strength f (Fig. 9.9)? The oscillator strength f is a quantum mechanical correction.

Let us first consider an absorption tube in which there is only absorption, described by κ and no reemission. Let the length be H and the optical thickness $\tau = \kappa_\nu H$: Then one has for the line dip:

$$R_\nu = 1 - e^{-\kappa_\nu H} \tag{9.93}$$

Fig. 9.9 Curve of grwoth. The
number of absorbing atoms
against the equivalent width is
plotted 1

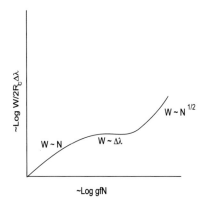

For a stellar atmosphere, we use the approximate formula:

$$R_\nu = \left(\frac{1}{\kappa_\nu H} + \frac{1}{R_c} \right)^{-1} \tag{9.94}$$

This formula states:

- For $\kappa_\nu H \ll 1$ you have absorption in an optically thin layer, and $R_\nu \approx \kappa_\nu H$.
- For $\kappa_\nu H \gg 1$ if one has absorption in an optically thick layer and $R_\nu \to R_c$, so the dip tends towards the limit for very strong lines.

For the *curve of growth*, the following parameters are plotted against each other: as abscissa $\log NHf + \text{const}$ and as ordinate $\log W_\lambda / 2 R_c \Delta \lambda_D$, where R_c is the maximum line depth. The curves or the line profiles show the following:

- With increasing NHf the line center approaches the maximum line depth R_c, here one has only radiation of the uppermost layers of the limit temperature T_0. The line becomes little broader at first, because the absorption coefficient decreases steeply with $\Delta \lambda$.
- If NHf continues to increase, then the damping wings also become essential. It is then

$$\kappa_\nu \approx NHf\gamma / \Delta \lambda^2 \tag{9.95}$$

One gets wide damping wings, the width of the line at the dip R_ν is determined by $\sqrt{NHf\gamma}$.

Let us now go to the consideration of equivalent widths, i.e., one integrates over the line profiles.

- For weak lines, W_λ increases linearly with NHf, one is in the linear part of the curve of growth.
- Flat part of the curve of growth: The equivalent width is equal to 2 to 4 times the Doppler width $\Delta\lambda_D$.
- For strong lines: $W_\lambda \approx \sqrt{NHf\gamma}$, one enters the damping region of the curve of growth.

The exact shape of the curve depends on α (formula 9.90); e.g., as the mean value for the metal lines of the solar spectrum, one has $\alpha = 0.03$.

9.6.2 Quantitative Spectral Analysis

A quantitative Spectral analysis is done in several steps:

1. Rough analysis: one calculates the whole atmosphere with constant mean values of T, P_e, H. Using the interpolation formula for the line depression, a curve of growth follows, and from the measured W_λ follows the product NH, i.e., the column density of absorbing atoms for the lower level of the transition. However, this requires f, γ to know.
2. The curve of growth is influenced by the Doppler width $\Delta\lambda_D$. It depends on T. In addition to thermal motions, turbulence velocities can also contribute to the Doppler width:

$$\Delta\lambda_D \propto (V_0^2 + \xi_{\text{turb}}^2)^{1/2} \qquad (9.96)$$

One must therefore determine ξ as an additional parameter. The easiest way to do this is to use lines that are located in the flat part of the growth curve, because that is where the dependence on $\Delta\lambda_D$ is greatest there.
3. One compares NH for energy levels with different excitation energies and for different ionization levels (Ca I, Ca II, ...)—then, according to Boltzmann and Saha, respectively, follow the temperature and the electron pressure. From this one then has the total numbers of particles, i.e. the abundance distribution of the elements.
4. From the abundance distribution of the elements follows the gas pressure P_G and from the hydrostatic equation

$$\frac{dp}{dr} = g\rho \qquad (9.97)$$

the gravitational acceleration g.
5. Now a fine analysis is built on top of the coarse analysis. One constructs a model of the stellar atmosphere using the equations of radiative transfer, etc., and sets plausible

values in it for T_{eff}, g and the chemical composition. Then one compares the model with observations and adjusts the model until it matches the observations.

9.7 Stellar Atmosphere Models

9.7.1 Comparison: Sun and Vega

Let us consider two different models:

- Sun (Table 9.3): G2V, $T_{eff} = 5\,770$ K, $g = 274$ m s^{-2}. The optical depth τ_0 refers to κ_λ at $\lambda = 500$ nm. $\bar{\tau}$ refers to Rosseland's mean value $\bar{\kappa}$.

  The chromosphere of the Sun, or, Fig. 9.10 the corona, can be examined, for example, with the lines of elements given in Table 9.4, respectively their different ionization levels. The higher the ionization, the higher the temperature necessary for it (Fig. 9.11).
- Wega (αLyr) (Table 9.5): A0V, $T_{eff} = 9\,400$ K, $g = 89$ m s^{-2}.

Table 9.3 Model atmosphere Sun Model atmosphere of the Sun *Kurucz,* 1979

$\bar{\tau}$	τ_0	T[K]	P_g[Pa]	P_e[Pa]
10^{-3}	1.1×10^{-3}	4485	3.46×10^2	2.84×10^{-2}
0.01	0.01	4710	1.29×10^3	1.03×10^{-1}
0.10	0.09	5070	4.36×10^3	3.78×10^{-1}
0.22	0.19	5300	6.51×10^3	6.43×10^{-1}
0.47	0.40	5675	9.55×10^3	1.34
1.0	0.84	6300	1.29×10^4	4.77
2.2	1.8	7085	1.52×10^4	2.13×10^1
4.7	3.5	7675	1.71×10^4	5.86×10^1
10	7.1	8180	1.89×10^4	1.27×10^2

Fig. 9.10 Spectrum of Vega (after *Bohlin, Gilliland*). Below 420 nm (= 4200 Å) measured (STIS, Space Telescope Imaging Spectrograph), above this value calculated according to a model atmosphere. Below 300 nm metals dominate, above 300 nm Balmer lines. The Balmer jump is clearly recognizable

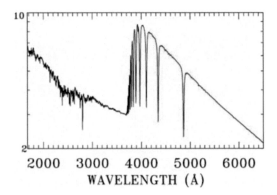

Table 9.4 Lines of elements in certain ionization levels, in which different heights of the chromosphere of the sun can be observed

Element	Height in 10^3 km	Temperature in 10^3 K
Si II	2	25
O II	2	45
Si III	2	70
Si IV	2	80
O III	2	100
O IV	2	200
O V	2.1	300
O VI	2.3	380
Si V	2.8	550
Si VI	6	1000
Mg X	16	1100
Fe X	200	2000

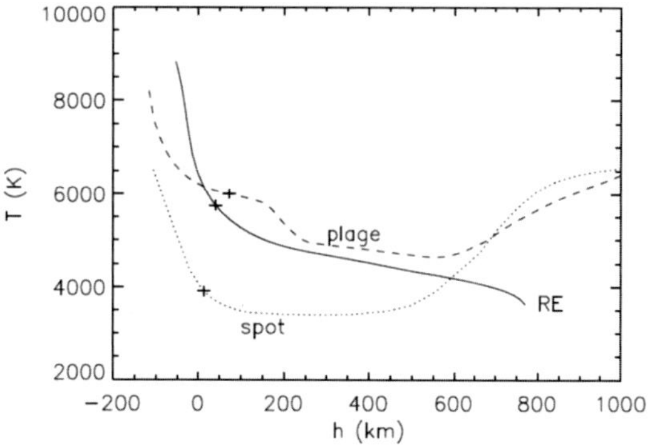

Fig. 9.11 Model of the solar photosphere and lower chromosphere after *Stuik, Bruls,* and *Rutten,* 1997. One can see the temperature profile for the atmosphere in a sunspot (spot), a plage, and in radiative equilibrium (RE)

Table 9.5 Model atmosphere of Wega according to *Kurucz,* 1979

τ_0	T[K]	P_g[Pa]	P_e[Pa]
0.6×10^{-3}	7140	6.52	4.31×10^{-1}
0.5×10^{-2}	7510	2.70×10^1	1.61
0.05	8150	9.13×10^1	7.33
0.11	8590	1.22×10^2	1.40×10^1
0.24	9240	1.53×10^2	2.94×10^1
0.53	10,190	1.79×10^2	5.81×10^1
1.3	11,560	2.12×10^2	9.21×10^1
3.6	13,480	2.99×10^2	1.40×10^2
11.5	16,000	5.81×10^2	2.77×10^2

With these models, it should be added that the altitude h can of course also be given in km. Then one always refers to the level $\tau = 1$, usually at a wavelength of 500 nm. For the sun often the *Harvard Smithsonian Reference Atmosphere* (HSRA) is used. This ranges at 500 nm from an optical depth of $\tau_{500} = 10^{-8}$ to $\tau_{500} = 25$.

Other models of the solar atmosphere are HOLMU *(Holweger, Muller)*, as well as the Models VAL-I, VAL-II and VAL-III *(Vernazza, Avrett, Loser,* 1977). An example is given in Fig. 9.11.

9.7.2   Numerical Solutions

We briefly sketch here how one can numerically solve the given systems of equations. If one introduces the quantity $\mu = \cos\theta$ then the following applies to the mean intensity:

$$J = \frac{1}{2} \int_{-1}^{1} I \, d\mu \qquad (9.98)$$

Where $\mu > 0$ is one hemisphere, $\mu < 0$ the opposite hemisphere of a star. Mostly it is assumed that $\mu > 0$ points to us. Decompose the above integral into:

$$J = \frac{1}{2} \int_{0}^{1} I^{+} \, d\mu + \frac{1}{2} \int_{-1}^{0} I^{-} \, d\mu \qquad (9.99)$$

and because of

$$J = \frac{1}{2} \int_{0}^{1} I^{+} \, d\mu + \frac{1}{2} \int_{0}^{1} I^{-} \, d(-\mu) \qquad (9.100)$$

finally follows the discretization of the expression for J:

$$J \approx \frac{1}{2} \sum_{j=1}^{m} a_j I_j^{+} + \frac{1}{2} \sum_{j=1}^{m} a_j I_j^{-} \qquad (9.101)$$

The rays of intensity I^{+} thus go outwards, the rays of intensity I^{-} go inwards. The quantities a_j are weighting functions of the integration. So one has approximated the integral by a sum:

$$\int_{x_1}^{x_n} f(x) \, dx \approx \sum_{1}^{n} a_i f(x_i) \qquad (9.102)$$

Examples can be found in the chapter "Mathematical Methods". We give here as an example of such an approximation:

$$\int_{x_1}^{x_5} f(x)dx \approx \frac{2\Delta x}{25}[7f(x_1) + 32f(x_2) + 12f(x_3) + 32f(x_4) + 7f(x_5)] \quad (9.103)$$

If one chooses the position of the points x_i well (even not equidistant), then much better results are obtained. Another method is that of spline interpolation.

Let us still consider the discretization of the transfer equation:

$$\mu\frac{dI}{d\tau} = I - S \quad (9.104)$$

For simplicity, we show the discretization for the monofrequency case. We discretize in n points τ_i where $i = 1, \ldots, n$. Furthermore we consider m angles μ_j. For the case of purely coherent scattering, the solution is of the transfer equation:

$$\mu\frac{dI}{d\tau} = I - \epsilon B - \frac{1-\epsilon}{2}\int_{-1}^{1} Id\mu \quad (9.105)$$

And the discretization gives:

$$\mu\frac{dI_\mu^+}{d\tau} = I_\mu^+ - \epsilon B - \frac{1-\epsilon}{2}\left[\sum_{j=1}^{m}a_jI_j^+ + \sum_{j=1}^{m}a_jI_j^-\right] \quad (9.106)$$

$$-\mu\frac{dI_\mu^-}{d\tau} = I_\mu^- - \epsilon B - \frac{1-\epsilon}{2}\left[\sum_{j=1}^{m}a_jI_j^+ + \sum_{j=1}^{m}a_jI_j^-\right] \quad (9.107)$$

This would be the discretization of the transfer equation at a given optical depth τ_i in a given direction μ_j.

In practice, one often computes with the Fautrier variables:

$$P_\nu(\tau_\nu, \mu) = \frac{1}{2}[I_\nu(\tau_\nu, \mu) + I_\nu(\tau_\nu, -\mu)] = \frac{1}{2}[I_j^+ + I_j^-] \quad (9.108)$$

$$R_\nu(\tau_\nu, \mu) = \frac{1}{2}[I_\nu(\tau_\nu, \mu) - I_\nu(\tau_\nu, -\mu)] = \frac{1}{2}[I_j^+ - I_j^-] \quad (9.109)$$

The average intensity is then:

$$J_\nu(\tau_i) = \int_0^1 P_\nu(\tau, \mu)d\mu \approx \sum_{j=1}^{m}a_jP(\tau_i, \mu_j) \quad (9.110)$$

9.8   Asteroseismology

With the help of asteroseismology one can look into the interior of stars. The method works in a similar way we can study the Earth's interior by seismology. from Earthquake waves.

9.8.1   Observations

The only information we receive and measure from stars is their radiation. Accurate photometry of stars shows small variations in brightness in many cases. So one studies the light curves of stars as a function of time. The small variations in brightness are due to oscillations of the star. Patterns of different magnitudes, as shown in Fig. 9.12, spread to different depths into the interior of a star: Waves with a large pattern extend deeper into the star's interior than waves that result in only a small surface pattern. In addition to photometry, one can also make velocity measurements; the individual parts of a star's surface oscillate relative to each other. These oscillations can be detected using the Doppler effect.

Why do such oscillation patterns occur? The stellar density increases inwards, which leads to a reflection of the wave fronts perpendicular to those in Fig. 9.12 The deeper lying part of a wave front is reflected. The lower lying part of a wave front comes rather into regions of higher temperature (T_2) and density than the higher (T_1). The speed of

Fig. 9.12 Vibration patterns propagating inward from the surface of a star. The larger the pattern, the deeper the layers in the star's interior, where they are reflected again

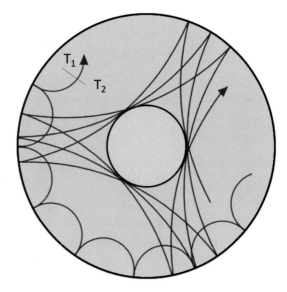

propagation of a sound wave is given by:

$$v \sim \sqrt{T/\rho} \tag{9.111}$$

Therefore, the lower lying part of the wave front propagates faster compared to the higher lying one and the wave is reflected upwards. The reflection at the surface is caused by the strong decrease in density. Stars therefore oscillate like resonant cavities. In order for oscillations to occur, any disturbance must be compensated by a counterforce (restoring force). In the case of the pressure the wave equation reads:

$$\Delta p = \frac{1}{c^2} \frac{\partial^2 p}{\partial t^2} \tag{9.112}$$

Ideally, these measurements are made from space. Two satellite missions of particular interest in this context are COROT (Convection, Rotation and Planetary Transits) and the Kepler mission. With COROT, about 120,000 stars were measured with a 30-cm telescope, with Kepler (telescope size 1 m) about 150,000 Stars.

The array of CCD camera used for imaging and photometry of stars with the Kepler satellite st shown in Fig. 9.13

9.8.2 Types of Waves

For Sun-like stars, there are three types of waves:

1. P-modes: These are generated by fluctuations in the pressure inside the stars. The local speed of sound determines their dynamics.

Fig. 9.13 The array of CCD cameras for the Kepler satellite. At the four corners, one can see much smaller CCD cameras used for guiding the telescope. Keplerconstantly observesBirminghamthe same Section of sky. NASA

2. G-modes: Also called "Gravity Modes"; they are caused by buoyancy in the interior of the star.
3. F-modes: These waves resemble surface waves on a body of water.

The oscillations are represented by the three parameters n (oscillation in radial direction), as well as l (denotes the degree) and m (denotes the azimuthal direction). In the analysis one considers:

- Large frequency intervals, $\Delta\nu$; in these intervals l is the same, but n changes by one value (overtone). $\Delta\nu$ gives information about the physics of the radial behavior of a star. If M is the total mass and R are the radius of the star, then

$$\Delta\nu \propto \left(\frac{M}{R^3}\right)^{1/2} \propto \rho \tag{9.113}$$

- The frequency of the maximum in the power spectrum of the oscillations is ν_{max}. This value is related to the acoustic cut-off frequency ν_{ac} of a star:

$$\nu_{ac} \propto MR^{-2}T_{eff}^{-1/2} \tag{9.114}$$

For example, from the analysis of the oscillations, one can determine the radius of a star.

Asteroseismology provides information about physical quantities such as pressure, temperature, etc. in the interior of the star.

In Fig. 9.14 you can see the power spectrum[7] of oscillations on the Sun.

In Fig. 9.15 the power spectra of oscillations of various stars are given. As an example, consider the stars α Centauri A and α Centauri B (see Fig. 9.15). For the Sun $\Delta\nu = 135\,\mu\mathrm{Hz}$, for α Centauri A is $\Delta\nu = 105\,\mu\mathrm{Hz}$. Since $\Delta\nu \approx \tau$ where τ is the time period within which a sound wave passes through the star, and $\Delta\nu \propto \rho^{1/2}$ gives a radius of $1.227\,R_\odot$ and a mass of $1.105\,M_\odot$. The star α Centauri B has a $\Delta\nu$ of $161\,\mu\mathrm{Hz}$. It must therefore be denser than the Sun. Its mass is $0.0934\,M_\odot$ and its radius $0.870\,R_\odot$.

9.9   Further Literature

We give a small selection of recommended further reading.
Theory of Stellar Atmospheres, I. Hubeny, D. Mihalas, Princeton Univ. Press, 2014

---

[7] See chapter on mathematical methods.

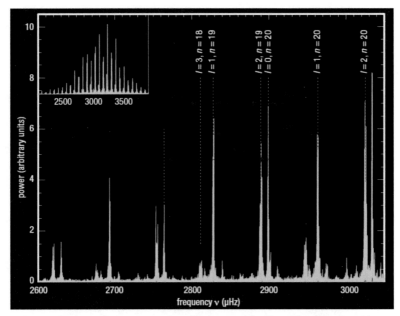

Fig. 9.14 Example of power spectrum of oscillations on the Sun. The distances Δv_0, Δv_1 resp. δ_{13}, δ_{02} are similar. BISON, Birmingham Solar Oscillations Network

An Introduction to Stellar Astrophysics, E. Böhm-Vitense, Cambridge Univ. Press, 1989
Radiative Transfer in Stellar and Planetary Atmospheres, Cambridge Univ. Press, 2020

Tasks

9.1 Calculate the wavelength of the Lyman-α-line.

Solution
$1/\lambda = R_H(1 - 1/4)$. So $\lambda = 121.168$ nm.

9.2 Calculate the mean square of velocity for (a) nitrogen atoms and (b) nitrogen molecules N_2 at $T = 800$ K.

Solution
$m_N = 14$ amu $= 14 \times 1.66 \times 10^{-27}$ kg.
   This gives us $v = \sqrt{3kT/m} = \sqrt{3 \times 1.38 \times 10^{-23} \times 800/(14 \times 1.66 \times 10^{-27})}$ and for (a) 1193 m/s, for (b) 846 m/s.

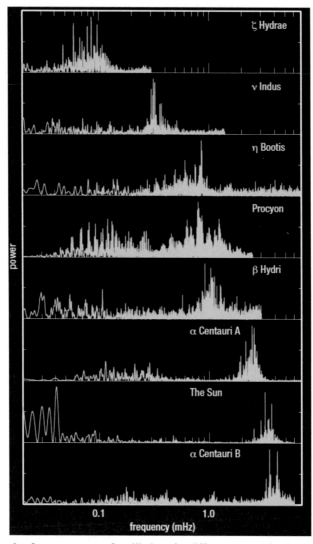

Fig. 9.15 Example of power spectra of oscillations for different stars. After H. Kjeldsen

9.3 Example Na: Ratio of singly ionized to neutral Na atoms for $T = 6000\,\text{K}$ and $P_e = 10\,\text{dyn/cm}^2$.

Solution

If $u_0 \approx g_{0.1} = 2$; $u_1 \approx g_{1.1} = 1$. For $\chi_0 = 5.14\,\text{eV}$. From this $N_1/N_0 = 4.4 \times 10^3$. If one calculates the whole for the ratio N_2/N_1 then one gets (high ionization energy of $\text{Na}^+ = \chi_1 = 47.29\,\text{eV}$): approx. $10\times^{-31}$.

So, if we take the values for the Sun, then of all the Na atoms, only the fraction $N_0/(N_0 + N_1) = 1/(1 + N_1/N_0) = 0.23 \times 10^{-3}$ in the neutral state.

9.4 Consider Fig. 9.5 At what wavelengths can one look particularly deep into the solar photosphere?

Solution
The lower the absorption, the deeper one can look, hence in the blue and at 1.6 μm in the IR.

9.5 What are the disadvantages of IR observations?

Solution
Poorer spatial resolution.

9.6 In the Sun, the coronal line CaXIII is observed very strongly. It is therefore 13 times ionized Ca. The ionization potential is 655 eV. From this, estimate the temperature of the corona.

Solution
$$kT \approx 655\,\text{eV} \approx 10^{-9}\,\text{erg}$$
$$T \approx \frac{10^{-9}}{1.4 \times 10^{-16}} \approx 7 \times 10^6\,\text{K}$$

Stellar Structure

10

In this chapter we cover the main equations describing the structure of a star. Furthermore, we trace the evolution of stars in the Hertzsprung-Russell diagram. The Voigt-Russell theorem states that the total stellar evolution is determined by the initial mass and the chemical composition.

10.1 Basic Physical Laws of Stellar Structure

Only a few equations are needed to describe the internal structure of a star. We start from the simplification that all physical parameters depend only on the distance r from the stellar center. So, for example, we can describe the temperature inside the star by the simple function $T(r)$. We therefore think of stars as homogeneous, isotropic, unflattened spheres of gas.

10.1.1 Hydrostatic Equilibrium

In a stable star gravity is balanced by the internal pressure:

Gravity (acting inward) = internal pressure (acting outward).

This state of equilibrium is called *hydrostatic equilibrium.*

© Springer-Verlag GmbH Germany, part of Springer Nature 2023
A. Hanslmeier, *Introduction to Astronomy and Astrophysics*,
https://doi.org/10.1007/978-3-662-64637-3_10

What happens if this equilibrium is not fulfilled? There are the two extreme cases:

- If the internal pressure falls away, the star collapses immediately as a result of gravity,
- without gravity, the star would expand.

We consider the simplest model of a star:

- Spherically symmetric gas sphere,
- homogeneous structure,
- the model should be static,
- no rotation,
- no magnetic fields.

From this follows: All physical parameters f depend only on the center distance r, thus become $f(r)$.

We briefly examine a case where such a simplification is not justified. The star *Vega* (α *Lyr*) is an example of a rapidly rotating star: $P_{\text{rot}} = 12.5$ Hours. Because of this rapid rotation, the temperature on Vega is different: at the equator, about 7600 K, at the poles about $10,000$ K. The star is oblate, and the poles are closer to the stellar center and therefore hotter.

The theorem of *Zeipel* states

$$T_{\text{eff}} \propto g_{\text{eff}}^{1/4} \tag{10.1}$$

where g_{eff} is the effective gravitational acceleration, i.e. the actual acceleration reduced by the effect of the centrifugal force .

Let r be the distance from the stellar center. Consider a thin shell of mass of thickness dr at the position r in the stellar interior. The mass per unit area is ρdr, the weight $-g\rho dr$. The weight is the inward gravitational force. The outward pressure is equal to the difference between the pressure P_i of the side of the mass shell facing the center and the pressure P_e:

$$P_i - P_e = -\frac{\partial P}{\partial r} dr \tag{10.2}$$

Thus we have (Fig. 10.1):

$$\frac{\partial P}{\partial r} = -g\rho \tag{10.3}$$

Fig. 10.1 Hydrostatic
equilibrium

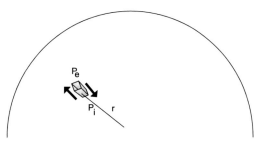

Fig. 10.2 On the derivation of
the mass continuity equation

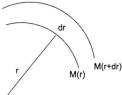

Let us set for $g = GM(r)/r^2$, then the condition for the *hydrostatic equilibrium* (Eulerian form):

$$\frac{dP}{dr} = -\frac{GM(r)\rho(r)}{r^2} \tag{10.4}$$

One can use instead of r is also the mass m inside the sphere with radius r as an independent variable, and one obtains then the Lagrangian Form:

$$\frac{\partial P}{\partial m} = -\frac{Gm}{4\pi r^4} \tag{10.5}$$

where $m = M(r)$.

Within a shell of thickness dr the mass is $dM(r)$ (Fig. 10.2):

$$\frac{dM}{dr} = 4\pi r^2 \rho(r) \tag{10.6}$$

This is known as *Mass continuity equation* and represents our second fundamental equation describing stellar structure. If R is the radius of the star, then the total mass is:

$$M = 4\pi \int_{r=0}^{r=R} \rho(r) r^2 dr \tag{10.7}$$

In Lagrangian form, the equation is:

$$\frac{\partial r}{\partial m} = \frac{1}{4\pi r^2 \rho} \tag{10.8}$$

We estimate the central pressure P_c for the Sun. $G = 6.67 \times 10^{-11} \text{N m}^2/\text{kg}^2$; $M_\odot = 1.989 \times 10^{30}$ kg, $R_\odot = 6.96 \times 10^8$ m. Equation 10.4 can be written approximately as:

$$\frac{dP}{dr} \approx \frac{P_{\text{surface}} - P_c}{R}$$

and the pressure at the surface is $P_{\text{surface}} \approx 0$.

From this, the mean density of the Sun follows to:

$$< \rho_\odot > = 3M_\odot/4\pi R_\odot^3 = 1410 \, \text{kg/m}^3$$

Let's set $r = R_\odot$ and $M(r) = M_\odot$ in the two basic equations, then we get:

$$P_c \approx GM_\odot < \rho_\odot > /R_\odot \approx 10^{14} \text{N/m}^2$$

Since 1 atm $= 1.01 \times 10^5$ N/m^2, we find $P_c = 10^9$ atm. Since the actual central density of the Sun is greater, the central pressure must also be greater.

10.1.2 Equation of Motion with Spherical Symmetry

We again consider a thin shell of mass with dm at the distance r from the center. The force f_P per unit area results from *Pressure gradients:*

$$f_P = -\frac{\partial P}{\partial m} dm$$

The gravitational force per unit area is :

$$f_g = -\frac{g \, dm}{4\pi r^2} = -\frac{Gm}{r^2}\frac{dm}{4\pi r^2}$$

Now, if the sum of both forces is not zero, then there is an acceleration of the mass shell:

$$\frac{dm}{4\pi r^2}\frac{\partial^2 r}{\partial t^2} = f_P + f_g$$

and by substituting and dividing by dm:

$$\frac{1}{4\pi r^2}\frac{\partial^2 r}{\partial t^2} = -\frac{\partial P}{\partial m} - \frac{Gm}{4\pi r^4} \qquad (10.9)$$

- If only pressure gradient → Outward acceleration , $\partial P/\partial m < 0$.
- If only gravity → Collapse.

From 10.9 hydrostatic equilibrium emerges if $\partial^2 r/\partial t^2 = 0$.
Now we investigate deviations from hydrostatic equilibrium:

- We assume the pressure vanishes. This gives us the so-called *free fall time* τ_{ff}. From the equations

$$\frac{1}{4\pi r^2}\frac{\partial^2 r}{\partial t^2} = -\frac{Gm}{4\pi r^4} \qquad g = Gm/r^2 \qquad |\partial^2 r/\partial t^2| \approx R/\tau_{\text{ff}}^2$$

we get the *free fall time*:

$$\tau_{\text{ff}} \approx \sqrt{\frac{R}{g\rho}} \tag{10.10}$$

- *Explosion time:* Here gravity is turned off, and we get:

$$\tau_{\text{expl}} \approx R\sqrt{\frac{\rho}{P}} \tag{10.11}$$

The speed of sound is the typical speed at which disturbances propagate in the stellar interior. The *speed of sound* is given by:

$$v_c \approx \sqrt{\frac{P}{\rho}} \tag{10.12}$$

Therefore τ_{expl} is of the same order of magnitude as it takes a sound wave to travel from the stellar center to the stellar surface (given by $\tau_s \approx R/v_c$).
In general, the formula for calculating the speed of sound in an ideal gas with an adiabatic exponent κ and molar mass M is:

$$c = \sqrt{\kappa \frac{P}{\rho}} = \sqrt{\kappa \frac{\Re T}{M}} \tag{10.13}$$

Let us make a rough estimate of the speed of sound at the surface of the Sun, $T = 6 \times 10^3$ K. For the molar mass, we set the value 0.029 kg/mol (actually the value for air), $\kappa = 1.4$, gas constant $\Re = 8.31\,\text{J mol}^{-1}\text{K}^{-1}$. We find 1500 km/s, which is very close to reality.

- The *hydrostatic time scale* is obtained by equating: $\tau_{ff} = \tau_{expl}$:

$$\tau_{hydr} = \sqrt{\frac{R^3}{GM}} \approx \frac{1}{2}\frac{1}{\sqrt{G\rho}} \tag{10.14}$$

For our sun $\tau_{hydr} = 27$ min. For red giants with $M \approx M_\odot$, $R \approx 100\, R_\odot$ the hydrostatic time scale become much longer: $\tau_{hydr} = 18$ d. For a compact White dwarf on the other hand with $M \approx M_\odot$, $R = R_\odot/50$ we get a very short time scale of $\tau_{hydr} = 4.5$ s.

Since in most cases the stars change on timescales that are very large compared to τ_{hydr} the assumption of a hydrostatic equilibrium is justified.

10.1.3 General Relativity

Effects of general relativity become important in the case of very strong gravitational fields: e.g. in the case of neutron stars. We will only sketch the derivation.

First, one starts from Einstein's field equations which shows the relation between matter (given by the so-called energy-momentum-tensor T_{ik}) and space curvature (given by the Ricci tensor R_{ik}) and the metric tensor g_{ik} which describes the distance between two points in space:

$$R_{ik} - \frac{1}{2}g_{ik}R = \frac{\kappa}{c^2}T_{ik} \qquad \kappa = 8\pi G/c^2 \tag{10.15}$$

Einstein's field equations:
Space-time geometry is on the left, energy/matter distribution is on the right.
Matter $\rightarrow$ Space curvature.

The metric tensor follows from the line element ds^2, which describes the distance between two points in space.

Example for line element in Euclidean space

$$ds^2 = dx^2 + dy^2 + dz^2 \tag{10.16}$$

Since the line element follows in general from

$$ds^2 = g_{ik}dx_i dx_k \tag{10.17}$$

one has in the Euclidean case:
  all $g_{ii} = 1$, and $g_{ij} = 0, if\, i \neq j$.

R is the curvature scalar and follows from R_{ik}. For an ideal gas the components of the energy-momentum tensor are

$$T_{00} = \rho c^2; \quad T_{11} = T_{22} = T_{33} = P \tag{10.18}$$

Let us assume for the line element in polar coordinates:

$$ds^2 = e^\nu c^2 dt^2 - e^\lambda dr^2 - r^2(d\theta^2 + \sin^2\theta d\phi^2) \tag{10.19}$$

From the line element we read the components of the metric tensor and from its derivatives one can calcclate the Ricci tensor. After a long calculation we find the *Tolman-Oppenheimer-Volkoff (TOV)* equation for hydrostatic equilibrium in general relativity:

$$\frac{\partial P}{\partial r} = -\frac{Gm}{r^2}\rho\left(1 + \frac{P}{\rho c^2}\right)\left(1 + \frac{4\pi r^3 P}{mc^2}\right)\left(1 - \frac{2Gm}{rc^2}\right)^{-1} \tag{10.20}$$

If we consider not too strong gravitational fields, then one can derive as approximation: One keeps in the expansion only terms which are linear $1/c^2$ are:

$$\frac{\partial P}{\partial r} = -\frac{Gm}{r^2}\rho\left(1 + \frac{P}{\rho c^2} + \frac{4\pi r^3 P}{mc^2} + \frac{2Gm}{rc^2}\right) \tag{10.21}$$

This is the *Post-Newtonian approximation.*

Fundamental to stable stars is the equation for hydrostatic equilibrium.

10.1.4 Equation of State

Our set of equations to describe stellar structure needs to be complemented by an equation of state. Let us suppose that the gas in a star satisfies the laws for an *ideal gas:*

$$P(r) = n(r)kT(r) \tag{10.22}$$

The pressure $P(r)$ thus depends on the particle density $n(r)$ (number of particles per m^3). k is the Boltzmann constant $k = 1.381 \times 10^{-23}$ J/K. One can also write:

$$n(r) = \frac{\rho(r)}{\mu(r)m_H}, \tag{10.23}$$

where $m_H = 1.67 \times 10^{-27}$ kg is the mass of the hydrogen atom. The equation of state of ideal gases is valid in the interior of the star as long as the interaction of neighbouring particles is small compared to their thermal (= kinetic) energy. The *Molecular weight* μ is equal to the atomic weight divided by the number of all particles (nucleus + electrons). If we assume complete ionization, then the molecular weight μ becomes for :

- Hydrogen: number of particles 2 (one proton, p, and one electron, e^-), therefore $\mu_H = 1/2$.
- For Helium: 3 particles ($2e^-$, nucleus), atomic weight = 4 (since in the nucleus 2 p and 2 n). Hence $\mu_{He} = 4/3$.

Very often one denotes with X the fraction of hydrogen, Y the fraction of helium, and Z the fraction of elements heavier than helium (such elements are often called metals in astrophysics). The average molecular weight is then :

$$\mu = [2X + (3/4)Y + (1/2)Z]^{-1} \approx 1/2 \tag{10.24}$$

We therefore get following *equation of state of ideal gases:*

$$P(r) = \rho(r)kT(r)/\mu(r)m_H \tag{10.25}$$

In the case of massive stars, the gas pressure is supplemented by the *Radiation pressure* (by momentum transfer of the photons):

$$P_{rad}(r) = \frac{a}{3}T^4(r) \tag{10.26}$$

$a = 7.564 \times 10^{-16}\,\mathrm{J\,m^{-3}\,K^4}$.

We estimate the central temperature of the sun. Using the values P_c and $< \rho_\odot >$:

$$T_c \approx \frac{P_c \mu m_H}{< \rho_\odot > k} = 12 \times 10^6\,\mathrm{K}$$

Modern computer models provide a central temperature of 14.7 million Kelvin. At such high temperatures, the gas behaves like a *plasma.* It consists of ions and electrons and is, on the whole neutral.

The total pressure is then:

$$P = P_g + P_{rad} \tag{10.27}$$

Consider a gas in a volume dV, which is completely ionized by pressure. n_e let be the number of free electrons. The velocity distribution of the electrons is given by a *Boltzmann distribution*, their mean kinetic energy is:

$$\bar{E}_{\text{kin}} = \frac{3}{2}kT \tag{10.28}$$

If (p_x, p_y, p_z) are the coordinates in momentum space, then:

$$f(p)dpdV = n_e \frac{4\pi p^2}{(2\pi m_e kT)^{3/2}} \exp\left(-\frac{p^2}{2m_e kT}\right) dpdV \tag{10.29}$$

Let us now assume that, n_e remain constant and T decreases. Then the maximum of the distribution function ($p_{\text{max}} = (2m_e kT)^{1/2}$) shifts to smaller values of p, and the maximum $f(p)$ becomes larger, since $n_e = \int_0^\infty f(p)dp$.

10.1.5 Degeneracy

The *Fermions* include particles with half-integer spin, such as electrons, but also other elementary particles such as quarks and nucleons (protons, neutrons). For these particles the *Pauli principle states:*
Each quantum cell of a 6-dimensional phase space.

$$(x, y, z, p_x, p_y, p_z) \tag{10.30}$$

must not contain more than two fermions, in our case electrons.
The volume of such a quantum cell is:

$$h^3 = dp_x dp_y dp_z dV$$

So if we consider a shell $[p, p + dp]$ in momentum space, then there are $4\pi p^2 dV/h^3$ quantum cells that contain no more than $8\pi p^2 dpdV/h^3$ electrons; therefore from quantum mechanics follows the condition:

$$f(p)dpdV \le 8\pi p^2 dpdV/h^3$$

The state in which all electrons have the lowest energy without violating the *Pauli exclusion principle* is that in which all phase space cells up to momentum p_F are occupied by two electrons; all other phase space cells $p > p_F$ are empty:

$$f(p) = \frac{8\pi p^2}{h^3} \qquad p \le p_F \tag{10.31}$$

$$f(p) = 0 \qquad p > p_F \tag{10.32}$$

From this then to be derived:

$$n_e dV = dV \int_0^{p_F} \frac{8\pi p^2 dp}{h^3} = \frac{8\pi}{3 h^3} p_F^3 dV \tag{10.33}$$

Thus, according to the Pauli exclusion principle, no more than two fermions differing in spin quantum number can occupy the same energy state. This is also called gas degeneracy. Because of the much lower mass, degeneracy occurs first in electrons. Thus, one may have the case where the electron gas is already fully degenerate, but the ion gas is not yet. In the case of degeneracy, the equation of state changes. In the case of complete degeneracy one distinguishes between

- *non-relativistic degeneracy,* $\rho < 2 \times 10^6 \, \mathrm{g\,cm^{-3}}$,

$$P = K_1 \rho^{5/3} \tag{10.34}$$

and

- *relativistic degeneracy,* $\rho > 2 \times 10^6 \, \mathrm{g\,cm^{-3}}$:

$$P = K_2 \rho^{4/3} \tag{10.35}$$

K_1, K_2 depend on the chemical composition.

> In the case of degenerate stellar matter, the density now depends only on pressure and no longer also on temperature.

Degeneracy can be expected for certain stars and in some cases electrons in other cases neutrons are degenerated: :

- Degenerate electrons: Red giants, white dwarfs.
- Degenerate neutrons degenerate electrons: Neutron Stars.

Degeneracy occurs at very high densities. Repulsion between electrons (neutrons) is a consequence of quantum mechanics (Pauli exclusion principle) and not electrical repulsion. In the case of degeneracy, the equilibrium condition is: Gravity = degenerate pressure. If the star receives more matter (through accretion), gravity increases; however, the degenerate pressure increases only slightly, and therefore the star shrinks.

The greater the mass of a degenerate star, the smaller its volume and therefore its radius.

10.1.6 Summary: Equation of State

We summarize for which physical parameters one has to calculate with which equation of state:

- Photon gas—Radiation pressure: . $\rho < 3.0 \times 10^{-23} \mu T^3$; pressure $P = 2.521 \times 10^{-15} T^4$.
- Ideal nondegenerate gas:
  $3.0 \times 10^{-23} \mu T^3 < \rho < 2.4 \times 10^{-8} \mu_e T^{3/2}$; pressure $P = 8.31 \times 10^7 \rho T / \mu$.
- Non-relativistic fully degenerate electron gas:
  $2.4 \times 10^{-8} \mu_e T^{3/2} < \rho < 7.3 \times 10^6 \mu_e$; Print $P = 1.004 \times 10^{13} \left(\frac{\rho}{\mu_e}\right)^{5/3}$
- Relativistic fully degenerate electron gas (white dwarfs):
  $7.3 \times 10^6 \mu_e < \rho \le 10^{11}$; Print $P = 1.244 \times 10^{15} \left(\frac{\rho}{\mu_e}\right)^{4/3}$.
- Degenerate neutron gas:
  $10^{11} \le \rho \le 10^{14}$; Pressure $P \approx 10^{10} (\rho)^{5/3}$.

10.2 Energy Transport

How is energy transferred inside a star? In principle, energy transport is possible by energy transport:

- thermal conduction,
- radiation,
- convection.

Heat conduction occurs through collisions. This type of energy transport works very well in solids (especially in metals); but less well in gases because their thermal conductivity is low.

10.2.1 Convection

Consider in Fig. 10.3 a bubble of gas that is supposed to move upward inside a star due to a random disturbance. As this travels a distance dr, its temperature changes from T_1 to T_2.

Fig. 10.3 Convection in a star. If the temperature change $(T_2 - T_1)$ of an upward moving gas bubble is smaller than that of the surrounding $(T_2' - T_1')$, it is lighter and rises further upwards, and one has convection

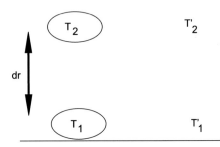

We assume here that during the ascent of the gas bubble there is no heat exchange with the environment. Then the temperature change of the gas bubble can be given by the *adiabatic temperature gradient*:

$$\frac{dT}{dr}\bigg|_{\text{ad}} \tag{10.36}$$

The surrounding matter changes its temperature from T_1' to T_2', and the temperature gradient is to be described by :

$$\frac{dT}{dr}\bigg|_{\text{rad}} \tag{10.37}$$

that is, by the radiation gradient. For the occurrence of convection in a star therefore applies the *Schwarzschild criterion*. If the temperature change of a volume element moving upwards due to a random perturbation is $|dT/dr|_{\text{ad}}$ and if the temperature gradient of the environment is equal to $|dT/dr|_{\text{rad}}$ then no convection occurs if holds:

$$\left|\frac{dT}{dr}\right|_{\text{rad}} < \left|\frac{dT}{dr}\right|_{\text{ad}} \tag{10.38}$$

In stars, convection occurs in different regions:

- Massive stars: the core is convective, the envelope is in equilibrium (Fig. 10.4, right). This is related to the extreme temperature dependence of energy production in the core, which implies a high radiation gradient. Convection in the core results in better mixing of the elements.
- Low-mass, cooler stars (e.g., the Sun): the core is in equilibrium, and the envelope becomes convective (Fig. 10.4, center). This is explained by the increasing number of layers from the surface inwards, in which elements such as hydrogen or helium are ionized, and therefore reduce the adiabatic gradient. In the case of the Sun, convection begins about 200,000 km below its surface.

Fig. 10.4 Convection in stars; in stars with more than 1.5 solar masses ($M > 1.5$) convection occurs in the core region

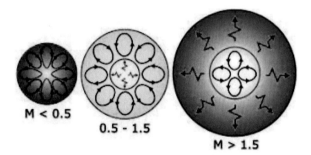

M < 0.5

0.5 - 1.5

M > 1.5

The cooler the stars, the deeper the convection zone reaches! Stars with less than 0.5 solar masses are fully convective.

In the Centre of the Sun the temperature is about 15 million Kelvin, and at the surface it is about 6000 Kelvin. In the radiative zone of the Sun, energy transport is by radiation. The formula for this can be derived by assuming a diffusion approximation. The free path length of the Photons l_{Ph} is given by:

$$l_{Ph} = \frac{1}{\kappa \rho} \tag{10.39}$$

It is only a few centimeters, so the diffusion approximation is a reasonable assumption ($l_{Ph} \ll R_\odot$).

The following estimate shows how long it takes for a photon to travel from the interior of the Sun to the surface and be emitted: The total solar radius is made up of all the partial distances $\mathbf{l_i}$, therefore:

$$\mathbf{L} = \sum_i \mathbf{l_i} = \mathbf{l_1} + \mathbf{l_2} + \dots$$

Since the partial distances are vectors pointing in arbitrary directions, holds:

$$< L^2 > = < l_1^2 > + < l_2^2 > + \dots + < L_N^2 >$$

therefore we get:

$$< L^2 > = N < l^2 >$$

Now $l \approx 1$ cm, $< L > = R_\odot = 7 \times 10^{10}$ cm and $< L^2 > \approx 10^{22}$ cm therefore $N \approx 10^{22}$ and from $t = < L^2 > /c = 3 \times 10^{12}$ s $= 10^5$ years.

Therefore, we now see photons (i.e., radiated energy) from the Sun produced by nuclear fusion in its interior about 100,000 years ago!

Let us consider diffusion in general: a concentration n of particles is said to depend on r. Let the mean free path length be l and its mean velocity v. Let j describe the diffusion flux of particles from sites of high concentration to sites of low concentration:

$$j = -\frac{1}{3}vl\frac{dn}{dr} \tag{10.40}$$

Now we put into this equation the parameters that describe our radiation field:

1. $n \to u = aT^4$, radiation density,
2. $v \to c$,
3. $l \to l_{Ph}$,
4. $j \to L_r/4\pi r^2$, radiation flux.

One can immediately see $dn/dr \to du/dr = 4aT^3 dT/dr$; $a = 4\sigma/c$, and resolved by the temperature gradient gives the diffusion approximation:

$$\frac{dT}{dr} = -\frac{3}{64\pi\sigma}\frac{\kappa\rho L_r}{r^2 T^3} \tag{10.41}$$

One can also derive this equation by the following reasoning: Take a thin shell, where the radiative flux is given by $F(r) = \sigma T^4(r)$. At the point $r + dr$ one has a temperature $T + dT$ and the flux is $F + dF = \sigma(T + dT)^4 \approx \sigma(T^4 + 4T^3 dT)$. dT is negative because the shell of mass is cooler on the outside than on the inside. The flux absorbed inside the shell is then:

$$dF = 4\sigma T^3(r)dT \tag{10.42}$$

This absorption comes from the *Opacity* (it describes quasi the transparency) of the stellar material:

$$dF = -\kappa(r)\rho(r)F(r)dr \tag{10.43}$$

On the other hand, the luminosity of the star is :

$$L(r) = 4\pi r^2 F(r) \tag{10.44}$$

and therefore

$$L(r) = -[16\pi\sigma r^2 T^3(r)/\kappa(r)\rho(r)](dT/dr) \qquad (10.45)$$

An exact treatment still gives the factor 4/3:

$$L(r) = -\frac{64\pi\sigma r^2 T^3(r)}{3\kappa(r)\rho(r)}\frac{dT}{dr} \qquad (10.46)$$

If one has a high opacity, then convection becomes dominant. The gradient is then:

$$\frac{dT}{dr} = \left(1 - \frac{1}{\gamma}\right)\frac{T(r)}{P(r)}\frac{dP}{dr} \qquad (10.47)$$

where $\gamma = c_p/c_v$ where c_p, c_V are the specific heat capacities at constant pressure and constant volume, respectively.

Let us determine the luminosity of the Sun from radiative transfer. For this we approximate $dT/dr \to -T_c/R_\odot$ and thus find the gradient -2×10^{-2} K/m. In the above equation we then set $r \to R_\odot$, $T(r) \to T_c$, $\rho(r) \to \rho_\odot$:

$$L_\odot = \frac{9.5 \times 10^{29}}{\kappa}\,\text{J/s}$$

Here we still have to define a suitable value for the opacity κ. κ is the effective cross-section per gas particle multiplied by the number of particles in 1 kg. A mass of 1 kg of completely ionized hydrogen contains 6×10^{26} protons and as many electrons. For electron scattering, the effective cross section is 10^{-30} m^2 for photoionization of hydrogen 10^{-20} m^2; photoionization as a source of opacity predominates in the solar interior, one has roughly:

$$10^{-3} \ll \kappa \leq 10^7$$

Thus our estimate is:

$$10^{22} \leq L_\odot \ll 10^{32}\,\text{J/s}$$

The mean value of 10^{27} J/s fits well to the measured value of 3.9×10^{26} J/s. This corresponds to an opacity of 2.4×10^3.

10.2.2 Opacity

The Opacity is a measure of the absorptivity in the stellar matter and therefore essential for energy transport.

It is composed of several components, which will be treated briefly.

Electron Scattering Once an electromagnetic wave passes an electron, the electron begins to oscillate and radiates. So the original radiation is attenuated, energy is transferred to the oscillating electron, and there is → Absorption.

The equation of motion of an electron subjected to an electric field **E** is:

$$m_e \left(\frac{d^2\mathbf{x}}{dt^2} + \gamma \frac{d\mathbf{x}}{dt} + \omega_0 \mathbf{x}^2 \right) = -e\mathbf{E} \tag{10.48}$$

γ is the damping constant. Electric field → excites electrons to oscillate.

Two limiting cases are obtained as solutions:

- $\omega \gg \omega_0, \gamma$ → Electron moves like a free electron, *Thomson effective cross section* σ_T

$$\sigma_T = \frac{8\pi}{3} \left(\frac{e^2}{4\pi \epsilon_0 m_e c^2} \right)^2 \tag{10.49}$$

- $\omega_0 \gg \omega, \gamma$: *Rayleigh scattering*

$$\sigma_R = \sigma_T \left(\frac{\omega}{\omega_0} \right)^4 \tag{10.50}$$

so scattering $\propto \omega^4$ or $\propto \lambda^{-4}$. In the Earth's atmosphere blue light is scattered more than red light. When starlight passes through an interstellar cloud, it becomes reddened.

The Thomson effective cross section is obtained as $\sigma_T = 6.65 \times 10^{-29}$ m^2, and when there are n_e electrons in the unit volume, then the absorption coefficient is:

$$n_e \sigma_T \tag{10.51}$$

It is found:

$$\kappa_\nu = 0.20(1 + X) \tag{10.52}$$

This *Thomson scattering* is independent of frequency, X is the fraction of hydrogen. This treatment neglects the momentum exchange between electrons and radiation, but this only becomes effective at very high temperatures. The momentum of the photons is $h\nu/c$, this is then partially transferred to the electrons after scattering $m_e v \approx h\nu/c$. The relativistic correction *(Compton scattering)* is effective when $v_e \approx 0.1c$.

Free-Free Transitions The electron are in thermal motion due to the temperatures. If such an electron now passes an ion, then the two charged particles can absorb radiation. The thermal velocity of the free electrons is $v \approx T^{1/2}$, and the time during which they absorb or emit, $\propto 1/v \approx T^{-1/2}$. *Kramers* derived the following relation:

$$\kappa_\nu \propto Z^2 \rho T^{-1/2} \nu^{-3} \tag{10.53}$$

Using Rosseland's mean, one obtains:

$$\kappa_{ff} = \rho T^{-7/2} \tag{10.54}$$

If one has a fully ionized gas (stellar center), then it holds *(Kramers)*:

$$\kappa_{ff} = 3.8 \times 10^{22}(1 + X)[(X + Y) + B]\rho T^{-7/2} \ \text{cm}^2 \, \text{g}^{-1}, \tag{10.55}$$

where $B = \sum_i X_i Z_i^2/A_i$ and A_i the atomic Mass numbers are.

Bound-Free Transitions Let us first consider bound-free transitions of a neutral hydrogen atom. In the ground state, the ionization energy is χ_0 ; it is ionized by a photon of energy $h\nu > \chi_0$. It follows, then:

$$h\nu = \chi_0 + \frac{1}{2}m_e v^2, \tag{10.56}$$

where v is the velocity of the released electron. If a_ν is the absorption coefficient per ion, $a_\nu = \kappa_\nu \rho/n_{ion}$ then $a_\nu = 0$, $v < \chi_0/h$ and $a_\nu > 0$, $v \geq \chi_0/h$. One obtains $a_\nu \propto \nu^{-3}$ for $v \geq \chi_0/h$. The so-called Gaunt *factor* is a quantum mechanical correction and occurs when the problem is treated exactly. Analogously, the matter continues for the first excited state, $a_\nu = 0$, $h\nu < \chi_1$ and $a_\nu \propto \nu^{-3}$ for $h\nu \geq \chi_1$ where χ_1 is the energy required to ionize a hydrogen atom from the first excited state. This is why the jagged shape of the absorption coefficient occurs.

Heat Conduction Like all particles, electrons can also transport energy by thermal conduction. Normally, their contribution to the total energy transport is negligible. The thermal conduction is proportional to the mean free path length l and in the non-degenerate case $l_{photon} \gg l_{particle}$.

Heat conduction becomes important for degenerate stellar material, i.e. in the interiors of far evolved stars as well as in white dwarfs. Here all quantum cells below the Fermi momentum p_F are occupied.

10.3   Energy Sources

Our sun has been radiating with almost unchanged luminosity for about 4.5 billion years. The question arose where this energy comes from. We will first cover classical sources of energy, such as the gravitational energy released during contraction, then nuclear fusion, which is the source of energy for most of a star's evolution.

A star continuously radiates energy. Therefore, stellar models are not static in the strict sense. Stars evolve. Let $\epsilon(r)$ be the rate of energy production related to the unit mass (J/s kg). In a strict sense ϵ also depends on T, P and the density, but for simplicity we write $\epsilon(r)$. For stellar structure, we assume : $\epsilon = 0$, except in the central regions where the energy is produced by thermonuclear fusion.

For the Sun, we obtain:

$$\epsilon_\odot \approx L_\odot/M_\odot = 2.0 \times 10^{-7} \, \text{J/s kg} \tag{10.57}$$

Inside a shell dr changes the luminosity by:

$$dL = 4\pi r^2 \rho(r)\epsilon(r)dr \tag{10.58}$$

In the following, we consider the various ways in which energy is generated.

Energy is released when the star contracts. Analogy: consider a rock falling to earth → Gravitational energy is converted into kinetic energy.

> **Virial Theorem**
> When a star slowly contracts, gravitational energy is released
>
> - half of which heats the star,
> - the other half is radiated.

Suppose we add the fraction $dM(r)$ to a mass $M(r)$, then the change in gravitational energy is:

$$dU = -\frac{GM(r)dM(r)}{r} \tag{10.59}$$

Let us integrate over all mass shells:

$$U = -\int_0^M G\frac{M(r)dM(r)}{r} = -q(GM^2/R) \tag{10.60}$$

Here q depends on the mass distribution in the sphere. If one has a uniform density, then $q = 3/5$. For most main sequence stars we can use $q = 1.5$.

Let us do an exercise. How long can the sun shine by contraction?

We calculate with

$$E \approx \frac{GM^2}{R} = \frac{6.67 \times 10^{-11}(2 \times 10^{30})^2}{7 \times 10^8} = 4 \times 10^{41}\,\text{J}$$

By comparison with Eq. 10.57: The Sun can only radiate at present luminosity for about 30 million years by releasing gravitational energy.

10.3.1 Thermonuclear Energy Production

Physical Preconditions

Prior to nuclear fusion process, the mass of the i nuclei involved would be $\sum M_i$. After fusion the resulting nucleus has a total mass $\sum M_p$. The mass of the fused nuclei is less than the mass of the original nuclei, and the missing amount, *Mass defect*, ΔM, is

$$\Delta M = \sum_i M_i - M_p \tag{10.61}$$

This missing mass is converted into energy according to Einstein's well-known formula:

$$E = \Delta M c^2. \tag{10.62}$$

If we consider as an example the *Hydrogen burning*. Here a helium nucleus is produced from four hydrogen nuclei:

$$4\,^1\text{H} \rightarrow {}^4\text{He} \tag{10.63}$$

- The total mass of $4\,^1\text{H}$ amounts to: $4 \times 1.0081 m_u$.
- The total mass of a ^{4}He-Kerns amounts to: $4.0029\,m_u$.

Therefore $\Delta m = 2.85 \times 10^{-2}\, m_u$ or 0.7% of the total mass is converted, which corresponds to an energy of $26.5\,\text{MeV}$. The following conversions are practical:

$$1\,\text{keV} \approx 1.16 \times 10^7\,\text{K} \tag{10.64}$$

$$931.1\,\text{MeV} \approx 1\,m_u \tag{10.65}$$

Let us apply this to the Sun. The Mass loss rate is $L_\odot/c^2 = 4.25 \times 10^{12}\,\text{g\,s}^{-1}$. So the sun loses 4 million tons per second. Let's assume that $1\,M_\odot$ is converted into He, then 0.7% corresponds to $1.4 \times 10^{31}\,\text{g}$ that is converted into energy and $\rightarrow$ the sun could life $3 \times 10^{18}\,\text{s} \approx 10^{11}\,\text{a}$.

Consider an atomic nucleus of mass M_{nuc}, mass number A, which contains Z protons of mass m_p and contains $(A - Z)$ neutrons of mass m_n. The binding energy E_B is then:

$$E_B = [(A - Z)m_n + Z m_p - M_{\text{nuc}}]c^2 \tag{10.66}$$

And the mean *binding energy* per nucleon f:

$$f = \frac{E_B}{A} \tag{10.67}$$

If one plots f vs. A, then one sees the following curve behavior:

- steep slope at the fusion of light elements,
- then flat slope up to the element ^{56}Fe,
- flat decrease from the element ^{56}Fe.

Accordingly, there are two ways to gain energy. Both have in common that the final product after fusion has a higher binding energy per nucleon than the initial products:

- up to the element iron: by fusion,
- elements heavier than Fe: by fission.

> In the fusion of hydrogen to helium, only 0.7% of the initial mass is converted to energy.

Let's take a closer look at fusion. Between two particles, which are charged with the same sign with the charges $Z_1 e$, $Z_2 e$ (Z denotes the *Nuclear charge number,* the number of protons in the nucleus) there is a *Coulomb repulsion:*

$$E_{\text{Coul}} = \frac{1}{4\pi \epsilon_0} \frac{Z_1 Z_2 e^2}{r} \tag{10.68}$$

One must bring the particles so close to each other that the short-range strong *Nuclear Forces* dominate over the long-range but weaker Coulomb repulsion forces. In this case, the interaction radius is:

$$r_0 \approx A^{1/3} 1.44 \times 10^{-15} \, \text{m} \tag{10.69}$$

And the *Coulomb barrier* then results to

$$E(\text{Coul}) \approx Z_1 Z_2 \, \text{MeV} \tag{10.70}$$

Inside the Sun, the temperature near the center is 10^7 K which corresponds to an energy of about one keV. Classically, nuclear fusion would thus be impossible in the stellar interior because of the too low temperatures. However, due to the *Tunnelling effect*[1] particles of lower energy can tunnel through the Coulomb barrier → fusion.

> Nuclear fusion in the stellar interior can only be explained by the quantum mechanical tunnel effect.

The probability that a particle tunnels through the Coulomb barrier is:

$$P(v) = e^{-2\pi \eta} \qquad \eta = \frac{1}{4\pi \epsilon_0} \frac{Z_1 Z_2 e^2}{\hbar v} \tag{10.71}$$

→ strong temperature dependence of thermonuclear reactions!

The thermonuclear reaction rates depend on: Number of particles n_j, n_k, cross section σ. The number of reactions per second is then $n_k \sigma v$ and if there are n_j particles are in the volume, it is :

$$r_{jk} = n_j n_k \sigma v \tag{10.72}$$

---

[1] G. Gamow, 1928.

The *Cross section* depends on v. Under normal conditions, particle velocities are distributed according to Maxwell-Boltzmann. Let the energy be

$$E = \frac{1}{2}mv^2 \tag{10.73}$$

and $m = m_j m_k/(m_j + m_k) \dots$ reduced mass. In the interval $[E, E + dE]$ we have thus:

$$f(E)dE = \frac{2}{\sqrt{\pi}} \frac{E^{1/2}}{(kT)^{3/2}} \exp^{-E/kT} dE \tag{10.74}$$

The averaged reaction probability is:

$$< \sigma v > = \int_0^\infty \sigma(E) v f(E) dE \tag{10.75}$$

Let X_i is the fractional mass of the particles, i.e.

$$X_i \rho = n_i m_i$$

and Q is the energy released per reaction, then the *energy generation rate:*

$$\epsilon_{jk} = \frac{1}{1 + \delta_{jk}} \frac{Q}{m_j m_k} \rho X_j X_k < \sigma v > \tag{10.76}$$

where $\delta_{jk} = 0$ if $j \neq k$ and $\delta_{jk} = 1$ if $j = k$.

Another effect is shielding by free electrons. Beyond a certain distance, the incoming particle senses a neutral conglomerate of a positively charged nucleus surrounded by a cloud of free electrons. A nucleus of charge Ze causes polarization in its environment: electrons of charge $-e$ are attracted, and their density n_e in the vicinity of the nucleus is greater. The other ions are repelled, and their density n_i is lower. We have therefore deviations of the n_e, n_i from the mean values $\bar{n}_e, \bar{n}_i$. For the potential Φ we find:

$$\Phi = \frac{Ze}{r} e^{-r/r_D} \tag{10.77}$$

Here r_D the *Debye radius,* which indicates the point at which the electrons start to shield the potential of the core. If $r \to 0$ then this potential changes to Ze/r. This also leads to a reduction of the Coulomb interaction and increases the probability of tunneling through the Coulomb barrier.

We now discuss the most important reactions. The superscript for the elements indicates the mass number, i.e., the number of protons and neutrons in the nucleus. ^{2}H is deuterium, i.e. a nucleus with one proton and one neutron, an isotope of hydrogen.

Hydrogen Burning The two hydrogen burning basic reactions are:

$$^1H + {}^1H \rightarrow {}^2H + e^+ + \nu \tag{10.78}$$

$$^2H + {}^1H \rightarrow {}^3He + \gamma \tag{10.79}$$

From here on, there are branching reactions:

- pp1:

$$^3He + {}^3He \rightarrow {}^4He + 2\,{}^1H \tag{10.80}$$

- Further:

$$^3He + {}^4He \rightarrow {}^7Be + \gamma \tag{10.81}$$

And from here, the branches:

- pp2:

$$^7Be + e^- \rightarrow {}^7Li + \nu \tag{10.82}$$

$$^7Li + {}^1H \rightarrow {}^4He + {}^4He \tag{10.83}$$

- pp3:

$$^7Be + {}^1H \rightarrow {}^8B + \gamma \tag{10.84}$$

$$^8B \rightarrow {}^8Be + e^+ + \nu \tag{10.85}$$

$$^8Be \rightarrow {}^4He + {}^4He \tag{10.86}$$

The Energy Production Rate ϵ of the pp process Is strongly temperature dependent and given by :

$$\epsilon \propto \rho T^5 \tag{10.87}$$

Hydrogen burning dominates at temperatures between 5 and 15×10^6 K.

CNO Cycle

Here, the carbon serves only as a catalyst. One has the following six reaction steps:

$$^{12}C + {}^1H \rightarrow {}^{13}N + \gamma \tag{10.88}$$

$$^{13}N \rightarrow {}^{13}C + e^+ + \nu \tag{10.89}$$

$$^{13}C + {}^1H \rightarrow {}^{14}N + \gamma \tag{10.90}$$

$$^{14}N + {}^1H \rightarrow {}^{15}O + \gamma \tag{10.91}$$

$$^{15}O \rightarrow {}^{15}N + e^+ + \nu \tag{10.92}$$

$$^{15}N + {}^1H \rightarrow {}^{12}C + {}^4He \tag{10.93}$$

Here the energy production rate is even more dependent on the temperature:

$$\epsilon_{CNO} \propto \rho T^{12...18} \tag{10.94}$$

> At lower temperatures, the pp chain predominates, and at higher temperatures, the CNO cycle predominates (Fig. 10.5).

The nascent Positron e^+ immediately annihilate with the electrons and radiate forming γ quanta. The neutrinos ν have a very small interaction cross section and can pass the star practically unhindered after their formation. In the process, they dissipate energy. In the interior of the sun, the fusion of a ^4He nucleus produces two Neutrinos, and a solar neutrino flux on Earth results. One measures a neutrino flux of 10^{15} neutrinos per m^2 per second.

Fig. 10.5 Energy production rates of pp and CNO reactions. The energy production by the CNO cycle predominates from about 18 Million K onwards

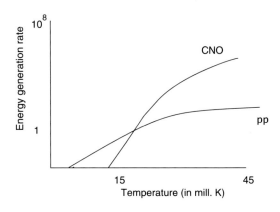

Helium Burning As soon as in the central region of the sun all hydrogen has fused into helium, the thermonuclear reactions cease. The temperature is still too low for further reactions to ignite. Only when it reaches 10^8, K by contraction (cf. virial theorem), helium burning begins:

$$^4\text{He} + {}^4\text{He} \rightleftharpoons {}^8\text{Be} \tag{10.95}$$

$$^8\text{Be} + {}^4\text{He} \rightarrow {}^{12}\text{C} + \gamma \tag{10.96}$$

$$^{12}\text{C} + {}^4\text{He} \rightarrow {}^{16}\text{O} + \gamma \tag{10.97}$$

The first reaction produces a ^8Be nucleus, which is unstable and decays after 10^{-1}s. Only if it reacts with a third ^4He nucleus within this short lifetime, a stable nucleus ^{12}C nucleus is formed. High densities are necessary for this. Some ^{16}O nuclei still react with ^4He and form ^{20}Ne.

The energy production rate is:

$$\epsilon_{\text{He}} \propto \rho^2 T^\nu \tag{10.98}$$

with $\nu = 20 \ldots 30$.

Carbon Burning Once He is burned and the temperature has increased high enough, carbon burning sets in at $5 \times 10^8 \ldots 10^9$ K and the following reactions occur:

$$^{12}\text{C} + {}^{12}\text{C} \rightarrow {}^{24}\text{Mg} + \gamma \tag{10.99}$$

$$\rightarrow {}^{23}\text{Mg} + n \tag{10.100}$$

$$\rightarrow {}^{23}\text{Na} + {}^1\text{H} \tag{10.101}$$

$$\rightarrow {}^{20}\text{Ne} + {}^4\text{He} \tag{10.102}$$

Oxygen Burning This starts at $T = 1.4 \times 10^9$ K:

$$^{16}\text{O} + {}^{16}\text{O} \rightarrow {}^{32}\text{S} + \gamma \tag{10.103}$$

$$\rightarrow {}^{31}\text{S} + n \tag{10.104}$$

$$\rightarrow {}^{31}\text{P} + {}^1\text{H} \tag{10.105}$$

$$\rightarrow {}^{28}\text{Si} + {}^4\text{He} \tag{10.106}$$

There are further reactions due to the capture of ^1H and ^4He.

Silicon Burning From $T \approx 2 \times 10^9$ K many reactions occur, the most important being the buildup of iron:

$$^{28}\text{Si} + {}^{28}\text{Si} \rightarrow {}^{56}\text{Fe} \tag{10.107}$$

This brings us to the end of the nuclear fusion chain. Further fusions no longer release energy, but consume it.

> We can explain the formation of all elements up to Fe by thermonuclear fusion processes in stellar interiors.

10.3.2 Neutrinos

As already mentioned, the cross section of neutrinos σ_ν with matter is very small. With an energy of E_ν one has

$$\sigma_\nu = (E_\nu / m_e c^2)^2 \times 10^{-44} \, \text{cm}^2$$

For neutrinos in the MeV range $\sigma_\nu \approx 10^{-44} \, \text{cm}^2$. This is by a factor 10^{18} smaller than the cross section for interactions between photons and matter. For a density of $\rho = n \mu m_u$[2] (let mean molecular weight be equal to 1), the mean free path length is:

$$l_\nu = \frac{1}{n \sigma_\nu} \approx \frac{2 \times 10^{20} \, \text{cm}}{\rho} \tag{10.108}$$

It follows:

- Normal stars: $\rho \approx 1 \, \text{g cm}^{-3}$ and $l_\nu = 100 \, \text{pc}$. Even if the density $\rho = 10^6 \, \text{g cm}^{-3}$ would be $l_\nu = 3000 \, \text{R}_\odot$.
- However, in a stellar collapse at the end of stellar evolution, the density can reach nuclear values, $\rho = 10^{14} \, \text{g cm}^{-3}$ and $l_\nu = 20 \, \text{km}$. Some of the neutrinos are then reabsorbed in the star, and one must take into account the energy transport of the neutrinos.

---

[2] $m_u = 1.67 \times 10^{-24}$ g.

Table 10.1 Thermonuclear processes in which neutrinos are released

$^1H + {}^1H \rightarrow {}^2H + e^+ + \nu$	pp	0.263 MeV
$^7Be + e^- \rightarrow {}^7Li + \nu$	pp2	0.80 MeV
$^8B \rightarrow {}^8Be + e^+ + \nu$	pp3	7.2 MeV
$^{13}N \rightarrow {}^{13}C + e^+ + \nu$	CNO	0.71 MeV
$^{15}O \rightarrow {}^{15}N + e^+ + \nu$	CNO	1.0 MeV

Neutrinos can occur in different flavours (electron-, muon- and tau-neutrinos), in nuclear fusion only the electron neutrinos are are important. A list of thermonuclear processes in which neutrinos are released is given in Table 10.1.

Furthermore, there are other processes that lead to the production of neutrinos:

1. Capture of electrons by protons – this happens at extremely high densities; let Z is the atomic number and A the atomic weight, then you have:

$$e^- + (Z, A) \rightarrow (Z - 1, A) + \nu$$

(Z, A) means an atom with charge number Z and the atomic weight A.

2. *Urca* process: electron capture and beta decay occur:

$$(Z, A) + e^- \rightarrow (Z - 1, A) + \nu$$
$$(Z - 1, A) \rightarrow (Z, A) + e^- + \bar{\nu}$$

3. Neutrinos by pair annihilation:

$$e^- + e^+ \rightarrow \nu + \bar{\nu}$$

This requires temperatures above 10^9 K.

4. Photoneutrinos:

$$\gamma + e^- \rightarrow e^- + \nu + \bar{\nu}$$

The analogue would be the scattering of a photon by an electron (Compton scattering).

5. Plasma neutrinos:

$$\gamma_{\text{plasm}} \rightarrow \nu + \bar{\nu}$$

Decay of a plasmon into a neutrino-antineutrino pair. The frequency of a plasma depends on whether it is degenerate or not. For non-degenerate plasma one has:

$$\omega_0^2 \frac{m_e}{4\pi e^2 n_e} = 1 \tag{10.109}$$

and for degenerate plasma:

$$\omega_0^2 \frac{m_e}{4\pi e^2 n_e} = \left[1 + \left(\frac{\hbar}{m_e c} \right)^2 (3\pi^2 n_e)^{2/3} \right]^{-1/2} \tag{10.110}$$

If an electromagnetic wave of frequency ω passes through a plasma and K is the wavenumber, then one has the following dispersion relation:

$$\omega^2 = K^2 c^2 + \omega_0^2 \tag{10.111}$$

The wave is thus coupled to the collective motions of the electrons, and only waves with $\omega < \omega_0$ can propagate. Multiplying the above equation by $h/2\pi$ then you have the square energy of a quantum, which behaves like a relativistic particle of rest energy $h/2\pi\,\omega_0$ called a *Plasmon*.

6. Neutrinos due to Bremsstrahlung. When an electron is decelerated in the Coulomb field of a nucleus, there occurs emission of a photon, which in turn can decay into a neutrino-antineutrino pair.

Nuclear fusion provides energy up to the element iron. These elements were formed inside the stars. All elements heavier than iron were formed by other processes, such as a supernova explosion.

10.4   Special Stellar Models

We consider two examples of simple stellar models, which help to simplify the extensive numerical calculations help.

10.4.1 Polytropic Models

In these models we set the following relation between pressure P and density ρ:

$$P = K\rho^\gamma \tag{10.112}$$

K is the polytropic constant and γ the polytropic exponent. Often one uses the *Polytropic index n*:

$$n = \frac{1}{\gamma - 1} \tag{10.113}$$

For a completely degenerate gas, this condition is satisfied by $\gamma = 5/3$, $n = 3/2$. Here one can calculate K, in other cases this is a free parameter.

Another special case would be a star with constant temperature (isothermal star):

$$\rho = \mu P/(\Re T_0)$$

Further special case: completely convective star. Here $\nabla = \nabla_{ad} = 2/5$ (∇ stands for temperature gradient), if one can neglect the radiation pressure and the star is completely ionized. Thus

$$T \approx P^{2/5}$$

and for an ideal gas with constant molecular weight $T \approx P/\rho$. Thus one has $\gamma = 5/3$, and K is also fixed again.

The first basic equation of the stellar structure can also be written as:

$$\frac{dP}{dr} = -\frac{d\Phi}{dr}\rho \tag{10.114}$$

where Φ is the gravitational potential. Furthermore one has *Poisson's equation:*

$$\frac{1}{r^2}\frac{d}{dr}\left(r^2\frac{d\phi}{dr}\right) = 4\pi G\rho \tag{10.115}$$

If we use our relation for polytropic stars, then the basic equation is:

$$\frac{d\Phi}{dr} = -\gamma K\rho^{\gamma-2}\frac{d\rho}{dr} \tag{10.116}$$

If $\gamma \neq 1$ then one can integrate this equation:

$$\rho = \left(\frac{-\Phi}{(n+1)K}\right)^n \tag{10.117}$$

At the surface $\Phi = 0$, $\rho = 0$. Substituting this into the Poisson equation, we have:

$$\frac{d^2\Phi}{dr^2} + \frac{2}{r}\frac{d\Phi}{dr} = 4\pi G\left(\frac{-\Phi}{(n+1)K}\right)^n \tag{10.118}$$

Now define:

$$z = Ar \qquad A^2 = \frac{4\pi G}{(n+1)^n K^n}(-\Phi_c)^{n-1}$$

$$w = \frac{\Phi}{\Phi_c} = \left(\frac{\rho}{\rho_c}^{1/n}\right)$$

The Poisson equation then goes into the *Lane-Emden* equations:

$$\frac{d^2 w}{dz^2} + \frac{2}{z}\frac{dw}{dz} + w^n = 0 \tag{10.119}$$

$$\frac{1}{z^2}\frac{d}{dz}\left(z^2 \frac{dw}{dz}\right) + w^n = 0 \tag{10.120}$$

From this we find the radial distribution of density:

$$\rho(r) = \rho_c w^n \qquad \rho_c = \left[\frac{-\Phi_c}{(n+1)K}\right]^n \tag{10.121}$$

For pressure we find:

$$P(r) = P_c w^{n+1} \qquad P_c = K\rho_c^\gamma \tag{10.122}$$

One can find a power series for $w(z)$ and finds:

$$w(z) = 1 - \frac{1}{6}z^2 + \frac{n}{120}z^4 + \ldots \tag{10.123}$$

For the following cases there is an analytical solution:

- $n = 0$

$$w(z) = 1 - \frac{1}{6}z^2 \tag{10.124}$$

- $n = 1$

$$w(z) = \frac{\sin z}{z} \tag{10.125}$$

- $n = 5$

$$w(z) = \frac{1}{(1 + z^2/3)^{1/2}} \tag{10.126}$$

Otherwise, one has to solve the Lane-Emden equation numerically.

We consider a polytropic model with index 3 for the Sun, $M = 1.98 \times 10^{33}$ g, $R = 6.96 \times 10^{10}$ cm. From a table one takes for $n = 3$: $z_3 = 6.897$ and $\rho_c/\bar{\rho} = 54.18$. With $\bar{\rho} = 1.41$ g cm^{-3} it follows for the central density $\rho_c = 76.39$ g cm^{-3}. Further more $A = z_3/R = 9.91 \times 10^{-11}$. From the relation

$$A^2 = \frac{4\pi G}{(n+1)K} \rho_c^{(n-1)/n}$$

follows $K = 3.85 \times 10^{14}$. Then from

$$P_c = K\rho_c^\gamma$$

the central pressure to $P_c = 1.24 \times 10^{17}$ dyn/cm^2. We assume the following chemical composition: $X \approx 0.7$, $Y \approx 0.3$. This gives an average molecular weight of $\mu = 0.62$. The central temperature follows from the ideal gas equation with $T_c = 1.2 \times 10^7$ K. The mass distribution is calculated from :

$$m(r) = \int_0^r 4\pi \rho r^2 dr = 4\pi \rho_c \int_0^r w^n r^2 dr \tag{10.127}$$

10.4.2 Homologous Equations

Very often one can solve problems in physics by starting from a known solution and then performing a transformation. Here we compare different stellar models with masses M, M', radii R, R' and consider so-called homologous points at which holds:

$$r/R = r'/R' \tag{10.128}$$

One speaks of homologous stars if holds $m/M = m'/M' = \xi$. The condition is then:

$$\frac{r(\xi)}{r'(\xi)} = \frac{R}{R'} \tag{10.129}$$

One introduces the following parameters: $x = M/M'$; $y = \mu/\mu'$; $z = r/r' = R/R'$; $p = P/P' = P_c/P_c'$; $t = T/T' = T_c/T_c'$; $s = l/l' = L/L'$. Now one can construct homologous main sequence stars:

$$\frac{dr}{d\xi} = c_1 \frac{M}{r^2 \rho}$$

$$c_1 = \frac{1}{4\pi}$$

$$\frac{dP}{d\xi} = c_2 \frac{\xi M^2}{r^4}$$

$$c_2 = -\frac{g}{4\pi}$$

$$\frac{dl}{d\xi} = \epsilon M$$

$$\frac{dT}{d\xi} = c_4 \frac{\kappa l M}{r^4 T^3}$$

$$c_4 = -\frac{3}{64\pi^2 ac}$$

and:

$$\frac{dr'}{d\xi} = c_1 \frac{M'}{r'2\rho'} \left[\frac{x}{z^3 d} \right]$$

$$\frac{dP'}{d\xi} = c_2 \frac{\xi M'2}{r'4} \left[\frac{x^2}{z^4 p} \right]$$

$$\frac{dl'}{d\xi} = \epsilon' M' \left[\frac{ex}{s} \right]$$

$$\frac{dT'}{d\xi} = c_4 \frac{\kappa' l' M'}{r'4 T'3} \left[\frac{ksx}{z^4 t^4} \right]$$

where $\rho/\rho' = d$; $\epsilon/\epsilon' = e$; $\kappa/\kappa' = k$. Thus, solving these equations yields multiple stellar models at once.

10.5   Further Literature

We give a small selection of recommended further literature.
Stellar Structure and Evolution, R. Kippenhahn, A. Weigert, Springer, 1996
Stars and Stellar Processes, M. Guidry, Cambridge Univ. Press, 2019

Stellar Interiors, V. Trimble, Springer, 2004
Introduction to Stellar Structure, W. J. Marciel, Springer, 2015

Tasks

10.1 At which point in the HRD are stars in hydrostatic equilibrium the longest?

Solution
Main sequence

10.2 Discuss why simplifications in stellar models are justified or what a consideration of rotation, magnetic field would change in the models!

Solution
Stars evolve very slowly, so stat. Model; Rotation $\rightarrow$ Flattening, magnetic field $\rightarrow$ Anisotropy, ...

10.3 Derive the classical condition for hydrostatic equilibrium from the TOV equation!

Solution
The solution is very simple: $c^2 \rightarrow \infty$.

10.4 Calculate the escape velocity of a white dwarf of 0.5 solar masses!

Solution
The radius of the object is 1.5 Earth radii. The escape velocity is defined as $v_e = \sqrt{2GM/R}$ and inserting the values yields $v_e = 3.7 \times 10^6$ m/s, i.e. about 1/100 of the speed of light!

10.5 At what temperature does Compton scattering reduce opacity?

Solution
We first consider Wien's law: $h\nu = 4.965\,kT$, and as soon as $T > 0.1m_e c^2/(4.965k)$, i.e. for $T > 10^8$ K the Compton scattering becomes important.

Stellar Evolution

11

In principle, there are the following stages in stellar evolution:

- Protostar,
- pre-main sequence evolution,
- main sequence,
- post-main-sequence existence.

The most important physical quantity that characterizes stellar evolution is the mass. In addition, the chemical composition plays a role. We have already mentioned the difference between stars of populations I and II. Population II stars contain significantly fewer metals (all elements heavier than He) than Population I stars.

11.1 Star Formation and Evolution

In this section we investigate the conditions under which a gas-dust cloud collapses. The evolution towards a protostar is then shown. Furthermore, we discuss the evolution of our Sun.

11.1.1 Protostars

Stars are formed by contraction from interstellar clouds consisting of gas and dust.

© Springer-Verlag GmbH Germany, part of Springer Nature 2023
A. Hanslmeier, *Introduction to Astronomy and Astrophysics*,
https://doi.org/10.1007/978-3-662-64637-3_11

Based on the virial theorem mentioned above, we know that half of the potential energy released during contraction is converted into thermal energy, i.e. it heats up the star, and the other half is radiated.

Let us consider the collapse of such a gas cloud. Let the mass of the cloud be M, the radius R, the total number of particles N, the mean particle mass $\overline{m}$ and the temperature T. The gravitational potential energy is thus:

$$U = -\text{const}\frac{GMN\overline{m}}{R} \tag{11.1}$$

The value of the constant depends on the internal matter distribution in the cloud. The kinetic energy per particle is on average:

$$E_{\text{kin}} = \frac{3}{2}kT \tag{11.2}$$

and per unit mass

$$E_{\text{kin}} = \frac{3}{2}\frac{kT}{\mu m_u}. \tag{11.3}$$

For a cloud of mass M is the kinetic energy thus given by Eq. 11.3, multiplied by the mass M or Eq. 11.2, multiplied by the number of particles N; this cloud will then contract when

$$U > E_{\text{kin}}. \tag{11.4}$$

This is called the *Jeans criterion*. The Jeans mass is:

$$M_J = \frac{3}{2}\frac{kT}{G\overline{m}}R \tag{11.5}$$

and from that the jeans density is obtained.

So only large masses can become gravitationally unstable.

We also see that especially cool regions of interstellar matter are relevant for star formation, otherwise the kinetic energy becomes too large.

In Fig. 11.1 protoplanetary disks in the Orion Nebula are shown.

One can also estimate how long it takes for a gas envelope that is not in hydrostatic equilibrium to collapse. Let us assume $\Delta E_{\text{kin}} = \Delta U$ (Virial theorem) and

$$1/2(dr/dt)^2 = Gm_0/r - Gm_0/r_0,$$

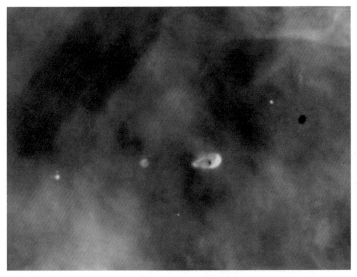

Fig. 11.1 Formation of protoplanetary disks in the Orion Nebula

then we find for *free fall time:*

$$t_{ff} = \int_{r_0}^{0} (dt/dr)dr = -\int_{r_0}^{0} \frac{dr}{\sqrt{Gm_0/r - Gm_0/r_0}}$$

$$x = r/r_0$$

$$t_{ff} = [r_0^3/(2Gm_0)]^{1/2} \int_{0}^{1} [x/(1-x)]^{1/2} dx \qquad (11.6)$$

$$x = \sin^2 \Theta$$

$$t_{ff} = \sqrt{\frac{3\pi}{32 G\rho}}$$

Before the temperature in the interior of a star is great enough to ignite nuclear reactions, one speaks of a *Protostar* or pre-main sequence evolution. The evolutionary paths for protostars in the HRD depend on their mass. There are four stages:

1. Collapse in free fall—particles do not collide with each other during free fall, internal pressure is zero.
2. The core regions collapse faster than the outer parts.
3. Once the core is formed, accretion from the shell occurs.
4. Only when the material surrounding the core is gone (due to radiation pressure) does the star become visible.

Fig. 11.2 Evolution of two protostars with (a) 1 $M_\odot$ and (b) 10 $M_\odot$ (dashed) to the main sequence. For case (a) the evolution takes 50 million years, for case (b) only 200,000 years

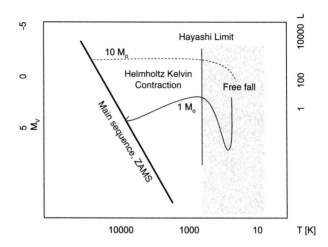

11.1.2 Collapse of a Sun-Like Star

We assume an interstellar cloud of sufficiently high mass. Collisions occur between the molecules during the collapse phase, and the dust radiates in the IR. This radiation can escape, and the cloud initially remains cool. But when the density of the core exceeds a critical value, then it becomes opaque (optical depth greater than 1)—the collapse of the core slows down, you have hydrostatic equilibrium, and a pre-main sequence star evolves. It takes a million years to get to this point. The luminosity of a star is given by:

$$L = 4\pi R^2 \sigma T^4 \tag{11.7}$$

In Fig. 11.2 the evolution is shown: Until the so-called Hayashi limit is reached, the star is fully convective, but gains luminosity by collapsing in free fall. Above a temperature of 2000 K dissociates H_2, so energy is consumed for this process, the luminosity of the star decreases. Above 10^4 K, hydrogen is ionized and the star becomes optically thick. Temperature and gas pressure increase, and after crossing the Hayashi limit, the contraction enters a Helmholtz-Kelvin phase (conversion of released gravitational energy into heat).

> The Hayashi limit separates the region of fully convective stars from no longer fully convective ones in the HRD.

In general: the luminosity is high because the radius R is still large. A Pre main sequence star glows due to contraction and accretion, the temperature hardly increases at first, and the luminosity decreases. Subsequently, the nucleus heats up and the opacity increases. A zone of radiative transfer develops, slowly moving from the inside to the outside. Then the evolution turns to the left in the HRD. If the temperature is high enough, the thermonuclear reactions ignite, one speaks of a *Zero Age Main Sequence Star, ZAMS*. Here only the outer envelope is still convective.

How long can a star stay on the main sequence? In principle, about 80% of its total lifetime. If the hydrogen content decreases in the interior due to nuclear fusion, then the temperature and density increase, and the star expands. Thus, the luminosity of the star also increases, and it evolves upward away from the main sequence. This can be estimated as follows: For luminosity, we have the mass-luminosity relation:[1]

$$L_*/L_\odot = (M_*/M_\odot)^{3.3} \tag{11.8}$$

The lifetime[2] of a star is given by:

$$t_*/t_\odot = (M_*/M_\odot)/(L_*/L_\odot) = (M_*/M_\odot)^{-2.3} \tag{11.9}$$

where $t_\odot = 10^{10}$, the main sequence lifetime of the Sun. Compare the main sequence lifetime of our Sun to that of a star of ten solar masses!

11.1.3 The Age of Stars

Stars in a cluster form at about the same time. Since massive stars evolve faster than low-mass stars (cf. Eq. 11.9), the main sequence of older star clusters clusters will no longer be fully occupied.

- In the HRD, the hot, massive, luminous stars are in the upper left.
- Hot, luminous massive stars have already moved away from the main sequence in older clusters.

Therefore, one can infer the age of a cluster, and thus of the stars in it, from the location of the turn off point from the main sequence.

In Fig. 11.3 one can see an HRD of two star clusters. It is clear that the star cluster M 67 must be somewhat younger,[3] because here are hotter stars still on the main sequence than

---

[1] Valid only for stars of the main sequence!

[2] Actually, the main sequence lifetime.

[3] Its age is given as about four billion years.

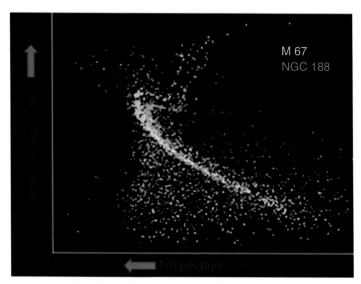

Fig. 11.3 Comparison of the HRD of two star clusters. From the position of the turn off point the age follows

in NGC 188. The star cluster NGC 188 is older than M 67 by a billion years and is one of the oldest open star clusters.

11.1.4 Evolution of a Star with One Solar Mass

As already mentioned above, a star with one solar mass reaches the zero-age main sequence, ZAMS, as soon as the pp chain ignites. After about ten billion years, its main-sequence existence ends, almost all the hydrogen in the core has been converted to helium, and the star expands slightly, increasing energy production as its temperature increases in the interior, and also increasing luminosity as its surface area increases. The nuclear reactions in the center eventually die out, but the fusion of hydrogen to helium continues in a shell around the core (shell burning). The radius of the star now increases considerably: the core contracts, so heat is produced (cf. virial theorem), and the hydrogen-burning shell heats up—more energy is produced, and the star expands. But this lowers the surface temperature, and thus the opacity increases, which leads to an increase in convection, which in turn is important for the mixing of the elements. The star evolves into a red giant and moves obliquely upward to the right in the HRD.

A *red giant* has the following structure: small dense core with $T \sim 50 \times 10^6$ K, degenerate electron gas in the core. So the gas pressure depends only on the density and not on T, the nucleus can thus resist the gravitational force even though there is no more fusion there. The contraction raises the temperature to 10^8 K and the triple-alpha process ignites.

Fig. 11.4 Evolution of a Sun-like star in the HRD from main sequence star to red giant and white dwarf. The main sequence stage (1) takes about 9 billion years, stages 2 to 4 are passed in a few 100 million years

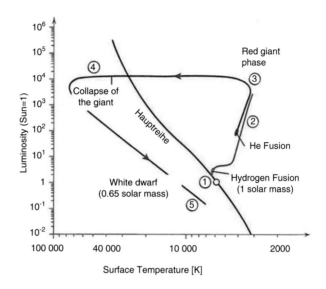

If this happens, then the heat spreads out through the very effective thermal conduction of the degenerate electrons. The entire nucleus then ignites, but since the matter is degenerate, only the temperature increases, the pressure remains constant, and the nucleus does not expand. This is called a *helium flash*. Only when the core temperature has reached 350 million K, the electrons are not degenerate and the core can expand and cool down. After this He flash, the stellar radius and thus its luminosity decreases slightly, and the star moves down and to the left in the HRD. When the He is consumed by the triple alpha process, its fusion occurs in a shell, and the star expands again. The electrons are degenerate again, and this time the core is enriched in carbon (Fig. 11.4).

The triple-alpha process is very strongly dependent on temperature, and thermal pulses occur that are in fact giant thermonuclear explosions. Such explosions happen every approx. 10^3 years and lead to luminosity changes of the star by up to 50% during some years. The star is located at the asymptotic giant branch in the HRD *(asymptotic giant branch, AGB)*. During these phases there are also very strong *stellar winds,* and in a few 1000 years the envelope is completely blown away; an expanding envelope forms around the star, heated by the hot core and excited to glow; this is called a *planetary nebula*. A very well-known example is the Ring Nebula, M57 (Fig. 11.5). The Ring Nebula is about 2000 light-years away and about 20,000 years old. The central star is a white dwarf, the luminosity is $15^{\text{m}}8$.

If the mass of a star is less than about 1 $M_\odot$, then the core temperature is not enough to start carbon burning. Within about 100,000 years, a *White Dwarf evolves*.

Fig. 11.5 The ring nebula
M57. The central star, a white
dwarf, can be seen in the center
(HST/NASA)

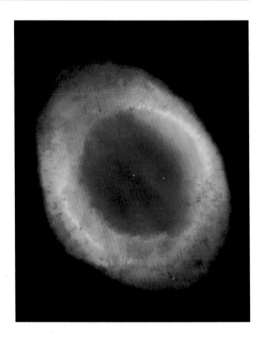

So the fate of our sun is as follows:

Evolution of our sun:
Main sequence star (about 10 billion years total) → Red giant (approx. 10^8
years)→ White dwarf.

11.2   Comparison of Stellar Evolution

An O5 star can reach a total age of about 5 million years, whereas an M0 star can reach
about 30 billion years if its mass is only 1/2 solar masses.

There are the following final stages in stellar evolution:

- White dwarfs,
- neutron stars, pulsars,
- black holes.

A detailed description of the final stages follows in the next sections.

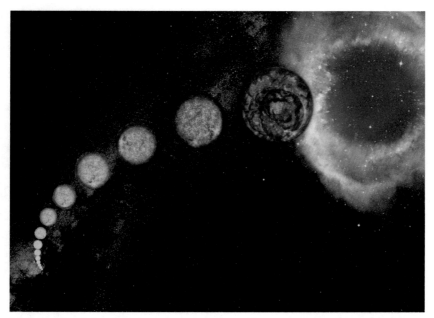

Fig. 11.6 Sketch of the evolution of low-mass stars from main sequence star (lower left) to planetary nebula (upper right). (ESO/Steinhöfel)

11.2.1 Low-Mass Stars

The evolution of low-mass stars is sketched in Fig. 11.6. At the end of the main sequence existence, the stars evolve into red giants and finally into a white dwarf. As they do so, the outer envelopes are slowly ejected; these envelopes can be seen glowing as planetary nebulae for a few thousand years.

11.2.2 Massive Stars

Massive stars burn elements down to iron. At the end of its evolution, the star has a shell-like structure (Fig. 11.7) with an iron core in the middle, followed by a shell of silicon burning, and so on. Once the mass of the iron core exceeds the Chandrasekhar limit mass ($1.4\ M_\odot$), there is an implosion of the core combined with the repulsion of the outer layers, which greatly enlarges the surface of the exploding star and therefore makes it very bright; a *Supernova* lights up.

Fig. 11.7 Structure of a
massive star at the end of its
evolution. As soon as the iron
core exceeds the
Chandrasekhar mass, a
supernova lights up

11.3   White Dwarfs

11.3.1 General Properties

They develop from Red Giants. As we have seen, in the final stages of stellar evolution
pulsations occur in which parts of the outer envelope are ejected. This results in the
formation of a planetary nebula (e.g. M57, Fig. 11.5).

Depending on the initial mass (the final mass leading to the formation of white dwarfs
is always below 1.4 solar masses) a distinction is made:

- Stars with $\leq 0.5\, M_\odot$ form He-white dwarfs, because the core temperature too low to
  ignite the helium.
- Stars with masses between 0.5–5.0 $M_\odot$ leave C-O stars behind.
- Stars with masses between 5–7 $M_\odot$ form O-Ne-Mg-rich white dwarfs.

The mass values given here refer to the initial mass of the star!

In the case of white dwarfs (WD, *white dwarfs*) the matter is so densely packed that the
electrons can no longer move freely, but form a degenerate electron gas.

Equilibrium state (hydrostatic equilibrium): The gravity of a white dwarf is compensated by the pressure of the degenerate electrons.

$\rightarrow$ However, this only goes up to 1.4 $M_\odot$ $\rightarrow$ *Chandrasekhar limiting mass.*

Chandrasekhar limiting mass: the final evolutionary stages of stars up to a mass of about 1.4 solar masses are white dwarfs.

One can easily establish a relationship between the Chandrasekhar limit mass, the radius and the mass of white dwarfs. In the case of complete non-relativistic degeneracy is:

$$P = K\rho^{5/3} \tag{11.10}$$

The condition for hydrostatic equilibrium gives:

$$P \approx M^2/R^4 \tag{11.11}$$

the density $\rho \approx M/R^3$ and therefore $P \approx M^{5/3}R^5$. One obtains:

$$R = \frac{4\pi K}{G(4/3\pi)^{5/3}M^{1/3}} \qquad R_{WD} \approx \frac{1}{M^{1/3}} \tag{11.12}$$

→ Therefore, the larger the mass of a white dwarf, the smaller its radius R.

White dwarfs glow by cooling, the thermal energy is given by:

$$E_{th} = \frac{3}{2}\frac{kT}{\mu m_u}M. \tag{11.13}$$

Consider a star with 0.8 $M_\odot$ and a temperature of 10^7 K then the thermal energy is 4×10^{40} J and we assume a luminosity of 10^{-3} $L_\odot$,[4] then the cooling time is τ_c:

$$\tau_c = E_{th}/L \approx 4 \times 10^{40}\text{J}/[(10^{-3})(3.8 \times 10^{26}\text{ J/s})] \approx 3 \times 10^9\text{ a}$$

The first white dwarf was found in 1862 by A. *Clark*: *Sirius* B, a hot but inconspicuous companion of Sirius (Fig. 11.8).

White dwarfs are divided into:

- DA: D stands for Dwarf, and A means a spectrum similar to that of an A star, i.e., hydrogen-rich.
- DB: spectrum with nebular lines.
- DC: predominantly continuous spectrum.

---

[4] Solar luminosity: $L_\odot \approx 3.86 \times 10^{26}$ J/s.

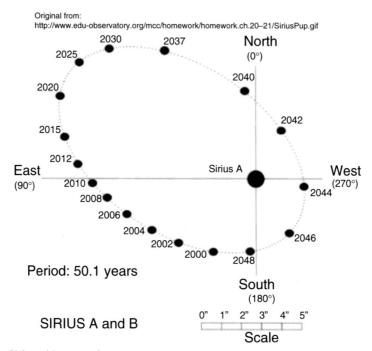

Fig. 11.8 Sirius with companion

The most reliable data are for DA stars, they lie shifted to the left parallel to the main sequence. The mean radius is 0.013 $R_\odot$, the mean mass 0.7 $M_\odot$ and density $10^9\,\mathrm{kg/m^3}$.

11.3.2 General Relativity and White Dwarfs

In the case of white dwarfs, the contribution of gravitational red shift becomes important, and the spectral lines appear shifted to the red. Photons of energy E have an equivalent mass of $E = mc^2$, and the gravitational field acts on this mass, leading to a decrease in its energy, or red shift, since photons of higher energy have a shorter wavelength than photons of lower energy. Let us first consider the classical case: The change of energy in the gravitational field is according to *Newton:*

$$\Delta(h\nu) = -GmM/R \tag{11.14}$$

Since the mass of the photon $m = E/c^2$ with $E = h\nu$, we obtain by substitution:

$$\Delta\nu/\nu_0 = -GM/c^2 R \tag{11.15}$$

Let us briefly consider the relativistic Doppler effect:

$$\frac{\Delta \nu}{\nu} = 1 - \frac{1}{\sqrt{1 - R_S/R}} \approx -\frac{GM}{Rc^2} \qquad R_S = \frac{2GM}{c^2} \qquad (11.16)$$

Where R_S is the Schwarzschild radius.

Thus, the relativistic Doppler effect depends on the ratio of the Schwarzschild radius R_S to the radius of the star R, and the effect increases for

- large masses,
- small compact objects (small value for R).

11.3.3 Magnetic Fields

In the formation of a white dwarf, the conservation of magnetic flux must be considered. This is the number of magnetic field lines multiplied by the area they penetrate. If now the star is compressed in the course of its evolution, the number of field lines remains the same, of course, but the surface area decreases, and therefore the magnetic field strength increases. One can show that the magnetic field strength of a White dwarf compared to the field strength of the Sun is:

$$B_{WD}/B_\odot = (R_\odot/R_{WD})^2 \qquad (11.17)$$

→ Extreme amplification of the magnetic field by compression of the star!

11.3.4 Brown Dwarfs

Unlike white dwarfs, brown dwarfs are not at the end of stellar evolution, but in their case hydrogen burning never ignited because the central temperatures were too low. The limits are not exactly definable, but one speaks of:

- Planets when $M < 0.002\ M_\odot$;
- Brown dwarfs, if $0.002\ M_\odot < M < 0.08\ M_\odot$.

For brown dwarfs near the 0.08 solar mass limit, there is a phase of deuterium burning (some 10,000 years).

The Hubble Space Telescope has been used to search for brown dwarfs. In this context, the object *Gliese 229* should be mentioned. The star is a double system, the main star a red dwarf and Gliese 229B a brown dwarf with more than 20 Jupiter masses. The companion is located at a distance of 40 AU from the main star.

11.4   Neutron Stars

11.4.1 Formation of Neutron Stars

For a contracting star at the end of stellar evolution whose mass is larger than the Chandrasekhar limit mass of 1.4 solar masses, the pressure of the degenerate electrons is no longer sufficient to resist the strong gravity. Matter is compressed to extremely high densities, and the inverse beta decay starts:

$$p^+ + e^- \rightarrow n + \nu \tag{11.18}$$

The protons and electrons combine to form neutrons; neutron gas is produced whose density reaches about 10^{17} kg/m^3. The neutrons form a degenerate gas, and a neutron star develops with a diameter of a few 10 km. In the interior there is a neutron liquid, in the outer regions there is a neutron superfluid and a crystalline surface (neutron lattice gas). In the outermost meters there exists an atmosphere of atoms, electrons, and protons, the atoms being mostly iron atoms.

Consider the gravitational red shift of a neutron star of 7 km radius (SI units used throughout):

$$\Delta\lambda/\lambda \approx GM/Rc^2 = \frac{6.67 \times 10^{-11} \times 2 \times 10^{30}}{9 \times 10^{16} \times 7 \times 10^3} \approx 0.2 \tag{11.19}$$

The structure of a neutron star is imagined as follows (Fig. 11.9):

- 15−16 km: The top layer consists of degenerate matter as in the white dwarf with an increase in density, from 10^7 to 4×10^{14} kg/m^3, above, of iron nuclei, and further down also of neutron-rich nuclei (e.g., gold, lead, uranium, ...).
- 11−15 km: Inner crust, the neutron-rich nuclei dissolve, and free neutrons appear in greater numbers, unable to decay in the relativistically degenerate electron gas surrounding them.
- At a distance of 11 km from the center there is a density of 200×10^{15} kg/m^3, the state of a strongly incompressible neutron liquid.
- Central region: density up to 400×10^{15} kg/m^3. Possibly the neutrons dissolve, and subnuclear particles such as free quarks could occur.

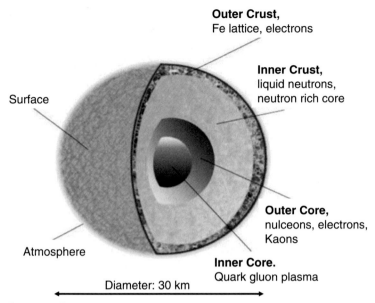

Fig. 11.9 Structure of a neutron star

As with the white dwarf, the diameter decreases with increasing mass, and there is a limiting mass for neutron stars analogous to the Chandrasekhar limiting mass, called the *Oppenheimer-Volkoff limit.*

> Calculations give a limiting mass for neutron stars in the range of three to four solar masses. Thus, even more massive stars evolve into black holes.

11.4.2 Pulsars

In 1967 the Hewish group wanted to study scintillation of radio sources on the sky with the help of a newly constructed radio telescope in Cambridge, England. Scintillation is the flashing of a radio source due to density fluctuations in the interplanetary plasma (caused by the solar wind) and in the interstellar medium. Very big surprise was then when extremely periodic radio signals were found. The Hewish group found a signal with the exact period of 1.33730113 s. This phenomenon was called *Pulsar*, although it will be shown below that the brightness change has nothing to do with pulsations of the star. In the meantime, more than 150 pulsars have been studied.

The regularity of the pulses is better than $1:10^8$. The energy of a pulse can vary strongly. Sometimes pulses fail. Typical pulsation durations range from a few 10 s to milliseconds.

Fig. 11.10 The Pulsar in the
Crab Nebula

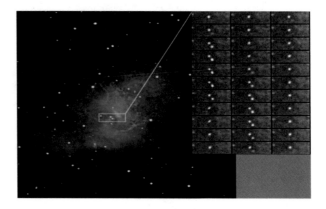

The primary pulses can still be decomposed into sub millisecond pulses. The first pulsars
were found at a frequency of 81.5 MHz. The pulsation times change in the long term, one
measures an increase of the periods around 10^{-8} s/a (seconds per year). This can only be
determined with atomic clocks, which have an accuracy of 10^{-10} s/a. The age of a pulsar
follows from its period divided by the rate of change of that period:

$$t_{\text{pulsar}} \approx P \left(\frac{dP}{dt} \right)^{-1} \tag{11.20}$$

A very well known pulsar is the Crab pulsar (Fig. 11.10): $P = 0.03$ s, $dP/dt = 1.2 \times 10^{-13}$ s/s, it therefore has age:

$$t_{\text{Pulsar}} \approx 10^{11} \text{ s} \tag{11.21}$$

indicating the approximate correctness of the formula. The supernova explosion that led
to the formation of the Crab pulsar occurred in 1054 AD.

Important in observation is the effect of *Dispersion:* if one examines a given pulsar at
lower frequencies, then the photons are slowed down by the electrons that are in the line
of sight of the pulsar. Longer wavelengths are slowed down more, and the electron density
in the line of sight is estimated from the observations. Conversely, if we know the average
electron density, we can determine the distance of the pulsar. Let us assume pulses of two
different frequencies f_1, f_2 are emitted at the time t_0 and arrive here at the times t_1, t_2. We
then receive:

$$t_1 - t_0 = d/v_1 \qquad t_2 - t_0 = d/v_2$$

Of course, we don't know t_0 but this is omitted in the case of:

$$t_2 - t_1 = (1/v_2 - 1/v_1)d \tag{11.22}$$

The velocities depend on the electron density, and if we know this, we can calculate the distance d. Interstellar matter, i.e., the matter between stars, does not have a constant density, a *Dispersion measure DM* has been introduced:

$$DM = \int_0^d n_e dl \tag{11.23}$$

and:

$$t_2 - t_1 = 2\pi e^2/m_e c(1/f_2^2 - 1/f_1^2)DM \tag{11.24}$$

$$D = (t_2 - t_1)/(1/f_2^2 - 1/f_1^2) \tag{11.25}$$

$$DM = 2\pi mcD/e^2 \qquad DM = 2.41 \times 10^{-16}D \tag{11.26}$$

Most pulsars are found at low galactic latitudes.

Furthermore the *Faraday rotation* has to be considered. The plane of polarization of linearly polarized radiation is rotated when it passes through a magnetic plasma. The Faraday rotation depends on:

- average electron density,
- mean magnetic field strength,
- λ^2 of radiation,
- Distance through the medium.

So one can measure the angle through which the plane of polarization is rotated at different wavelengths and then say something about above quantities.

How Do Pulses Occur?

To explain them, one needs:

- a rapidly rotating neutron star, which has a high rotational energy E_{rot} and
- a dipolar magnetic field, which transforms rotational energy into electromagnetic energy. In Fig. 11.11 a model of an oblique rotator is shown.

Pulsars are rapidly rotating neutron stars.

Fig. 11.11 The oblique rotator model of a pulsar; charges are accelerated in the strong magnetic field, and focused synchrotron radiation is produced

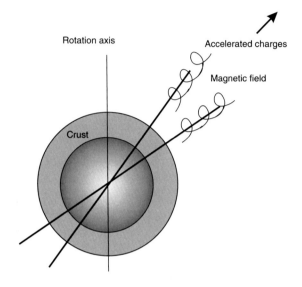

Suppose our Sun collapses into a neutron star with radius R. Let us examine the expected magnetic field strengths! Let $R_{ns} = 73$ km, it holds:

$$B_{ns} = B_\odot (R_\odot / R_{ns})^2 \approx 10^6 \text{ T}$$

This is purely hypothetical, since our Sun will evolve into a white dwarf.

The magnetic axis is inclined with respect to the axis of rotation. The rotation creates an electric field by induction, and this accelerates particles of the crust. The particles thus accelerated (mainly electrons) emit *Synchrotron radiation*. The torque of the accelerated particles slows down the rotation, and therefore the slower pulsars rotate, the older they are.

Let us consider rotation for a moment:

$$\frac{v^2}{R} = \frac{GM}{R^2} \tag{11.27}$$

This is the stability condition (centrifugal force must be less than or at most equal to gravity). The rotation period is:

$$P = 2\pi R / v$$

From the above equations we obtain a typical density for neutron stars of. $\rho = 4 \times 10^{16} \text{ kg/m}^3$.

As we have seen, there are very many binary stars and multiple systems. In 1974 *Hulse* and *Taylor* analyzed a pulsar already known by then and found a period of 7.75 h, which can be explained by the orbital motion of two components. Therefore it is a double pulsar. This object (PSR 1913+16) is 5 kpc from us, and the semi-axis of the double system is only as large as the radius of the Sun. The masses are 1.4 $M_\odot$ and 1.3 $M_\odot$.

Pulsars and neutron stars are formed in supernova outbursts. The pulsar in the Crab Nebula emits pulses with an energy of 10^{28} W from the optical to the X-ray range. Here a deceleration of its rotation by 4×10^{-13} s/s or 10^{-5} s/a has been measured. The rotational energy provides the energy budget for the nebula surrounding the pulsar. There is a conversion of rotational energy into kinetic energy and finally into radiation energy of the nebula. The rotational energy is (I is the moment of inertia):

$$E_{\text{red}} = \frac{1}{2} I \omega^2 \tag{11.28}$$

$$\omega = 2\pi / P \tag{11.29}$$

$$I = \frac{2}{5} M R^2 \tag{11.30}$$

Let us assume that all the rotational energy is converted into radiant energy:

$$\frac{d E_{\text{rad}}}{dt} + \frac{d E_{\text{rot}}}{dt} = 0 \tag{11.31}$$

Now we put in:

$$\frac{d E_{\text{rot}}}{dt} = \frac{1}{2} \frac{d}{dt} (I \omega^2)$$

$$= \frac{1}{2} \frac{d}{dt} \left[\frac{2}{5} M R^2 \left(\frac{2\pi}{P} \right)^2 \right]$$

$$= \ldots = -\frac{8}{5} \pi^2 M R^2 P^{-3} \frac{dP}{dt}$$

and since $L = d E_{\text{rad}}/dt = -d E_{\text{rot}}/dt$, we get:

$$L = \frac{8}{5} \pi^2 M R^2 P^{-3} \frac{dP}{dt}$$

$$\frac{dP}{dt} = \frac{5}{8\pi^2} \frac{L P^3}{M R^2}$$

We determine the rate of pulse changes for the Crab pulsar. We use to estimate $M = 1\, M_\odot$, $R = 10$ km and $L = 10^{31}$ W and $P = 1$ s:

$$\frac{dP}{dt} = \frac{5}{8\pi^2} \frac{10^{31}}{2 \times 10^{30} (10^4)^2} = 10^{-8} \text{s/s}$$

For the Crab pulsar: $P \approx 0.03$ s, therefore one has:

$$dP/dt = 10^{-13}\,\text{s/s}$$

This agrees with the observations.

11.5 Supernovae

11.5.1 Classification

> In a supernova, the star explodes and the outer shell is ejected.

Supernovae (SN) reach absolute magnitudes of -16^{M} to -20^{M} and can thereby increase the brightness of an entire galaxy. From historical records is known, for example, the SN from the year 1054, which was observed by Chinese astronomers. It was so bright that the star could be seen in the daytime sky. The brightness progression of a SN shows a rapid rise to maximum and then a drop of two to three magnitudes within a month, before a slower decline in brightness. The radiant energy released during an SN explosion is 10^{44} J. The neutrinos carry off much more energy still. The first neutrinos originating from a SN to be recorded were those from SN 1987A. The collapse of a star leads to the release of gravitational energy (be $R = 15$ km):

$$E_{\text{grav}} = \frac{GM^2}{R} \approx \frac{(6.67 \times 10^{-11})(2 \times 10^{30})^2}{1.5 \times 10^4} \approx 2 \times 10^{46}\,\text{J} \qquad (11.32)$$

There are two types of supernovae:

- Type I: They occur in elliptical and in spiral galaxies, it is a white dwarf that explodes due to sudden onset of carbon fusion. The white dwarf accretes matter from a companion. Once the Chandrasekhar limit is reached, the star collapses and a type I supernova is formed.

  Another form of a type I supernova is possible in a compact binary star system where both components have evolved into a white dwarf. The masses move around the common center of mass, this is an accelerated motion, and accelerated masses radiate gravitational waves according to general relativity (cf. Electrodynamics: Accelerated charges radiate electromagnetic waves). As a result, the orbital angular momentum decreases, and the two components approach each other until they merge. A type Ia supernova is formed, and the onset of carbon detonation is likely to rupture the star completely, leaving no remnant star (e.g., neutron star). The light curves of SN Ia are very similar.

Type I is divided into type Ia, b, c. In general, type I lacks hydrogen lines. Spectra of SN Ib resemble those of SN Ia near the maximum, but later those of SN II.

- Type II: occur exclusively in spiral galaxies; massive (10 to 100 solar masses) stars at the end of their evolution. The detonation is caused by gravitational collapse. Inside, an iron core is formed in a highly evolved star, which collapses into a neutron star.

At the end of the evolution of a massive star, an inert Fe core remains, which produces no energy; neutrinos escape and dissipate energy. Density increases, protons and electrons form neutrons and neutrinos. Matter around the nucleus impacting the nucleus at 15% of the speed of light causes the nuclear mass to increase. The nucleus collapses as soon as the Chandrasekhar mass is reached. There is no counterforce to gravity at this moment, as the pressure of the degenerate electrons is released ;

The neutron densities become so high that the nucleus becomes incompressible, a repulsive matter wave is formed, which propagates outward as a shock wave . This shock wave causes the actual explosion, the inner region compresses further and forms a neutron star or a black hole (Table 11.1).

The spectra of both types show emission lines that are often accompanied by short-wavelength absorption components, so-called P-Cygni lines $\rightarrow$ expanding gas shell (absorption lines are produced in the shell that moves toward the observer). One measures ejection velocities of up to 2×10^4 km/s, higher for SN I than for SN II. In SN II one observes similar lines as in novae, Balmer lines, He, metals (Ca II, Fe II) and later forbidden lines like [OI] and [OIII].

Brightnesses:

- SN Ia: $M_{\mathrm{B,max}} = -19^\mathrm{m}$; in the first 20 to 30 days after maximum the brightness decreases by two to three magnitudes; light curves are very similar $\rightarrow$ Standard candles for distance determination!

Table 11.1 Comparison of type Ia and type II supernovae

	Ia	II
Cause	White dwarf in double star	Massive star
Spectrum	No H	H
Max. bright.	brightness $1^\mathrm{m}5 >$ type II	
Light curve	Sharp maximum	Broader maximum
	All have the same brightness	Diverse brightness
Occur in	All galaxies	Only spiral galaxies
Expansion	10,000 km/s	5000 km/s
Radio emission	–	Available

- SN Ib/c: Brightness at maximum $1^{m}5$ lower than SN Ia.
- SN II: stronger dispersion of maximum brightness; $M_{B,max} = -17 \ldots - 18^{m}$.

11.5.2 Nuclear Synthesis During a SN

Towards the end of its evolution a star with a mass between 10 and 20 solar masses has a shell-like structure: C, He, and H shells. The Fe core contracts and its temperature increases. At 10^9 K occurs *Photodisintegration of* Fe:

$$^{56}Fe + \gamma \rightarrow 13\,^4He + 4n \tag{11.33}$$

This reaction is endothermic and requires about 100 MeV. The nucleus loses energy and contracts more rapidly. The following reactions lead to the formation of a degenerate neutron gas:

$$^4He \rightarrow 2p + 2n \tag{11.34}$$

$$p + e^- \rightarrow n + \bar{\nu} \tag{11.35}$$

The upper layers also fall inward, heat up, and nuclear fusion begins. This happens explosively, and the outer layers are repelled. So many energetic neutrons are formed, which can be absorbed by the heavy nuclei. There is an *r*-process *(rapid)* and an *s*-process *(slow),* where the terms "rapid" and "slow" refer to beta decay, respectively:

$$n \rightarrow p + e^- + \bar{\nu} \tag{11.36}$$

This takes about 15 min. In the *r* process, neutron capture occurs faster than beta decay. For example, the *r*-process leads to $^{56}Fe + n \ldots$ the ^{61}Fe. This is only stable for about 6 min, and if during this time the *s process* neutrons are captured, then emerges:

$$^{56}Fe \rightarrow ^{56}Co + e^- + \nu \tag{11.37}$$

For type II supernovae only the *r* process plays a role. In red giants, nucleosynthesis after the *s*-process plays an important role.

For both type I and type II supernovae, the major source of energy in emission is radioactive decay: ^{56}Ni decays with a half-life of 6.1 days to ^{56}Co, and this decays with a half-life of 77.3 days to the stable ^{56}Fe.

Fig. 11.12 Crab Nebula M1.
A supernova remnant, distance
6300 Ly. (Credit: NASA, ESA,
S. Beckwith (STScI), and The
Hubble Heritage Team
STScI/AURA)

11.5.3 Observed Supernovae

Several hundred supernova outbursts have been observed to date, and several dozen per year through surveys. Because of their high luminosity, these outbursts are observable not only in our Galaxy, but even in distant galaxies; at the time of greatest brightness, a supernova even outshines an entire galaxy. In our Milky Way only three supernova outbursts have been registered in the last 900 years:

- 1054 (Remnant = Crab Nebula, see Fig. 11.12),
- In 1572 *Tycho Brahe* saw a supernova,
- In 1604, Kepler observed a supernova in the constellation Ophiuchus.

Let us do some calculation. Kepler observed a supernova whose brightness was about the same as Jupiter's during its opposition. How far away was this supernova from us?

Assuming it was a type I SN, then $M = -19$; the apparent brightness of Jupiter is -2^m5. Therefore, it follows from the distance modulus:

$d \approx 20{,}000$ pc $= 65{,}000$ light years.

In 1987 a supernovae was observed in the large Magellanic cloud[5], Supernova1987A (the "A" stands for the first supernova of 1987). It could be seen with the naked eye in the southern sky (light curve, Fig. 11.13). Before the outburst was detected, neutrinos were

---

[5] A dwarf galaxy, belongs to our galaxy.

Fig. 11.13 Light curve of
supernova 1987 A. The time is
given in Julian day count

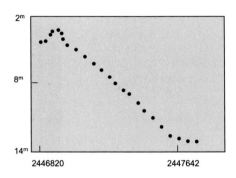

found with a neutrino detector in Kamioka, Japan, which must have been emitted from
SN 1987 A. The neutrinos are produced at the enormous nuclear densities of 10^{14} kg m^{-3}
the neutrinos are still scattered in the core region and therefore leave the star with a delay.
The neutrino signals were received a few hours before the first optical observation of the
outburst. The neutrino burst corresponds to a total energy of 10^{45}–10^{46} J. Only 1% of the
gravitational binding energy is released as optical radiation and as kinetic energy of the
ejected envelope. Such observations are also essential for the question of the neutrino rest
mass. The observations suggest an upper limit for the neutrino rest mass of 10 to 30 eV
c^{-2}.

The discovery was made on February 24, 1987, and in July light echoes from small
matter rings 140 and 400 pc before the supernova were detected. From July to November
the brightness decreased exponentially, which is related to the mean decay time of the
^{56}Co agrees.

On average, one supernova outburst can be expected in 50 years per galaxy.

The evolution of light echoes around object V838 is shown in Fig. 11.14. This
originated from a nova outburst (early 2002) of an object in the constellation Monoceros
(Unicorn) at a distance of 6.1 kpc.

11.6   Black Holes

The existence of objects whose gravity is so strong that not even light can escape was
already suspected by Newton. In principle, all objects could become a black hole if they
were suitably compressed. Quantum physics shows that even black holes evaporate over
very long periods of time.

Fig. 11.14   Evolution of the light echo around V838 (Source: Hubble Space Telescope)

11.6.1 General

We consider an object whose escape velocity

$$v_{esc} = \sqrt{\frac{2GM}{R}} \tag{11.38}$$

equal to the speed of light c; so we get for the radius of a black hole:

$$R = \frac{2GM}{c^2} \tag{11.39}$$

For such an object, nothing can escape, not even light particles, hence the name black hole. The radius of a black hole therefore depends on its mass M and is called *Schwarzschild radius*. Matter falls into a black hole on a spiral path.

From the above formula, the hypothetical Schwarzschild radius for our Sun is 3 km.

A hypothetical trip into a black hole yields that if the black hole has a mass of ten solar masses, at 3000 km from the singularity, you will be torn apart by the enormous tidal forces. 10^{-5} s after passing the Schwarzschild radius, one arrives in the singularity. For an outside observer, the spacecraft suffers an ever-increasing red shift as it flies in, and time passes slower and slower. The light is red shifted until you can no longer detect the signals.

As we have seen, above a mass larger than the Chandrasekhar limit mass, the pressure of the degenerate electrons is not sufficient to stop the gravitational collapse at the end of the star's evolution. As the mass increases, a neutron star is still possible. As the star contracts, the gravitational field at its surface becomes stronger and stronger, and with it relativistic effects, which roughly depend on the ratio of the Schwarzschild radius to the actual radius of an object. Finally when the mass is big enough, light can no longer escape, the escape velocity is equal to the speed of light, and thus everything is trapped by the gravitational field, you have a set of events from which no escape is possible, and this is called a black hole. The limit from which no escape is possible is called the *Event horizon.*

At the center of a black hole, there is a singularity. No known laws of physics exist there anymore, and nothing can be predicted. Such singularities are beyond our knowledge, since they are separated from us by the event horizon. But there are also solutions according to general relativity, which would allow an astronaut to avoid a collision with this singularity, he could instead fall into a *Wormhole* which would mean that he would come out at a completely different location in the universe. Such journeys through space and time have a disadvantage: the solutions are extremely unstable, the slightest disturbance would lead to a fall into the singularity.

The mathematical description using the Schwarzschild metric yields three solutions:

- Black holes;
- White holes: opposite of black holes, only matter, energy flows out; violate 2nd law of thermodynamics;
- Wormholes: also called *Einstein-Rosen-Bridges* connect different parts of the universe.

Rotating black holes are described by a Kerr metric. In addition to the event horizon, there is a so-called ergosphere, which envelops the event horizon; within the ergosphere, matter cannot be kept stationary.

The formation of a black hole results in the emission of *gravitational waves.* These extract energy from the system. Gravitational waves emitted when two black holes collide were directly detected for the first time in 2015.

Gravitational waves are emitted by all accelerating moving masses.

Gravitational waves (ripples in space-time) are also produced by the motion of the Earth around the Sun, energy is thus extracted from the system, but very little, and the effect here is extremely small. This is different, for example, with the pulsar PSR 1913+16, where we have two neutron stars orbiting each other. They lose energy by emitting gravitational waves and thus spiral towards each other.

The size and shape of a black hole depend only on its mass and rotation, but not on other parameters such as chemical composition, etc.

11.6.2 Candidates for Black Holes

There are many candidates for black holes: e.g. the system *Cyg X-1* (Fig. 11.15). This is a powerful X-ray source in the sky (brightness in the range 2–11 keV: 2×10^{30} W, distance: 2.5 kpc). It is a binary star system in which matter is blown away from one component and spirals in an accretion disk towards an unseen companion (which is likely to be a black hole due to its large mass, 16 solar masses), heating it enough to emit X-rays. The mass of the other star is 33 solar masses. The blue supergiant shows periodic Doppler shifts in the absorption lines (period five days), which is interpreted as motion around the system's center of mass. The X-ray intensity varies in the range of 0.001 s, indicating a very compact X-ray source.

Fig. 11.15 Cygnus-X1: blue supergiant with black hole companion (Adapted from Chandra/Harvard)

11.6.3 Quantum Theory of Black Holes

The Area of a black hole cannot decrease, similar to that Entropy of a closed system.

Classically, black holes cannot emit radiation, yet there is so-called *Hawking radiation*. The emission is smaller, the larger the mass of the black hole is. According to quantum theory, there are quantum fluctuations even in vacuum, and particle/antiparticle pairs are created. One speaks of virtual particles.

Heisenberg's uncertainty principle states:

$$\Delta t \, \Delta E \geqq \frac{h}{4\pi} \qquad E = hf \tag{11.40}$$

A virtual particle of the energy range ΔE therefore has a lifetime of range Δt. The energy of a photon pair is $2\Delta E$.

Calculate the lifetime of a virtual photon pair of orange light ($f = 5 \times 10^{14}$ Hz).

$$\Delta t = \frac{1}{8\pi f} = 8 \times 10^{-17} \, \text{s}$$

Energy cannot be created out of nothing, one of the partners of a particle/antiparticle pair has positive energy, the other negative energy. The energy of real particles is always positive. A real particle that is close to a mass has less energy than one that is far away from that mass, because energy must be expended to keep it away from the mass. The gravitational field inside a black hole is so strong that even a real particle can get negative energy there. In this way, a virtual particle with negative energy can also fall into a black hole and become a real particle or antiparticle—it does not need to annihilate with its partner. Its partner can also fall into the black hole or escape from the vicinity of the black hole as a real particle or antiparticle. We as observers from outside then have the impression that a particle is emitted from the black hole. The smaller the black hole is, the shorter is the distance for a particle with negative energy to become a real particle. Small black holes therefore radiate more intensely. Therefore, the smaller the mass of a black hole, the higher the temperature. If the mass becomes extremely small, there is a final evaporation and a violent burst of radiation (Table 11.2).

Bekenstein, Hawking et al. showed that a black hole has a non-vanishing temperature which is calculated from:

$$T = \frac{hc^3}{16\pi^2 kGM} \approx 10^{-7} \frac{M_\odot}{M} [\text{K}] \tag{11.41}$$

Table 11.2 Various astrophysical objects; escape velocity, v_e, Schwarzschild radius, R_S (hypothetical for Earth, Sun and white dwarf)

Object	$M[M_\odot]$	Radius [km]	v_e	R_S
Earth	$1/3 \times 10^{-5}$	6357	11.3	9 mm
Sun	1	7×10^5	617	2.9 km
White dwarf	0.8	10^4	5000	2.4
Neutron star	2	8	2.5×10^5	5.9
Galact. Core	5×10^6	?	?	15×10^6

and the energy of the radiation is:

$$E = \frac{hc^3}{16\pi GM} \tag{11.42}$$

Hawking temperature: the larger the mass, the slower the black hole evaporates.

Suppose an Earth mass (5.3×10^{26} kg) be a black hole. What would be the energy of its radiation and at what frequency could it be observed?

Solution:

$$E = hc^3/(16\pi GM) = 8.9 \times 10^{-25} \, \text{J}$$

and because of

$$E = hf$$

follows a frequency of $f = 1.35$ GHz.

The temperature of a black hole with a solar mass is only 10^{-7} K. This is much less than the temperature of the background radiation (see section Cosmology) which is 2.7 K. At present, such black holes "warm up". However, as the universe continues to expand, at some point the temperature of the background radiation would drop below that of the black holes, and they may cool down. For a black hole with a solar mass, it takes 10^{66} yr for it to evaporate. However, could be very small black holes that were formed during the Big Bang and are now evaporating.

11.6.4 Accretion

As we have seen, there are different phases of stellar evolution in which accretion plays an important role. According to *Zel'dovich* the luminosity of a star due to accretion of matter:

$$L \approx \Phi \frac{dM}{dt} \approx 2 \times 10^{31} \left(\frac{\Phi}{0.1c^2} \right) \left(\frac{M}{M_\odot} \right)^2 \left(\frac{10^4}{T} \right)^{3/2} N \text{ erg/s} \qquad (11.43)$$

Here Φ is the gravitational potential near the surface of the star, T and N are temperature and particle density of the gas, respectively.

Let us consider two cases:

- White dwarf with $M \approx M_\odot$ and $R = 10^9$ cm, $\Phi = 10^{-4} c^2$. One finds $L \approx 10^{28} N$ erg/s. For corresponding values of N is then given by L, and from this we can deduce the temperature of $T = 500$–2000 K.
- Neutron star: Also let $M \approx M_\odot$, $R \approx 10^6$ cm; the radiation occurs mainly in the UV (15...900 nm).

11.7 Gamma Ray Bursts

Gamma Ray Bursts (GRB), were first observed around 1970. The discovery was made by chance with the Vela satellites, which were supposed to monitor the ban on nuclear testing. Since then, well over 2000 GRBs have been recorded.

11.7.1 Properties of GRB

The *bursts* come from random celestial directions (Fig. 11.16). There is no concentration to the galactic plane → GRBs could be from the Galactic halo or even from the Oort cloud.

The duration of the bursts ranges from fractions of a second to minutes. In order to reveal the origin of the bursts we need distances but distance determination was not possible until radio or optical sources could be identified.

Energy release: within seconds as strong as some 10^4 Supernova explosions.

With the Burst and Transient Source Experiment Burst (BATSE[6]) aboard the Compton Ray Observatory, it was possible to register this gamma ray radiation and determine its position within seconds. On January 23, 1999, only 22 s after observing a GRB with a robotic telescope in New Mexico, an optical image was obtained of the region of the sky

---

[6] Removed from Earth orbit by NASA in 2000.

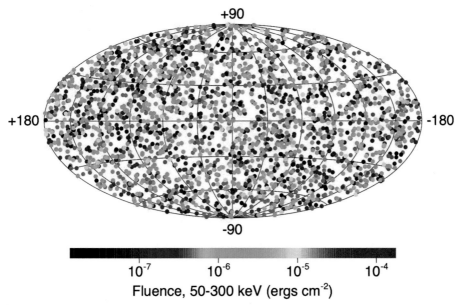

Fig. 11.16 Distribution of GRBs in the sky (galactic coordinates, the galactic equator runs in the middle). (Source: M. Briggs)

where the GRB was observed. Within 25 s a dramatic increase in brightness was observed. Spectra were then taken using the Keck telescope, and the object had a red shift of $z = 1.6$, which corresponds to a distance of 3000 Mpc.

GRBs have also been associated with mass extinctions of animal and plant species in Earth's history *(mass extinction)*.

11.7.2 Explanation of GRB

There are several theories:

- Collapse of a massive star $\rightarrow$ a black hole with an extreme magnetic field arises, one also speaks of a Hypernova.
- Merging of two neutron stars or neutron star + black hole.
- Bursts are directed like in a pulsar. This would lead to an overestimation of the released energy.
- Gravitational lensing effect enhances a less intense burst. Between GRB source and observer is a massive object which acts as a lens .

- Magnetar: Neutron star with extremely strong magnetic field. Starquakes occur from time to time by interaction of the magnetic field with the crust $\rightarrow$ Gamma rays. During the outburst of object SGR 1900 +14 on August 27, 1998, gamma rays striking the Earth's atmosphere raised the number of ions in the ionosphere at night to daytime levels.

11.8 Variable Stars

Variable stars are interesting for several reasons. On the one hand, they are usually at the end of stellar evolution, and often, in addition to pulsations, envelope expulsion occurs. On the other hand, some groups of variable stars are important standard candles by which distances can be determined. One knows their actual luminosity, and by comparison with the easily measurable apparent brightness follows the distance.

11.8.1 General

The term "variable star" in astrophysics always refers to stars whose brightness changes. As we saw in the section on the Sun, it too is strictly speaking a variable star, but the types of stars discussed here are variable to a much greater extent. If we examine the light curve of a variable, we can derive two parameters:

- Period of the change in brightness P;
- Amplitude of the brightness change A.

The designation is made within a constellation with large Latin letters R, S,...Z and then continuing with RR, RS,...ZZ and AA, respectively,...QZ as well as simply with V and a number. The first star found to be variable is the star Mira (*o* Ceti) (1596 by *Fabricius* discovered). In general, a distinction is made between:

- Pulsating Variables : giants or supergiants of all spectral classes; the cause of the change in brightness is more or less periodic pulsations of the atmosphere.
- Eruptive Variables: often stars of low luminosity; there are random eruptions of gas.
- Eclipsing Variables: The cause here is the mutual eclipsing; some representatives of these groups are very close binary stars, so that there is also an exchange of matter.

Pulsation Mechanism

Stars pulsate when they are not in hydrostatic equilibrium.

If a star expands due to increased gas pressure, the matter density decreases until the point of hydrostatic equilibrium is reached, gravity dominates again and the star contracts. In both cases, you have overshooting, above the corresponding equilibrium points. In this process, energy dissipation occurs, and normally pulsations therefore come to a rapid halt.

Thus, as in the case of pulsating stars, in order to sustain pulsations over long periods of time, one needs a mechanism that can compensate the dissipation. Opacity plays a significant role in energy transport. If the opacity is large, the radiation cannot escape and the star appears faint. If the star is compressed at the time of greatest opacity, then the excess radiation (cf. virial theorem) is stored and exerts an additional pressure $\rightarrow$ Mechanism to maintain the pulsations (κ-mechanism).

Pulsation variability: Increase in opacity upon compression as singly ionized helium absorbs UV radiation, becoming doubly ionized. The He^+-Ionization zone is cooler than surrounding regions because energy is consumed to ionize. Thus, one has a region of instability.

There is a simple relation between the density of the star and its pulsation period. Assume that the matter falls freely onto the star after expansion; then Kepler's law applies to this gas:

$$\frac{P^2}{R^3} = \frac{4\pi^2}{GM} \tag{11.44}$$

Thereby P the period of the pulsation, R the radius of the star. Because of

$$P^2 \approx R^3/M \qquad M \approx \bar{\rho}R^3 \tag{11.45}$$

one gets:

$$P^2 \approx R^3/(\bar{\rho}^3 R^3) \approx 1/\bar{\rho} \tag{11.46}$$

If a star radiates like a black body, then:

$$L \propto R^2 T_{\text{eff}}^4 \tag{11.47}$$

Fig. 11.17 Location of the different types of variable stars in the HRD

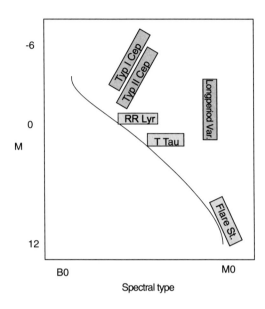

If we observe a pulsating star at two different times 1 and 2, then the ratio of the luminosities is given by L_1 and L_2 at these times:

$$\frac{L_1}{L_2} = \left(\frac{R_1}{R_2}\right)^2 \left(\frac{T_1}{T_2}\right)^4 \tag{11.48}$$

Luminosity changes are therefore associated with temperature and radius changes (pulsations).

In addition to regularly variable stars, there are also stars whose variability is not strictly periodic; these irregularly variable stars include ZZ-Ceti stars (hydrogen-rich white dwarfs with periods between 3 and 30 min), BY-Draconis stars (late-type dwarf stars), and L-type variable stars (slowly variable, amplitudes up to 2^m, giants and supergiants). The location of different types of variable stars in the HRD is outlined in Fig. 11.17.

11.8.2 Pulsation Variable

Cepheids

The light variation of Cepheids occurs strictly regularly with a period between 1 and 50 days (Fig. 11.18). They are very bright supergiants of types F to K. That is why they can still be observed at great distances (more than 300 have been found, for example, in our neighboring galaxy about 2.5 million light-years away, the *Andromeda galaxy*.

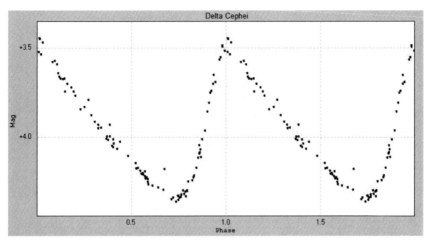

Fig. 11.18 Light curve of the star δ Cephei

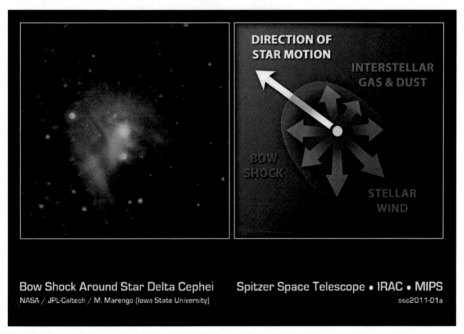

Fig. 11.19 Shock front around δ Cephei (Spitzer Telescope)

The star δ Cephei has a period of 5.37 days and changes its diameter by 2 million km during this time. Using the Spitzer IR telescope, a shock front has been detected resulting from the star's motion through the interstellar medium. Furthermore, strong stellar winds have been observed (Fig. 11.19).

Around 1950, it was found that there were two groups of Cepheids:

- δ-Cephei stars: They occur in the galactic plane, called Cepheids of the population I.
- W-Virginis Cepheids: They occur in the halo of the Galaxy or in the Galactic center, called Cepheids of population II.

The amplitudes are less than 2^m, larger in the blue than in the red. The cause of the change in brightness is pulsation, the radius changes being about 10% for the δ-Cepheids, for the W-Virginis-Cepheids about 50%. The largest diameter occurs during the descent phase of brightness, and the temperature is greatest when the brightness is highest. By measuring the Doppler shift of the spectral lines, one can determine the velocities; from the integration of the radial velocity curve, the change in radius follows.

When this group of variables was detected in the Magellanic Clouds , it was found that there is a relationship between their apparent brightness and period. Since all the stars in this dwarf galaxy neighboring us are about the same distance away, this means that there is a relationship between the periods and luminosities of these objects, the *period-luminosity relation*. The luminosity and absolute magnitude are related and follow from the period P:

$$\delta\mathrm{Ceph\,Pop\,I} : M = -1.80 - 1.74\log P \tag{11.49}$$

$$\mathrm{W\,Vir\,Pop\,II} : M = -0.35 - 1.75\log P \tag{11.50}$$

> So once you have determined the period of these objects, you know their absolute brightness and thus the distance.

This is a very important method of distance determination for nearby Galaxies.

RR Lyrae Stars

They often occur in globular clusters. Globular clusters are spherically arranged collections of some 10^4 Stars distributed in a halo around a galaxy. In them one finds the oldest stars. The periods are $<1^d$, spectral types B8...F2, amplitudes around 1^m. Sometimes several periods overlap. The absolute magnitudes are fairly constant $\rightarrow$ Standard candles

$$M_{vis} = +0.5 \pm 0.4. \tag{11.51}$$

In the HRD they are located at the gap in the horizontal branch where there can be no stable stars.

δ-Scuti Stars

This group is also called dwarf Cepheids. They are pulsating giants (F) with very short periods $0.^d5\ldots0.^d2$ and small amplitudes.

β-Canis-Majoris Stars

Spectral type: B1... B2, III...IV. Periods between 3 and 6 h. Multiple periods always occur. The amplitudes are only a few hundredths of a magnitude.

Mira-Variables

Periods between 80 and 1000 days; striking are the very large amplitudes between two and more than four magnitudes. In terms of light curve shapes, the following types are distinguished:

- α: ascent steeper than descent, minima wider than maxima.
- β: almost symmetrical.
- γ: irregularities in light curves, humps, etc.

Mira stars belong to spectral classes M, S, or C. The enormous changes in brightness result from the strong variation in absorption bands and not from changes in temperature directly.

From the measurements of the radial velocities of emission and absorption lines, it follows that the envelopes are expanding at about 10 km/s.

In addition, these stars lose about 10^{-8}–10^{-6} $M_\odot$/year in mass (Fig. 11.20). Some Mira stars show maser emission in the radio region at a wavelength of 18 cm from OH.

Let us consider the prototype *Mira (oCeti)* The maximum radius occurs at the minimum luminosity and is about $320\,R_\odot$, and the minimum radius occurs at the brightness maximum and is about $220\,R_\odot$, the brightness varies between $2.^m0$ and $10.^m1$. The pulsation period is 331 days. The star itself has a white dwarf as its companion.

11.8.3 Semi-regular Variables

Periods range from 30 to 1000 days, amplitudes are usually less than one to two magnitudes. There is either a good mean period, or the period is disturbed by irregularities. The following subgroups are distinguished:

- SRa: Red giants, M, C, S; smaller amplitude than Mira stars, otherwise the same.
- SRb: Also M, C or S. The periodicity is interrupted by completely irregular phases; example: AF Cyg.
- SRc: Supergiants of intermediate type (G8...M6); example: μ Cep.
- SRd: yellow giants and supergiants (F...K).

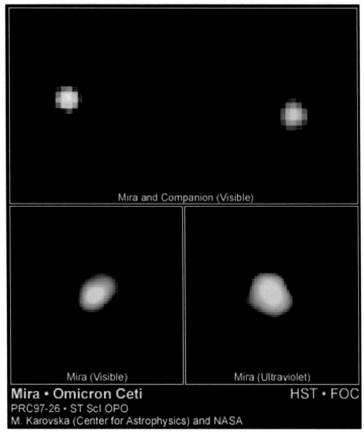

Fig. 11.20 HST image of the variable star Mira. Mira is seen to be losing mass in UV light (GALEX (Galaxy Evolution Explorer))

RV Tauri Stars
Type F...K, I, II. regular alternation of shallow and deep minima. Periods between 30 and 150 days. The amplitudes are up to three magnitudes.

α^2-Canum Venaticorum Stars, Ap Stars
These stars have the following features: (a) strong magnetic fields (measured by Zeeman effect, 0.1 to 1 T), (b) abnormally strong lines of rare elements like Si, Mn, Cr, Sr, Eu. Individual line groups change their intensity, which leads to variations in brightness of about $0.^m1$. Periods range from 1 to 25 days, magnetic fields are variable, which is explained by:

- Oblique rotator: Magnetic axis does not coincide with axis of rotation of star.
- Activity cycle similar to Sun, but orders of magnitude stronger.
- Huge magnetic spots on surface, change by rotation.

About 10–15% of late B and A stars belong to this group. The elemental anomalies (e.g. Os I, Pt II, Pm) can be explained by neutron irradiation or by complex diffusion processes.

Metal Line Stars, Am Stars
Cooler than the Ap stars; mostly members of binary systems, rotational velocity <100 km/s; strong lines of the Fe group; about 10% of the A stars belong to this group.

11.8.4 Eruptive Variables

Novae and Nova-Like Variable Stars
These objects are sometimes called *cataclysmic variables*. They are hot dwarf stars whose brightness increases by seven to 20 magnitudes within a short time (hours to months). After a short maximum the brightness returns to the original value in the course of years to decades. About the Praenovae is only known that they are A-subdwarfs. Sometimes there is a slight increase in brightness up to 1.5 magnitudes before the outburst. One divides the novae into:

- Na: rapid nova, very steep rise. Decline by 3^m in less than 100 days.
- Nb: slow nova, descent around 3^m in more than 100 days.
- Nc: extremely slow nova, many years at maximum.
- Nr (Nd): recurrent nova, recurrent bursts of brightness. Example: TCrB; outburst 1866, 1946 at $\Delta m = 8.6$.
- Nl (Ne): Nova-like variable.

The spectra for Na and Nb are also divided into Q0 to Q9. The typical progression of these two types looks like this:

- steep slope: 7–10^m within one day;
- short standstill, possibly even decline before the maximum;
- steep final rise to maximum, Q0;
- main maximum, Q1;
- first descent around $3.^m0$; first spectrum similar to an F supergiant (Q2), then broad, vigorous emission of H and ionized metals (Q3). Later H, OII, NII, NIII (Q4, Q5);
- Transitional stage, nebular spectrum, NII, forbidden line [OIII] (Q7);
- Postnovae.

Knowing the time of brightness decline by three magnitude classes in days (t_3), then one can determine the absolute maximum brightness from the following empirical formula:

$$M_{max} = -11^M + 2.5 \log t_3 \qquad (11.52)$$

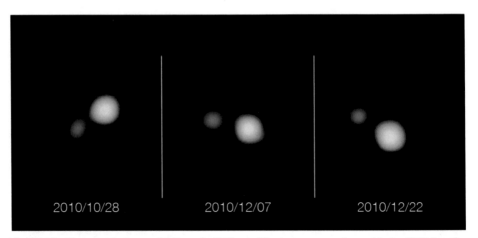

Fig. 11.21 The symbiotic star SS Leporis (17Lep). The image was taken with the VLT interferometer. A red giant orbits a hotter companion. The stellar images were coloured according to the temperatures (VLTI/ESO)

Novae are close binary stars with a hot, blue component (white dwarf). The cooler red component, which has less mass, gives off matter to the white dwarf. For the white dwarf its temperature and density at the bottom of the atmosphere increases, a hot spot forms, and thermonuclear reactions can begin. During the eruption, about 10^{38} J is released.

In Fig. 11.21 the *symbiotic double star* SS Lep is shown. There is an exchange of matter between the components. The object is about 270 pc from us, and the orbital period is 260 days. The size of the orbit is about 0.005". Such observations are only possible with the VLT-PIONIER (Precision Integrated Optics Near-infrared Imaging ExpeRiment). The light from the four VLT telescopes is made to interfere with each other; this results in a greater baselength, which in turn increases the telescope resolution.

About 200 novae have been observed in our Galaxy. It is estimated that there are about 50 novae per year in our Galaxy. Postnovae are often surrounded by expanding nebulae.

In August 1975 a nova was observed in the constellation Cygnus (maximum brightness $1^{m}.8$).

Among the Nl stars are the S Doradus stars, the γ-Cassiopeiae stars, Z-Andromedae stars, P-Cygni stars and others. Among the dwarf novae (DN) one counts the U-Geminorum stars. They all have weak fluctuations in brightness in common, and then suddenly there are outbursts between two and six magnitudes within a few days. The longer the pause between the eruptions, the more violent they are (pent-up energy).

R Coronae Borealis Stars

These are supergiants; the brightness remains constant for months or even years. Then within days the brightness decreases by several magnitudes, amplitudes up to 7^m. Almost all of them are located in regions of interstellar matter, nebular patches. Based on the blue shift of the spectral lines, we know that matter is ejected from the Star, carbon condenses, and soot clouds eclipse the star. The expansion velocity is about 60 km/s and the mass loss rate $\dot{M} \approx 10^{-5} M_\odot$/year.

T Tauri Stars

These are pre-main sequence stars with masses between 0.2 and two solar masses. They have an extended convection zone. Some are thought to have large spots on their surfaces (based on brightness changes when the spots move due to stellar rotation). Emission lines of H and ionized Ca indicate an active chromosphere. Forbidden lines similar to nebulae are also found in some T Tauri stars. This indicates circumstellar material. Observations in X-rays show very strong variations (up to factor 10) within one day. These are probably huge flare outbursts in their photospheres. The observed excess of IR radiation can be explained simply: The dust/gas cloud surrounding it absorbs the star's shortwave radiation and re-emits in the IR. One observes strong stellar winds (10^{-7}–$10^{-8} M_\odot$/year).

Our Sun also went through a T Tauri stage, which had a significant impact on the formation of the planets' primordial atmospheres. During this stage, it emitted an amount of X-ray radiation equal to a factor of 10^3 stronger than today.

Flare Stars

In dwarf M stars, flares are observed with released energies between 10^{21} and 10^{27} J. The increase in brightness by up to six magnitudes occurs within a few seconds to minutes; they are also called UV Ceti Stars. Such stars are probably very common, but the probability of detection is low because of their low luminosity.

RS Canum Venaticorum Stars

Are found in binary star systems. Orbital periods are around seven days. The stars exert strong tidal interactions on each other. The radio flares observed in these types are 10^5–10^6 times stronger than for the Sun. The radio spectra are polarized and non-thermal, suggesting synchrotron radiation. One also detects a brightness modulation with the rotation period, suggesting giant stars pots.

11.8.5 Peculiar Stars

The suffix p in the spectral type indicates a special feature, e.g. anomalous metal abundance.

Wolf-Rayet Stars
Stars of very high luminosity, expanding atmosphere, extremely broad emission lines. Temperature about 30,000 K, radii between 3 and 25 $R_\odot$, masses 10–20 $M_\odot$. The emissions come from the expanding envelope. One distinguishes the following types:

- WC: strong C lines,
- WN: strong N lines.

Be and Shell Stars
Shells can form around Be stars due to the high rotational velocities that characterize these stars.

11.8.6 Planetary Nebulae

Due to their nebular appearance, which resembles that of a planet in the telescope, this misleading name was introduced for white dwarfs that have ejected a luminous gas envelope. Again, there is an expanding atmosphere, and H-lines as well as He-lines are observed. The strongest lines are the forbidden lines of the elements O and Ne. Such forbidden lines arise as a result of the low gas density. Atoms can be excited at metastable levels, since collision, which would contribute to depopulation of the level, is very unlikely. The forbidden lines of [O III] are at 500.7 and 495.5 nm, respectively. The forbidden nitrogen line [NII] is at 658.4 nm.

The hot star at the center heats the gases to about 10,000 K. The gas temperature increases as one moves away from the center. High-energy photons are absorbed less often than low-energy photons. In the outer nebular regions, the low-energy photons have already been absorbed, and the remaining high-energy photons cause the temperature increase.

In general, Mira stars are thought to be the precursors of planetary nebulae.

A prominent example of a planetary nebula is the Cat's Eye Nebula, NGC 6543 (Fig. 11.22). It is located at a distance of 1500 light-years in the constellation Dragon. The inner part of the nebula has only 20″ extension, the outer part extends over 6.4 arcminutes and was formerly ejected by the very old central stare. The inner part has a temperature of 8000 K and a particle density of 5000 particles per cm³, the outer part is much thinner and has a temperature of about 15,000 K. The 80,000 K hot central star loses about 20 trillion tons per second due to stellar winds, which is about $3 \times 10^{-7}\,M_\odot/\text{year}$.

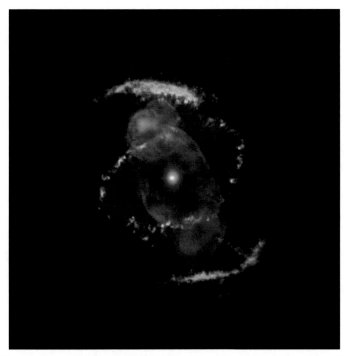

Fig. 11.22 The Cat's Eye Nebula, NGC 6543, an example of a planetary nebula. The image was taken with three filters: Red (Hα, 6563 nm), blue (neutral oxygen, 630 nm), and green (ionized nitrogen, 658 nm). (HST image)

Other well-known examples of planetary nebulae are the Dumbbell Nebula (M27, distance about 1400 light-years, diameter 3 light-years) and the Ring Nebula (M57, distance about 2300 light-years, diameter about 0.9 light-years, age about 20,000 years). In total, more than 1500 planetary nebulae are known in the Milky Way. Compared to the several 100 billion stars in the Milky Way, this is not much, but these nebulae only shine for a few 10,000 years. .

11.9 Stellar Activity

In addition to the stars with very strong variabilities discussed above, it is now possible to determine activity cycles, spots, etc. for "normal" stars.

11.9.1 Stellar Activity and Convection

The activity of the Sun can be explained by a dynamo process, where

- rotation,
- magnetic fields,
- convection zone

interacts. Magnetoacoustic waves are generated in the region of the convection zone, which heat up the chromosphere and the corona, among others.

From the theory of stellar structure we know: Stars of later spectral type than F0 ($T \approx 6500$ K) have a convection zone that extends to the surface. The convection zone extends deeper into the stellar interior the later the spectral type. This can be explained simply. Convection occurs when the adiabatic temperature gradient of an element moving upwards due to a random perturbation is smaller than the radiation gradient of the surroundings. If there is ionization of hydrogen H^+ or helium (in this case He^+, is singly ionized He, and He^{++}, i.e., doubly ionized He), then the radiative gradient increases and the adiabatic gradient decreases, respectively, which favors convection. For cooler stars the surface temperature is lower, the zone above which hydrogen is ionized extends deeper into the stellar interior than for hotter stars.

The chromospheric activity of a star can be determined by measuring the Ca-II H and K emission lines. Long series of measurements then allow the stellar activity cycle to be determined. In 1957 *Wilson* and *Bappu* found that the width of the Ca-II emission lines is a function of the absolute luminosity, *Wilson-Bappu effect* → Method of distance determination. The width of a line can have various causes (rotation of the star, magnetic fields, turbulence)—turbulence is the most important here.

Skumanich found that rotational velocity and chromospheric activity of the Stars decrease with age.

The decrease in stellar rotation can be given as follows:

$$\Omega_{eq} \approx t^{-1/2} \qquad (11.53)$$

Ω_{eq} is the angular velocity at the star's equator, t is the age of the star. One can only determine the equatorial velocity if the inclination of the rotation axis is also known (Fig. 11.23). *Gyrochronology* is the method of determining the age of a star from its rotation rate. This is calibrated at the sun.

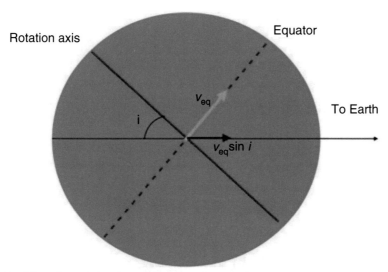

Fig. 11.23 When determining the rotation rate of a star, the mostly unknown inclination of the rotation axis must be taken into account. Thus, apparently slowly rotating stars may nevertheless rotate rapidly if their axis is only slightly inclined

Table 11.3 Summary young stars/old stars

	Young stars	Old stars
Activity	Large amplitudes	Low amplitudes
Activity cycle	Irregular	Regular
Rotation	Rapid	Slow
Chrome. activity	High	Low

From the rotation rate of stars one can infer their age.

In young, rapidly rotating stars, for example, huge flare outbursts are observed in the X-ray region. A summary is given in Table 11.3

Out flowing stellar winds with magnetic fields dissipate angular momentum to the interstellar medium, leading to the slower rotation rates observed in old stars.

Mass losses can be explained by blue-shifted star absorption components of strong lines (Ca II H and K lines, Fig. 11.24). Mass-loss rates can be extracted from (after Reimers, 1975):

$$\dot{M} = 4 \times 10^{-13} \frac{L}{gR} \tag{11.54}$$

Where $\dot{M}$ in solar masses/year, L, g and R in units of the sun.

red

blue

Fig. 11.24 P-Cygni profile. The component shifted to blue is from an envelope moving towards the observer (stellar wind)

Example
cool super giants: $\dot{M} \approx 10^{-7}$–10^{-5}, Sun $\dot{M} \approx 2 \times 10^{-14}$.

The X-ray luminosity also depends on the age and rotation of the star.

Another interesting observation is that of coronae in O and B stars. In their spectra one observes highly excited atoms (N V, O VI) as well as strong X-ray emissions. To explain the existence of these coronae, a mechanism other than the hydrogen convection zone is needed, since these stars show no convection near the surface. Here, compressions in the stellar wind are assumed to be heated by shock waves.

11.9.2 Mass Loss of Stars

It is very easy to show that outer layers of stars (coronae) are not stable. We give here the derivation already developed by Parker in 1958. Let us replace in the hydrostatic equation the density ρ by

$$\rho = \frac{\mu m_H}{k} \frac{P}{T}$$

then The hydrostatic equation:

$$\frac{dP}{dr} = -\frac{GM}{r^2} \frac{\mu m_H}{k} \frac{P}{T} \tag{11.55}$$

At the lower boundary of the corona let $r = r_0$ and $T = T_0$. The heat flux through a surface $4\pi r^2$ be

$$4\pi r^2 K \frac{dT}{dr} \tag{11.56}$$

The thermal conductivity K of a Plasma is

$$K \propto T^{5/2} \tag{11.57}$$

and one obtains

$$r^2 T^{5/2} \frac{dT}{dr} = \text{const} \tag{11.58}$$

or the solution:

$$T = T_0 \left(\frac{r_0}{r}\right)^{2/7} \tag{11.59}$$

Thus

$$\frac{dP}{P} = -\frac{GM\mu m_H}{kT_0^{2/7}} \frac{dr}{r^{12/7}} \tag{11.60}$$

and with $P = P_0$ at the position $r = r_0$

$$P = P_0 \exp\left[\frac{7GM\mu m_H}{5kT_0 r_0}\left[\left(\frac{r_0}{r}\right)^{5/7} - 1\right]\right] \tag{11.61}$$

→ If $r \to \infty$, the pressure remains finite and does not vanish. The asymptotic value of P is larger than the typical pressure of the interstellar medium.

How are stellar winds driven? A distinction is made between:

- Thermally driven stellar winds: high temperature drives the wind; example: corona of the Sun.
- Radiation pressure: Its magnitude can become as large as gravity. This mechanism works for massive stars. A star remains stable as long as:

$$\frac{GM}{R^2}\rho > \frac{L}{4\pi R^2}\frac{\rho\kappa}{c} \tag{11.62}$$

It follows the Eddington Limit:

$$L < \frac{4\pi cGM}{\kappa} \tag{11.63}$$

- Stars that are rotating rapidly loose mass as well.

Currently, the Sun's mass loss is: $10^{-14} \, M_\odot/\text{year}$.
Red Giants: Low surface gravity, therefore stronger stellar winds.
Mass loss at late stages of stellar evolution leads to the formation of planetary nebulae.

11.10 Further Literature

We give a small selection of recommended further reading.
The Life of Stars, G. Shaviv, Springer, 2010
Theory of Stellar Structure and Evolution, D. Prialnik, Cambridge Univ. Press, 2009
Stars and Stellar Evolution, K. de Boer, W. Seggewiss, EDP, 2008
Stellar Evolution Physics, I. Iben, Cambridge Univ. Press, 2013
Solar and Stellar Magnetic Activity, C. Schrijver, C. Zwaan, Cambridge Univ. Press, 2000
Solar and Stellar Activity Cycles, P. Wilson, Cambridge Univ. Press, 2005

Tasks

11.1 Calculate the free fall time for the above cloud with $R = 10^{15}$ m to a radius with $R = 10^{11}$ m.

Solution
20,000 years.

11.2 Estimate the Jeans mass for an interstellar cloud 100 light-years in diameter, its temperature 30 K. $R = 100 \times 10^{16}$ m

Solution
$$M > \frac{3}{2} \frac{1.38 \times 10^{-23} \times 100 \times 10^{16} \times 30}{6.67 \times 10^{-11} \times 1.6 \times 10^{-27}} \approx 10^{34} \, \text{kg} \approx 10^4 M_\odot$$

11.3 Calculate the relativistic Doppler effect (a) for the Sun, (b) for a white dwarf with $0,8 \, M_\odot$, $R = 0.01 \, R_\odot$.

Solution
Substituting the values for the Sun gives $c\frac{\Delta}{\nu} = 2.117 \times 10^{-6}$...and this corresponds to a Doppler effect of... $c\frac{\Delta\nu}{\nu} = 635$ m/s.
The values for the white dwarf are: $\frac{\Delta\nu}{\nu} = 1.7 \times 10^{-4}$...and this corresponds to a Doppler effect... $c\frac{\Delta\nu}{\nu} = 58.8$ km/s.

11.4 Calculate the magnetic field strength of a white dwarf with R = 7000 km.

Solution
Assume the values for the Sun: $B_\odot = 10^{-4}$ T, $R_\odot = 7 \times 10^5$ km. Then you get $B_{WD} = 1$ T.

11.5 How can brown dwarfs be found? Why are they so difficult to observe?

Solution
If companion in a binary system, due to motion of the primary; low luminosity.

11.6 At what wavelength do you observe light emitted at 600 nm from a neutron star of about 1 solar mass and radius 7 km?

Solution
One observes the radiation at 720 nm.

11.7 What is the apparent brightness of a type Ia supernova in the Andromeda Galaxy?

Solution
The Andromeda Galaxy is 2.5 million light-years away from us, this corresponds to $2.5 \times 10^6/3.26$ pc. If one substitutes into the formula for the distance modulus $M = -19$:
$m = -19 + 5\log d - 5 \rightarrow 5.^m3$. So you could just see a supernova exploding in the Andromeda Galaxy with the naked eye under very good conditions.

11.8 At what distance would a SN Ia have to explode so that it surpasses the full moon in brightness?

Solution
A little less than 250 pc = 815 Ly.

11.9 Show that the energy released in a nova outburst is equivalent to the thermal energy content of a thin shell of 5×10^6 K the mass $10^{-3} M_\odot$ is equal to.

Solution
$E = 3/2[kTM/m_u] = 3/2[1.38 \times 10^{-16} \times 2 \times 10^{33} \times 10^{-3} \times 5 \times 10^6/1.66 \times 10^{-24}]$

11.10 Show that the mass of a black hole is approx. 2×10^{19} kg must amount to in order to glow deep red. Could you actually see this radiation?

Solution
No, you calculate the energy!

11.11 Show the mass at which black holes could annihilate today.

Solution
Solution: If a black hole had a mass of $10^{-9} M_\odot$, then it would explode today ($T > 2.7$ K).

11.12 Visually, the luminosity variation of a Mira stars is, say, 1:100 (how many magnitudes would that correspond to?). Bolometrically, the variation is only 1:2, which corresponds to a $\Delta T \sim 500\,°C$ corresponds. Verify that.

Solution

Approach: compare the Planck curves.

Interstellar Matter

12

In this section, we examine the interstellar matter that is dispersed between the stars and is particularly important for star formation. If you look up at the sky with the naked eye, on a clear moonless night, in northern geographic latitudes especially well in summer and autumn, you can see the Milky Way. This delicately glowing band that spans the sky is made up of about 300 billion stars, including our sun. However, the Milky Way (Galaxy) appears torn in some areas → interstellar dust absorbs the light from stars behind it.

12.1 Discovery, General Properties

Even a glance at the sky with the naked eye shows that there are dark regions in the Milky Way that appear to be starless. Nevertheless, at first astrophysicists did not get the idea that these might be huge dust clouds absorbing the light from stars behind them.

12.1.1 Discovery of Interstellar Matter

Interstellar dust is particularly conspicuous for its absorption of starlight. If you look at the distribution of stars along the Milky Way band in the sky, you can clearly see star-poor regions. These are the locations of large extended dust clouds.

In the spectrum of stars, interstellar matter was be detected only about 100 years ago. If we examine spectral lines of double stars, they are shifted to blue when one component approaches us as a result of the motion around the main component, and to red when it moves away. In some double stars, however, lines were discovered which showed no such shifts: particularly striking are the absorption lines of the Ca^+ (393 and 396 nm) and of the Na^+ (589, 590 nm); these come from interstellar gas located between the star and the observer.

© Springer-Verlag GmbH Germany, part of Springer Nature 2023
A. Hanslmeier, *Introduction to Astronomy and Astrophysics*,
https://doi.org/10.1007/978-3-662-64637-3_12

Interstellar matter is composed of gas and dust.

12.1.2 Composition of Interstellar Matter

Interstellar matter is composed of two components:

- Gas: atoms, molecules, ions; this is visible when excited to glow.
- Dust: solid particles; appears either as a dark cloud or luminous as a reflection nebula.

Interstellar matter is diffusely distributed and often glows spectacularly: well-known examples are the *Horsehead Nebula* or the *North American Nebula* (Fig. 12.1). These gas nebulae are usually caused to glow by young stars, mainly in the red light of the hydrogen line Hα.

The composition is similar to that of stellar matter, i.e. predominantly hydrogen. The average density is 10^{-24} g/cm^3. But since it fills a huge space, its mass is about 3–10% of the total mass of our Milky Way, the Galaxy. The ratio of gas to dust is 100:1.

Fig. 12.1 The North American Nebula, NGC 7000, a luminous H-II region. It is about 2000 light-years away from us and difficult to observe because it is very extended (120 by 100 arcminutes). Hanslmeier/Kohl, Pretal

12.2   Interstellar Dust

The observation of dust is difficult, because firstly the concentration is mostly low and secondly no spectral lines can be seen. Detection is only possible via:

1. Extinction,
2. Reddening,
3. Polarization,
4. Reflection of starlight.

12.2.1 Extinction

As we mentioned at the beginning, there are *dark clouds,* which largely attenuate the light from stars behind them. Counts of extra-galactic stellar systems provide basic information about extinction in the Milky Way. Here the use of Galactic coordinates is instructive: the Galactic equator is defined by the plane of the Milky Way. Within ± 10 degrees of Galactic latitude, few extragalactic systems are seen because this is where the extinction is greatest. The dust is therefore highly concentrated towards the Milky Way plane. A_V is the extinction contribution in the visual, then:

$$m_V - M_V = A_V + 5^{\mathrm{m}} \log(r) - 5^{\mathrm{m}} \tag{12.1}$$

where r is the distance in [kpc]. Empirically, we find:

- $A_V = 1 \ldots 2^{\mathrm{m}}/\mathrm{kpc}$, when there are dark clouds,
- $A_V = 0.^{\mathrm{m}}3/\mathrm{kpc}$ on average in the plane of the galaxy.

An object at a distance of 10 kpc is therefore attenuated by up to about ten magnitudes!

The extinction is wavelength dependent, the short wavelength light is attenuated more than the long wavelength light. One speaks of an interstellar reddening and defines the *color excess* as:

$$E(\lambda_1, \lambda_2) = (m_{\lambda_1} - m_{\lambda_2}) - (m_{\lambda_1} - m_{\lambda_2})_0 \tag{12.2}$$

where m_λ is the observed brightness and $m_{\lambda 0}$ is the brightness that would result if the extinction contribution were zero. On the right hand side of Eq. 12.2 there is the difference between observed color and intrinsic color. The intrinsic color of a star can be determined by comparing it to other stars of the same spectral type.

One can determine the extinction by *Star counts* by comparing the field under investigation with a reference field.

In the so-called *Wolf diagram* one enters the number of stars at an apparent brightness (Fig. 12.2). One counts the stars per square degree in the brightness interval $[m - 1/2, m +$

Fig. 12.2 In the Wolf diagram, one plots the number of stars against their apparent brightness. Without absorption, one observes a uniform increase to fainter stars; if there is a dark cloud in between, this is delayed by the amount ϵ

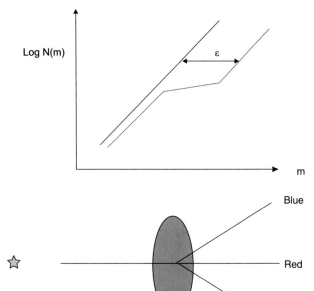

Fig. 12.3 The wavelength dependence of interstellar absorption. Blue is scattered more strongly than red

1/2] in the region of a dark cloud and a comparison field. If all stars had the same absolute magnitude M, a dark cloud, which lies in the range of the distances r_1, r_2 would reduce the increase, i.e. cause an extinction by the amount of Δm and thus the star numbers $N(m)$ in this region decreases noticeably. Thus, one can roughly indicate the distance of the dark cloud. Many dark clouds are no more than 100 pc away from us. Possibly there is a connection of the two large dark cloud complexes in Taurus and Ophiuchus via our solar system.

Furthermore, one can still investigate the dependence of interstellar absorption as a function of wavelength (Fig. 12.3) can be investigated. The interstellar absorption is very low in the IR and radio range, but it increases strongly in the optical and UV. A broad maximum occurs at a wavelength of 220 nm. This absorption band is mainly caused by Graphite particles.

12.2.2 Scattering

Actually the interstellar extinction is not due to absorption but to *scattering*. In the scattering theory, the ratio wavelength λ to particle diameter d is crucial:

- $d/\lambda \to 0$: diameter of the particles $\ll$ wavelength. *Rayleigh scattering,* is proportional to λ^{-4}.

- $d/\lambda \gg 1$: Here the scattering is independent of wavelength.
- $d/\lambda \approx 1$: This is found for interstellar dust grains. From the relation it follows that the grains are some 10^2 nm in size, and the scattering is proportional to λ^{-1}.

The chemical composition of the dust is determined from the extinction curve; graphite particles are found, as indicated above, as well as silicates.

> A typical interstellar gas cloud is characterized by the following quantities: $A_V \approx 1^m$, density one particle per $100\,\mathrm{m}^3$, diameter about 10 pc and temperature about 100 K.

Where do the dust particles come from? From the outflowing stellar envelopes, condensations, soot flocs, etc.

12.2.3 Polarization

Light from distant stars is partially polarized. *Hiltner* and *Hall* found in 1949 a correlation between the degree of polarization the interstellar reddening and the interstellar extinction. The electric field strength vector preferably oscillates parallel to the Galactic plane.

If $I_{\parallel}$ is the intensity of the light oscillating parallel to the plane of polarization and $I_{\perp}$ is the intensity of the light oscillating normal to the plane of polarization, then one defines the degree of polarization P as follows:

$$P = \frac{I_{\parallel} - I_{\perp}}{I_{\parallel} + I_{\perp}} \tag{12.3}$$

The polarization is also given in magnitude classes:

$$\Delta m_P = 2.^m5 \log(I_{\parallel}/I_{\perp}) \tag{12.4}$$

One has a correlation of P found with the color excess:

$$\Delta m_P \approx 2.^m17 P \qquad \Delta m_P \approx 0.19 E(B, V) \tag{12.5}$$

Such a polarization can be easily explained by assuming that the scattering particles are elongated and their axes have a statistical preferred direction in space. For fixed λ the ratio d/λ is larger for light oscillating parallel to the major axis than for light oscillating perpendicular to it. So you have different extinctions for the two directions of oscillation.

The alignment of the particles is done by the galactic magnetic field[1] in which the para- or diamagnetic dust particles arrange themselves with the axis of their smallest moment of inertia on average preferably parallel.

Information about the interstellar polarization is obtained by measurements of stars at different Galactic longitudes and latitudes.

Reflection Nebulae
They are always found as diffuse Dust nebulae in the vicinity of early type (hot) stars. Thereby the dust particles reflect the light of these stars (e.g. *Pleiades Nebula*). The extent is about 1 pc, and the luminous stars are mostly B types. Radiation from the star has blown away a shell. Reflection nebulae are always sites of star formation.

12.3   Interstellar Gas

As already mentioned the ratio of gas to dust is 100:1. Hydrogen is the most important element, accounting for 70% of the mass and 90% of the particles. Interstellar hydrogen can occur in two forms:

- Neutral hydrogen, H-I,
- ionized hydrogen, H-II.

12.3.1 Neutral Hydrogen

Because of the low temperatures of the interstellar gas, almost all hydrogen atoms are in the ground state according to the Boltzmann formula, and if anything, these Lyman series lines absorb in the UV ($\lambda < 122$ nm). Further interstellar absorption lines such as Na, Ca are observed .

In 1951, the 21-cm line was discovered by *(van de Hulst)*. Consider a neutral Hydrogen atom; the spins of electron and proton have two settings: (a) F = 1, both are parallel, (b) F = 0, both are antiparallel. At the transition from F = 1 to F = 0, radiation of the frequency or wavelength

$$\nu = \Delta E / h = 1420 \, \text{MHz} \tag{12.6}$$

$$\lambda = c / \nu = 21 \, \text{cm} \tag{12.7}$$

is released. The transition is in Fig. 12.4 illustrated.

---

[1] Strength approx. 10^{-10} T.

Fig. 12.4 The 21-cm line of neutral hydrogen arises from the transition from parallel nuclear and electron spin (left) to antiparallel nuclear and electron spin (right)

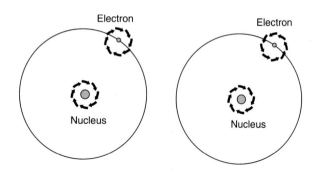

This transition is a *forbidden transition;* the *metastable level* F = 1 is long-lived without collision. A denotes the reciprocal lifetime of an excited state. In the forbidden transitions A is by 10^{-6}-fold smaller than in normal transitions. A is proportional to v^3 i.e. it is by the factor 10^{-17} smaller than for optical transitions. The lifetime of the excited state is $t = 1/A = 3.5 \times 10^{14}$ s. Under laboratory conditions during t many collisions occur, so that $F = 1$ depopulates by collision transitions. An equilibrium between the two hyperfine structure levels 1 and 0 is established according to the Boltzmann formula. The statistical weights are $g = 2F + 1$, and we find for the ratio $N_1/N_0 \approx 3$, since the exponential terms add up to 1. The density of all H atoms is $N_H \approx N_1 + N_0$ and therefore $N_1 = \frac{3}{4} N_H$.

Let us still calculate the emitted intensity, assuming the optically thin case here: $\tau_v = \kappa_v l \ll 1$:

$$I_v = \kappa_v \frac{2v^2 kT}{c^2} l, \tag{12.8}$$

since we use the Rayleigh-Jeans approximation for the Kirchhoff-Planck function. This follows from the Planck formula:

$$I_v = \frac{2hv^3}{c^2} \frac{1}{e^{hv/kT} - 1} \tag{12.9}$$

$$hv/kT \ll 1 \qquad e^x = 1 + x \text{ if } x \ll 1 \tag{12.10}$$

$$\frac{2hv^3}{c^2} \frac{1}{e^{hv/kT} - 1} \rightarrow \frac{2hv^3}{c^2} \frac{1}{1 + \frac{hv}{kT} - 1} \tag{12.11}$$

The absorption coefficient of the 21-cm line is obtained due to $N_0 = 1/4N_H$ to:

$$\kappa_v = \frac{1}{4\pi\epsilon_0} \frac{\pi e^2}{mc} f \frac{hv_0}{kT} \phi(v) \frac{1}{4} N_H \tag{12.12}$$

The stimulated emission can be taken into account by the factor $(1 - \exp(-hv_0/kT)) \approx hv_0/kT \ll 1$. Where an optically thick layer is reached, this leads to a maximum intensity value of :

$$I_\nu = B_\nu(T) = 2\nu^2 kT/c^2 \tag{12.13}$$

and one obtains here a temperature of 125 K. $\phi(\nu)$ is the function for the line profile and in this case only determined by the Doppler effect: In the volume unit there are $N(V_r)dV_r$ H atoms with radial velocities $[V_r, V_r + dV_r]$ and one finds:

$$\frac{N(V_r)dV_r}{N_H} = \phi(\nu)d\nu = \phi(\nu)\frac{\nu_0}{c}dV_r \tag{12.14}$$

The 21-cm line is broadened by the Doppler effect, the shock broadening is negligible (why?) and also the natural line width ($= A/2\pi$). The Doppler broadening is caused by the thermal motion of the radiating atoms with a velocity of about 1 km/s. The line is broadened by the Doppler effect. Furthermore the line is broadened by observing different cloud complexes in the line of sight. All stars and also the interstellar matter participate in the galactic rotation motion. Therefore, clouds that are at different distances from the galactic center also have different rotational velocities.

Optical depth $\tau = 1$ is not reached with the interstellar matter, since applies:

$$\tau = n_H \sigma \, \Delta l \tag{12.15}$$

$\sigma = 10^{-18}\,\mathrm{cm}^2$ is the absorption cross section per atom for the 21-cm radiation, and Δl is the path length along the line of sight. The number of atoms per cm^3 is $n_H = 1$. Without the broadening, already at a distance of less than 1 kpc the value $\tau = 1$ and the gas would be opaque for this radiation.

Given a purely thermal line broadening, one can immediately determine the kinetic temperature of the cloud:

$$\frac{mv^2}{2} = \frac{3}{2}kT \tag{12.16}$$

and from:

$$\frac{\Delta\nu}{\nu} = \frac{v}{c} \tag{12.17}$$

one obtains at a $\Delta\nu = 6 \times 10^3\,\mathrm{Hz}$ a kinetic temperature of about 20 K. If H-I clouds are located in front of strong radio sources, one can sometimes observe the 21-cm line in absorption.

The observation of the 21-cm line gives the distribution of neutral hydrogen in interstellar matter.

12.3.2 Emission Nebulae, H-II Regions

As we have seen Hydrogen is largely neutral and detectable only with the 21-cm line in the radio region (H-I region). Particularly spectacular, however, are the luminous H-II regions, also known as the *emission nebulae*. Famous examples are the Orion Nebula (Fig. 12.5), the North American Nebula, etc. The radiation from these nebulae does not come from their own energy sources, but is produced by the stars embedded in them.

The radiation of an emission nebula consists of:

- Continuum,
- strong emission lines,
- nebular lines.

The continuum arises from free-bound transitions; recombination of ionized hydrogen and helium occurs. Free-free radiation from free electrons decelerated by Coulomb fields of protons is also important.

The emission lines are again recombinations of hydrogen atoms and helium atoms, but here the electron does not end up in the ground state, but at a level n. Then there are cascade-like transitions from $n \rightarrow n - 1 \rightarrow n - 2 \rightarrow n - 3 \dots$ accompanied by emission of a photon each time. The Balmer lines are formed during the transitions to the first excited level. It is important to take into account the low density of interstellar matter. The atoms are hardly disturbed by neighboring particles during the transitions, and so energy transitions with high n are possible, which could never occur under laboratory conditions. Example: Transition in the hydrogen atom from $n = 110 \rightarrow n = 109$ corresponds to a wavelength in the radio range at 6 cm.

Nebular lines: Long time their assignment was uncertain, and it was thought that in the emission nebulae there are elements unknown on Earth (e.g. called nebulium). However, they are forbidden transitions of ions such as

O^+, O^{++}, N^+, and others.

These ions have levels of low excitation energies (1 eV), which are metastable. Excitation to these levels occurs by inelastic electron collisions. Due to the low density of interstellar matter, the lifetime of the metastable levels is shorter than the time between two collisions, and therefore a radiative transition occurs.

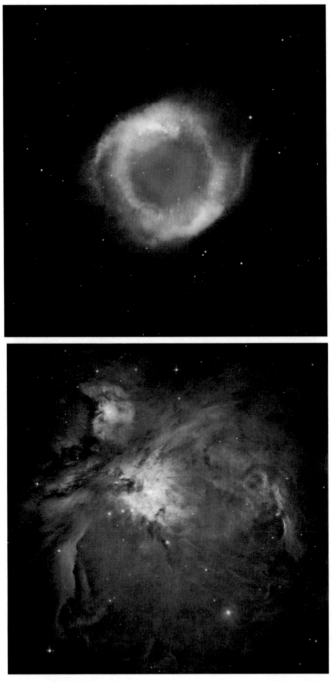

Fig. 12.5 Bottom: Orion Nebula, M42, dist. 1500 Ly. Composite on Spitzer telescope and HST image; top: Helix Nebula, distance 650 Ly, Spitzer telescope. (Source: NASA/JPL-Caltech/T. Megeath [University of Toledo] & M. Robberto (STScI)/ Credit: NASA/JPL-Caltech/K. Su [Univ. of Arizona])

The intensity follows from

$$I = \text{const} \int n_e^2 ds, \tag{12.18}$$

where s is the line of sight and $n_i n_e = n_e^2$. The emission measure is then:

$$EM = \int n_e^2 ds \tag{12.19}$$

One gives the length of the line of sight in pc, the number of particles per cubic centimeter. Then one finds values for EM from $10 \, \text{cm}^{-6}$ for just visible objects up to $10^9 \, \text{cm}^{-6}$.

One can easily estimate the extent of an H-II region. To ionize the neutral hydrogen, one needs $13.595 \, \text{eV}$ or a radiation with $\lambda < 91.2 \, \text{nm}$. Therefore the temperature of the stars, which should excite the H-II region to glow, must be sufficiently high ($T_{\text{eff}} > 2 \times 10^4 \, \text{K}$), i.e. spectral type B1 or earlier. We will therefore find H-II regions only in the vicinity of stars of very early types. The number of N_{Lc} quanta[2] determines the visible extent of the nebula, where here the quanta by $\lambda < 91.2 \, \text{nm}$ are meant, in the Lyman continuum. s_0 is the distance from a star up to which a hydrogen gas of density n_H can be kept ionized. This is counteracted by the recombinations, and αn_e is the number of recombinations per unit time with α as recombination coefficients. One finds ($n_e \approx n_H$):

$$N_{\text{Rek}} = \frac{4\pi}{3} s_0^3 n_H \alpha n_e \tag{12.20}$$

The equilibrium condition is: $N_{\text{Lc}} = N_{\text{Rek}}$, and we thus get the Radius of the *Strömgren sphere* (Table 12.1):

$$s_0 = \left(\frac{3 N_{Lc}}{4\pi \alpha N_H^2} \right)^{1/3} \tag{12.21}$$

This thus gives how far the extension of an H-II region can be. If we relate this to $n_H = 1/\text{cm}^3$ then we have:

Each L_{Lc}-photon must give up energy $13.6 \, \text{eV}$ to lead to ionization. The rest goes into electron kinetic energy, and this is on average more than was previously removed from the electron gas during recombination. So we have a heating mechanism. The most important cooling processes are then the radiation processes at the forbidden nebular lines. Their excitation energy comes from the kinetic energy of the free electrons during the elastic collisions. Thus, an equilibrium occurs when the kinetic temperature of the gas is about $10^4 \, \text{K}$. The H-II regions are 100 times hotter than the H-I regions.

---

[2] The subscript LC stands for Lyman Alpha Continuum.

Table 12.1 Radius of the Strömgren sphere for $n_H = 1/cm^3$ as a function of spectral type

Spectral type	B1 V	B2 I	O9 V	O 9 I	O5 I
R_S/pc	5	15	50	100	150

Expanding shock fronts form at the boundaries of the H-II regions. This limits their visibility duration. Further expansion leads to a decrease in density, and the visibility duration is less than 10^6 years.

12.3.3 Special Emission Nebulae

To explain the nebulae discussed so far, one only needs high-energy photons (from a hot star) in a previously neutral interstellar gas.

With the *planetary nebulae* as well as in the envelopes around novae one has up to 5 times the temperature compared to the normal emission nebulae, and also the forbidden lines appear more strongly. Due to the expansion of the envelope, planetary nebulae can exist for about 10^4 years. However, nova envelopes can only be for seen about 100 years because they expand more rapidly.

The emission nebulae around *Wolf-Rayet stars* are formed by strong stellar winds accompanied by relatively large mass loss. The envelopes of these stars have temperatures between 10,000 K and 30,000 K.

Supernova explosions also result in the repulsion of an envelope, which expands at velocities between 5000 and 10,000 km/s. This results in the formation of a shock front because these velocities are above the local speed of sound. At the shock front the temperature is 10^6 K and one finds thermal Bremsstrahlung observed in the X-ray region. Synchrotron radiation observed in the optical and radio domains originates from accelerated charges. Magnetic fields are compressed and strong turbulence occurs. The lifetime of such structures is 10^5 years, and the diameter can be up to 50 pc.

Also of interest are *molecular clouds*. Normally molecules are decomposed by the interstellar radiation field (photo dissociation) resp. also by energy-rich cosmic radiation. So molecules can only form in areas protected from this radiation. This is possible in areas with strong dust absorption. Thus e.g. H_2, CO, ... is formed under protection by dust.

Compared to an atom, a molecule has two more degrees of freedom: 1. rotation around the center of gravity, 2. counter-oscillation of the atomic nuclei. But also these degrees of freedom are quantized; at transitions between two rotation- or oscillation-levels one has small energy-differences, therefore radiation occurs in centimeter- and decimeter-radio-range. The particle density varies between 10^2 and $10^6/cm^3$. Clouds containing methyl alcohol or formic acid have also been discovered. Because of their relatively high density, molecular clouds become gravitationally unstable, making them ideal sites of Star formation. In extreme cases temperatures of up to 10^{10} K have been measured. This can be explained by a pumping mechanism, the particles arrive at a metastable level which is strongly overpopulated. The transition to the ground state then occurs by induced emission, i.e. the incident radiation is extremely amplified. This *Maser effect* is mainly found in the molecules OH, H_2O, SiO. A similar process also occurs in variable stars (Mira stars). Such maser sources are very compact and only a few AU in size → Evolution to a protostar.

12.3.4 Light Echoes

The star V838 Mon V838 is a variable star, but very irregular. In 2002 an outburst occurred, the luminosity increased to $6\times 10^5 \, L_\odot$. Light echoes were observed, the light being reflected from dust envelopes further and further out in time (Fig. 11.14). These were not previously visible, giving the impression that the star is exploding as a whole. The time that elapses between the eruption and the observation of the echo provides information about the size of the illuminated dust cloud. The distance of the star from the earth is 10 kpc.

The whole thing is reminiscent of a nova, however it was found that the star was only expanding and had not ejected a shell.

12.4 Cosmic Rays

12.4.1 Discovery

During a Balloon flight *V. Hess* noted in 1912, that the electrical conductivity of the atmosphere increases with altitude. He was only able to prove this by an ionizing explain cosmic *rays*. This refers to atomic nuclei, electrons and positrons, which propagate at very high speeds of up to 0.9 c. Helium and other heavier atomic nuclei represent only 9% of all cosmic ray particles.only 2% are electrons.

12.4.2 Composition and Origin

It is interesting to note that there is a greater abundance of the light elements Li, Be and B than in stars. These are formed by the breakup, *spallation*, of C, N and O nuclei in collision with protons in interstellar space. The total energy we receive from cosmic rays on Earth is comparable to the total energy of starlight incident on us. The big question, however, is where do these particles come from? Light propagates in a straight line, so it is easy to trace where the light particles come from. The particles of cosmic rays are charged, so they are deflected by magnetic fields (Earth's magnetic field, heliosphere, galactic magnetic field), they fall in on a spiral path. Because of the relatively strong galactic magnetic field, it is certain that most particles cannot come from outside the galaxy. Only particles with very high energy could come from outside our galaxy. Possible sources for these particles are supernova explosions.

Before the radiation hits the Earth's atmosphere, we speak of the *primary component* of cosmic rays. The primary particles of cosmic rays produce secondary particle showers in the Earth's atmosphere.

12.4.3 Magnetic Fields and Charged Particles

Let us briefly consider the deflection of cosmic rays in a homogeneous magnetic field of flux density **B**. A particle of charge e, the rest mass m_0 and velocity v perpendicular to **B** moves moves on a circle with *Cyclotron frequency:*

$$\omega_c = \frac{e\mathbf{B}}{m} = \frac{1}{\gamma}\frac{e\mathbf{B}}{m_0} \tag{12.22}$$

where by it follows from the theory of relativity that the moving mass m with rest mass m_0 is related as follows:

$$m = \gamma m_0 = \frac{m_0}{\sqrt{1 - \left(\frac{v}{c}\right)^2}} \qquad p = mv \tag{12.23}$$

$$E = mc^2 \tag{12.24}$$

This follows from equating the centrifugal force mv^2/r_c with the Lorentz force evB. For the *Gyration radius r_c* get we:

$$r_c \approx \frac{E}{ceB} \tag{12.25}$$

or if one calculates E in GeV and r_c in pc:

$$r_c \approx 1.08 \times 10^{-16}\frac{E}{B} \tag{12.26}$$

(Thereby B given in T). The Earth's magnetic field causes, that relatively low-energy charged particles reach the Earth's surface only in certain zones around the geomagnetic poles. However, this so-called broad effect disappears for high energies (>10 GeV).

12.4.4 Solar Activity and Cosmic Rays

Upon entering the solar system, cosmic rays interact with the interplanetary magnetic field controlled by the Sun *(heliosphere)*. Therefore, the intensity of cosmic rays changes with two periods:

- 27-day cycle because of synodic solar rotation,
- 11-year cycle because of solar activity.

Cosmogenic isotopes such as ^{14}C, produced by cosmic rays are therefore anti-correlated with solar activity. When the solar activity is high, the influence of the interplanetary magnetic field is large, and few particles can enter the heliosphere from outside the solar system.

Particles coming from outside the solar system are called GCR Galactic*(galactic cosmic rays)*; neutral particles that also come from outside the solar system but are ionized by solar radiation are called ACR*(anomalous cosmic rays)*. Finally, there is the SEP *(solar energetic particles)*.

12.5 Further Literature

We give a small selection of recommended further literature.
Physics of the Interstellar and Intergalactic medium, B. T. Draine, Princeton Univ. Press, 2011
The Orion Nebula, C.R. O'Dell, Belnap Press, 2003
Astrophysics of Gaseous Nebulae and Active Galactic Nuclei, D.E. Osterbrock, G.J. Ferland, Univ. Science, 2005
Atomic Processes in Gaseous Nebulae, A. Prozesky, Lap Lambert, 2012.

Tasks

12.1 A hypothetical example: The radius of our Milky Way is 50,000 Ly. Assuming (which is false) that the interstellar matter is distributed spherically in the Galaxy, what would be its total mass?

Solution
1 Ly $= 10^{13}$ km $= 10^{18}$ cm. The volume of the galaxy is: $\frac{4\pi}{3}r^3 \approx 600 \times 10^{66}$ cm^3. This multiplied by the above mean density gives a mass of 3×10^{11} $M_\odot$ (solar mass $= 2 \times 10^{33}$ g).

12.2 The largest radio telescope that can be steered in all directions is located in Effelsberg and has a diameter of $D = 100$ m. Calculate the resolving power when observing the 21-cm line.

Solution
The resolving power is $\Theta = 206,265\lambda/D \; [''] = 206,265 \times 21/10,000[''] \approx 7'$.

The Galaxy

13

The word *Galaxy* comes from the Greek and means Milky Way. For the ancient Greeks the Milky Way consisted of namely milk poured out by the goddess Hera. If we write *Galaxy* we mean our Milky Way, *galaxy* means any giant star system.

It has been known for about 100 years that there are extragalactic systems outside our Milky Way. However, details can only be studied in the case of the Milky Way. On the other hand, global statements (e.g. about the shape of a galaxy) can be made more easily with extragalactic systems. That's why a comprehensive study of the Galaxy is important for research of extragalactic systems and vice versa (analogue to relation solarphysics (details) $\longleftrightarrow$ physics of stars).

In order to make statements about the structure of the Galaxy, it is first necessary to know the true spatial distribution of the stars. For this reason, in this chapter we first give a summary of the most important methods of distance determination.

13.1 Methods for Determining Distances

Information about the dimension of the Milky Way can be obtained if the distances of the objects are known. We therefore discuss here the relevant distance determination methods.

13.1.1 Trigonometric Methods

Trigonometric Parallax
Let a be the mean Earth-Sun distance and r is the distance of a star, then

$$\sin \pi = \frac{a}{r} \tag{13.1}$$

© Springer-Verlag GmbH Germany, part of Springer Nature 2023
A. Hanslmeier, *Introduction to Astronomy and Astrophysics*,
https://doi.org/10.1007/978-3-662-64637-3_13

Fig. 13.1 Due to the motion
of the Sun at 20 km/s in the
direction of the constellation
Hercules, the base b becomes
longer

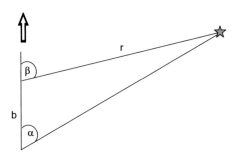

Thereby the parallactic angle π is the angle at which (as seen from the star) the radius of
the Earth's orbit appears. Due to the high precision of satellite measurements, parallaxes
can be determined with up to 0.001″ accuracy. The most accurate Earth-based parallax
measurement was made on a pulsar with a radio telescope (interferometer); the parallax
of this object was 0.4 milliarcseconds (0.0004″). This gives the distance r of this object to
$r = 1/\pi = 2400$ pc. The GAIA mission, launched in 2013, provides parallaxes to within
20 microarcseconds! For 20 million stars, one will obtain parallaxes with an error less than
1%!

Secular Parallax
Our Sun is moving at 20 km/s towards the constellation Hercules, and the distance travelled
b in the direction of Hercules becomes longer and longer with time (Fig. 13.1). The
distance results then from the sine theorem $r/\sin\alpha = b/\sin(180° - [\alpha + 180° - \beta])$:

$$r = \frac{b \sin \alpha}{\sin(\beta - \alpha)} \tag{13.2}$$

Stellar Cluster Parallax
The members of a star cluster move in the same direction to the vanishing point. If Θ is
the angle between the connecting line and the direction of motion (vanishing point), μ the
proper motion of the star (in ″ per year), v_r the radial velocity in km/s, then (Fig. 13.2):

$$v = v_r / \cos \Theta \qquad v_t = v \sin \Theta \tag{13.3}$$

The positional displacement of a star in the sky is called its proper motion proper *motion
proper motion*); this depends not only on the tangential velocity v_t also depends on the
distance of the star, and from this follows $v_t = \mu r = \mu/\pi$ and finally:

$$\pi'' = 4.74 \frac{\mu''}{v_r \tan \Theta} \tag{13.4}$$

The angle μ denotes the proper motion in the sky as seen by the observer (given by v_t and
distance r). The factor of 4.74 is obtained by dividing the Earth's orbital velocity by 2π.

Fig. 13.2 Method of stellar cluster parallax. An observer B sees a group of stars moving with velocity $\mathbf{v}$ moves in a direction with angle Θ moved on the sky

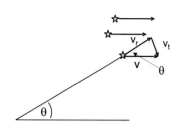

The method of stellar cluster parallax is shown in Fig. 13.2 explained. The proper motion μ is given in $''$ per year (a) to state.

Dynamic Parallax

For visual double stars, one can use this method if the major semi major axis of the orbit is in $''$ and the masses are is known. From the 3rd Kepler's law it follows:

$$\frac{a^3}{P^2} = M_1 + M_2 \tag{13.5}$$

and finally:

$$\pi = \frac{a''}{a[\text{AU}]} = \frac{a''}{[P^2(M_1 + M_2)]^{1/3}} \tag{13.6}$$

Whereby M in solar masses (hypothetically, one can start from $M = 2$), P the orbital period in years and a in AU.

Diameter Method

If the true diameter D is known (in AU) and the apparent diameter d'' then one has:

$$\pi'' = d''/D \tag{13.7}$$

Photometric Parallax

Here one must take into account the absorption coefficient γ of the interstellar matter; one assumes that the absolute brightness M of a star is known, m is the easily measured apparent magnitude:

$$m - M = 5 \log r - 5 = -5 \log \pi - 5 \tag{13.8}$$

Spectroscopic Parallax

This is to be taken from the spectrum. Problems with this are: Classification error, natural scattering within a given absolute magnitude, calibration of the HRD.

13.1.2 Photometric Standard Candles

Parallax from the Color-Brightness Diagram, CBD

Let us assume we have an HRD or a CBD of s star cluster, then by superimposing this diagram with a diagram in which the absolute values are known, one obtains:

→ the distance modulus m-M (from shift of the ordinate) resp.

→ the reddening (shift the abscissa).

Cepheids, RR-Lyr

Here we are dealing with pulsation-variable stars. The period of the brightness change can be easily determined. From the period-luminosity relation follows M (absolute magnitude) and from the measured apparent magnitude m the distance r ($m - M = 5 \lg r - 5$).

There are two classes of *Cepheids:*

- Classical Cepheids (δ-Cephei stars), Population-I Cepheids: young stars, close to the Galactic plane, in young clusters. Near the Galactic plane, extinction is strong, so observe in the near IR, e.g., K-band at $\lambda = 2.4\,\mu$m.
- W-Virginis stars, Population-II Cepheids: low-mass, low-metal stars, usually in the halo of the Galaxy, in globular clusters, or near the Galactic center.

Advantage: Cepheids are bright, therefore one has here a large range for distance determination and the method even works for Cepheids in extragalactic systems.

The RR-Lyr stars also belong to population II, they are metal-poor and can be found in the halo, in globular clusters and in the galactic bulge; the magnitude is between 0.5 and 1.0 absolute magnitude.

Novae, Supernovae

For Novae the following empirical relation exists:

$$M_{V,\max} = -9.96 - 2.31 \log m_r \tag{13.9}$$

This equation describes a relationship between the absolute maximum brightness and the rate of decrease (m_r) during the first two magnitudes. Thereby m_r is given in magnitude units per day. At maximum a nova can reach $M_V \approx -10$.

For supernovae, one assumes:

Type I:$-18.^M7$

Type II: $-16.^M5$

A supernova can become as bright as an entire galaxy. So this method has a long range!

Similar Total Brightness for Similar Objects

Here we assume, all globular clusters, H-II regions etc. have the same total brightness. It is of advantage not to select the brightest objects, because just these ones could be anomalously bright.

Interstellar Absorption Lines

If the interstellar gas is uniformly distributed, then the equivalent width of the lines of the Ca^+ and the distance r are related by:

$$r(\text{kpc}) = 3.00W(K) \tag{13.10}$$

($W(K)$ denotes the equivalent width of the K-line of the Ca^+) and for the D lines of the Na:

$$r(\text{kpc}) = 238(W(D_1) + W(D_2))/2 \tag{13.11}$$

with $W(D_i)$ as the equivalent width of the sodium D_i-line., $i = 1, 2$

Wilson-Bappu Effect

Let us examine in detail the profile of the H and K lines of ionized Ca. The profile show different parts (see Fig. 13.3):

- Broad absorption line, H_1, K_1; it arises from cool gas in the atmosphere of the star.
- In the middle of the profile an emission, H_2, K_2, caused by re-emission in the higher atmosphere.
- Central absorption, H_3, K_3, caused by cool gas.

Fig. 13.3 The width of the central absorption, W_3 ...superimposed on the emission, is useful in determining the absolute brightness of cool stars. . .

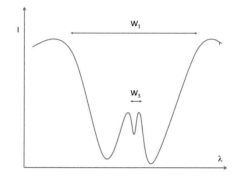

There is a correlation between the absolute brightness M_V of a star and the equivalent width of this central absorption, this is the Wilson-Bappu effect:

$$\frac{dM_V}{d \log W_3} = \text{const} \tag{13.12}$$

13.2   The Structure of Our Milky Way

What are the major components of the Galaxy, what information can we gain from stellar counts and their distribution? In addition, we discuss the closer neighborhood structure of the Sun in the Milky Way.

13.2.1 Rough Structure

The rough structure of the Galaxy can be given as follows:

- Disc, diameter about 50 kpc;
- central thickening, galactic bulge;
- Halo, approximately spherical distribution of stars and globular clusters around the disk.

The galactic center *(GC)* is orbited by the Sun, and its distance from the GC is, according to the official definition of the IAU (International Astronomical Union) (the true value is 8.1 kpc):

$$R_0 = 8.5 \, \text{kpc.} \tag{13.13}$$

Some important data of the components are given in Table 13.1.[1] In this table, the mass M given in $10^{10} \, M_\odot$, the luminosity (meaning the one actually observed) L_B in $10^{10} \, L_\odot$, the diameter in kpc, the scale height in kpc, the *Velocity dispersion* σ_z perpendicular to the galactic plane in km/s. The *Metallicity* [Fe/H] indicates logarithmically the fraction of Fe compared to that in the Sun: Fe $= -1$ means only 1/10 of the solar abundance of Fe. We can already clearly see the structure of the Galaxy from this table. The scale height is the characteristic thickness of the respective component.

---

[1] From P. Schneider, Extragalactic astronomy and cosmology, Springer.

Table 13.1 Important data for the components of the Galaxy. DM denotes Dark Matter

	Neutr. gas	Thin disk	Thick disk	Bulge	Stell. halo	DM halo
M	.0.5	6	0.2–0.4	1	0.15	
L_B	–	1.8	0.02	0.3	0.1	0
M/L_B	–	3		3	~1	
diam.	50	50	50	2	100	>200
Form	$e^{-h_z/z}$	$e^{-h_z/z}$	$e^{-h_z/z}$	Bar	$r^{-3.5}$	$\frac{1}{(a^2+r^2)}$
Scale h.	0.13	0.33	1.5	0.4	3	2.8
σ_z	7	20	40	120	100	–
[Fe/H]	>0.1	−0.5 to −0.3	−1.6 to −0.4	−1 to 1	−4.5 to −0.5	–

The Sun, and thus the solar system, orbits the center of the Galaxy at a distance of about 8 kpc.

13.2.2 Galactic Coordinates

The brightest part of the Galaxy (Fig. 13.4) and thus the center lies in the constellation of *Sagittarius* the anticenter lies in the constellation *Auriga* (Wagoner). Already *G. Galilei* recognized in the seventeenth century, that the bright clouds, viewed with a telescope, dissolve into many individual stars. 1781 *Messier* published a catalogue of non-stellar objects. In it one finds numerous gas nebulae, star clusters, etc., which belong to the Galaxy, but also extragalactic objects. The catalogue contains about 100 objects.

Our galaxy is a highly oblate system, and our Sun is located far from the center.

To better describe the distribution of objects in the Galaxy, galactic coordinates have been introduced (Fig. 13.5):

- Galactic equator; along it the galactic longitude *l* is measured eastward from the center,
- galactic latitude *b*, north and south of the equator. So the galactic north pole is located at $b = +90°$.

The galactic *north pole* (Fig. 13.5) has the coordinates in the equatorial system (RA, Decl mean right ascension and declination, epoch 2000).

Fig. 13.4 Section of the Milky Way with dark clouds (source: pa/imageBROKER/Markus Obländer)

Fig. 13.5 The galactic coordinates

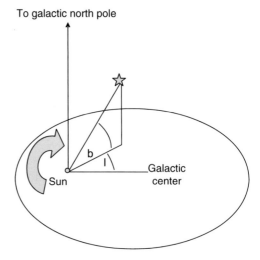

RA = 12h51m, Decl = +2°7,7′ (constellation Coma Berenices).

The galactic *Center* lies at

RA = 17h46m, Decl = −28°56′ (constellation Sagittarius).

Another coordinate system often used is cylindrical coordinates. R denotes the distance from the center, z the height above the disk, and Θ the angular distance of an object in the disk from the position of the Sun, viewed from the Galactic center. The distance of an object from the GC is then $\sqrt{R^2 + z^2}$. Since the matter distribution is approximately axially symmetric, one can apply to the density, for example:

$$\rho = \rho(R, z) \tag{13.14}$$

so there is no Θ-dependence.

13.2.3 Distribution of the Stars

The easiest way to obtain information about the shape of the galaxy is to examine the spatial distribution of the stars. If their distribution is uniform, then the directions in which we see more stars are equivalent to the directions in which the Galaxy continues to expand. Let ω be the solid angle, r is the distance; the solid angle takes an area A:

$$A = \omega r^2 \tag{13.15}$$

Now let's examine the volume between r and a distance dr further away:

$$dV = dA\,dr = r^2 dr\,d\omega \tag{13.16}$$

Let $n(r)$ be the stellar density (number of stars in this unit volume), $N(r)$ be the number of stars in the volume under consideration:

$$N(r) = n(r)V = n(r)r^2 dr\,d\omega \tag{13.17}$$

Let's assume that all the stars are equally bright, that is, they have the same absolute magnitude M. $r(m)$ be the distance of the stars with apparent magnitude m:

$$N(m) = \frac{4}{3}\pi r^3(m)n \tag{13.18}$$

Let's take only stars of absolute brightness M (if, for example, one chooses those with the same spectral type). Then one finds:

$$\log r = (m - M + 5)/5 = 0.2\,m + \text{const} \qquad r = 10^{0.2\,m + \text{const}} \tag{13.19}$$

and:

$$\log N(m) = 0.6\,m + C \qquad\qquad (13.20)$$

So this equation describes the number $N(m)$ of stars at a given magnitude as a function of their apparent magnitudes. The deviations from this law result from

- non-uniform distribution,
- interstellar Absorption.

> Absorption by galactic dust complicates the observation of extragalactic objects in the optical range. This occurs in the galactic latitude range $|b| < 10°$ known as *zone of avoidance*.

In the constellation Cassiopeia the absorption in the visible is very high, and therefore one was surprised to find here in 1994 with a 25 m radio telescope the dwarf galaxy Dwingeloo 1,[2] which is only 10 million light-years away from us. It also has a companion galaxy, Dwingeloo 2. However, these galaxies are not likely to belong to the Local Group, which includes the Milky Way.

An important quantity is also the *dynamic pressure*. The stars move randomly perpendicular to the disk, one therefore gets a finite thickness of a population, as in a thermal distribution. The dynamical pressure determines the *scale height* (cf. barometric altitude Formula):

→ the greater the dynamic pressure, the greater the velocity dispersion. σ_z the greater the scale height h.

Example
Solar neighborhood

- Stars younger than 3 Gyr[3]: $\sigma_z = 16$ km/s, $h \approx 250$ pc;
- Stars older than 6 Gyr: $\sigma_z = 25$ km/s, $h \approx 350$ pc.

---

[2] Named after a Dutch radio telescope.
[3] 1 Gyr = 1 gigayear, 10^9 Years.

13.2.4 Galaxy: Components

One finds the following components of the galaxy:

- *Galactic bulge:* central thickening; strong extinction up to 28^m in the visual! However, there are some windows, e.g. *Baade window* (1951 *Baade* found at galactic coordinates $l = 0.9°$, $b = -3.9°$ about the size of a full moon; Baade's window also contains the globular cluster NGC 6522), through which one can look into the galactic center (especially in the IR). The bulge has the shape of a bar, the main axis points $30°$ away from us, and the scale height is 400 pc. There is a relation between the surface brightness I and the distance from the center R, called *De-Vaucouleurs profile:*

$$\log\left(\frac{I(R)}{I_e}\right) = -3.33\left[\left(\frac{R}{R_e}\right)^{1/4} - 1\right] \tag{13.21}$$

$I(R)$ is an area brightness and is measured in $L_\odot/pc^2$, R_e the effective radius, half of the luminosity is emitted within R_e. Another form is:

$$I(R) = I_e\exp(-7.669[(R/R_e)^{1/4} - 1]) \tag{13.22}$$

and with the de Vaucouleurs profile, there is a relationship between the rms radius, luminosity and surface brightness:

$$L = \int_0^\infty dR 2\pi R I(R) = 7.215\pi I_e R_e^2 \tag{13.23}$$

The *Metallicity* is always calibrated at the sun.

Note: In astrophysics, all elements heavier than He are called metals.

Example
Metal-poor object has $[Fe/H] = -2 \rightarrow$ 2/100ths of the solar metallicity.
The metallicity of the stars in the bulge is between -1 and 1 (mean $= 0.2$), so there is also a young population, about $10^8 M_\odot$ neutral gas. The mass of the bulge is $10^{10} M_\odot$.

- *Center:* with central core, about 5 kpc across. Sun is about 8 kpc from center and is north of galactic plane.

- Flat disk: diameter 50 kpc, thickness 1 kpc. This is divided into:

    1. young thin disk: greatest amount of dust and gas; also currently star forming. Scale height 100 pc.

    2. Old thin disk: scale height about 325 pc.

    3. thick disk: scale height 1.5 kpc.

  Molecular Gas is important for star formation, the scale height is only 65 pc. Observations of gas in the Galaxy are mainly made by the 21-cm line of neutral hydrogen (H-I) as well as by emissions of CO. Dust can be observed by extinction and by discoloration and emits in the far IR (FIR). The IRAS and COBE satellites have been used to study the Galactic dust distribution. The temperatures are between 17 and 21 K. Gas and dust are concentrated to the galactic plane. The gas disk is bent *(warped)*, which may be due to tidal interaction with the gravitational field of the *Magellanic Clouds.*[4]

- *Halo:* The galaxy is surrounded by a halo containing mainly globular clusters (about 150 are known) and field stars. There are two populations of *globular clusters:*

    1. old globular clusters: [Fe/H] < -0.8, spherically distributed around GC;

    2. globular clusters with [Fe/H] > -0.8; flatter distribution, similar to thick disk.

  Most globular clusters are located at a distance $r < 35$ kpc from GC, but some at $r > 60$ kpc. They may have been captured by the Magellanic Clouds.

  The extent of this halo is about 50–70 kpc.

  One can also find neutral H- and high velocity clouds. These do not participate in galactic rotation. The *Magellanic Stream* is an H-I emission and follows the Magellanic Clouds as they orbit the Galaxy, and is due to tidal interaction between the Milky Way and the Magellanic Clouds.

- *Spiral arms:* Inside the disk are young objects arranged in spiral arms. The disk rotates around the center, with a rotational speed at the location of the Sun of 250 km/s and an orbital period of 200 million years. From this, the total mass of the system can be estimated to be about 2×10^{11} $M_\odot$. The Galaxy's total mass brightness is -20^{M}.

The concentration of different objects to the Galactic plane looks as follows:

- strong concentration: A stars, novae, T Tauri stars, galactic nebulae, planetary nebulae;
- medium concentration: F stars, Mira stars, dark clouds;
- weak concentration: G-, K-stars;
- no concentration: RR-Lyr stars, globular clusters.

There is a problem (bias) regarding the concentration in such a statistics: No or even a weak concentration show objects that a) are not concentrated on the galactic plane even in reality or b) are very faint and only observable near the Sun (Table 13.2).

---

[4] These are two dwarf galaxies gravitationally bound to the Milky Way.

Table 13.2 Number of stars up to a certain apparent magnitude m

m	$N(m)$	m	$N(m)$
6	3000	14	12×10^6
8	32,000	16	55×10^6
10	270,000	18	240×10^6
12	1.8×10^6	20	945×10^6

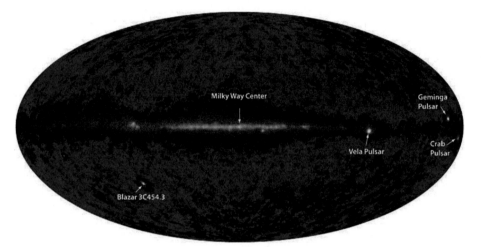

Fig. 13.6 Gamma-ray emission map of the sky. The Galactic plane is seen, as well as several objects, including the Crab Nebula and Geminga (source: NASA)

13.2.5 Local Solar Environment, Local Bubble

Within a 10-pc neighborhood of the Sun one finds: 254 stars; no supergiant, no giant, two subgiants, eight white Dwarfs, 244 main sequence stars. There are no O, B stars, and only four A stars.

The *Local Bubble (Local Bubble)* is a kind of empty space in the interstellar medium around the Sun. The diameter is 300 Ly, the density of the neutral hydrogen gas is 0.05 atoms. cm^{-3}. Thus, the density is only 1/10 of the average density of the interstellar medium in the Galaxy (0.5 atoms cm^{-3}). This void was created by the explosions of supernovae during the past $10-20$ million years. In the constellation Gemini there is a supernova remnant called Geminga (Fig. 13.6). The object was detected by the X-ray satellite ROSAT in 1991 and shows a periodic X-ray emission with a period of 0.237 s. So it is a relatively slowly rotating pulsar. It was even claimed to have found an Earth-like planet around this pulsar, orbiting at a distance of 3.3 AU with a period of 5.1 years.

More candidates for these supernovae could be exploded stars in the *Pleiades* star cluster. To visualize the distribution of stars, let's consider the following two models:

- Let's assume that our Sun is the size of a cherry. In this model the distance Sun-Pluto would be 60 meters. However, the distribution of stars would then be such that there would be a cherry in approximately every European capital.
- If the scale is 1:1 392,000,000, the solar diameter is about 1 mm. The average stellar distances in the galactic disk would then be around 30 km, in open clusters 2.5 km, in globular clusters 600 m, and in the center of the Galaxy 70 m. Thus we see that even with the very dense stellar concentration in the galactic center collisions between stars are very unlikely.

13.2.6 Stellar Statistics

In stellar statistics , one assumes the number of stars in a brightness interval: $m \pm 1/2$ relative to one square degree in the sky. If space is Euclidean and uniformly filled with stars, and if interstellar absorption can be neglected, then the number of stars is given by $\log(N(m)) = 0.6m + \text{const.}$

The basic equation of stellar statistics includes:

- Density function $D(r)$,
- Luminosity function $\phi(M, r)$ which gives the fraction of stars of absolute luminosity M in the interval $M, M + dM$,
- $A(r)$ is the interstellar absorption.

Then the equation is:

$$N(m) = \omega \int_0^\infty D(r)\phi(r, M)r^2 dr \tag{13.24}$$

The Absorption can be proportional to $\csc b$. If one looks from the Sun towards the galactic plane, the absorption is stronger than if one looks perpendicular to the galactic equator.

Seen from the outside, the following picture of the Galaxy results: The Galaxy is a spiral system; this follows from the distribution of interstellar matter and early stars. One knows with certainty three Spiral arms: Perseus arm, Orion arm (Sun), Sagittarius arm.

13.3 Star Populations and Density Waves

The distribution of young and old stars gives us an insight into the evolution of the Galaxy. One question is how the spiral structure comes about. The spiral structure is maintained by density waves. First, however, we deal with stellar populations. We can divide stars into different generations according to their ages.

13.3.1 Star Populations

In the principle we distinguish between two stellar populations:

- *Population II* (Table 13.4): This includes all stars that are further away from the galactic plane (Table 13.3).
- Population *I:* Stars in this population are concentrated towards the galactic plane, and are therefore also referred to as the *Disk population.* Here the metal abundance is significantly higher.

Population I stars are found in open clusters, O, B associations, and near interstellar gas and dust clouds (Table 13.3).

Going from the galactic halo to the spiral arms, the abundance of heavy elements increases. From this we can infer the origin of the Galaxy: it formed from a cloud consisting mainly of H and He, from which metal-poor Population II stars formed.

> Population II stars are old, metal-poor stars.

Table 13.3 Star populations: population I

	Extreme pop. I	Old pop. I
Type. objects	Interest. matter O, B, T Tauri, young open clusters class. Ceph.H-II regions	Sun A, Me Giants older open clusters
Age [10^9 a]	<0.1	0.1–10
Conc. to.		
Gal. center	None	Low
Galact. orbit	Circular	Nearly circular

Table 13.4 Stellar populations: population II

	Disc pop. II	Intermediate pop. II	Halo pop. II
Type. objects	Planet. nebula gal. Novae RR Lyr$_{P<0.4\,d}$	Long period. mutable.	Globular cluster RR Lyr$_{P>0.4\,d}$ Pop. II ceph
Age [10^9 a]	$\leq$10	$\approx$10	$\geq$10
Conz. to.			
Gal. center	Strong	Strong	Strong
Galact. Orbit	High e	Eccentr.	High e

Fig. 13.7 On the density wave theory

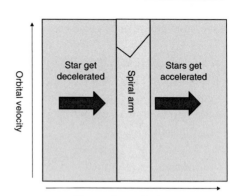

13.3.2 Density Waves, Spiral Structure

The *spiral arms* of the galaxy are caused by dynamic processes. If one had only Kepler orbits of the stars, then the arms would wind up after a short time. But since many spiral galaxies are observed, there must be a process whereby the arms are constantly being created.

In the *density wave theory*, it is assumed that spiral arms are formed by rotating density waves.

The galactic disk is originally unstable to disturbances. These disturbances, which propagate much like a sound wave, grow and attract matter along spiral paths, but they rotate only 1/2 times as fast as the disk. This means that matter passes through such a density wave and gas passing through the disk is brought and compressed in spiral structures (Fig. 13.7). This is why we find young stars of population I near the spiral arms. The density wave model goes back to *Lin* and *Shu* . A density wave rotating slower than matter causes potential minima where stars and interstellar matter reside. The spiral pattern thus moves through the matter and remains. The gas that concentrates there defines the spiral arms and is constantly consumed for star formation. An analogue to a density wave is a truck driving slowly in one lane on a highway. Behind him, a traffic jam forms, which however always consists of different cars.

Altogether, therefore, the following picture of the Galaxy is: A spherical halo with a diameter of about 100 kpc; there are old stars of population II and the *Globular clusters.* Their orbits have large eccentricities and are highly inclined, the orbital period around the galactic center is about 10^8 a. Globular clusters can be divided into two groups with respect to their orbits:

- random motions, without rotation around the galactic center,
- flatter distribution and rotation around the center at about half the rotational speed that would normally be the orbital velocity at that point.

In the halo area there are also the *Magellanic Clouds,* which are dwarf galaxies and companions to our galaxy, as well as gas. Near the galactic plane, the density of stars increases greatly. About half the mass of the galaxy is within the orbit of the Sun.

13.4 Rotation of the Galaxy

From the rotational behavior of the Galaxy, conclusions can be drawn about the mass distribution in and outside the Galaxy.

13.4.1 Radial and Tangential Motion

The velocity vector of a star can be decomposed into two components:

- *Radial component:* This is directed away from the observer at red shift and towards the observer at blue shift; measurable by the Doppler effect:

$$\frac{\Delta \lambda}{\lambda_0} = \frac{v_r}{c} \tag{13.25}$$

Thereby λ is the measured wavelength of a line and λ_0 its rest wavelength. When analyzing the observations, one must still take into account the motion of the Earth around the Sun, which occurs at 30 km/s. The accuracy of measurement is 10 m/s.
Tangential component: The motion of a star on the apparent celestial sphere is as *proper motion* μ observed. The proper motion of a star comes from the tangential component v_t. This, of course, depends on the distance of the star for a given velocity. If the stars are very far away, then they can be taken as reference stars. The largest measured proper motion has *Barnard's star* with $10''$/year. Other examples of stars with high proper motions are: *61 Cygni* ($5,22''$/year); *Arcturus* ($2,28''$/year).

Fig. 13.8 Distance
determination using a star
cluster as an example. Over
time, the angular extent of the
cluster as seen from us
decreases

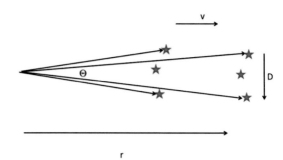

The velocity in space is given by:

$$\mathbf{v}^2 = v_r^2 + v_t^2 \qquad \tan\Theta = \frac{v_t}{v_r} \tag{13.26}$$

The method of star cluster parallax has been tested on the open star cluster of the *Hyades*
(contains about 40 stars). The radial velocity of the cluster is 39.1 km/s, and it moves to a
convergence point which is at $RA = 95.3°$, $Dekl = 7.2°$. One finds a $m - M = 3.23$.
This value can be used to calibrate the HRD. After all, we have shown earlier that the
HRDs of known objects and globular clusters can be made to coincide by a vertical
shift that is due to the distance. The open cluster *Hyades* was used for calibration.
With the HIPPARCHOS-satellite it was possible to obtain even more precise position
determinations and to improve the calibration. The distance of the Hyades is 44.3 pc. The
distance scale was extended to the star clusters *Praesepe* ($d = 159$ pc) and $h + \chi\,Persei$
($d = 2330$ pc).
A simple method of determining the distance of the Hyades is shown in Fig. 13.8. If we
assume that all the stars in the cluster are moving radially away from us, then over time
the angular extent of the cluster decreases. The distance r then follows from:

$$\Theta = D/r \tag{13.27}$$

$$\dot\Theta = -D\dot r/r^2 = -\dot r\Theta/r \qquad \dot r = v \qquad \rightarrow r = \ldots \tag{13.28}$$

The method works for the following reasons:

- Hyades are close enough,
- the members of the cluster are moving in unison, so $\ldots \dot D = 0$.

13.4.2 Galactic Rotation, LSR

Viewed from the Galactic North Pole, the rotation is clockwise.

In the cylindrical coordinate system R, Θ, z the velocity components are:

$$U = \frac{dR}{dt} \qquad V = R\frac{d\Theta}{dt} \qquad W = \frac{dz}{dt} \tag{13.29}$$

The galaxy rotates differentially thus not like a rigid body. One defines a fictitious system of rest LSR *(local standard of rest)*; this LSR Local Rest is valid for a) location of the sun, b) circular orbit of the sun. For the velocity components one then has:

$$U_{LSR} = 0 \qquad V_{LSR} = V_0 \qquad W_{LSR} = 0 \tag{13.30}$$

with V_0 as the orbital velocity at the location of the Sun. The peculiar velocity v of an object is its velocity relative to the LSR, and the components are:

$$\mathbf{v} = (u, v, w) = (U - U_{LSR}, V - V_{LSR}, W - W_{LSR}) = (U, V - V_0, W) \tag{13.31}$$

$\mathbf{v}_\odot$ is the motion of the sun relative to the LSR. For this vector one finds:

$$\mathbf{v}_\odot = (-10, 5, 7)\,\text{km/s} \tag{13.32}$$

i.e. there is inward and upward motion away from the galactic plane. The orbit of the Sun around the Galactic center is not exactly circular, and therefore one has a motion relative to the local rest system of 19.5 km/s in the direction of the constellation Hercules ($l = 56$ deg, $b = 23$ degrees). At the celestial sphere the sun moves to the *Apex;* this has already been determined in the year 1783 by *Herschel.* Stars in solar neighborhood move randomly; if the Sun were at rest in the LSR, then statistically all these velocities would add up to zero. Stars on a great circle 90 degrees from the apex resp. *antapex* have the largest proper motions, and thus one can fix these two points.

If we look at the velocity dispersion of stars, we find a correlation with metallicity or age. The oldest stars have the largest peculiar velocities. For the orbital velocity of the LSR we get the value:

$$V_0 = V(R_0) = 220\,\text{km/s} \tag{13.33}$$

From centrifugal acceleration = gravity acceleration

$$V^2/R = GM/R^2 \tag{13.34}$$

follows mass of galaxy within R: $8.8 \times 10^{10}\,M_\odot$. The orbital period of the sun is then :

$$P = \frac{2\pi R_0}{V_0} = 230 \times 10^6\,\text{a} \tag{13.35}$$

Fig. 13.9 Principle of
measuring galactic rotation
velocity

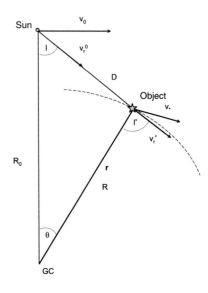

13.4.3 Galactic Rotation Curve

We want to answer the question: How does the orbital velocity (of a star) change as a
function of the distance from the center, $V(R)$? This is called the galactic rotation curve.
For simplicity, let us assume that the motion occurs in a disk. Let Θ the angle between the
sun and the object as seen from the GC (galactic center). Then

$$\mathbf{r} = R \begin{pmatrix} \sin \Theta \\ \cos \Theta \end{pmatrix} \qquad \mathbf{V} = \dot{\mathbf{r}} = V(R) \begin{pmatrix} \cos \Theta \\ -\sin \Theta \end{pmatrix} \tag{13.36}$$

From Fig. 13.9 also follows

$$\mathbf{r} = \begin{pmatrix} D \sin l \\ R_0 - D \cos l \end{pmatrix} \tag{13.37}$$

Thus, (sine theorem $D/\sin \Theta = R/\sin l$):

$$\sin \Theta = (D/R) \sin l$$

$$\cos \Theta = (R_0/R) - (D/R) \cos l$$

Suppose that $\mathbf{V}_\odot \approx \mathbf{V}_{\mathrm{LSR}} = (V_0, 0)$. From Fig. 13.9 one finds:

$$v_r = v_{r*} - v_{r\odot} = v_* \sin l_* - v_\odot \sin l$$

$$v_t = v_{t*} - v_{t\odot} = v_* \cos l_* - v_\odot \cos l$$

and

$$R \sin \Theta = D \sin l$$

$$R \cos \Theta + D \cos l = R_0$$

hence

$$v_r = R_0 \left(\frac{v_*}{R} - \frac{v_\odot}{R_0} \right) \sin l$$

$$= R_0 (\Omega - \Omega_0) \sin l$$

$$v_t = R_0 \left(\frac{v_*}{R} - \frac{v_\odot}{R_0} \right) \cos l - D \frac{v_*}{R}$$

$$= (\Omega - \Omega_0) R_0 \cos l - \Omega D$$

where the angular velocity

$$\Omega(R) = \frac{V(R)}{R} \tag{13.38}$$

is. The values with index 0 are for the Sun. The relative velocity is therefore:

$$\Delta V = \begin{pmatrix} R_0(\Omega - \Omega_0) - \Omega D \cos l \\ -D\Omega \sin l \end{pmatrix} \tag{13.39}$$

and the radial and tangential components of this motion are respectively:

$$v_r = \Delta V \begin{pmatrix} \sin l \\ -\cos l \end{pmatrix} = (\Omega - \Omega_0) R_0 \sin l \tag{13.40}$$

$$v_t = \Delta V \begin{pmatrix} \cos l \\ \sin l \end{pmatrix} = (\Omega - \Omega_0) R_0 \cos l - \Omega D \tag{13.41}$$

The distance of an object from the GC follows from:

$$R = \sqrt{R_0^2 + D^2 - 2R_0 D \cos l} \tag{13.42}$$

Now we still examine the rotation curve near Ω_0, i.e. for $D \ll R_0$. Then by the approach:

$$\Omega - \Omega_0 \sim \left(\frac{d\Omega}{dR}\right)_{R_0} (R - R_0) \tag{13.43}$$

$$v_r = (R - R_0) \left(\frac{d\Omega}{dR}\right)_{R_0} \sin l$$

However, since $\Omega(R) = V(R)/R$ is, it still follows:

$$R_0 \left(\frac{d\Omega}{dR}\right)_{R_0} = \frac{R_0}{R} \left[\left(\frac{DV}{dR}\right)_{R_0} - \frac{V}{R}\right] \sim \left(\frac{dV}{dR}\right)_{R_0} - \frac{V_0}{R_0}$$

and the components are:

$$v_r = \left[\left(\frac{dV}{dR}\right)_{R_0} - \frac{V_0}{R_0}\right] (R - R_0) \sin l$$

$$v_t = \left[\left(\frac{dV}{dR}\right)_{R_0} - \frac{V_0}{R_0}\right] (R - R_0) \cos l - \Omega_0 D$$

Since $|R - R_0| \ll R_0$, follows $R_0 - R \approx D \cos l$. Thus:

$$v_r \approx AD \sin 2l \qquad v_t = AD \cos 2l + BD \tag{13.44}$$

and the *Oort's constants* are:

$$A = -\frac{1}{2} \left[\left(\frac{dV}{dR}\right)_{R_0} - \frac{V_0}{R_0}\right] \tag{13.45}$$

$$B = -\frac{1}{2} \left[\left(\frac{dV}{dR}\right)_{R_0} + \frac{V_0}{R_0}\right] \tag{13.46}$$

Thus, the radial and tangential velocity fields relative to the Sun exhibit a sinusoid with period π. Here the components v_r, v_t are out of phase by $\frac{\pi}{4}$.

We thus obtain the double wave (Figs. 13.10 and 13.11), where the velocity components are shifted with respect to each other. Today one calculates with $A = 14$ and $B = -12$.

Fig. 13.10 Stars closer to the center of the Galaxy rotate faster than the Sun (∗) around the center; stars farther out rotate more slowly (**a**). The motion relative to the Sun is shown in ((**b**). In (**c**) and (**d**) are the radial components and the tangential components of the velocity vector, respectively

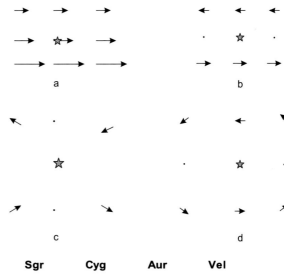

Fig. 13.11 Double wave of galactic rotation. Dashed line is radial velocity; solid line is v_t

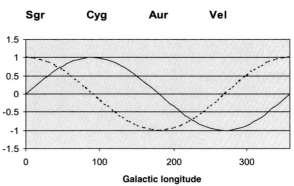

According to the recommendation of the International Astronomical Union (1985) the following applies $R_0 = 8.5$ kpc, $\Theta_0 = 220$ km/s.

We derive Oort's constants for a Keplerian orbit around a mass $M_G = 1.5 \times 10^{11} M_\odot$:

$$\omega^2 R = \frac{GM_G}{R^2} \tag{13.47}$$

$$\omega(R) = \sqrt{\frac{GM_G}{R^3}} \tag{13.48}$$

then:

$$d\omega/dR = -(3/2)(GM_G)^{1/2}R^{-5/2} = -3\omega/2R \tag{13.49}$$

and one finds:

$$A = (3/4)\omega_0 = 19\,\text{km/s kpc} \tag{13.50}$$

$$B = A - \omega_0 = -6.5\,\text{km/s kpc} \tag{13.51}$$

Here one has $R_0 = 8.5\,\text{kpc}$. The deviations from the values derived earlier are easily explained by the fact that our Galaxy is not a point mass.

The measurements show:

$$A = (14.8 \pm 0.8)\,\text{km s}^{-1}\,\text{kpc}^{-1} \tag{13.52}$$

$$B = (-12.4 \pm 0.6)\,\text{km s}^{-1}\,\text{kpc}^{-1} \tag{13.53}$$

Now we treat briefly the two cases:

- $R < R_0$ If one looks at the intensity profile of the 21-cm-line, one sees several maxima, one receives radiation from clouds of different distances to us or to the galactic center. For a cloud at the tangential point (Fig. 13.12) the largest radial velocity directed away from us is:

$$v_{r,\text{max}} = 2 A R_0 (1 - \sin l) \tag{13.54}$$

If one knows R_0 then it follows that the distance of the cloud to the galactic center (tangential point method).
- $R > R_0$ Here one needs objects whose distances can be determined directly (Cepheids, ...). From l, D then follows R. Each object with known D and v_r provides a measuring point for the galactic rotation curve. From this one finds that for $R > R_0$ the curve does not decrease outward (would be logical, since gas and star density decrease outward!).

Fig. 13.12 Tangential point method. The different gas clouds are assumed to move on circular orbits around the galactic center. A cloud at the tangential point then shows the largest radial velocity

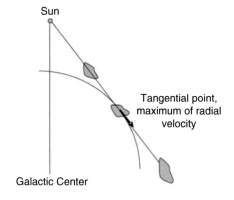

Sun

Tangential point, maximum of radial velocity

Galactic Center

Fig. 13.13 Distribution of
neutral hydrogen measured by
21-cm radio emission

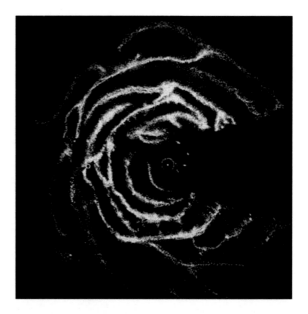

The Milky Way thus contains a further matter component, which dominates of the
visible mass for $R > R_0 \rightarrow$ Dark matter.

The spiral structure follows from the distribution of neutral hydrogen and CO emissions
(Fig. 13.13).

13.5   Dark Matter in the Milky Way

The existence of dark matter follows from the rotational velocity curve (Fig. 13.14). In a
Keplerian motion, one would expect: $V \approx R^{-1/2}$, i.e., the rotational velocity decreases
outward.[5] However, one observes that V(R) is constant for $R > R_0$. This is found for all
spiral galaxies.

---

[5] Cf: solar system; Mercury moves faster around the Sun than, say, Mars.

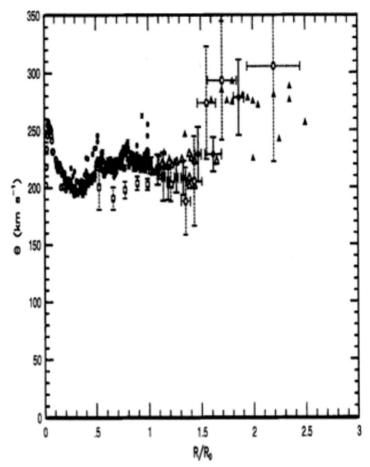

Fig. 13.14 Rotation curve of the Galaxy as a function of distance from the Galactic center. Open and filled circles mean determination from H-I tangential point method, crosses CO measurements, open squares planetary nebulae, filled squares H-II regions, etc. (adapted from Honma, Sofue, 1997, PASJ, 49, 453)

13.5.1 The Nature of Dark Matter

The main question is: why is Dark matter dark, i.e. it does not emit electromagnetic radiation. There are two completely different dark matter explanations.:

Astrophysical Explanation Compact objects, low-mass and therefore faint stars, brown dwarfs or final stages of stellar evolution such as black holes or white dwarfs. These objects are grouped together under the term MACHO: Massive Compact Halo Objects. For example M-dwarfs, would be visible only in the IR because of their low temperature. Measurements in the near IR, however, showed a too low density of such objects. White dwarfs (WD) have another problem: they are at the end of the evolution of low-mass stars, and the universe is too young (about 13.7 billion years) to have produced enough such objects that transformed from white dwarf to dark dwarfs. The explanation with neutron stars (mass $>1,4$ solar masses) or even black holes also has difficulties. Massive stars produce heavy elements during nuclear fusion, which are released into interstellar space during a supernova explosion. The galaxy thus becomes enriched in heavier elements over time $\rightarrow$ chemical evolution of the Galaxy. However, the observed amount of heavy elements suggests that the total number of supernova explosions is less than the number of black holes and neutron stars needed to explain dark matter. Brown dwarfs (mass $< 0,08\, M_\odot$) could contribute at least 10% of normal mass to dark matter.

Explanation from Particle Physics One assumes that dark matter is made up of unknown particles. Here one distinguishes between HDM, *Hot Dark matter* and *Cold Dark matter, CDM*. The two attributes *hot* resp. *cold* mean fast or slow motion of particles (temperature is a measure for speed of particles). For HDM, neutrinos with low mass would be most likely, but this would have serious consequences for the formation of the universe: Top-Down-Evolution, so first galaxy-clusters formed , then galaxies, etc. However, it is found that some galaxy clusters are only in the process of forming, so more a bottom-up scenario is favoured.

Unknown elementary particles that could be candidates for CDM are WIMPs *(weakly interacting massive particles)*. Candidates for these are photinos and neutralinos. These are particles predicted by supersymmetry (also called SUSY), i.e. for every Fermion (matter particle) there is a a Boson (Interacting particles). Candidates are Photinos or Neutralinos. Other particles postulated are Axions that have a very low mass.

13.5.2 Galactic Microlensing

As early as 1986 Paczynski proposed to find MACHOs by microlensing experiments. Light is deflected by massive particles in a gravitational field; a beam of light that is at a distance ξ at a point mass M, is deflected by the angle α:

$$\alpha = \frac{4GM}{c^2\xi} \tag{13.55}$$

(this equation applies only to weak gravitational fields).

If D_{ds} denotes the distance between the lens (in our case MACHO) and the source, D_s the distance from the observer to the source, and D_d the distance of the observer to the lens, then applies to the deflection:

$$\alpha = \frac{D_{\mathrm{ds}}}{D_s}\frac{4GM}{c^2 D_d} \tag{13.56}$$

Multiple images are possible, and there is also a magnification or gain effect.

Thus, if the halo consists of compact objects, a distant source should be "lensed" by one of these MACHOs from time to time, and whenever the MACHO moves between the observer and the source a gain in brightness of the lensed object should be observable. Stars of the Magellanic Clouds are ideal for this purpose. The light curves would have to be symmetric as well as achromatic (independent of wavelength, this is not the case for normal variable stars), and since the probability is small, only one microlensing event per source should be observed.

Results of surveys so far showed 20 events in the direction of Magellanic Clouds and about 1000 in the direction of the Bulge.

General:

Our Galaxy contains about $2-4\times10^{11}$ Solar masses of visible matter; dark matter could exceed this value by up to a factor of 5!

13.6   Galactic Center

The most important information about the Galactic center is obtained from radio and IR observations. Since the GC can be observed from "close up", it is an important prototype for the centers of extragalactic systems.

13.6.1 Definition of the Center

The definition seems difficult at first: center of gravity of the system, place around which all stars orbit etc.

Radio observations reveal a complex structure – central disk of H-I gas, about 100 pc to 1 kpc. From the rotational velocity, it follows a mass of $3 \times 10^7 \, M_\odot$ (within 2 kpc). There are radio filaments extending perpendicular to the Galactic disk, and supernova remnants. In the Baade window discussed earlier, microlensing surveys were performed in the optical domain.

In the central region there are also many globular clusters and gas nebulae. In the X-ray region one detects emission from hot gas as well as X-ray binaries.

The inner 8 pc contain the radio source Sgr A. This consists of several components:

- circumstellar molecular ring, torus at $2\,\text{pc} < R < 8\,\text{pc}$. The ring has a strikingly sharp inner edge. This sharp edge is explained by an energetic event in the galactic center within the last few 10^5 years.

- Sgr A East: non-thermal, synchrotron radiation. Probably supernova remnant, age between 100 and 5000 years.
- Sgr A West: thermal source, H-II region, spiral structure.
- Sgr A*: strong compact radio source, near the center of Sgr A West. Extension less 3 AU, shines in the millimeter and centimeter range. Other galaxies also often show a compact radio source at their center. Therefore it is assumed that Sgr A* is the Center of our Milky Way.

13.6.2 Central Star Cluster and Black Hole

In the K-band ($\approx 2\,\mu$m) one finds a star cluster concentrated on Sgr A*. Due to the high stellar density it can come to close star encounters here within 10^6 years $\rightarrow$ thermalization. However, the motion of these stars shows a deviation from thermalization and indicates a central mass concentration. From the studies of the kinematics of these stars (see Fig. 13.15), one concludes a point mass of approx. $(2.87 \pm 0.15) \times 10^6\,M_\odot$ concentrated to an area of less than 0.01 pc. This is a supermassive black hole, SMBH. *(supermassive black hole)*. However, this dominates the rotation curve only up to 2 pc!

Fig. 13.15 Orbits of stars near the Galactic center during a 15-year period. From the motion around the center, the mass of the supermassive black hole can be estimated (photo: Keck Telescope)

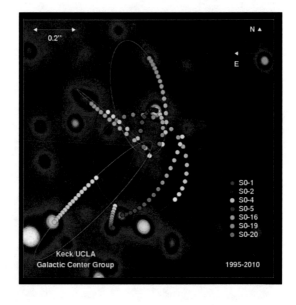

The dynamics of stellar motions near the galactic center suggest a supermassive black hole there.

13.7   Evolution of the Galaxy

First, we will again discuss the problem of the spiral structure of the Galaxy, then we will briefly discuss the age and magnetic field of the Galaxy.

13.7.1 Theories to the Origin of the Spiral-Arms

For the origin of the spiral arms of the galaxy, the *density-wave-theory* can be consulted.

Other considerations for the origin of the spiral structure go e.g. from *tidal interactions* of the Galaxy with neighboring dwarf galaxies (Magellanic Clouds, Fig. 13.16) or from the collision of two galaxies (as distance between galaxies within a galaxy cluster is only

Fig. 13.16 The Large Magellanic Cloud (LMC); together with the Small Magellanic Cloud, the two dwarf galaxies are companions of our Milky Way. First mentioned by *Al Sufi* around 964, then by *Magellan* during his circumnavigation of the world in 1519

Table 13.5 Companions of the Milky Way; d is the distance, D is the approximate diameter, year is the year of discovery

Designation	d [10^3 Ly]	Year	M_V	D[10^3 Ly]
Sagittarius Sgr	50	1994	−13.4	>10 (?)
Large Magell. Cloud (LMC)	160	–	−18.1	20
Small Magell. Cloud (SMC)	180	–	−16.2	15
Ursa Minor (UMi)	220	1954	−8.9	1
Sculptor (Scl)	260	1938	−11.1	1
Draco (Dra)	270	1954	−8.8	0.5
Sextans (Sex)	290	1990	−9.5	3
Carina (Car)	330	1977	−9.3	0.5
Fornax (For)	450	1938	−13.2	3
Leo II	670	1950	−9.6	0.5
Leo I	830	1950	−11.9	1

about 20-fold of their diameter, collisions between galaxies are relative frequent; e.g. our Milky-Way will collide within some 10^7 years, with a recently discovered Dwarf Galaxy).

Table 13.5 gives an overview of neighbouring dwarf galaxies.

13.7.2 Age of the Galaxy and Magnetic Field

The *age* of the galaxy can be estimated by the age of the oldest halo Objects in it. One finds values between 13 and 14 billion years, which corresponds approximately to the age of the universe. From the distribution of the objects of the halo one can estimate that the galaxy originated from a 100 kpc large cloud. The free fall time of such a cloud is 10^8a. Initially globular clusters formed, and the remaining matter formed a flat disk. As the density of the disk increased, there were repeated so-called *bursts of* star formation. The most massive stars did not survive long and enriched matter with heavier elements.

The *galactic magnetic field* follows mainly from the *Faraday rotation* (Rotation of the plane of polarization when linearly polarized radiation passes through a magnetic plasma). Furthermore, one can study the Zeeman effect at the 21-cm line. Particles of cosmic rays move on spiral paths around magnetic field lines, and where the field lines converge, the particles are accelerated (this happens, for example, in the Earth's magnetic field for charged particles from the Sun).

The Galactic magnetic field, reconstructed from more than 40,000 individual measurements is shown in Fig. 13.17. The magnetic field has different polarity above and below the galactic disk. The measurement was made by analyzing the polarization of radio radiation from distant galaxies. Faraday rotation twists the plane of polarization. The origin of the galactic magnetic field is assumed to be a *dynamo mechanism* as we already know it (compare solar dynamo or dynamo in the interior of the earth). The field lines run parallel

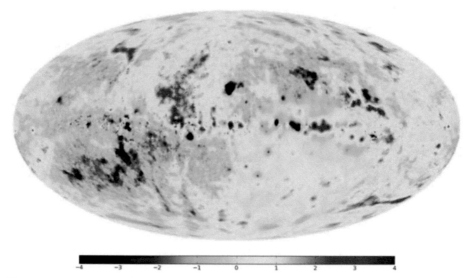

Fig. 13.17 The galactic magnetic field (MPIFA/MPG)

to the galactic plane around the GC. One also sees small-scale magnetic field structures, which are associated with turbulent gas-movements .

13.8   Further Literature

We give a small selection of recommended further reading.
Visual galaxy: The Ultimate Guide to the Milky Way and Byond, C. Hadfield, National Geographic, 2019
The Milky Way, B.J. Bok, P.F. Bok, Harvard Books on Astronomy, 1981

Tasks

13.1 Calculate the apparent brightness of a supernova at a distance of ten million light years.

Solution
Assume the absolute magnitude -18^M and use the formula f.d. distance modulus.

13.2 We estimatet estimate Milky Way Mass the mass of the Milky Way, M_G, from. The sun orbits around the center of the galaxy with a period of 240×10^6 years at a distance of 8.5 kpc from the galactic center.

Solution

Kepler's third law: $M_G + M_\odot = \frac{R^3}{P^2}$. Since 1 pc = 2.1 × 10^5 AU: $M_G + M_\odot \approx M_G =$ (1.8 × 10^9)^3/(2.4 × 10^8)^2 = (5.8 × 10^{27})/(5.8 × 10^{16}) ≈ 10^{11} $M_\odot$

13.3 Find the velocity of the sun around the galactic center.

Solution

The velocity is given by the relation: $v_\odot^2/R_\odot = GM_G/R_\odot^2$

13.4 What is the tangential velocity of a star that has a proper motion of 0.1″/year and is 100 pc away from us?

Solution

47.4 km/s (from formula 13.4).

13.5 Calculate the deflection of light for $M = M_\odot$, $R = R_\odot$.

Solution

For the solar limb one finds $\alpha = 1.74″$.

Extragalactic Systems

14

Galaxies are the largest building blocks of the universe. However, it has long been unknown whether they are independent systems of stars or clouds of gas within our Milky Way galaxy.[1] So they were also called spiral nebulae because very many have a spiral structure. It was not until 1924 that their true nature was recognized: At that time the astronomer *Hubble,* succeeded in resolving the nearest galaxies into individual stars and determining their distance on the basis of Cepheids (in M31, Andromeda Galaxy, Fig. 14.1).

14.1 Classification

There are several types of galaxies, which differ in appearance and content of metals. However, this is not an evolutionary sequence. We start this chapter with an overview of galaxy catalogues. In these catalogues one finds a lot of unprocessed data material, thus also material for new research.

14.1.1 Catalogues

Galaxies can be found in all catalogues that do not focus on special objects, e.g. in the Messier catalogue[2]. The objects in this catalog are relatively bright and therefore popular amongst amateur astronomers. M31 (Fig. 14.1) is the Andromeda Galaxy, M33 is the Triangulum Nebula, M51 is called the Whirlpool Galaxy, etc. The first part of the Messier

---

[1] Controversy between *Shapley* and *Curtis,* Great Debate, 1920.

[2] *Ch. Messier,* 1730–1817.

© Springer-Verlag GmbH Germany, part of Springer Nature 2023
A. Hanslmeier, *Introduction to Astronomy and Astrophysics,*
https://doi.org/10.1007/978-3-662-64637-3_14

Fig. 14.1 The Andromeda galaxy M 31 with the two companions M 32 and NGC 205. The distance to us is 2.4 million light years

catalogue was published in 1774, there are 110 objects in the extended version. In the NGC (New General Catalogue) one finds 8000 Objects. It was published by Dreyer in 1888. In the IC (Index Catalogue, extensions of the NGC, 1895 IC I and 1908 IC II) further galaxies are listed.

There are also catalogues specifically for galaxies such as the Shapley-Ames Catalogue or the Reference Catalogue of Bright Galaxies (de Vaucouleurs). The third edition, RC3, contains 23,000 galaxies. Up to the blue brightness of 15^m5 and a diameter $>1'$ and a red shift of less than 15,000 km/s the RC3 contains 11,900 galaxies and seems to be almost complete here. The Shapley-Ames catalogue was made in 1932 and contains 1249 objects up to magnitude 13^m2. In 1982 for this Catalogue the Data updated (Revised Shapley-Ames Catalog [RSA] of *Tammann* and *Sandage*).

PGC (Principal Galaxies Catalogue), published in 1989, contains more than 70,000 galaxies. Other modern surveys are the MGC (Millennium Galaxy Catalogue). The Sloan Digital Sky Survey (SDSS), (Fig. 14.2) consists of several parts, which are also still being extended: SDSS-I (2000−2005), SDSS-II (2005−2008), SDSS-III 2008−2014, and SDSS-IV 2014−2020, with more than a million galaxies recorded. From their spectra, one can infer the distance. The 2.5 m Apache Point telescope was used to create the inventory (Fig. 14.3). The project was funded primarily by the A.P. Sloan Foundation.

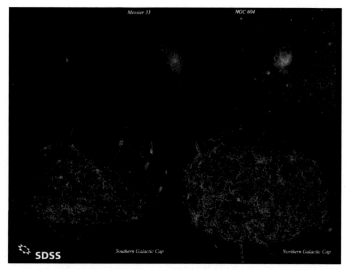

Fig. 14.2 The SLOAN digital sky survey

Fig. 14.3 The 2.5 m Sloan Foundation telescope. SLOAN Found

A 150 megapixel CCD camera was used for the survey, which covers 1.5 square degrees of sky at a time. Some projects specifically conducted during this survey are:

- SEGUE: Sloan Extension for Galactic Understand and Exploration. Here one can find spectra of more than 230,000 stars, their ages, radial velocities, etc. This should help to better understand the structure of the Milky Way.
- APOGEE: Apache Point Observatory Galactic Evolution Experiment. Similar to SEGUE, but also observations in the near infrared (1.6 μm). With this one can penetrate the intergalactic dust.
- BOSS: Baryon Oscillation Spectroscopic Survey Baryon. Spectra of luminous galaxies up to red shift 0.7 and 160,000 quasars at $z \sim 2.0$. As we show in the chapter on cosmology, red shift is a measure of distance. With these measurements, a picture of the large-scale structure of the universe emerges.
- MaNGa: Mapping Nearby Galaxies. High-resolution spectroscopy of nearby galaxies. One places an entire fiber optic bundle on a galaxy and then simultaneously obtains spectra at all locations in the fiber optic bundle. In Fig. 14.4 shows the principle. The individual optical fibers are represented by circles and several of them cover a Galaxy, in this example, 500 million light-years away from us. .

14.1.2 Hubble Classification

According to *E. Hubble*, galaxies are divided into three categories (Fig. 14.5):

- elliptical galaxies,
- spiral galaxies,
- irregular systems.

In the case of spiral galaxies, a further distinction is made between normal spirals (Fig. 14.6) and barred spirals (Fig. 14.7), in which the nucleus is bar-shaped. Hubble thought that this was an evolutionary sequence for galaxies, but this is wrong. In the *Hubble scheme*, the E galaxies are often referred to as early types and the irregulars as late types, but this is purely historical and has nothing to do with their evolution.

A very well known example of a spiral galaxy seen from above M51 seen spiral galaxy is the Whirlpool Galaxy, M51 (Fig. 14.6).

Galaxies are collections of up to several 100 billion stars and are the largest building blocks of the universe.

Fig. 14.4 SLOAN: project MaNGa. Multiple fiber optic bundles guide light from a galaxy to a spectrograph. The diameter of the glass fibers is 2 arcseconds. SLOAN Found

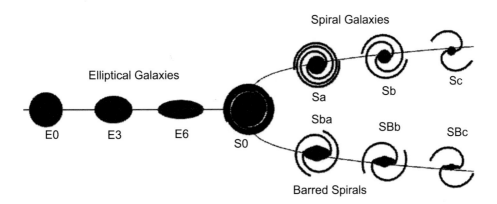

Fig. 14.5 The Hubble classification scheme of galaxies

Fig. 14.6 The so called whirlpool galaxy M51, distance about 30×10^6 Ly. An example of a spiral galaxy seen from above (Credit: NASA, ESA, S. Beckwith [STScI], and The Hubble Heritage Team STScI/AURA)

14.1.3 Active Galaxies

Besides these normal galaxies, however, there are other interesting types. In the case of normal galaxies, the radiation can be understood as a superposition of the spectra of their stars. Since the temperatures of the stars are between $\approx 10^3 \ldots \approx 10^4$ K, it follows that the resulting Planck curve has a relatively well-defined maximum, normal galaxies radiate in a narrow range of wavelengths $\rightarrow$ optical, near IR (NIR) (Fig. 14.8).

Fig. 14.7 NGC 1300, a barred spiral; distance about 18.8 Mpc. HST image

In the case of *active galaxies* a large fraction of the luminosity comes not from the stars (from thermonuclear fusion) but from gravitational energy released, by the infall of matter into an *supermassive black hole*. Therefore, such galaxies are very luminous in the radio and/or X-ray range. In many spectra of active galaxies one also finds very strong emission lines, and from their widths follow velocities of some 10^3 km/s. The galaxy M82 shows activity caused by a collision with another galaxy (Fig. 14.9).

Quasars Here the luminosity comes from a very compact region and can be up to $10^{13} L_\odot$ thus exceeds the brightness of the galaxy by a factor of 100.

Starburst Galaxies In normal galaxies, the rate of star formation is about 3 $M_\odot$/year. The rate of star formation in a galaxy can be determined from Balmer lines of H, which mostly consist of H-II regions around young hot stars. For *starburst galaxies* the *rate of star formation* is 100 $M_\odot$/year or even more. Many young stars radiate in the blue or UV, but there are also large dust and molecular clouds in these galaxies. These are excited to glow by the short-wavelength radiation of young, hot stars in the IR and submillimeter range → *ultra luminous infrared galaxies*, ULIRGs. Many of these ULIRGs interact with other galaxies.

Active galaxies are mostly galaxies with extremely bright nuclei.

Fig. 14.8 Galaxy NGC 4013,
a galaxy seen from the edge.
Distance: 55 million Ly
(Source: Hubble Space
Telescope)

14.1.4 Other Classifications of Galaxies

There is also the classifications from *De Vaucouleurs* using three criteria:

1. Spiral structure: *E*—elliptical, *S, Im*..., so similar to Hubble.
2. Bar: *A*—no bar, *B*—Bar, *AB*—Intermediate,
3. Ring: *s*—no ring (so-called s-shape), *r*—ring, *sr*—intermediate class.

The *Yerkes-Morgan scheme* looks like this:

1. Spectrum: *a*—A stars dominate, etc. to *k*, where K stars dominate.
2. Shape: *B* barred spiral, *D* rotationally symmetric, *E* elliptical, *Ep* elliptical with dust absorption, *I* irregular, *L* low surface brightness, *N* small bright core, *S* spiral.
3. Inclination: 1 galaxy seen from above, ... 9 galaxy seen from the edge.

Fig. 14.9 The galaxy M82, distance about 12 million Ly. Its shape was changed by interaction with the neighboring galaxy M81 about 500 million years ago. Composite of several images: X-ray data (CHANDRA, colored blue), IR (Spitzer telescope), and hydrogen emissions (HST, orange). (Source: NASA/JPL-Caltech/STScI/CXC/UofA/ESA/AURA/JHU)

Example   Andromeda galaxy, M31, has the type: kS5.

14.2   Discussion of the Individual Types

In this chapter we discuss the individual galaxy types in detail. Furthermore, we briefly discuss our two companion galaxies, the Magellanic Clouds, as well as population synthesis.

14.2.1   Elliptical Galaxies, E

These are roughly divided into:

- Normal ellipticals: gE—Giant ellipticals, E—normal luminosity, cE—compact ellipses. The absolute luminosity is: $-23 < M_B < -15$. The masses lie between 10^8 and

$10^{13}\,M_\odot$. The expansion is defined with the value D_{25}. This is the extent at which the brightness has decreased to $25^m/\mathrm{arcsec}^2$ (mostly in the blue, B). For these types $D_{25} = 1$–$200\,\mathrm{kpc}$. To the group of the normal ellipses one counts still the S0 (spindle galaxies).

- Dwarf ellipses: dE, cE; lower surface brightness, lower metallicity. Brightness values range from $-13 < M_B < -19$ and $1 < D_{25} < 10$.
- cD: extremely luminous, at the center of dense galaxy clusters (ex: M87). The values are $-22 < M_B < -25$; $300 < D_{25} < 1000$. They have a diffuse envelope, i.e. large M/L ratio.
- BCD, Blue compact dwarfs: blue color (B-V = 0, …, 0.3), contain relatively much gas. $-14 < M_B < -17$; $D_{25} < 3$.
- Dwarf spheroids: dSPH's *(dwarf spheroidals)* Low brightness, can only be observed at relatively short distances (within Local Group). $-8 < M_B < -15$; $10^7 < M_\odot < 10^8$; $0.1 < D_{25} < 0.5$.

From the brightness profile, one can tell if the halo region is important. If we plot the brightness in magnitudes per square arc second against the distance from the center (in $('')^{1/4}$), one finds a *de Vaucouleurs law* for normal E, i.e., a linear relation. For cD, for large r a deviation, a light excessis found $\rightarrow$ Halo luminous. cD is found in the center of massive galaxy clusters, so here the environment is important.

The elliptical galaxies are red (except BCD). There is gas and dust, but the contents is smaller than in spiral galaxies.

Signs for that : X-ray emission of hot gases ($10^7\,$K), Hα emission lines ($10^4\,$K) and cold gas ($10^2\,$K, H-I 21-cm emission, and CO molecular lines). In general: metallicity increases inward.

The dynamics of elliptical galaxies is important to explain why they are oblate. One determines from:

- Doppler shift of absorption lines $\rightarrow$ Rotational velocity v_{rot},
- Doppler broadening of the lines $\rightarrow$ velocity dispersion σ_v of the stars.

One finds that $v_{\mathrm{rot}} \ll \sigma_v$, i.e., rotation cannot explain the ellipses. How does this result in a stable elliptical distribution of stars without rotation?

Could it be that, as a result of close encounters between two stars, a *Thermalization* of orbits occurs? The *relaxation time* is:

$$t_{\mathrm{relax}} = t_{\mathrm{cross}}\frac{N}{\ln N} \tag{14.1}$$

t_{cross} is the characteristic time of a star to traverse the system $\approx 10^8\,$yr. If one sets the number of stars $N \approx 10^{12}$ then it follows that $t_{\mathrm{relax}} \gg \tau$, with τ is the age of the universe. The stars therefore behave like a collisionless gas. Elliptical galaxies are pressure-stabilized; their shape can be explained by an anisotropic stellar distribution in velocity space.

There is also evidence of complex evolution: boxiness (i.e., nearly rectangular shape), nuclei that rotate in opposite directions, shells, ripples. These shapes are due to interactions with other galaxies.

> Elliptical galaxies can be relatively compact and low-mass, but they can also be very massive. They are poor in interstellar matter.

14.2.2 Spiral Galaxies

The types Sa and SBa are also called early-type spirals, again it should be emphasized that this designation has nothing to do with evolution. If one goes from the early-type spirals to the late-type spirals, one finds the following trends:

- the ratio L_{bulge}/L_{disk} decreases,
- spiral arms open up,
- spiral arms become more structured, bright clumps = H-II regions.

In the brightness profile one has to distinguish between bulge and disk:

- bulge: de-Vaucouleurs profile,
- disk: exponential profile.

Freeman found that the central surface brightness μ_0 is almost the same for different galaxies. However, there are also LSBs, *Low Surface Brightness Galaxies*.

To determine the *rotation curves* (i.e. $v_{rot}(R_c)$) i.e. the rotation velocity as a function from the distance to the galaxy centre, the inclination of the disk must be taken into account i.e. the orientgation to the line of sight.. This inclination angle can be estimated from the observed axial ratio of the disk. One also measures the extent of the H-I disk, which is usually larger than that of the stellar disk. Similar to our Milky Way, one does not find a drop of the rotation curves towards the outside. From this follows again a *halo* of dark matter. One can use the following estimation:

$$v_{lum}^2 = GM_{lum}(R)/R \tag{14.2}$$

This is a rotation curve due to luminous matter M_{lum}. The mass of dark matter then simply follows from the difference between v_{lum} and v, the observed velocity.

$$M_{\text{dark}}(R) = \frac{R}{G}[v^2(R) - v_{\text{lum}}^2(R)] \tag{14.3}$$

The greater the luminosity, the steeper increases $v(R)$ in the central region, as well as the maximum rotational velocity v_{max}. That is, the mass increases with luminosity. For example, the Sa types have $v_{\text{max}} \approx 300$ km/s, Sc have $v_{\text{max}} \approx 175$ km/s and Ir only 70 km/s. It is interesting that galaxies with $v_{\text{rot}} < 100$ km/s do not have a spiral structure. So there exists possibly a minimum halo mass for the formation of the spirals? The spiral arms consist of young stars and H-II regions and are more conspicuous in the blue. As pointed out at the previous section, they are not made of matter, because otherwise they would unwind after a few rotations, but they are stationary *Density waves*. Density is increased by 10–20% compared to ambient, and this compresses molecular clouds and stimulates star formation.

Supernova remnants expand out of the disk, matter is transported into the halo, and a corona is observed in X-rays.

From 21-cm observations, in 1977 *Tully* and *Fisher* concluded: There is a relation between the maximum rotation speed and the luminosity of galaxies.

$$L \propto v_{\text{max}}^{\alpha} \tag{14.4}$$

And $\alpha \approx 4$. The longer the wavelength, the smaller the scattering, because at long wavelengths radiation less affected by dust absorption. If one knows the rotation speed, then the luminosity follows and by comparison with the apparent brightness the distance of a galaxy. Interesting here are measurements in the 21-cm line; the Doppler width of this line corresponds to $2v_{\text{max}}$.

For elliptical galaxies there is the *Faber-Jackson relation:* If σ_0 means the velocity dispersion at the center, then:

$$L \approx \sigma_0^4 \tag{14.5}$$

In Fig. 14.8 you can see a Spiral Galaxy edge on.

Density waves explain the maintenance of spiral structure in S galaxies.

14.2.3 Irregular Galaxies

About 3% of all galaxies (apparent observed distribution, not corrected for distance) belong to this type. They have neither an elliptical structure nor a spiral structure. One distinguishes: (a) Irr I: There is some structure, according to de-Vaucouleurs there are the subgroups Sm (hint of a spiral structure) and Im (no spiral structure). (b) Irr II: No structure. Another group are the dIrr, which are irregular dwarf galaxies.

The dIrr are probably very important for cosmology. The metallicity is very low, the gas content very high. They may be the *first galaxies* that were formed in the early universe.

The masses are between 10^8–10^{10} $M_\odot$ and the diameters between 1 and 10 kpc. The blue luminosities are between –13 and –20. The best known examples of an irregular galaxy are the *Magellanic Clouds*. However, these are now classified as SBm, meaning barred spirals of irregular shape.

14.2.4 Distribution Among the Types

It is important to distinguish between the observed distribution and the true distribution. In the observed distribution, bright galaxies are preferred, while in a distribution that takes into account, for example, all galaxies within a given volume of space, their brightness does not matter. In Table 14.1 the observed (i.e. apparent) distribution and the true distribution are given.

> The true distribution of the galaxies shows that at least half of all galaxies are of the Irr type.

14.2.5 Integral Properties and Diameters

The photometric properties can be summarized as follows: Elliptical galaxies are redder than S and these again redder than Irr. The S types are redder the more pronounced the nucleus. The E-types are dominated by population II and old population I. The S-types

Table 14.1 Comparison between observed and true distribution of galaxies. For the true distribution, all galaxies up to a certain distance are considered

	Spiral gal. (%)	Ellipt. galax. (%)	Irr. systems (%)
Observed distribution	77	20	3
True distribution	33	13	54

have old population I in the core and young population I in the spiral arms. If one knows the photometric diameter a_{rad} (angular diameter in rad) as well as the distance of the galaxy d, then the linear diameter is:

$$s = a_{rad}d \qquad (14.6)$$

The definition of the angular diameter of a galaxy comes from the definition of an isophote level above the sky background. As mentioned earlier, elliptical galaxies can become largest.

The *luminosity* of a galaxy follows from the radiation flux and its distance. However, the luminosity must be corrected for:

- Absorption in our Milky Way $\rightarrow$ a function of galactic latitude.
- Absorption within the galaxy itself. For E galaxies this correction is small.
- Correction for galaxy inclination i: A galaxy that is seen from above *(pole on)* appears brighter than one seen from the edge *(edge on)*. For the B-brightness there is the following correction:

$$A_B(i) = 0,7 \log \sec(i) \qquad (14.7)$$

The inclination results from

$$\cos^2 i = \frac{(b/a)^2 - \alpha}{1 - \alpha^2} \qquad (14.8)$$

Where b/a is the observed *axis ratio* and α the axial ratio for a galaxy of the same type seen from the edge.

- K-correlation: the universe is expanding, therefore galaxies are moving away from us, cosmological red shift occurs.

For the mass determination the *virial theorem* should be mentioned:

$$2 < E_{kin} >= - < E_{pot} > \qquad (14.9)$$

Let r_h be the radius of a galaxy within which half the mass is located, and $< v^2 >$ the mean square of the velocity of the proper motion of the stars, M is the total mass, then the virial theorem becomes:

$$< v^2 >= 0.4\,GM/r_h \qquad (14.10)$$

An important quantity in this context is the *Luminosity function* of Galaxies. This indicates how the members are distributed according to their luminosity. For example, let $\Phi(M)dM$

is the number density of galaxies with an absolute luminosity from the range $[M, M+dM]$. Then the total density is:

$$\rho_{ges} = \int_{-\infty}^{\infty} \Phi(M)dM \tag{14.11}$$

The global distribution of galaxies is given by a so-called. *Schechter luminosity function* given. There exists a luminosity L^* above which the distribution decays exponentially, and for small L there is a parameter α which describes the slope.

14.2.6 The Magellanic Clouds

One knows the *large magellanic cloud, LMC*, and the *small magellanic cloud, SMC*.

Due to their proximity to us, the LMC can be resolved into single stars and searched for e.g. Cepheids . This then gives their distance very simply. Another method will be briefly discussed. In 1987 a supernova (SN1987A) exploded in the LMC. The progenitor star had previously ejected matter by stellar winds. This now came to glow as a result of the SN explosion. Since one can well assume that the stellar winds were nearly isotropic, a circular ring should light up. Since it appears elliptical, one can infer the angle of inclination with respect to the line of sight. The ring was seen after a time R/c excited to glow, i.e. after the actual explosion. However, we do not observe the glow uniformly; the part of the ring closer to us glows earlier. We can therefore determine the true diameter of the ring from the light travel time delay and the angle of inclination, and the distance by comparing it with the easily measured apparent diameter. The value for SN1987A gives 51.8 kpc.

The diameter of the LMC is 7.7 kpc, and it contains about 10^{10} Stars. The SMC is about 64 kpc from us, the diameter is 3.1 kpc, and it contains 2×10^9 stars.

> The two Magellanic Clouds are connected to each other and to the Milky Way by thin hydrogen gas $\rightarrow$ Magellanic Current.

14.2.7 Population Synthesis

Aim   To understand the spectrum of galaxies as a superposition of stellar spectra.

The distribution of stars changes as their evolution on the main sequence depends on mass. The massive stars leave the main sequence after only a few million years.

> The spectral energy distribution of a galaxy indicates how star formation and evolution have taken place.

Let us assume that stars form from a molecular cloud. The mass distribution of the stars cannot be predicted theoretically; high-mass and low-mass stars form. The *Initial Mass Function, IMF* denotes the initial mass distribution during star formation. One often takes the Salpeter IMF, M is the mass of a star:

$$\Phi(M) \approx M^{-2.35} \tag{14.12}$$

This equation states: less massive than low-mass stars are formed. It is not clear whether this depends on

- Metallicity,
- mass of the galaxy.
- Good approximation for stars with $M > 1\ M_\odot$, flatter for low-mass stars.

The star formation rate, the mass of gas converted into stars per unit time is:

$$\Psi = -\frac{dM_{\text{gas}}}{dt} \tag{14.13}$$

The mass of the interstellar gas decreases as stars form. Since massive stars already after a few million years at the end of their evolution exploded to a supernova the metal content Z has increased over time. Due to stellar winds, ejection of matter (e.g. in planetary Nebulae, supernovae), the metal content continues to increase.

One can describe the total spectral luminosity of a galaxy at time t by a convolution of the star formation rate with the spectral energy distribution of a stellar population.

Isochrones are locations of the same time in the HRD.

- Early stage: spectrum and luminosity of a population are dominated by most massive stars, intense UV radiation; after $\approx 10^7$ years flux below 100 nm is strongly reduced, after $\approx 10^8$ years no longer present. Brightness in the NIR increases, massive stars have evolved to red giants.
- After several 10^7 years an edge appears at 400 nm: Opacity in stellar atmospheres changes (singly ionized Ca and Balmer discontinuity of H).
- $10^8 < 10^9$ Jahre Emission in the NIR remains strong, after 10^9 years Red Giant Stars (RGB) $\rightarrow$ NIR; after 3×10^9 years UV emission increases, as blue stars are on the horizontal branch after AGB phase, and there arealso white dwarfs.
- Between 4 and 13×10^9 years there is hardly any evolution of the spectrum.

14.3 Supermassive Black Holes

In the previous chapter it was shown that there is a Super massive Black Hole (SMBH) at the centre of the Milky Way. Evidence for such SMBHs in galaxies are mainly active galactic nuclei, AGN *(active galactic nuclei)*. Matter falls into the Black hole, releasing energy in the process. How do normal galaxies differ from AGNs?

14.3.1 Detection of SMBHs

If the escape velocity is $v_{esc} = \sqrt{2GM/r}$ equal to the speed of light, we get the Schwarzschild radius[3] r_s:

$$r_s = \frac{2GM}{c^2} \qquad (14.14)$$

The hypothetical Schwarzschild radius for our Sun is 3 km, for a SMBH in the galactic center $r_s \approx 10^7$ km. The galactic center is 8 kpc away from us, so the angular radius is $6 \times 10^{-6''}$. The resolving power of a radio telescope depends on the spacing of the antennas used to synthesize the signal. Through radio observations *(Very Long Baseline Interferometry, VLBI)*, one will arrive at an angular resolution of this order.

> The space occupied or influenced by a SMBH is very small, so direct detection is difficult.

Let us assume that there is a mass concentration of $M_\bullet$ at the center and the characteristic velocity dispersion is σ. We see an influence of a SMBH for distances smaller than

$$r_{BH} = \frac{GM_\bullet}{\sigma^2} \qquad (14.15)$$

on the kinematics of the stars and the gas. Let D the distance of the galaxy, then find:

$$\Theta_{BH} = \frac{r_{BH}}{D} \approx 0.1'' \left(\frac{M_\bullet}{10^6 M_\odot}\right) \left(\frac{\sigma}{100 \text{ km/s}}\right)^{-2} \left(\frac{D}{1 \text{ Mpc}}\right)^{-1} \qquad (14.16)$$

This is where the atmospheric perturbation-free observation with the Hubble Space Telescope comes into play. Within r_{BH} the velocity dispersion must therefore increase in the presence of an SMBH.

---

[3] *K. Schwarzschild,* 1916, Solution of Einstein's field equations for a point mass.

Example In the case of the galaxy NGC 4258, compact water maser sources have been measured using VLBI techniques. In the field of a point mass with $M_\bullet = 3.5 \times 10^7 \, M_\odot$ the deviations from a Kepler rotation are less than 1%. Instead of an SMBH, one could also imagine an ultra compact star cluster, but this would not be stable for long. From the existence of SMBHs in the Galaxy and in AGNs, the assumption of an SMBH follows as a very likely explanation.

14.3.2 SMBHs and Galaxy Properties

One has strong evidence for SMBHs in about 40 galaxies so far. Is there an influence of mass $M_\bullet$ on properties of the galaxy? One finds a correlation between $M_\bullet$ and (i) the absolute bulge brightness, (ii) the velocity dispersion:

$$M_\bullet \approx M_{\text{bulge}}^{0.9} \tag{14.17}$$

$$M_\bullet \approx 1.2 \times 10^8 M_\odot \left(\frac{\sigma_e}{200 \, \text{km/s}} \right)^{3.75} \tag{14.18}$$

As an example of a galaxy with one of the first discovered SMBHs, M87, which is the centre of the "*Virgo Cluster of Galaxies*". This elliptical galaxy is located about 20 Mpc away from us. The Hubble Space Telescope was used to study the central region of this galaxy. Evidence of a very rapidly rotating accretion disk was found. The measured velocities in the accretion disk fit very well with theoretical models of such disks around SMBHs. Further evidence for a central SMBH in M87 is an unusually high stellar concentration towards the center and gas clouds rotating at very high velocities. Furthermore, jets propagating at 0.99 c are observed. These jets are produced by the clustering of extremely hot matter outside the event horizon of the SMBH at the strong magnetic fields. It has also been observed with the CHANDRA X-ray satellite telescope that eruptions cause ripple-like propagating disturbances. Incidentally, the galaxy M87 also has a large number of globular clusters in the Halo area.

With the Hubble Space Telescope it could also be shown that the *Andromeda galaxy* M31 has an SMBH of $140 \times 10^6 \, M_\odot$. About 200 blue stars only 200 million years old were found in the vicinity of the SMBH within a narrow distance range of one light year. These orbit the SMBH in about 500 years at a speed of 2000 km/s.

14.4   Active Galaxies

In normal galaxies, the radiation is thermal. Characteristic for active galaxies are:

- non-thermal processes,
- synchrotron radiation, masers,
- thermal processes with extremely high energies.

In the spectrum, one usually observes a strong component of the emission. In the case of emission lines, the forbidden lines (indicated by square brackets, [ ]) should also be noted. For synchrotron radiation to occur, a magnetic field and relativistic electrons are needed. For the flux here applies:

$$F(v) = F_0 v^{-\alpha} \tag{14.19}$$

and from this:

$$\log F(v) = -\alpha \log v + \text{const} \tag{14.20}$$

In a log-log plot of the spectrum, the spectral index α shows up as an increase.

14.4.1 Active Galactic Nuclei

AGNs *(active galactic nuclei)* are characterized by:

- high luminosity ($> 10^{37}$ W),
- non-thermal emission lines,
- the flux is excessive in the UV, IR and radio regions compared to that of normal galaxies,
- a small region that is variable,
- large contrast between core brightness and outer brightness,
- jets; in images, they appear to explode,
- sometimes broad emission lines.

Seyfert Galaxies
In 1943 *C. K. Seyfert* found six galaxies which are of type S and show broad emission lines in the spectrum. Furthermore they have an unusually bright nucleus. Today almost 100 such objects are known. More than 90% of all Seyfert galaxies are of type S, the rest elliptical. Conversely, it can be said that about 1% of all spiral galaxies are Seyfert galaxies. It is assumed that all galaxies go through a Seyfert stage.

The following areas are distinguished: At the center is a small region that emits photons that strike two different gas regions farther out:

- A zone in which broad emission lines are formed; these are variable, with variabilities ranging from a few weeks to months. This results in an extension of these regions of about 1/10 light-year.
- A zone in which narrow lines arise that are thought to be 10^3-times more extended.

Since no forbidden lines appear in the first zone, the density must be $10^{13\ldots15}$ ions per cm^3 and thus the mass 30...50 $M_\odot$. One can divide the Seyfert galaxies into two types:

1. Type I: shows extreme Doppler broadening of the hydrogen lines (5000–10,000 km/s). The forbidden lines are only moderately broadened (200–400 km/s).
2. Type II: only narrow lines compared to type I, 200–400 km/s.

Some Seyfert galaxies have been found to change type within a few years. This could be caused by a dust cloud moving in front of the small area where the broad lines come from. Thus a type I Seyfert galaxy becomes a type II.

The total energy output of Seyfert galaxies is $10^{37\ldots38}$ W, brightness can vary during a few days to months. Many Seyfert galaxies are members of a binary system, and tidal effects therefore play a role in explaining their activity (Fig. 14.10).

BL Lacertae Objects

The following characteristic properties apply to this group of galaxies, which were initially mistakenly thought to be ordinary variable stars (hence the name, which is actually typical of variable stars):

- fast change of brightness in the radio, IR and visible range,
- no emission lines,
- non-thermal continuous radiation,
- strong and rapidly variable polarization,
- appear star-like,
- emissions vary rapidly and randomly (sometimes by a factor of up to 100).

Usually you have synchrotron radiation in active galaxies in the UV region. There is gas near the nucleus, which is ionized, and recombination gives rise to the emission lines.

About 40 BL Lac objects are known. Some of them show signs of a surrounding galaxy. Some are in clusters of galaxies. Since no spectral lines are seen, their distance determination is very difficult. In some cases, where the continuum is faint, emission lines shine through, and from the red shift their distance follows.

Fig. 14.10 M77, distance 60 million light years. A type II Seyfert galaxy

14.4.2 Radio Galaxies

Typically, galaxies are observed with a luminosity of 10^{33} W. The radio galaxies radiate much brighter in the radio range with up to 10^{37} W. Two groups are distinguished here: extended radio galaxies and compact Radio Galaxies.

- The extended radio galaxies often appear as double galaxies connected by arcs (distance in the range of one Mpc). The radio components are always outside the optically visible galaxy and much larger. The galaxy is usually located in the center of the radio sources.
- The compact radio galaxies are again very small (a few light-years) in terms of the extent of the radio source, which is localized in their cores.

We consider examples:

M87 (Fig. 14.11) has a radio source 1.5 light months across at its core. Protruding from the core is a jet that is about 6000 light-years long and has a luminosity of 10^{34} W. The entire galaxy as well as the jet emit X-rays. With the VLA and the Einstein satellite knots in the jet were detected. With the HST one has found an accretion disk with 65 Ly diameter oriented perpendicular to the jet. From the rotation of the matter in this disk one can infer the mass inside: $3 \times 10^9 \, M_\odot \rightarrow$ SMBH (previous section). Such jets are characteristic for many elliptical galaxies in their cores as well as almost all radio galaxies.

Fig. 14.11 M87 with matter
jet produced by bunching in the
strong magnetic field (HST
image and NASA sketch)

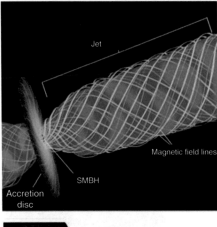

Cygnus A is one of the strongest radio sources in the sky. Here you can see the typical double structure of a radio galaxy. Two arcs are observed, from which, in the radio region, about 10^{38} W come, each of these arcs, "lobes", being about 17 kpc in diameter. The galaxy is located in the center and is of type E with a dust cloud in the center. An active core is observed. An overlay of images in the optical, X-ray, as well as radio regions is shown in Fig. 14.12 see.

Centaurus A: is the closest active galaxy to us, distance $d = 10 \times 10^6$ Ly and similar to Cyg A. Within this galaxy is another (Fig. 14.13). In Fig. 14.14 we see an overlay of the optical image with dark lines suggesting radio emission and an infrared image (ISO satellite), respectively. The IR data indicate the dust of a spiral galaxy, while Cen A itself is an elliptical galaxy is. The gravity of the elliptical galaxy is likely responsible for maintaining the spiral structure. Matter flows along the spiral arms into a central supermassive black hole, thus producing the radio emission of the elliptical galaxy.

Very Long Baseline Interferometry, VLBI
Since the resolution in the radio range is low due to the long wavelengths, attempts are being made to combine many radio telescopes into one (section on Astronomical Instruments). A new project in these efforts is the EVN (European VLBI Network), in which a total of 18 antennas are operated. One observes per year in four sessions, each of which lasts about three to five weeks.

Fig. 14.12 The radio galaxy Cygnus A, about 600 million light years from us. Here optical images (Hubble telescope) were overlaid with X-ray images (CHANDRA) and radio images (NSF, NRAO)

Fig. 14.13 Detailed view of Centaurus A, taken with the *wide field planetary camera* of the HST

Figure 14.15 shows a superposition of several images. In the background is a *Hubble Deep Field (HDF)*-image of distant galaxies, above a radio map of the Westerbrok radio telescope (contours). The HDF image was taken with the CFHT telescope. Because of the large baseline, these radio images from the VLBI (right) have three times the resolving power of the HST (Hubble Space Telescope). They were based on the 100-

Fig. 14.14 Centaurus A: a radio contour map is superimposed on the optical image

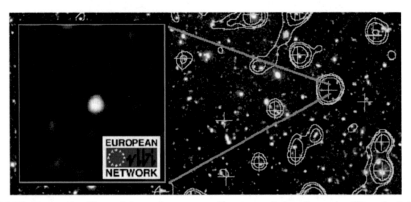

Fig. 14.15 Extremely distant faint radio sources. Contours of data from the Westerbork Radio Telescope are inscribed, which in turn are superimposed on an HDF (CFHT)

m radio telescope at Effelsberg, the 76-m Lovell telescope in the UK, the 70-m NASA DSN antenna near Madrid, and six other radio telescopes. The images show extremely small radio sources (diameter around 600 light years). The origin of the radio emission from these compact sources is interpreted by the assumption of a supermassive black hole (several 10^6 solar masses).

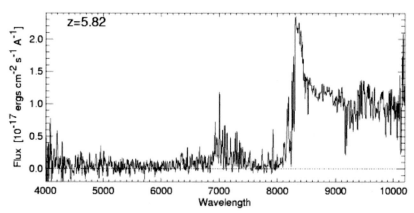

Fig. 14.16 The Lyman-α-emission line from a quasar with $z = 5.82$. On the left, the emission is superimposed by absorption (Lyman-α-Forest). Note: due to the high red shift the lines are in the near IR!

14.4.3 Quasars

By 1960 there were several catalogues of Radio sources, and attempts were made to identify these radio sources with stars or galaxies. *Th. A. Matthews* and *A. R. Sandage* found in 1960 a star-like object at the position of the radio source 3C48 (3rd Cambridge catalogue). It was referred to as *quasi-stellar object* or quasar for short, QSO (Quasi Stellar Object). The spectrum showed very broad emission lines, but they could not be identified at first. *M. Schmidt* identified then in 1963 the emission lines of the object 3C273. He found out that they are quite ordinary lines, *Balmer lines*, but highly redshifted (0.158) (e.g. Fig. 14.16). In this Figure we see Lyman lines shifted to the near IR.

The red shift is defined by the Doppler effect red shift quasars:

$$\frac{\Delta\lambda}{\lambda_0} = \frac{v}{c} = z \tag{14.21}$$

Thus, for the red shift, the relationship is:

$$v = cz \tag{14.22}$$

A red shift of $z \geq 1$ would result in the object moving away from us at faster than light speed.

A high red shift is common to all quasars (values range from 0.06 to >6.0). Here one must apply the relativistic formula for the Doppler effect:

$$z = \frac{\Delta\lambda}{\lambda_0} = \sqrt{\frac{1 + v/c}{1 - v/c}} - 1 \tag{14.23}$$

> At velocities close to the speed of light, values of red shift greater than 1 are obtained using the relativistic Doppler formula.

In addition to broad emission lines, which are strongly red shifted, many quasar spectra also have absorption lines. To observe absorption lines, one needs (a) a source of continuous emission and (b) between this source and the observer a medium which absorbs. The red shifts of the absorption lines often differ from those of the emission lines. One has to keep in mind that quasars are very far away from us and absorption can occur somewhere between the quasar and the observer. According to their absorption lines, quasars can be divided into:

- Quasars with broad absorption lines: originitae from gas is in the vicinity of the quasar, interstellar clouds that are strongly accelerated and expanding at up to 0.1 c.
- Quasars with sharp lines: Here the velocity differences between emission lines and absorption lines are only 3000 km/s. You can also see lines of metals.
- Quasars with sharp metal lines.
- *Lyman Alpha Forest:* Here you have lines of low velocity dispersion. Example: Quasar PHL 938 has in emission a red shift of 1.955, in absorption however 1.949, 1.945 and 0.613. Light of a quasar with $z = 2$ can cross up to 20 galaxy clusters on its way to the observer and thus produce such lines. 60% of the lines of the H-Lyman-Forest are explained in this way, the rest could indicate matter that has not clustered into galaxies.

From the observed changes in the brightness of quasars, which occur within days to a year, one can estimate their extension: Let the period of the change be t, then the radius of the object follows R from:

$$R \leq ct \tag{14.24}$$

There have also been attempts to explain the high red shift of the Quasars which, via the Hubble relation, also means a great distance, by other effects but in the case of the quasars it is well established that the high redshifts suggest that they are indeed extremely distant. However, since they are very bright (some of 13th magnitude), up to 1000 times more energy is produced in them than in the brightest galaxies, and this in the a small region (a few light-days). The continuous spectrum is produced by synchrotron radiation. Electrons orbit in a magnetic field, i.e. they are accelerated. They radiate and thus move slower. Thus one can estimate that per year about 10^{43} J are emitted. It is assumed that in the center of a quasar there is a supermassive black Hole with masses up to $10^{7\ldots9}\,M_\odot$. Passing stars are torn apart by the tidal effect, and the matter forms an *Accretion disk* as in the case of the X-ray double sources discussed earlier. Ionized gas flows out perpendicular to the plane of rotation of the accretion disk as a jet. Calculations indicate that about one solar mass per year is needed to explain $10^{12} L_\odot$.

Let us estimate what amount of energy is released when 1 $M_\odot$ falls onto a black hole of $M = 10^9 M_\odot$. Suppose the mass falls from infinity down to $R = 3 \times 10^9$ km.

$$E_{\text{pot}} = -\frac{GMm}{R} \tag{14.25}$$

You get $E = 10^{47}$ J. But the conversion of this released energy to radiation is not 100%, matter in the accretion disk is also accelerated, etc.

In the case of the quasar 3C 273, the propagation of a gas bubble along the jet emanating from the nucleus has been detected at faster-than-light speeds (Fig. 14.17).

This can be easily understood by assuming that the jet is only slightly inclined with respect to the observer's line of sight (Fig. 14.18). Suppose that in a jet, masses of matter are ejected at nearly the speed of light. The jet has an angle of 8 degrees to the line of sight. The jet is ejected from the nucleus N and reach point A in 101 years. It should reach point B in 100 years. So the angle between A and B is 8°, but this is equivalent to 14 light years. But the light at B arrived one year earlier than the light at A. The light reaches us as observers after several billion years. First the light from B reaches us and after one year the light from A. We therefore have the impression that the source has moved from B to

Fig. 14.17 Quasar (appears star-shaped) 3C273 with a jet. NASA

Fig. 14.18 For the explanation of superluminal velocities

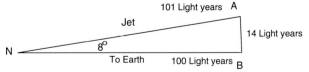

A within one year, covering the distance of 14 light years! Therefore we wrongly interpret the velocity $v = 14c$.

In strong gravitational fields, light gets deflected. This has been proven by measuring the *deflection of light rays* of stars during a total solar eclipses.

One has found *double quasars* with identical properties: The double quasars 0957+561A and 0957+561B are only $6''$ apart, their spectra are almost identical and have $z = 1.41$. In fact these are optical images of the same quasar produced by an intervening galaxy. Such observations of *Gravitational lensing* are particularly important for several reasons:

- Proof of General Relativity.
- The quasar is farther away than the galaxy, hence evidence that their red shifts are really to be interpreted "cosmologically", i.e. indicate their distances.
- Cool gas in the vicinity of the galaxy produces the absorption lines.

Figure 14.16 shows the spectrum of a quasar with $z = 5.82$. Quasars must therefore have formed very early in the universe. In the range between $z = 2$ and $z = 4$ the quasar density is 100 times larger than today.

Let us consider a simple example. Let D_{ls} the distance from the source (quasar) to the lens (galaxy), D_l the distance from the lens to the observer, and D_s the distance from the source to the observer. Then one finds the following expression for the deflection θ (Fig. 14.19) if the source is point-like:

$$\theta = \sqrt{\frac{D_{ls}}{D_l D_s} \frac{4GM}{c^2}} \tag{14.26}$$

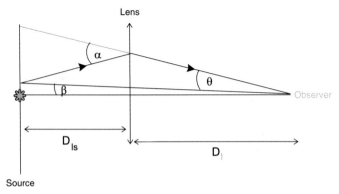

Fig. 14.19 On the formation of a gravitational lens. Light from a source (e.g., a distant quasar) passes through a lens (e.g., a galaxy) and is deflected, resulting in multiple images of the source

Table 14.2 Relative luminosity of different objects in different wavelength ranges

Object	X-ray	Optical	Radio
Milky way	1	1	1
Radio galaxy	100–5000	2	2000–2×10^6
Seyfert galaxy	300–70.000	2	20–2×10^6
Quasar 3C273	2.5×10^6	250	6×10^6

For a circular lens, the magnification (ratio of the image area to the area of the source) is:

$$\mu = \frac{\theta}{\beta}\frac{\mathrm{d}\theta}{\mathrm{d}\beta} \tag{14.27}$$

Table 14.2, the relative luminosity of various objects is given for comparison.

Microquasars

Quasars are essentially AGNs, *Microquasars* are of much smaller length scale. They are binary stars whose compact component is a black hole. A massive star (10–30 $M_\odot$) loses mass to the black hole. The accretion rate is $10^{-4}\,M_\odot$ per year. The accretion disk heats up strongly $\rightarrow$ X-ray emission. Microquasars are thus X-ray binaries (XRBs, *X-ray binaries*). One distinguishes between the LMXBs (low mass ..), HMXBs *(high mass ...)*, BHXBs *(black hole...)*, ULX *(ultra luminous sources)*. A well known source is e.g. Cyg X-1, SS433.

14.4.4 Galaxies with High Red Shift

Since 1996 many galaxies with $z = 3$ are known. Narrow band filters are used to find such objects. Suppose a galaxy has $z = 3$. It then shows much less light at $\lambda \leq 480$ nm $\approx$ (1 + 3) 121.6 nm than at longer wavelengths. At wavelengths below 360 nm ($\approx$ (1 + 3) 91.2 nm), there is almost no radiation any more, because the absorption of photons by ionization is even more efficient than the L_α absorption. Such a galaxy is therefore invisible in the U-filter, and one has a U-dropout. In practice, one compares the brightness in three filters and finds galaxies strongly red shifted like this.

14.4.5 Blazar

To be classified as a blazar, an object must meet the following conditions.:

- The object must appear to be point-like; some blazars have faint nebulae around them, but almost all of the radiation comes from the blazar.
- There are no absorption lines in the spectra.

- Visible light is mostly partially polarized.
- Its radiation varies much more rapidly and with greater amplitude at all wavelengths than a normal quasar.

Blazars are interpreted by a gas jet originating from material in the vicinity of a black hole; the jet points almost exactly in the direction of the Earth. In a gamma-ray energy range between 30 MeV and 30 GeV, 60 blazars could be found. These gamma rays can only originate in relativistic jets.

14.5 Galaxy Clusters

The galaxies arrange themselves in galaxy clusters. These in turn are usually part of so-called superclusters. The superclusters are therefore the largest structures in the universe.

14.5.1 The Local Group

Our Milky Way belongs with about 20 other galaxies to the so-called *Local Group*. In this group there are only three large spiral galaxies, the Andromeda Galaxy M31, the Triangle Galaxy M33 as well as our own system. The other members are dwarf galaxies like the large and the small Magellanic Cloud (Table 14.3, the Draco dwarf galaxy is given for comparison).

14.5.2 Abell Catalogue of Galaxy Clusters

Abell has divided galaxy clusters into regular (showing spherical symmetry) and irregular. Examples of galaxy clusters can be found in Figs. 14.20 and 14.21. The regular clusters contain only E or S0 systems and have many members (more than 1000). The irregular systems include all types of galaxies. Our Local Group also belongs to this type of galaxy cluster. They contain few members. The Local Group spans about 1 Mpc. A neighbouring cluster is the Fornax- and contains about 16 systems. The Coma Cluster cluster extends

Table 14.3 Data of some members of the local group

Object	Type	M_V	Distance (kpc)	Mass ($M_\odot$)
M 31	Sb	−21.1	690	3×10^{11}
Milky way	Sb or Sc	−21	–	4×10^{11}
M 33	Sc	−18.9	690	4×10^{10}
LMC	Irr I	−18.5	50	6×10^9
SMC	Irr I	−16.8	60	1.5×10^9
Draco	dE3	−8.6	67	10^5

Fig. 14.20 A galaxy cluster, distance 142 Mpc, image diameter about 120 kpc. In the foreground is the giant elliptical galaxy ESO 325-G004, Abell S0740 (Source: Hubble Space Telescope)

over nearly 10 Mpc and contains 100 galaxies brighter than $M = -16$. The closest large galaxy cluster is the Virgo Cluster which stretches across $12°$ in the sky and is about 16 Mpc away from us. Its diameter is 3 Mpc. From the Palomar Sky Survey Abell has found 2700 clusters. Examples of galaxy clusters are shown in Table 14.4.

Important for the understanding of galaxy clusters is the luminosity function. This indicates how many galaxies there are at a certain brightness per volume unit in space. One enters in a diagram as abscissa a magnitude (brightness) interval and as ordinate the log of the number of galaxies.

Fig. 14.21 The galaxy cluster Abell 2218. Note the gravitational lensing effect in the image. Hubble Space Telescope

Table 14.4 Some galaxy clusters

Designation	Expansion [°]	Del. [Mpc]	Radius [Mpc]	L [10^{11} $L_\odot$]
Ursa. major	0.7	270	1.31	7.1
Virgo	12	19	1.07	12
Canes venatici	19	8	1.23	1.5
Coma	4	113	2.6	49
Pisces	10	66	0.47	4.2

14.5.3 Galaxy Collisions

Comparing the distribution of galaxies in a Cluster of Galaxies with galaxy diameter, we find that the average distance between galaxies is less than 100 times their diameter. The distribution of planets in the solar system, on the other hand, looks different: The distances between planets are roughly the 10^5 of their diameter. fFr the distribution of stars in a galaxy we get the factor 10^7. From this estimate it is immediately clear that collisions between galaxies may occur relatively frequently, whereas during such collisions stars themselves collide extremely rarely. However, in such a collision the shape of the affected galaxies changes due to the tidal interaction. Here, for example, the group of cD galaxies are to be mentioned, which are characterized by:

- widely extended halos (up to 1 Mpc),
- sometimes multiple nuclei,
- at the center of clusters of galaxies.

Since more than 50% of cD galaxies have multiple nuclei, it can be assumed that they were formed by cannibalism.

Galaxy clusters are defined according to *Bautz* and *Morgan* as follows:

- Type I: The cluster is dominated by a massive cD galaxy.
- Type II: Contains predominantly elliptical galaxies.
- Type III: Is divided into IIIE, the cluster contains few spiral galaxies, and IIIS, the cluster contains many spiral galaxies.

> Galaxies occur in galaxy clusters, collisions between galaxies occurred repeatedly throughout the evolution of the universe.

The Milky Way will collide with the Andromeda Galaxy in about 4 billion years.

While the probability of collision between stars during a galaxy collision is extremely low, compressions of interstellar gas occur, greatly increasing the rate of star formation.

14.5.4 Super Cluster

By counting the galaxies in the *Palomar Sky Survey* several million galaxies were found. A two-dimensional image shows that these are often connected by chains and that there is a higher degree of structure. The local super cluster is dominated by the Virgo cluster and also contains the Local Group. Other super clusters such as Hercules or the one in the Hydrus-Centaurus region have been found. In between, however, there are also empty regions referred to as *Voids* (e.g. Bootes Void). The super clusters are not spherical, but curved, often pancake-like filaments. All rich clusters are within super clusters, and probably more than 90% of all galaxies are in super clusters. The voids are spherical and contain no bright galaxies. There is also matter between the clusters, intergalactic matter.

From a statistical investigation of the distance of galaxies, derived from their red shift and compared with other methods of distance determination, it is shown for galaxies of our environment that there is a systematic deviation, which can be explained by a *Great Attractor*. The local group falls into the center of the local Supercluster (near Virgo), which in turn moves at 570 km/s towards $l = 307°$, $b = 9°$. The gravitational potential required for this requires 5×10^{16} stars. This Great Attractor lies in the plane of the Galaxy, where interstellar absorption is very strong and it is therefore difficult to observe.

The galaxy cluster Abell 2218 is shown in Fig. 14.21.

WINGS stands for Wide Field Nearby Galaxy Cluster Survey. This inventory contains 77 galaxy clusters distributed across the northern and southern skies. Spectroscopic studies were made with the 4.2 m William Herschel telescope on La Palma. The instrument used is the 2dF Multfiber Spectrograph. Up to 150 glass fiber cables, each 1.6 arcseconds in diameter, were used.

We give as an example of WINGS spectra in Fig. 14.22 some galaxy spectra. The positions of typical lines are plotted. Note how increasing red shift z affects on the position of the lines.

14.5.5 Special Galaxy Clusters

We give some examples of special galaxy clusters to illustrate the dynamics in the Universe. The so-called bullet cluster is actually two colliding clusters of galaxies. Between them is empty space, but computer models show that dark matter must be present (see Fig. 14.23). The distance is 3.7 billion light years (GLy). In X-ray light, the X-ray satellite CHANDRA can be used to see the emission of the colliding gas components of both galaxies. This gives the distribution of baryonic matter. The collision may have occurred about 150 million years ago. Background objects are subject to gravitational lensing by the two galaxy clusters $\rightarrow$ distribution of dark matter. This is shown in Fig. 14.23 by means of the Contour Plots shown.

Another example of a rather chaotic galaxy cluster is Abell 520 (Fig. 14.24). The large composite image shows the mass distribution of baryonic (visible) and dark matter, as well as a composite of visible light and X-ray images. The distribution of dark matter in this cluster is unusual; it is not concentrated at the edges of the cluster, but is heavily concentrated toward the center.

> The Milky Way belongs to the local group, which in turn is part of the local supercluster.

14.6   Further Literature

We give a small selection of recommended further reading.
Galaxies: Birth and Destiny of our Universe, G. Schilling, Firefly Books, 2019.
Galaxy Formation, M.S. Longair, Springer, 2008

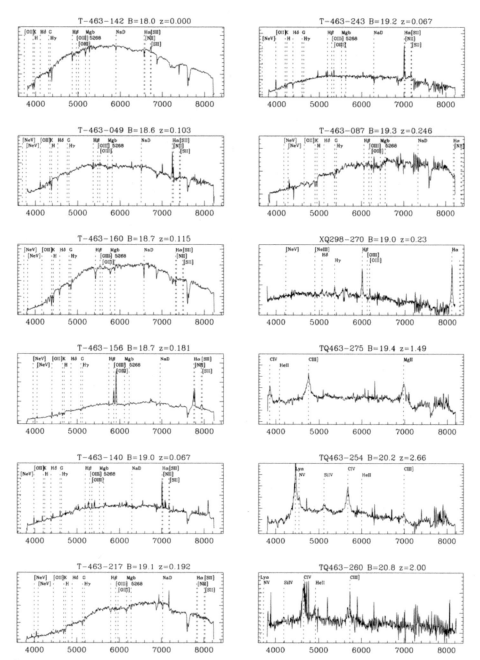

Fig. 14.22 Spectra from the WINGS project. z stands for red shift. The position of typical spectral lines is plotted (Credit: 2dFGRS Team)

Fig. 14.23 Bullet cluster with contours resulting from the distribution of mass; from this follows the distribution of dark matter. NASA, Hubble Space Telescope

Galactic Dynamics, J. Binney, S. Tremaine, Princeton Series in Astrophysics, 2008
Gas Accretion onto Galaxies, A, Fox, R. Dave, Springer, 2017.
Astrophysics of Gaseous Nebulae and Active Galactic Nuclei, D.E. Osterbrock, G.J. Ferland, Univ. Science Books, 2005
Physics of Active Galactic Nuclei at all Scales, D. Alloin, R. Johnson, P. Lira, Springer Lec. Notes in Physics, 2006
Fifty Years of Quasars, M. D'Onofrio, P. Marziani, J.W. Sulentic, Springer, 2012.
Gravitational Lensing of Quasars, A. Eigenbrod, EPFL Press, 2011
The Formation and Disruption of Black Hole Jets, I. Contopoulos, D. Gabuzda, N. Kylafis, Springer, 2014

Tasks

14.1 As an average value, one can calculate a diameter of 15 kpc for a galaxy. Determine the angular extent of the Andromeda Galaxy, whose distance is 600 kpc!

Solution
More than full moon diameter.

14.2 Calculate the wavelength at which Hα-line of hydrogen looks for the quasar 3C273 given above, $z = 0.157$.

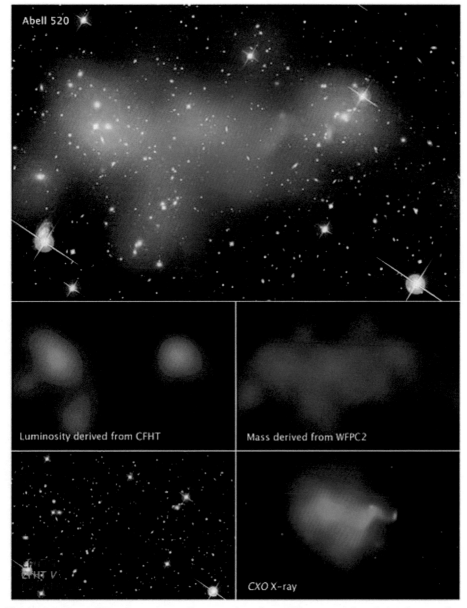

Fig. 14.24 The A520 galaxy cluster. Images taken from different wavelength bands as well as composite (large image above). NASA, CHANDRA

Solution

The wavelength is changed from 656.3 nm to 656.3 nm + 656.3 × 0.157 nm shifted.

14.3 With what speed does a quasar move with $z = 2$ away from us?

Solution

$z = 2 = \Delta\lambda/\lambda_0 = [(1 + v/c)/(1 - v/c)]^{1/2} - 1$ and

$(1 + v/c)/(1 - v/c)) = 3^2 = 9 \qquad v/c = 8/10 = 0.8$

Cosmology

15

How big is the universe, how old is it, how did it come into being, what does its future development look like? *Cosmology* is concerned with such questions. The first philosophical reflections on the structure of the universe can be found already in the ancient peoples.

Distance determinations play a major role in understanding the structure of the universe, and here we mention three milestones:

Copernicus, Kepler: were concerned with distances and dynamics of the bodies of the solar system;

Bessel: First measurement of a stellar parallax. Bessel used the star 61 Cygni, which showed the largest known proper motion at that time: $4.1''$/year in right ascension and $3.2''$/year in declination. In 1838 he determined the parallax of this star: $0.3''$. This measurement showed definitively: Our solar system is only a tiny part of the universe.

Hubble: From the Period-Luminosity Relationship of the Cepheids, he was able for the first time to determine the distance of the "Andromeda Nebula" and classify it as an extragalactic system.[1]

> Only since about 100 years we know that there are many other galaxies beside our galaxy.

---

[1] He reported at the annual meeting of the American Astronomical Society, 1925, "Cepheids in Spiral Nebula."

© Springer-Verlag GmbH Germany, part of Springer Nature 2023
A. Hanslmeier, *Introduction to Astronomy and Astrophysics*,
https://doi.org/10.1007/978-3-662-64637-3_15

Modern cosmology has contributed significantly to the development of general relativity and modern particle physics.

15.1   Expansion of the Universe

In this section, observational results relevant to cosmology are discussed. These can then be used to build a model of the formation and evolution of the universe.

15.1.1 View into the Past

In general: The objects (galaxies) are very far away from us; recently there has been a great increase in knowledge due to the development of new detectors (CCD, CMOS) and large telescopes (> 8 m). The objects are at a distance D by us. The light propagates with finite speed c, and therefore we see a source at distance D today in a state in which light propagated by a time interval Δt earlier than today:

$$\Delta t = \frac{D}{c} \tag{15.1}$$

> The present state of the universe is observable only in the near vicinity. We are, so to speak, looking into the past

If we observe galaxies at a distance of 10^{10} light years, we see light from a time when the age of the universe (about 13.7 billion years) was only about 1/3 of today's!

Suppose we are in a Euclidean space at the position $\mathbf{r} = 0$ at the time $t = t_0$. Since light propagates at finite speed, we only see space-time points for which:

$$|\mathbf{r}| = c(t_0 - t) \tag{15.2}$$

i.e. only this part of the universe is observable for us.

15.1.2 Olbers Paradox

Olbers (1758–1840) arose the question: Why is it dark at night? Let's assume, n_* is the number density of stars and n_* is constant in space and time. The number of stars within in a spherical shell of radius r and thickness dr is $n_* 4\pi r^2 dr$. Each star occupies a solid angle $\pi R_*^2/r^2$ so the stars in the spherical shell occupy the solid angle

$$d\omega = 4\pi r^2 dr n_* R_*^2 \pi/r^2 \tag{15.3}$$

i.e. r truncates out. So the sky would be completely filled with star discs and would be bright (shine like a body with a temperature of approx. 10^4 K).

Therefore, the universe (i) cannot be infinitely extended, (ii) cannot have existed for an infinitely long time.

15.1.3 Galaxy Counts

If one counts galaxies with a luminosity $n > L$ then the number of galaxies within a spherical shell of radius r and thickness dr is:

$$n(> L)4\pi r^2 dr \tag{15.4}$$

galaxies. We further assume that the luminosity function is constant in space and time. The relation between flux S and luminosity L is

$$L = 4\pi r^2 S \tag{15.5}$$

and one finds for the number:

$$N(> L) = \int_0^\infty dr 4\pi r^2 n(> L)4\pi r^2) \tag{15.6}$$

Because of $L = S4\pi r^2$, $r = \sqrt{L/(4\pi S)}$ and $dr = dL/(2\sqrt{4\pi L S})$ we find

$$N(> L) = \int_0^\infty \frac{dL}{2\sqrt{4\pi L S}} \frac{L}{4\pi S} n(> L)$$
$$= \frac{1}{16\pi^{3/2}} S^{-3/2} \int_0^\infty dL \sqrt{L} n(> L) \tag{15.7}$$

We see: Independently of the luminosity function, the following holds for the count of galaxies in the universe:

$$N(> L) \propto S^{-3/2} \tag{15.8}$$

However, this contradicts observations.

> The universe cannot be Euclidean, infinite and static.

15.1.4 The Redshift of the Galaxies

Let v be the measured Radial velocity of a galaxy, d their distance. *Hubble* found around 1926 that galaxies are moving away from us, the velocity being proportional to the distance of the galaxy.

The *Hubble relation*

$$v = d\,H_0 \tag{15.9}$$

states that (a) galaxies are moving away from us (except for those in our galaxy cluster) and (b) the escape velocity of galaxies is greater the farther they are from us (distance d). If one introduces the redshift z, then holds:

$$v = cz \qquad d = cz/H_0 \tag{15.10}$$

If all galaxies are moving away from us, you might think we are at the center of the universe. But the explanation is quite simple. Galaxy movements reflect the expansion of the universe. Imagine a balloon with galaxies marked as black dots. When the balloon is inflated, i.e. the universe expands, then you observe from any given point that the other points move away. Hubble's law applies to all galaxies in the universe.

> The redshift of the galaxies is explained by the *Expansion* of the universe.

Einstein published his Field Equations in 1915. The solution of the field equations led to non static universe. At that time nothing was known about an expanding universe. Therefore he introduced the *Cosmological constant* introduced to keep the solutions static. This constant was also necessary because a universe consisting of mass would automatically contract if it were static.

Another problem is the exact determination of the Hubble constant H_0. In 1929 Hubble gave a value of $530\ \mathrm{km\,s^{-1}\,Mpc^{-1}}$. 1958 one was with a value of 75, later there were two variants with (a) 50 and (b) 100. Measurements with the HST and WMAP come to a value accepted today as valid from:

$$H_0 = (69.7 \pm 4.9)\ \mathrm{km\,s^{-1}Mpc^{-1}} \tag{15.11}$$

A galaxy at a distance of 1 Mpc is therefore moving away from us at about 70 km/s.
One also writes:

$$H_0 = 50h \qquad h = 1.4 \tag{15.12}$$

15.1.5 The Age of the Universe

From the expansion, one can estimate an age of the world; let us consider the units of the quantities of the Hubble relation:

- Hubble law: $V = RH$
- Unit of velocity: $[v] = \text{km/s}$
- Unit of distance: $[R] = \text{Mpc}$
- Unit of Hubble constant: $[H] = [\frac{\text{km/s}}{\text{Mpc}}]$
- $\rightarrow [1/H] = s$

The reciprocal of the Hubble constant provides an estimate of the age of the universe.

$$\tau_0 = r/v = 1/H_0 = \frac{1}{h50\,\text{km s}^{-1}\,\text{Mpc}^{-1}} = 20 \times 10^9 h^{-1}\text{a} \tag{15.13}$$

So let's put in $h = 1.4$ (modern value of the Hubble constant), then the universe would be approx. 13.65×10^9 years old. This is consistent with the observed oldest globular clusters, whose age is 12 Ga.[2]
Of course, this assumes a uniform expansion of the universe. To a first approximation, this is given. However, there are two important exceptions:

- *Inflationary phase:* roughly between 10^{-35} and 10^{-33} s after the big bang the universe expanded extremely strong by a factor of 10^{26}.
- If one compares the present expansion of the universe with that of earlier epochs, it turns out: the universe is expanding at an accelerated rate. This is explained by the presence of *Dark energy.*

---

[2] 1 Ga = 1 Gigajahr so 10^9 a.

15.1.6 Homogeneity and Isotropy

The Hubble radius of the universe is:

$$R_H = \frac{c}{H_0} = 2997h^{-1}\mathrm{Mpc} \qquad (15.14)$$

There are structures in the universe:

- Galaxies,
- galaxy clusters,
- Superclusters.

The largest structures are the super clusters, with dimensions up to about $100h^{-1}$ Mpc. This is still small compared to the Hubble radius, so the universe can be considered homogeneous and isotropic.

> Although there are structures in the 13.65 billion year old universe, it can be simplistically described as homogeneous and isotropic.

This assumption greatly simplifies the complex calculations.

15.1.7 Methods of Distance Determination

In order to make statements about the structure and distribution of objects in the universe, it is necessary to know the distances. Most of the methods for distance determination have already been discussed in detail and shall only be listed here:

- Cepheids, RR Lyrae stars (from the period of brightness change follows the distance),
- globular clusters (standard candles),
- Novae (standard candles),
- supernovae (standard candles),
- Tully-Fisher relationship (rotational behavior of galaxies indicates their luminosity),
- luminosity of *planetary nebulae:* They emit 15 % of the light in the 500.7-nm-line ([OIII]). This part can be filtered out with a narrow band filter and then you have practically only the planetary nebulae of a galaxy. The absolute brightness then follows from the empirical Relationship:

$$M = -2.5\log(F_{500.7} - 13.74) \qquad (15.15)$$

Table 15.1 Methods of distance determination applied to members of the Virgo cluster. The uncertainty of the method is given in absolute magnitudes M

Method	Uncertainty [M]	Del. [Mpc]	Range [Mpc]
Cepheids	0.16	14.9 ± 1.2	20
Novae	0.40	21.1 ± 3.9	20
Planet. Nebula	0.16	15.4 ± 1.1	30
Globular cluster	0.40	18.8 ± 3.8	50
Tulley-Fisher	0.28	15.8 ± 1.5	> 100
Supernovae Ia	0.53	19.4 ± 5.0	> 1000

In this equation $F_{500.7}$ denotes the Radiation flux at 500.7 nm. Cepheids and RR Lyrae stars were identified in the closer galaxies (distance below 100 million light-years). With the HST, the horizon within which such objects can be seen directly has been substantially extended (Table 15.1).

15.2 Newtonian cosmology

In the universe, only gravitational force and electromagnetic force act on large length scales. Electric forces neutralize, leaving only gravity, and Einstein developed General Relativity in 1915, relating gravity to space-time curvature. Newton's theory is valid on small length scales, and one can consider simple world models.

15.2.1 Expansion

Although we have the impression that all galaxies are moving away from us—due to expansion of space—this does not mean that we occupy an excellent position in the universe. Because of

$$\mathbf{v} = H_0 \mathbf{r} \tag{15.16}$$

is valid for any other galaxy, which is at a distance $\mathbf{r_1}$ relative to us and has the velocity $\mathbf{v_1}$ relative to us:

$$\mathbf{v} - \mathbf{v_1} = H_0(\mathbf{r} - \mathbf{r_1}) \tag{15.17}$$

Therefore, a Hubble relation exists there as well: $\mathbf{v_1} = H_0 \mathbf{r_1}$.

We assume a homogeneous sphere expanding radially. At time $t = t_0$ the coordinate is $\mathbf{x}$ and it changes over time to:

$$\mathbf{r}(t) = a(t)\mathbf{x} \tag{15.18}$$

$a(t)$ ... cosmic scale factor. It holds $\mathbf{r}(t_0) = \mathbf{x}$ and thus $a(t_0) = 1$. The quantity t_0 can be chosen arbitrarily, and we set $t_0 =$ today. The expansion rate is given by

$$\mathbf{v}(\mathbf{r}, t) = \frac{d}{dt}\mathbf{r}(t) = \frac{da}{dt}\mathbf{x} = \dot{a}\mathbf{x} = \frac{\dot{a}}{a}\mathbf{r} = H(t)\mathbf{r} \tag{15.19}$$

So the expansion rate is (cf. Hubble constant!):

$$H(t) = \frac{\dot{a}}{a} \tag{15.20}$$

and to the time $t = t_0$ one has $H_0 = H(t_0)$.

15.2.2 Equation of Motion

If one considers a spherical shell with radius x at time t_0. For any t the radius is $r(t) = a(t)x$, and the mass is:

$$M(x) = \frac{4\pi}{3}\rho_0 r^3 = \frac{4\pi}{3}\rho(t)a^3(t)x^3 \tag{15.21}$$

The mass density of the present universe is ρ_0, and it holds:

$$\rho(t) = \rho_0 a^{-3}(t) \tag{15.22}$$

If we consider a particle on the surface of a sphere, we get an inward gravitational acceleration:

$$\ddot{r}(t) = -\frac{GM(x)}{r^2} = -\frac{4\pi G}{3}\frac{\rho_0 x^3}{r^2} \tag{15.23}$$

and if we substitute: $r(t) = xa(t)$ it follows that

$$\ddot{a}(t) = \frac{\ddot{r}(t)}{x} = -\frac{4\pi G}{3}\frac{\rho_0}{a^2(t)} = -\frac{4\pi G}{3}\rho(t)a(t) \tag{15.24}$$

15.2.3 Conservation of Energy

We multiply Eq. 15.24 by $2\dot{a}$, and because of $d(\dot{a}^2)/dt = 2\dot{a}\ddot{a}$ and $d(-1/a)/dt = \dot{a}/a^2$ is:

$$\dot{a}^2 = \frac{8\pi G}{3}\rho_0\frac{1}{a} - Kc^2 = \frac{8\pi G}{3}\rho(t)a^2(t) - Kc^2 \tag{15.25}$$

(Kc^2 is a constant of integration). Multiplying this equation by $x^2/2$ and find:

$$\frac{v^2(t)}{2} - \frac{GM}{r(t)} = -Kc^2\frac{x^2}{2} \tag{15.26}$$

The constant K is proportional to the total energy of the system. One can see immediately:

- $K < 0$ The right side of (15.25) positive, $da/dt > 0$ today, and remains positive for all times, universe expands eternally.
- $K = 0$: right side always positive, but for $t \to \infty$ goes $da/dt \to 0$.
- $K > 0$: right side of (15.25) becomes zero if $a = a_{max} = (8\pi G\rho_0)/(3Kc^2)$, expansion stops, and contraction occurs.

The special case $K = 0$ at which the universe expands forever corresponds to a critical density ρ_{cr}:

$$\rho_{cr} = \frac{3H_0^2}{8\pi G} = 1.88 \times 10^{-29}h^2 \, \text{g/cm}^3 \tag{15.27}$$

The density parameter Ω_0 is

$$\Omega_0 = \frac{\rho_0}{\rho_{cr}} \tag{15.28}$$

and one has: $K > 0 \to \Omega_0 > 1$.

Observations reveal: The density $\Omega_* \approx 0.01$ present due to stars and luminous matter is thus much too low (Fig. 15.1).

Fig. 15.1 Expansion of the
universe for three values of k

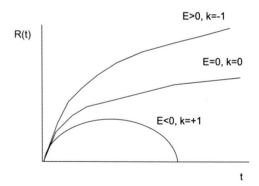

> There is a critical matter density. This determines whether the universe expands
> forever, or collapses again after a certain expansion phase.

15.3 Theory of Relativity

The theory of relativity developed by *A. Einstein* is divided into the special theory of
relativity (1905), which describes the transformation between two reference systems
moving with uniform velocity, and the general theory of relativity (1915), which specifies
the transformation between reference systems moving arbitrarily against each other (and
also includes gravity, as a curvature of space and time).

15.3.1 Special Theory of Relativity

Maxwell's equations of electrodynamics showed that electromagnetic waves (light)
propagate. Since waves need a medium to propagate, it was assumed that light propagates
in a medium called the *ether* and scientists tried to measure the motion relative to the
ether. In 1887 there was the famous experiment of *Michelson*, to determine the motion
of the Earth relative to the aether, but the result was negative. The result showed that all
observers had the same speed of light c. Consider being in a train that moves with a velocity
v. Then one would have expected:

– The person in the train measures for the propagation of light the speed c
– the speed of light an observer measures outside the moving train should be $c + v$ (in
  thel ight propagatipn direction of the train's motion).

However, this is not the case. Both observers measure the same speed of light c!

Let us consider two reference frames f, f'. One measures the following coordinates:

- in the system f: x, y, z, t
- in the system f' : x', y', $z,'$, t'.

The movement in x-direction is given by a classical *Galilean transformation:*

$$\Delta x' = \Delta x - v \Delta t \tag{15.29}$$

$$\Delta y' = \Delta y \tag{15.30}$$

$$\Delta z' = \Delta z \tag{15.31}$$

$$\Delta t' = \Delta t \tag{15.32}$$

However, it also follows that:

$$c' = c - v \tag{15.33}$$

That is, the speed of light would not be constant in all systems, and thus a contradiction to the experiment of *Michelson*. However, the *Lorentz transformation* solves this problem.

Let the Lorentz factor be:

$$\gamma = \frac{1}{\sqrt{1 - \frac{v^2}{c^2}}} \tag{15.34}$$

The Lorentz transformation (again, we assume only motion in the x-direction) is then:

$$dx' = \gamma(dx - v dt) \tag{15.35}$$

$$dy' = dy \tag{15.36}$$

$$dz' = dz \tag{15.37}$$

$$dt' = \gamma \left[dt - \frac{v}{c^2} dx \right] \tag{15.38}$$

Thus we have the concept of *Space-time.* The space-time interval between two events is:

$$ds^2 = dx^2 + dy^2 + dz^2 - c^2 dt^2 \tag{15.39}$$

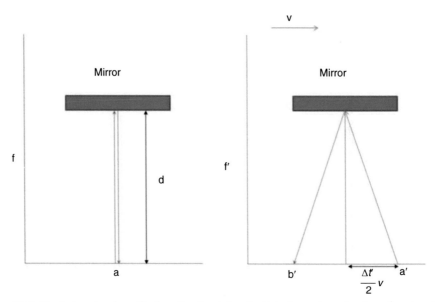

Fig. 15.2 To derive the time dilation. On the right, the light beam is emitted at a' and arrives, because of the motion of the system with the velocity v, at b'

Note: Between any two intervals is ds^2 independent of the motion of the observer, is a scalar invariant quantity. We denote the intervals as follows:

- $ds^2 < 0$ time-like,
- $ds^2 > 0$ space-like,
- $ds^2 = 0$ zero.

In relativity, one has the concept of a space-time continuum.

Wee derive from Fig. 15.2 the time dilation: On the left we have the system of rest, light is returned to a emits, reflects at b reflects and travels the distance $2d$ within the time Δt back. On the right, a ray of light is emitted at a' but arrives at b' because the system is moving with the speed v to the right – therefore light travels the distance $2d'$. We have therefore:

System at rest

$$\Delta t = \frac{2d}{c} \tag{15.40}$$

Moving system

$$\Delta t' = \frac{2d'}{c} = 2\frac{\sqrt{d^2 + \left(\frac{1}{2}v\Delta t'\right)^2}}{c} \tag{15.41}$$

Because of $\Delta t'^2 = 4d^2/c^2 + 4v^2\Delta t'^2/(4c^2)$ and $2d/c = \Delta t$ we get $\Delta t = \Delta t'\sqrt{1 - v^2/c^2} = \Delta t'/\gamma$.

A moving clock (system $'$) goes slower than a stationary $\rightarrow$ Time dilation.

$$\Delta t' = \gamma\,\Delta t \tag{15.42}$$

Analogous to that the *Length contraction* is obatined:

$$\Delta x' = \gamma\,\Delta x \tag{15.43}$$

For the addition of velocities we find: Let there be an observer B$'$ in a system which moves with respect to the observer B with the velocity v in the x-direction. The observer B$'$ measures the velocity of a body in its system to be u' the following velocity results for the observer B:

Velocity of the object measured in the moving system, is u'_x , and u_x is the velocity of the object measured in the system at rest. Then holds:

$$u_x = \frac{u'_x + v}{1 + \frac{u'_x v}{c^2}} \tag{15.44}$$

For energy and momentum holds:

$$E = \gamma mc^2 \qquad p = \gamma mv \tag{15.45}$$

Energy and mass are equivalent:

$$E = mc^2 \tag{15.46}$$

No object, no wave, i.e. no information can be transmitted faster than the speed of light. In order to even reach this speed, an infinite amount of energy must be expended. An object, which moves with superluminal speed from A to B, violates the *Causality Principle* (cause would then occur after effect).

Fig. 15.3 The space-time
diagram. Because of the
principle of causality, the world
lines can only go from the past
cone to the future cone. The
two 45°-lines mean $v = c$,
photons with rest mass $= 0$
move along these lines. The
space-like area is forbidden,
because here $v > c$ would be

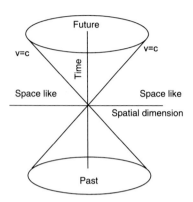

> An event is represented by three space and one time coordinate.

In the space-time diagram, a space coordinate is plotted against the time-space-time cone
(Fig. 15.3).

15.3.2 Four Vectors, Transformations

Figure 15.4 shows the Minkowski diagram: Two coordinate systems $K(x, ct)$ and
$K'(x', ct')$ move against each other with the speed v. The events in the systems have the
coordinates (x_1, y_1, z_1, t_1) as well as (x'_1, y'_1, z'_1, t'_1). Let us now apply the transformation
equations:

$$x = \frac{x' + vt'}{\sqrt{1 - v^2/c^2}}, \qquad y = y', \qquad z = z', \qquad t = \frac{t' + v(x'/c^2)}{\sqrt{1 - v^2/c^2}} \tag{15.47}$$

and calculate

$$L^2 = x^2 + y^2 + z^2 - c^2t^2 = \dots \rightarrow \qquad L = L' \tag{15.48}$$

The length of the four-vector is independent of the coordinate system.
The general transformation of a four-vector is ($\beta = v/c$):

$$A_1 = (A'_1 + \beta A'_4)\gamma, \qquad A_2 = A'_2, \qquad A_3 = A'_3 \tag{15.49}$$

$$A_4 = (A'_4 + \beta A'_1)\gamma \tag{15.50}$$

Fig. 15.4 The Minkowski diagram. Two coordinate systems K(x, ct) and K' (x', ct') move against each other. The following applies $tan\Theta = v/c$

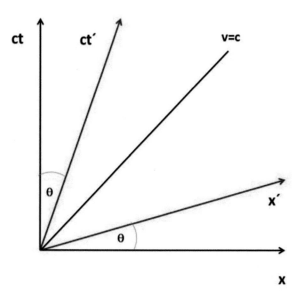

Examples of four vectors:

- Momentum 4-vector

$$(p_x, p_y, p_z, E/c) \tag{15.51}$$

with $E =$ Energy.

- The potential 4-vector in electrodynamics: consists of a vector potential (A_x, A_y, A_z) and as the 4th component the scalar potential Φs:

$$(A_x, A_y, A_z, \Phi) \tag{15.52}$$

From this one can deduce the *relativistic Doppler effect*. The energy is the fourth component of the four-vector $(\mathbf{p}, E/c)$ and transforms as follows:

$$E = \gamma(E' + vp'_x) \tag{15.53}$$

and therefore

$$E = \frac{E' + (Ev/c)\cos\Theta'}{\sqrt{1 - v^2/c^2}} = E'(1 + \beta\cos\Theta')\gamma \tag{15.54}$$

and because of $E = hv$ becomes

$$v = v'(1 + \beta\cos\Theta')\gamma \tag{15.55}$$

Fig. 15.5 Space-time diagram around a black hole. The event horizon is approached from the right. R_S. At this (at location 2) the escape velocity is $v = c$ and inside $r < R_s$ it is greater than c (e.g., at location 3). Space and time are practically reversed

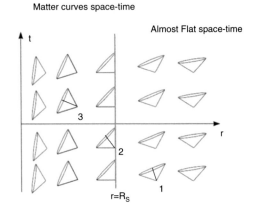

Two special cases:

- Source moves radially to detector, it emits light waves of frequency v_0; the frequency measured by the detector is then:

$$v = v_0 \sqrt{\frac{1 - \beta}{1 + \beta}} \tag{15.56}$$

- Transverse Doppler effect: the relative motion of the light source is perpendicular to the line that lies between the source and the observer:

$$v = v_0 \sqrt{1 - \beta^2} \tag{15.57}$$

$\beta = v/c$. The *transverse Doppler effect* is a consequence of time dilation. The space-time diagram around a black hole is given in Fig. 15.5. At the event horizon given by the Schwarzschildraius, space and time coordinates get inversed.

15.3.3 General Theory of Relativity

In principle the structure of space depends on the line element ds^2. This gives the distance between two points in the four-dimensional space-time continuum. In the simplest case of Euclidean space the distance between two points is (x^1, x^2, x^3) and $(x^1 + dx^1, x^2 + dx^2, x^3 + dx^3)$ in *Cartesian coordinates* is:

$$ds^2 = (dx^1)^2 + (dx^2)^2 + (dx^3)^2 \tag{15.58}$$

In special relativity, you have require a four-dimensional *Minkowski space,* and the line element is:

$$ds^2 = -(dx^0)^2 + (dx^1)^2 + (dx^2)^2 + (dx^3)^2 \tag{15.59}$$

Note: Sometimes one writes the time coordinate as positive and the three space coordinates as negative, sometimes vice versa.

The most general form of the line element is:

$$ds^2 = g_{ik} dx^i dx^k \tag{15.60}$$

Here one assumes the *Einstein's sum convention:* indices occurring twice on one side of an equation are automatically summed, so here over i, k. g_{ik} means *metric tensor,* it determines the metric of the space.

> **The metric tensor describes the metric of the space.**

The metric given by g_{ik} and describing space-time is dynamically connected with matter. Matter curves space-time. Riemannian geometry describes this mathematically, namely the geometric properties of space-time are intrinsic, i.e. one does not need a super space to explain the curvature.

15.3.4 Matter and Space-Time Curvature

Einstein's field equations describe the connection between the matter distribution (described by the tensor T_{ik}) and the curvature properties of space (described by the Ricci tensor R_{ik}):

$$R_{ik} - \frac{R}{2} g_{ik} + \Lambda g_{ik} = -\kappa T_{ik} = -\frac{8\pi G}{c^4} T_{ik} \tag{15.61}$$

The Ricci tensor is obtained by summation over n obtained from the Riemann tensor:

$$R_{ik} = R^n_{ink} \tag{15.62}$$

The curvature scalar is found from:

$$R^n_m = g^{in} R_{km} \qquad R = R^n_n \tag{15.63}$$

Fig. 15.6 A point is
represented in the x–y-system
by coordinates (A_x, A_y) and in
the oblique x'-y' system by its
contra variant (A^1, A^2) and
covariant components (A_1, A_2)

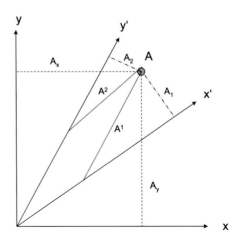

In an orthogonal coordinate system, there are only two components of a vector (Fig. 15.6, x–y system), whereas in a non orthogonal coordinate system there are the contravariant components (written with index above) and the covariant Components (written with index below) (Fig. 15.6, x'-y'-system).

If one wishes to determine the derivative of a vector or tensor field, one must take the difference of two vectors sitting at different points in space time. However, two such vectors cannot be compared directly because they are elements of different tangent spaces. One vector must first be transported parallel along a curve to the location of the other vector.

In principle, the Riemann tensor can be derived as follows: At each space-time point, the gravitational field is made to vanish → flat spaces. Now we investigate how these are related according to special relativity. Thus, the gravitational field is made to disappear by coordinate transformations. There is a simple analogy for this: in a free-falling elevator on Earth, the gravitational field disappears. If the elevator has a very weak gravitational field, then one can neglect the inhomogeneity of the field.

If one now transfers this to a curved space, a tangential surface can be defined on a curved spherical surface to every point.

- Curvature of the spherical surface ⇔ Space curvature;
- Tangent surface ⇔ flat space.

The problem is how to describe the relationship between different tangent surfaces in space → Covariant derivatives of a vector field.

The energy-momentum tensor is:

$$(T^{ik}) = \begin{pmatrix} w & \frac{S_x}{c} & \frac{S_y}{c}\ddot{a} & \frac{S_z}{c} \\ \frac{S_x}{c} & G_{xx} & G_{xy} & G_{xz} \\ \frac{S_y}{c} & G_{yx} & G_{yy} & G_{yz} \\ \frac{S_z}{c} & G_{zx} & G_{zy} & G_{zz} \end{pmatrix}$$

(15.64)

w—energy density, (S_x, S_y, S_z)—energy-current density, G_{ij}—stress tensor—whose diagonal components correspond to the pressure exerted by a radiation field.

In cosmology, one usually uses the following for the energy-momentum tensor:

$$T^{ik} = \left(\rho + \frac{P}{c^2}\right)u^i u^k - Pg^{ik}$$

(15.65)

where $u^i u^k$ is the quadratic velocity, P the pressure (radiation field), ρ is the mass density. This corresponds to the tensor for a fluid field.

> Galaxies are therefore taken to be elements of an ideal cosmic fluid, The pressure is assumed as isotropic.

15.3.5 Metric of the Space

If one knows the metric, then one can derive the curvature tensor from it.

For this one needs the *Christoffel symbols, affine connections*

$$\Gamma^i{}_{k\ell} = \frac{1}{2}g^{im}\left(\frac{\partial g_{mk}}{\partial x^\ell} + \frac{\partial g_{m\ell}}{\partial x^k} - \frac{\partial g_{k\ell}}{\partial x^m}\right) = \frac{1}{2}g^{im}(g_{mk,\ell} + g_{m\ell,k} - g_{k\ell,m})$$

Here $g_{k\ell,m}$ denotes a derivative of the component $g_{k\ell}$ with respect to component m. From this one finds:

$$R^\rho{}_{\sigma\mu\nu} = \partial_\mu\Gamma^\rho{}_{\nu\sigma} - \partial_\nu\Gamma^\rho{}_{\mu\sigma} + \Gamma^\rho{}_{\mu\lambda}\Gamma^\lambda{}_{\nu\sigma} - \Gamma^\rho{}_{\nu\lambda}\Gamma^\lambda{}_{\mu\sigma}$$

(15.66)

with $\partial_\mu = \partial/\partial x^\mu$.

So we outline the whole procedure:

- Give some metric of space-time $ds^2 = \ldots$
- → Calculate Christoffel symbols

- • → Calculate the Riemann tensor
- • → Insert this into Einstein's field equations ...

Motion in the Gravitational Field

Motion occurs along a geodesic line in Riemannian space; the equation of motion, according to special relativity, is given by ($d\tau = ds$):

$$\frac{d^2x^\mu}{ds^2} = -\Gamma^\kappa_{\mu\nu}\frac{dx^\mu}{ds}\frac{dx^\nu}{ds} \qquad (15.67)$$

> Basic statement of Einstein's field equations: Matter distribution determines space-time curvature.

Schwarzschild Metric

The Schwarzschild metric applies to a very simple situation: just one mass M in space-time, outside $T_{ik} = 0$

$$ds^2 = \left(1 - \frac{R_s}{r}\right)c^2dt^2 - \frac{dr^2}{1 - (R_s/r)} - r^2(d\theta^2 + \sin^2\theta d\phi^2) \qquad (15.68)$$

with

$$R_s = \frac{2GM}{c^2} \qquad (15.69)$$

as *Schwarzschild radius*. Application: non-rotating black hole.

For practice, let us consider the components of the metric tensor in the Schwarzschild metric. These are obtained from the factors that precede the squares of the coordinate differentials, so for example:

$$g_{00} = \left(1 - \frac{R_s}{r}\right) \qquad (15.70)$$

$$g_{11} = \frac{1}{1 - R_s/r} \qquad (15.71)$$

Kerr Metric

If one also takes the angular momentum (rotation) into account, the result is the Kerr metric.

This metric can be used to understand various special processes in the vicinity of black holes:

- the generation of jets,
- the quasi-periodic oscillations in stellar black holes,
- the relativistic broadening of emission lines,
- the *Lense-Thirring effect* in accretion disks, *frame-dragging*.

Any rotating mass virtually drags the local reference frame along with it *(frame dragging)*. The more mass rotating in small space, the larger this effect. It was found by *Lense* and *Thirring*.[3] The area within which a particle is forced to co-rotate is called *Ergosphere*.

Here again an analogue to electrodynamics becomes visible. One can deduce the magnetic field **B** from a vector potential **A** :

B = curlA.

Let us conceive of the space-time continuum as three-dimensional subspaces with constant time. The lapse function mediates from one of these hyper surfaces to the next later time, the shift function from one location to another on the hyper surface → gravitomagnetic field. A mass flow generates a gravitomagnetic field, just as an electric current generates a magnetic field in its vicinity.

In April 2004, the satellite Gravity Probe B was launched to demonstrate the Earth's space-time curvature and the Lense-Thirring effect using gyroscopes. Due to the rotating space-time of the Earth, the gyroscopes start a precession motion. The effect is additive and increases per revolution around the Earth. The Lense-Thirring effect after one year of revolutions around the earth amounts to $42 \times 10^{-3\prime\prime}$. It is orders of magnitude smaller than the geodetic precession caused by space-time curvature.

The gravitomagnetic dynamo caused by the Lense-Thirring effect is of great importance for astrophysics: in the vicinity of rotating black holes, tube-shaped magnetic fields are generated. By accretion then the jets are formed as well as the also observed quasiperiodic oscillations, which coincide with the Lense-Thirring frequency (produced by gyroscopic motion). They lie in the Microquasars in the kHz range.

Robertson-Walker Metric

This line element is used to describe the expanding universe. $a(t)$ is the cosmic scale factor. We assume a homogeneous, isotropic universe → Universe with constant curvature. The curvature is given by the curvature parameter k.

- $k = 0$ Euclidean, circumference of the circle $2\pi r$, area of the circle $r^2\pi$.
- $k > 0$ spherical or elliptic; circumference of the circle $< 2\pi r$, area $< r^2\pi$.
- $k < 0$ hyperbolic; circumference of the circle $> 2r\pi$, area $> r^2\pi$.

---

[3] In 1918.

The line element here is

$$ds^2 = c^2dt^2 - a(t)^2 \left[\frac{dr^2}{1 - kr^2} + r^2(d\theta^2 + \sin^2\theta d\phi^2) \right] \tag{15.72}$$

We roughly sketch the derivation of this metric. What is the equation of a hypersurface of constant curvature in a higher dimensional space? Solution:

$$x_1^2 + x_2^2 + x_3^2 = \frac{1}{k}R^2 \tag{15.73}$$

In Euclidean space, the metric is:

$$ds^2 = dx_1^2 + dx_2^2 + dx_3^2 \tag{15.74}$$

If one substitutes from (15.73) the expression for x_3 then follows:

$$ds^2 = dx_1^2 + dx_2^2 + \frac{(x_1 dx_1 + x_2 dx_2)^2}{\frac{R^2}{k} - x_1^2 - x_2^2} \tag{15.75}$$

and with $x_1 = r \cos\Theta$, $x_2 = r \sin\Theta$:

$$ds^2 = \frac{R^2 dr^2}{R^2 - kr^2} + R^2 r^2 d\Theta^2 \tag{15.76}$$

Robertson-Walker metric: One substitutes R by $a(t)$, calculate in 3 coordinates, then comes the term $R^2 \sin^2\theta d\Phi^2$ to it.

15.3.6 Friedmann-Lemaître Equations

If one takes the Robertson-Walker-metric as a basis, then from the field equations follow the Friedmann-Lemaître equations[4]:

$$\left(\frac{\dot{a}}{a}\right)^2 = -k\frac{c^2}{a^2} + \frac{8\pi G}{3}\rho + \frac{c^2}{3}\Lambda \tag{15.77}$$

$$2\frac{\ddot{a}}{a} + \left(\frac{\dot{a}}{a}\right)^2 = -k\frac{c^2}{a^2} - \frac{8\pi G}{c^2}P + c^2\Lambda \tag{15.78}$$

---

[4] *Friedmann 1922, Lemaître 1927.*

ρ contains the density of matter as well as radiation and also dark matter. From the two equations, by subtraction, we get:

$$\frac{\ddot{a}}{a} = -\frac{4\pi G}{3} \left(\rho + \frac{3P}{c^2} \right) + \frac{c^2}{3} \Lambda \tag{15.79}$$

A cosmological constant $\Lambda > 0 \rightarrow$ means repulsive acceleration, counteracts gravity. The static world model results from the Friedmann-Lemaître equations by $\dot{a} = 0, k = \Lambda a^2, \Lambda = 4\pi G\rho/c^2$.

One can attribute to Λ an equivalent mass density ρ_Λ

$$\Lambda = \frac{8\pi G}{c^2} \rho_\Lambda \tag{15.80}$$

and then the Friedmann-Lemaître equations read:

$$\left(\frac{\dot{a}}{a} \right)^2 = -k\frac{c^2}{a^2} + \frac{8\pi G}{3}(\rho_M + \rho_\Lambda) \tag{15.81}$$

$$\frac{\ddot{a}}{a} = -\frac{4\pi G}{3}\left[(\rho_M + \rho_\Lambda) + 3\left(\frac{P}{c^2} - \rho_\Lambda \right) \right] \tag{15.82}$$

15.3.7 The Cosmological Constant and Vacuum Energy

Quantum field theories are concerned with a unified description of fields and particles. As we know from quantum physics, properties like energy and momentum of particles are quantized. In the so-called second quantization, the fields are also quantized.

- First quantization: properties of particles quantized, e.g. energy, momentum.
- Second quantization: fields describe interactions between particles; these fields are also quantized.

According to the *Quantum Field Theory* the vacuum consists of fluctuating matter fields and virtual particles, and it can have a non-zero energy density. This provides an explanation for the cosmological constant:

$$\rho_\Lambda = \rho_{\text{vac}}$$

The cosmological constant corresponds to the energy density of the vacuum. The associated pressure is negative $P_{\text{vac}} = -\rho_{\text{vac}}c^2$.

The negative pressure of the vacuum energy can be made plausible: Consider the first law of thermodynamics. A change of the internal energy dU is equal to (i) sum of the heat supplied or dissipated and (ii) the work done pdV. An increase in volume means an output of work, hence the negative sign. The internal energy is therefore $U \propto$ volume V. Thus, if as a result of an increase in volume the internal energy is increased, because of

$$dU = -PdV \tag{15.83}$$

P to be negative.[5]

Three components can be given for the density:

$$\rho = \rho_m + \rho_r + \rho_{\text{vac}} = \rho_{m+r} + \rho_{\text{vac}}, \qquad P = P_r + P_{\text{vac}} \tag{15.84}$$

density of radiation (ρ_r) and density of matter (ρ_m) were combined and the dust was considered to be free of pressure. The pressure of a gas is determined by the thermal movement of the particles. Air molecules move at about $v \approx 300$ m/s, the pressure $P \ll \rho v^2$ i.e. the pressure is gravitationally insignificant. Such is the case in the universe today, *pressureless matter.* If the thermal velocity is equal to the speed of light, then you have the limiting case of radiation. An example of this is cosmic microwave background (CMB) radiation. Also particles with vanishing rest mass belong to it as well as particles whose thermal energy is much larger than the rest energy:

$$kT \gg mc^2 \tag{15.85}$$

The radiation pressure is:

$$P_r = \frac{1}{3}\rho_r c^2 \tag{15.86}$$

If ρc^2 is the energy density, then

$$\frac{d}{dt}(c^2 \rho a^3) = -P\frac{da^3}{dt} \tag{15.87}$$

Let us now consider the evolution of density:

$$\rho_m(t) = \rho_{m,0}a^{-3}(t) \tag{15.88}$$

$$\rho_r(t) = \rho_{r,0}a^{-4}(t) \tag{15.89}$$

$$\rho_{\text{vac}}(t) = \rho_{\text{vac}} = \text{const} \tag{15.90}$$

---

[5] For an adiabatic volume change dV the work is $dU = -PdV$.

The index "0" stands for today, $t = t_0$. Note the a^{-4} dependence of the radiation flux:

- As with matter, the number density of photons changes proportionally to a^{-3}.
- Photons are furthermore red shifted by cosmic expansion, $\lambda \propto a$, $E = hc/\lambda$; $E \propto a^{-1}$.

One can write the dimensionless density parameters for matter, radiation and vacuum as:

$$\Omega_m = \frac{\rho_{m,0}}{\rho_{crit}} \qquad \Omega_r = \frac{\rho_{r,0}}{\rho_{crit}} \qquad \Omega_\Lambda = \frac{\rho_{vac}}{\rho_{crit}} \tag{15.91}$$

and

$$\Omega_0 = \Omega_m + \Omega_r + \Omega_\Lambda \tag{15.92}$$

$\Omega_m > 0.3$ if one considers the galaxies and their halos. The energy density of radiation today is very small (photons from the CMB as well as neutrinos from the early universe).

$$\frac{\rho_r(t)}{\rho_m(t)} = \frac{\rho_{r,0}}{\rho_{m,0}} \frac{1}{a(t)} = \frac{\Omega_r}{\Omega_m} \frac{1}{a(t)} \tag{15.93}$$

Radiation and dust had the same energy density when

$$a_{eq} = 4{,}2 \times 10^{-5} (\Omega_m h^2)^{-1} \tag{15.94}$$

amounted to.

Because of $\rho = \rho_{m+r} = \rho_{m,0} a^{-3} + \rho_{r,0} a^{-4}$ becomes:

$$H^2(t) = H_0^2 \left[a^{-4}(t)\Omega_r + a^{-3}(t)\Omega_m - a^{-2}(t)\frac{Kc^2}{H_0^2} + \Omega_\Lambda \right] \tag{15.95}$$

Now we put in the values for today:

$$H(t_0) = H_0 \qquad a(t_0) = 1 \tag{15.96}$$

→ Integration constant:

$$K = \left(\frac{H_0}{c}\right)^2 (\Omega_0 - 1) \approx \left(\frac{H_0}{c}\right)^2 (\Omega_m + \Omega_\Lambda - 1) \tag{15.97}$$

Table 15.2 Characteristic parameters of the universe

Age	$13.7 \pm 0.5 \times 10^9$ Years
Diameter	96×10^9 Light years
Mass	10^{53} kg
Number of galaxies	10^{11}
Number of particles	$\approx 10^{79}$
Number of photons	10^{88}
Present temperature	2.75 K
Average matter density	2.3×10^{-26} kg/m^3

$\rightarrow$ K stands for the curvature of the Space:

- $K = 0$ Space Euclidean, flat.
- $K > 0$: two-dimensional analogue is spherical surface. Radius of curvature is $1/\sqrt{K}$, sum of angles in a triangle is greater than 180 deg.
- $K < 0$: hyperbolic space, sum of angles in a triangle is less than 180 degrees.

Note: General relativity states that matter is associated with the space time continuum, but nothing about the topology. If the universe possesses a simple topology, it is in the case of $K > 0$ finite, in the case $K \leq 0$ infinite. However, it is always unbounded, even a spherical surface is a finite but unbounded space.

One still defines the *Deceleration parameter*:

$$q_0 = -\frac{\ddot{a}a}{\dot{a}^2} = -\frac{1}{2}\Omega_m - \Omega_\Lambda \tag{15.98}$$

From this we see: If $\Omega_\Lambda = 0$ then the expansion is slowed down (due to gravity). If Ω_Λ is large enough, then q_0 can become positive, i.e. accelerated expansion (Table 15.2).

One takes the following values today:

$$\Omega_m = 0.3 \qquad \Omega_\Lambda = 0.7 \tag{15.99}$$

15.3.8 Gravitational Waves

The gravitational waves predicted by general relativity have already been detected in double pulsars. They only occur in when masse get accelerated (cf. electromagnetism: accelerated charges emit electromagnetic waves). They are most likely to be detected in:

- rapidly orbiting objects, e.g. double pulsars,
- non-rotationally symmetric, very rapidly rotating objects,
- asymmetric collapse or explosion of a massive object.
- Merging of massive objects such as neutrons stars, black holes.

Consider a binary star system with masses M_1, M_2 with period P, large semi-axis a and reduced mass $\mu = M_1 M_2 / (M_1 + M_2)$; $M = M_1 + M_2$. Then, for the luminosity of the gravitational wave, we find (*Peters* and *Matthews*):

$$L = \frac{32}{5} \frac{G^5}{c^{10}} \frac{\mu^2 M^3}{a^5} L_0 f(e) \tag{15.100}$$

where is:

$$L_0 = c^5 / G = 3.63 \times 10^{59} \, \text{erg/s} \tag{15.101}$$

and

$$f(e) = \frac{1 + \frac{73}{24} e^2 + \frac{37}{96} e^4}{(1 - e^2)^{7/2}} \tag{15.102}$$

with e for the orbital eccentricity. In practical units this gives:

$$L \approx 3.0 \times 10^{33} \text{erg/s} \left(\frac{\mu}{M_\odot} \right)^2 \left(\frac{M}{M_\odot} \right)^{4/3} \left(\frac{P}{1h} \right)^{-10/3} f(e) \tag{15.103}$$

For a deformed neutron star Gravitational wave emission with mass M, radius R, rotation period P, moment of inertia $I = 2/5 M R^2 / 5$ and ellipticity ϵ a luminosity due to gravitational wave emission can also be calculated from:

$$L = \frac{32}{5} \frac{G}{c^5} I^2 \epsilon^2 \left(\frac{2\pi}{P} \right)^6 \tag{15.104}$$

Gravitational waves are perturbations of space-time that propagate at the speed of light. A sensitive interferometer system (Fig.15.7) can measure these extremely small changes in position due to the space-time ripples. Thus, in September 2015, the first direct experimental evidence of gravitational waves was obtained, almost exactly 100 years after Einstein postulated them. The measured signal came from two black holes that merged. The distance of the objects to us was about 500 Mpc. A length change $h = 10^{-20}$ (strain) was detected by LIGO (Laser Interferometer Gravitational Observatory).

15.4   Dark Energy, Accelerated Expansion

15.4.1 Observations

Type Ia Supernovae , as already discussed, are very well suited as standard candles up to distances of 500 Mpc. Their true brightness is in a narrow range except for a few outliers. With the help of the HST, Cepheids have been found in nearby galaxies, calibrating the

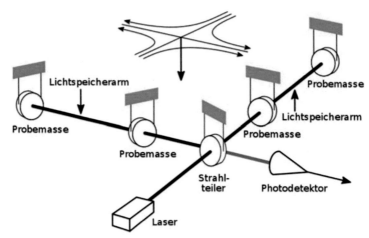

Fig. 15.7 Gravitational wave detector using two interferometer arms, LIGO concept, laser interferometer gravitational observatory. LIGO

magnitudes of SN Ia. There have been two major surveys, the Supernovae Cosmology Project (SCP) and the High Z Supernovae Search Team (HZT). These groups were able to find 70 type Ia supernovae.

The measurements are practically only possible with $\Omega_\Lambda \neq 0$ to explain it. So the cosmological constant, originally introduced to maintain a static universe, now describes the accelerated expansion of the universe.

Today's universe is expanding faster than it used to.

Another clear indication is the anisotropy of cosmic microwave radiation.

The early universe was extremely hot, the atoms were completely ionized, there were only:

- free electrons,
- atomic nuclei, mostly protons.

Because of the many free electrons, photons were extremely scattered, so the universe was opaque.

When, as a result of expansion, the universe cooled to about 3000 K, recombination occurred. The electrons recombined to form atoms, and the universe became transparent.

At the time of recombination, which was about 400 000 years after the Big Bang, the universe was anisotropic in some places. There were condensations, gravitational potential wells. Photons falling into such potential wells gain energy. If a photon falls into a potential well, it has to expend energy to get out, the same energy it gained falling in: δE_1. However, as the universe expands, the potential well, while the photon is in it, has become shallower, it therefore requires less energy, δE_2 to get out. Therefore $\delta E_2 < \delta E_1$, the photon therefore gains energy (which is lost to the potential well due to expansion). This is called *Sachs-Wolfe effect*[6] and through this the cosmic background radiation became anisotropic in some places.

15.4.2 Dark Energy

In 2001 the satellite WMAP[7] could find further evidence for the existence of dark energy. Dark energy opposes gravity in its effect, i.e. it is antigravitational (repulsive). Therefore, there is an accelerated expansion of the universe, in agreement with the type-SNIa observations. Dark energy accounts for about $70-74\%$ of the energy of the universe.

This dark energy is to be distinguished from dark matter. Baryonic matter makes up only between 2–5% of the matter of the universe, and the dynamics of galaxy clusters show that the sum of baryonic matter and dark matter cannot exceed about 30% (Fig. 15.8). Dark matter consists of particles that do not radiate electromagnetically, or at least radiate extremely weakly, and interact only weakly. Candidates for this would be the *(weakly interacting particles)* WIMPs and super symmetric particles.

What does dark energy consist of?

- vacuum-energy (see previous section): the quantum vacuum can be explained by the *Casimir effect*. Between two metal plates certain modes are missing—outside the plates they exist. This creates a quantum pressure that pushes the plates together.
- Effect of a scalar field, quintessence. Associated with this field are extremely light elementary particles elementary ($10^{-82}m_e$).
- String theories; the universe consists of more than four dimensions. At large distances, gravitational interaction weakens, and space expands more.

---

[6] R.K. Sachs, A.M. Wolfe, 1967.

[7] Wilkinson Microwave Anisotropy Probe.

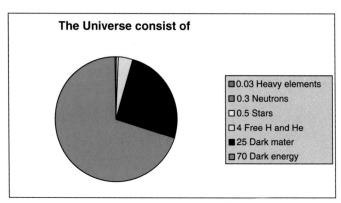

Fig. 15.8 Composition of the universe. The figures are in %

- Topological defects, imperfections as a result of spontaneous symmetry breaking in the early universe during the inflationary phase.
- Phantom energy: According to this theory, which was proposed in 2003, the universe would tear apart after 10^{50} years including all particles within *(big rip)*. However, this would already be noticeable by ultra-energy particles in cosmic rays, which have not been observed. Interestingly, this Big Rip could be observed: First, the most distant galaxies would disappear, and then successively closer and closer ones.

15.5   The Early Universe

15.5.1 Big Bang: Observational Hints

Expansion of Space

We described in the previous section the expansion of the universe, which had to start from a singular point. This is a first hint to the *Big Bang*. When this took place can be given by the world age, with the scale factor $a \rightarrow 0$. Theoretically, it could be that $\dot{a}$ was zero at some point in the past, i.e. a took a minimal value. However, this can be ruled out by observation: $\Omega_m > 0.1 \rightarrow a_{\min} > 0.3$. Since moreover a is related to the red shift z :

$$z_{\max} = \frac{1}{a_{\min}} - 1 \tag{15.105}$$

there should be a maximum red shift. In the meantime, however, one has observed quasars with $z > 6$ observed!

Background Radiation

A further indication of a hot big bang was the detection of the *Cosmic Background Radiation, CBR* detected in 1965 by *Penzias* and *Wilson*[8] . It corresponds to a Planck radiation curve of a body with a temperature of about 3 K, so its maximum lies in the microwave range. The background radiation is isotropic and originates from the time when the redshift was $z \approx 1000$, a relic from the *Recombination Era of* the universe (whose age was then about 400,000 years). When the universe was younger than 400,000 years the temperature was high and there were only free particles, atomic nuclei and electrons. Because of expansion the universe cooled (adiabatic cooling), and when temperature dropped to 3000 K, electrons were captured by (mainly) hydrogen nuclei, and the universe became transparently $\rightarrow$ decoupling of radiation and matter.

> The decoupling of radiation and matter occurred at a time when the universe was about 400,000 years old. The Cosmic Background radiation also dates from this time.

As early as 1945 *Gamow*, when studying thermodynamics in the radiation-dominated early universe, suggested that there must have been primordial nucleosynthesis, and *Alpher* and *Herman* postulated the existence of a 5-K background radiation.

Primordial Nucleosynthesis

Another important clue is primordial nucleosynthesis. The simplest process is that of Deuterium *D* formation:

$$p + n \rightarrow D + \gamma \tag{15.106}$$

Here, the temperature must be lower than the binding energy E_b of the deuterium so $kT \ll E_b$; at higher temperatures the formed D is destroyed again by photodissociation, only when $T_D \approx 8 \times 10^8$ K during the first $t \approx 3$ min the formation of D is predominant. D then forms into He^4, where the binding energy of 28 MeV is even higher, thus even more stable against photodissociation. Other fusion reactions will be discussed later.

> In total, in the first three minutes after the Big Bang, the following elements were formed: 75% hydrogen, 24% helium, and 1% lithium.

---

[8] They were awarded the Nobel Prize in Physics in 1978.

15.5.2 Sunyaev-Zel'dovich Effect

Compton Effect

We start with the photoelectric effect. *Hallwachs* discovered that the energy of the electrons triggered by the photoelectric effect is determined only by the frequency of the triggering light, their number only by the intensity of this light. Einstein assumed that only certain energy contributions $h\nu$ are available for electron triggering, which leads to the concept of photons and is of the same amount of energy that is exchanged between the oscillator and the radiation field according to Planck. Photons have the energy $h\nu$, move in vacuum with the speed of light c and have zero rest mass and thus momentum $p = h\nu/c = h/\lambda$. *Compton* found the effect named after him in 1922: Monochromatic X-rays are scattered by matter with an increase in wavelength. The wavelength of the scattered light is larger, the larger the scattering angle θ Is. For the backward scattering $\theta = \pi$ he found a wavelength increase of 0.0485 Å, independent of the irradiated wavelength. This effect can be interpreted as a collision process between the X-ray photon and the electron of the scattering matter; by conservation of energy and momentum this process is completely described. One finds after long calculation:

$$\lambda' - \lambda = \frac{2h}{mc} \sin^2 \frac{\theta}{2} \tag{15.107}$$

Inverse Compton Effect

Due to the free electrons in the universe there is a change in intensity ΔI_ν:

$$\frac{\Delta I_\nu}{I_\nu} = \frac{yx\exp^x}{\exp^x - 1} \left(x \frac{\exp^x + 1}{\exp^x - 1} - 4 \right) \tag{15.108}$$

with $x = \frac{h\nu}{kT}$ and the Compton parameter:

$$y = \int \frac{kT_e}{m_e c^2} \sigma_T n_e dl \tag{15.109}$$

n_e—density of free electrons, T_e—their temperature, m_e—their mass, σ_T—the Thomson cross section, l—the path length of the photons. The effect is also called *Sunyaev-Zel'dovich effect*. The measurements show $y \leq 1.5 \times 10^{-5}$. As photons pass through a cluster of galaxies, scattering of photons occurs, and some gain energy (a kind of inverse Compton effect). For example, if one observes at a wavelength of 1 cm, the gain is 0.05%. Thus, one observes a deficit of background cosmic photons, since some have been shifted to higher energies as a result of Compton scattering (Fig. 15.9).

Fig. 15.9 Spectrum of the 3-K
background radiation
(indicated above by two Planck
functions); due to the
Sunyaev-Zel'dovich effect
there is a shift to higher
energies

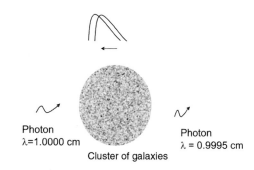

Photon
λ=1.0000 cm

Photon
λ = 0.9995 cm

Cluster of galaxies

R is the radius of a galaxy cluster and σ_T the Thomson effective cross section, then one finds for the resulting temperature deviations:

$$\frac{\Delta T}{T} = \frac{4Rn_e kT_e \sigma_T}{m_e c^2} \tag{15.110}$$

The 3-K background radiation is isotropic and exhibits deviations only on very small scales, which were measured with COBE (Cosmic Background Explorer) and WMAP very accurately(Fig. 15.10) :

- 10^{-3}: Fluctuations of this magnitude show an angular dependence of the temperature. This can be represented with the help of a dipole distribution:

$$T(T_0)\left(1 + \frac{v}{c}\cos\theta\right) \tag{15.111}$$

This dipole asymmetry can be interpreted as a motion of the Earth relative to the co-moving reference frame, which follows from the expansion of the universe and in which the background radiation appears isotropic. One gets a velocity of 350 km/s in the direction of the galactic coordinates $l = 264.25° \pm 0.33°$, $b = 48.22° \pm 0.14°$.
- With an accuracy of 10^{-5} the background radiation exhibits temperature fluctuations which no longer satisfy a simple dipole distribution. These are interpreted as fluctuations in the early universe and are therefore of great importance for the structure formation in the early Universe.

In mid-May 2009, the space probe PLANCK together with the probe HERSCHEL (observes in the IR region) was launched. It is located at a distance of 1.5 million km from Earth, at the Lagrangian point L2, and the entire sky has been mapped in the microwave range. Thus one looks back to the young universe, when it was about 400,000 years old (Fig. 15.11).

Fig. 15.10 Background
radiation measured by COBE.
(**a**) One can clearly see the
dipole asymmetry (amplitude
10^{-3}), (**b**) with emission from
the Galaxy (between 0.3 and 1
cm), (**c**) adjusted; the dark
regions are 0.0002 K warmer
than the bright ones

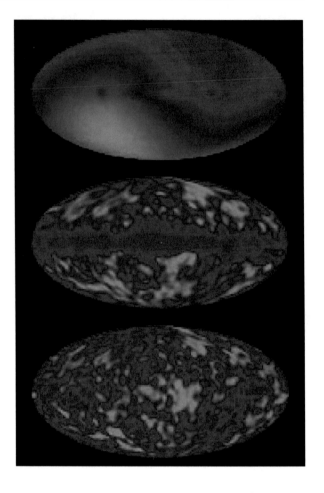

At $z = 1\,100$ there were no stars or galaxies; when the universe was 500 million years
old it already consisted of 27% dark and normal matter and 73% dark energy. Anisotropies
in the range of 10^{-6} K were found. These were shaped by the first young galaxies.

The deviations of the CMB in the range of $10^{-3}\ldots 10^{-6}$ K can be expanded into an
angular distribution by spherical surface functions:

$$\frac{\Delta T(\theta, \phi)}{T} = \sum_{l=1}^{\infty} \sum_{m=-l}^{l} a_{lm} Y_{lm}(\theta, \phi) \tag{15.112}$$

The power spectrum can then be given as follows:

$$C_l = <|a_{lm}|^2 > \tag{15.113}$$

Thus, for each l we get the strength of the l-th moment. Now plot l vs. $\Delta T[\mu K]$: $\rightarrow$
so-called angular power spectra, i.e. the intensity distribution of the CMB in the sky is

Fig. 15.11 The extreme deep field taken with the Hubble Space Telescope shows galaxies up to a distance of slightly more than 13 billion light-years ($z = 10$). One finds about 5000 galaxies on it (NASA/HST)

expanded into multipoles. At $l \approx 220$ the first acoustic peak occurs, at $l \approx 540$ the second. Due to the Sunyaev-Zel'dovicheffect, CMB photons were scattered by protogalaxies (Compton scattering), and these scattering effects modify the power spectra.

Note: Angular distances in the CMB of 1 degree correspond to the size of the horizon at the time of decoupling. Particles farther than 1 degree away in the CMB could not interact, anisotropies on these scales indicate perturbations that existed before the inflationary Phase (next chapter).

Another property of the CMB is polarization. So the background radiation must have been scattered by electrons. Now it was just mentioned that the CMB comes from the recombination era, so there were no more free electrons, so where did the scattering come from? This comes from the reionization era of the universe. The neutral cosmos was re-ionized by the first hot stars. So we get two important pieces of information from the background radiation:

- Temperature fluctuation → Anisotropies
- Polarization.

15.5.3 Acoustic Oscillations

Let us consider an inhomogeneous mass distribution . This leads to an inhomogeneity of the space $\rightarrow$ can be represented by Fourier expansion as a superposition of many periodic functions. Each periodic potential can be assigned a very specific oscillation frequency. At the time of recombination these oscillations cease. A temperature distribution is generated which depends on the state at the time of decoupling.

First Peak in the Power Spectrum First Frequency reaches its maximum. It is the angle which describes the acoustic horizon. This is the distance that the acoustic wave can just traverse until the moment of decoupling.

The *second peak* is an extremum of the wave with the double frequency and so on. From this one can determine the baryon part of the total matter in the universe. Baryons shift the equilibrium position of the oscillations by their mass.

From the *third peak* follows the radiation driving. If the radiation density is known, then follows the the fraction of dark matter.

15.5.4 Formation of Particles

In Table 15.3 the rest masses of some important particles are given.
Very often energies are given with the help of temperature. If one calculates with

$$E \approx kT$$

and the energy unit $1\,\mathrm{eV} = 1.60 \times 10^{-19}\,\mathrm{J}$ from E, then

$$\frac{1\,\mathrm{eV}}{k} = \frac{1.60 \times 10^{-19}\,\mathrm{J}}{1.38 \times 10^{-23}\,\mathrm{J/K}} \tag{15.114}$$

and

$$1\,\mathrm{eV} = 1.1605 \times 10^4\,k\,\mathrm{K} \tag{15.115}$$

Table 15.3 Rest masses of some important particles

Particle	Designation	Mass in MeV/c^2
Proton	m_p	938.3
Neutron	m_n	939.6
Electron	m_e	0.511
Muon	m_μ	140

Calculate which energy to a temperature of 10^{12} K corresponds to! Solution: from $E = kT$ follows $E = 1.38 \times 10^{-23} \times 10^{12} = \ldots J = \ldots eV$
Solution: 100 MeV.

Let us consider the universe at $T = 10^{12}$ K which corresponds to 100 MeV. From Table 15.3 it follows that all the baryons we know today were already there. The WIMPS,[9] which possibly represent a part of the dark matter, have rest-masses > 100 GeV. Except for the WIMPs, all Particle are in equilibrium. There are the following reactions:

$$Compton - Scattering$$

$$e^{\pm} + \gamma \leftrightarrow \quad e^{\pm} + \gamma$$

$$Pair\ ₅generation,\ Annihilation$$

$$e^{+} + e^{-} \leftrightarrow \quad \gamma + \gamma$$

$$Neutrino - Antineutrino - Scattering$$

$$\nu + \bar{\nu} \leftrightarrow \quad e^{+} + e^{-}$$

$$Neutrino - Electron - Scattering$$

$$\nu + e^{\pm} \leftrightarrow \quad \nu + e^{\pm}$$

The mass of the neutrinos is less than 1 eV. These reaction rates are influenced by the number density $n \approx a^{-3} \approx t^{-3/2}$ and the cross section $\sigma \approx a^{-2}$. At early times the particles were in equilibrium, the reaction rates were greater than the expansion rates.

For $T < 10^{10}$ K the neutrinos are no longer in equilibrium with the other particles; a decoupling, or freezing out of the neutrinos occurs, the universe is 1 s old.

Once the temperature has fallen below 5×10^{9} K ($kT \approx 500$ keV), electron-positron pairs can no longer be created efficiently.

15.5.5 Quarks and Quark-Gluon Plasma

In the Standard Model of Particle physics Quarks and Leptons are the fundamental building blocks of matter. Quarks are point-like fermions with spin 1/2 and third-numbered elementary charge of $+2/3$ or $-1/3$. Furthermore, they have the property of Color charge. Quarks exist in six, *flavors,* and their masses and charges are given in Table 15.4.

Protons and neutrons are not elementary particles, but are made up of *Quarks*. All particles that occur in nature are without *color*: therefore in combinations *bgr* (blue-green-red) or their antiparticles.

---

[9] Weakly interacting massive particles.

Table 15.4 Properties of quarks

Name	Mass	Charge	Antiparticle	Charge
u (up)	5 MeV	+2/3	$\bar{u}$	+2/3
d (down)	10 MeV	−1/3	$\bar{d}$	−1/3
s (strange)	200 MeV	−1/3	$\bar{s}$	+1/3
c (charmed)	1.5 GeV	+2/3	$\bar{c}$	−2/3
b (bottom)	4.7 GeV	−1/3	$\bar{b}$	+1/3
t (top)	180 GeV	+2/3	$\bar{t}$	−2/3

In addition to these six quarks there are the antiquarks. Hadrons are divided into baryons (consisting of three quarks) and mesons (consisting of two quarks with color and anticolor). Examples:

- Protons: uud,
- neutrons: udd.
- π−Meson: $u\bar{d}$, $\bar{d}$ means antidown quark.

Other particles consisting of more than three quarks have since been found such as the tetra-quark or the penta-quark.

The interaction between quarks takes place through exchange particles , the so-called gluons.

Under normal conditions quarks are *confined (confinement,* chiral symmetry broken), only above 150 MeV quarks are quasi free (*deconfinement,* chiral symmetry restored). Then a quark-gluon plasma is formed. This plays a role in the formation of the universe, in quark stars, in the interior of neutron stars and in Magnetars.

In the early universe, the phase transition to quark-Gluon plasma occurred at 170 MeV or $T = 10^{12}$ K. Free quarks were found in the laboratory with the RHIC (Relativistic Heavy Ion Collider, Brookhaven National Laboratory, Long Island, USA) in March 2003 experiments with gold-deuteron collisions were performed, the small deuteron (atomic nucleus of deuterium) shoots through the much larger gold atoms and quarks are torn out.

15.5.6 Particle Generation

Consider an elementary particle of m . At some point in the evolution of the universe, there was a time when valid:

$$kT \approx mc^2 \tag{15.116}$$

Prior to this time, a collision between two photons resulted in the creation of a particle-antiparticle pair. Once T below mc^2/k sank, particle pairs could no longer form, and particles and antiparticles annihilated each other, provided the density and thus the collision frequency were large enough. One can also subdivide the elementary particles into:

1. Leptons: Neutrinos, electrons, muons, tauons and their antiparticles.
2. Hadrons:
   (a) Baryons: Protons, neutrons, hyperons (unstable).
   (b) Mesons.

The critical temperature for Hadrons is 10^{12} K. Before this time, therefore, one speaks of the *Hadron era* of the universe. One has the following sections (t... Age of the universe):

1. $T \approx 10^{12}$ K, $t \approx 10^{-4}$ s: Pair annihilation of muons; muon neutrinos and their antiparticles decouple.
2. $T < 10^{11}$ K, $t > 0{,}01$ s: The mass difference between neutrons and protons (1.3 MeV $\approx T = 1{,}5 \times 10^{10}$ K) begins to produce more protons and fewer neutrons. The following processes are important in this process:

$$n + \nu_e \leftrightarrow p + e^- \qquad n + e^+ \leftrightarrow p + \bar{\nu}_e \tag{15.117}$$

$$n \leftrightarrow p + e^- + \nu_e \tag{15.118}$$

The number ratio $n : p$ is determined by the temperature:

$$N_n / N_p = \exp(-1.5 \times 10^{10}/T) \tag{15.119}$$

3. At a temperature of $T \approx 10^{10}$ K when the uniserse had an age of $t \approx 1$ s the electron neutrinos and the antineutrinos decouple.
4. $T \approx 5 \times 10^9$ K, $t \approx 4$ s: the electrons and positrons annihilate each other; neutrino cooling fixes the $n : p$ ratio. One then has only the β decay:

$$n \rightarrow p + e^- + \bar{\nu}_e \tag{15.120}$$

5. $T \approx 10^9$ K; $t \approx 10^2$ s: nuclear synthesis begins, and ^{4}He, deuterium (^{2}H) and ^{3}He, ^{7}Li are produced. For example:

$$p + n \leftrightarrow {}^2\text{H} + \gamma \tag{15.121}$$

$$^2\text{H} + {}^2\text{H} \leftrightarrow {}^3\text{He} + n \tag{15.122}$$

$$^3\text{He} + n \leftrightarrow {}^3\text{H} + p \tag{15.123}$$

$$^3\text{H} + {}^2\text{H} \leftrightarrow {}^4\text{He} + n \tag{15.124}$$

As the universe expands, cooling occurs. Heavy particles were created at higher temperatures, light particles at lower temperatures. A temperature of 10^{12} K corresponds to 100 MeV. All baryons known to us were already present.

15.6   Symmetry Breaking in the Early Universe

We first consider the four fundamental forces that describe the interactions between particles today. It turns out that the weakest of these forces, gravity, determines the large-scale structure of the universe, i.e. the shape of galaxies, clusters of galaxies, stars, planets, and so on. Attempts are being made to attribute these forces to one force in what are called GUTs, grand unified theories. String theories and quantum loop gravity extend the standard model of particle physics and can also explain gravity.

15.6.1 The Four Forces of Nature

In the nineteenth century, three basic forces of nature were known: gravitation, electricity, and magnetism. *Maxwell* was able to show that electricity and magnetism can be unified to electromagnetism. Atomic physics and quantum physics led to the discovery of further basic forces (Table 15.5).

The *strong force* (strong interaction) is stronger than gravity by a factor 10^{40}, but acts only at very short distances holding atomic nuclei together. The *electromagnetic force* (EM) is long-range, but can be attractive (between different charges) and repulsive (between charges of the same sign), and only for charged particles. Therefore, the electric forces in the universe neutralize each other.

The *weak interaction* is responsible for radioactive decay and the neutrino interaction. As an example, the decay of a free neutron in Fig. 15.12 is shown in a Feynman diagram (Fig. 15.12). In these diagrams the x-axis is the time.

Table 15.5 The four fundamental forces; K means coupling strength, R the range

Force	K	R [m]	Transducer	Acts on
Strong	1	10^{-15}	Gluons	Quarks
EM	1/137	∞	Photons	All charged Particles
Weak	1×10^{-5}	10^{-18}	W^+, W^-, Z^0	Quarks & Leptons
Gravity	10^{-40}	∞	Gravitons	Particles with mass

Fig. 15.12 For example, the electroweak interaction explains the decay of a free neutron (consisting of udd quarks) to a proton (consisting of uud quarks)

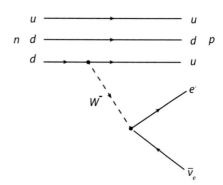

The *Gravity* is by far the weakest interaction, and yet it dominates the structure of the universe. There are two reasons for this:

- The range is unlimited,
- Between two masses in the universe there is only attraction and no repulsion (but cf. electromagnetic interaction).

It's assumed, all four basic forces result of one single force by so called spontaneous symmetry breaking.

Electromagnetic, weak interaction and strong interaction are described in GUTs *(grand unified theories)*. There is experimental evidence for this assumption of unification of the four forces of nature: About 40 years ago it was discovered that the electromagnetic force combines with the weak interaction to form the electroweak force. In the so-called *Standard Model* the interaction can be calculated. In 1979 *Glashow, Salam* and *Weinberg* received Nobel Prize for their theory of the electromagnetic and the weak interaction, 1984 discovered *Rubbia* and *Simon* the From W and Z particles predicted by this model.

Particles can be divided into:

- Bosons: integer spin. $s = 0, 1, 2, \ldots$
- Fermions: half-integer spin; $s = 1/2, 3/2, \ldots$

Fermions bosons are subject to the Pauli Principle The bosons, on the other hand, do not and can act as transmitters of forces.

→ In physics, forces are transmitted by so-called exchange particles. Analogy: How can two ice skaters repel each other? By throwing snowballs at each other; they are attracted by each throwing the snowballs in the opposite direction. In this case the snowballs act as interacting particles.

The photons are the transmitters of the electromagnetic force. If one has thus an electron bound to a proton, photons are constantly exchanged here to transmit the electromagnet attraction. In the weak interaction, these are the W^+-,W^--, Z^0-particles, each of which has 100 times the mass of the proton. In the case of the strong force, the transducers are the Gluons.

The leptons are fermions subject to the electroweak interaction as well as gravity, but not the strong force. These particles include electrons and neutrinos with their corresponding antiparticles.

15.6.2 The Early Universe

So far we have studied the evolution of the early universe from a point in time of 10^{-4} s after the big bang, when its temperature was 10^{12} K. The physics of these processes can be understood and verified today in particle accelerators. For even earlier phases, however, higher energies are required and we must rely on theoretical considerations..

At an age of $t \approx 10^{-6}$ s, at which the temperature was $T \approx 10^{13} K$, the quarks annihilate with their antiparticles, you have a soup of quarks and leptons. The electromagnetic force and the weak interaction combine to form the electro weak force. So a phase transition occurs. The theory for this was developed by Weinberg and Salam at CERN with the help of accelerator experiments (W and Z particles were detectee).

Up to 10^{-35} s After the Big Bang, the electroweak force was united with the strong interaction—GUTs *(grand unified theories)*. One of the predictions of the GUTs is the decay of the proton within 10^{31} years.

> At the ever-increasing temperatures of the early universe, a unification of the electromagnetic force with the weak force occurred, and when the age of the universe was only 10^{-35} s the electroweak force unified with the strong force (GUT).

In Fig. 15.13 the behavior of the coupling constants of the four fundamental forces as a function of energy is plotted. One can see how the constants approach each other at high energies.

This splitting or phase transition can be thought of as the phase transition of ice to liquid water. Ice is in a higher state of symmetry (crystals) than water because water, being a liquid, can take any form.

15.6.3 Inflationary Universe

In 1981 A. *Guth* has developed the concept of the *inflationary universe*. When the strong interaction separated from the weak and electromagnetic, a phase transition (symmetry breaking) occurred when the universe was about 10^{-35} s old. This filled the universe with energy, the so called vacuum energy (or false vacuum energy). Thus for about 10^{-32} s gravity became repulsive, and this resulted in extreme expansion of the universe (by a factor of 10^{26}). As soon as the phase transition was completed, the normal evolution continued (Fig. 15.14).

So the energy density of the vacuum transformed into matter and radiation. The theory of the inflationary universe is needed to explain peculiarities of the universe:

- Horizon problem: Consider the universe in the microwave range. Two opposite points in the sky can never have been physically related, but inflationary expansion can explain this.
- Flatness problem: The density parameter Ω_0, which describes the ratio of actual matter density to critical matter density (the density required to maintain a closed universe), is

Fig. 15.13 At higher energies, the coupling constants of the four forces (s strong, w weak, em electromagnetic and g gravitational) approach each other

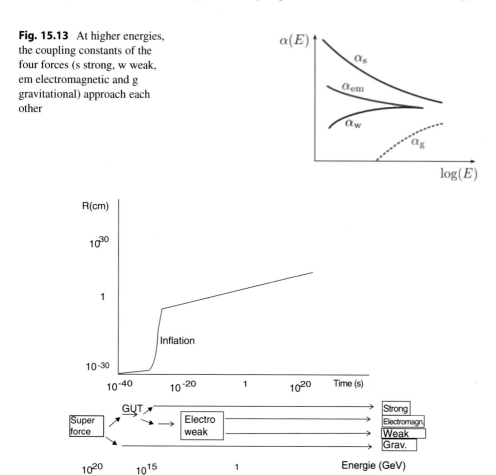

Fig. 15.14 Evolution of the universe

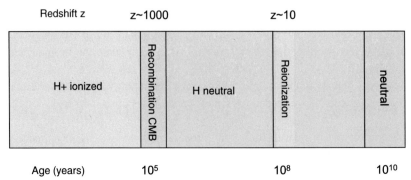

Fig. 15.15 Evolution of the universe: At $z = 1000$ the universe becomes transparent, from this time comes the cosmic background radiation (CMB). Then the universe was neutral (mainly neutral H). Due to massive, luminous stars and quasars, a new ionization at $z = 10$ took place when the age was about 10^8 years

between 0.03 and 2. But already 1 s after the big bang was $1 - \Omega_0$ was only 10^{-15}, so the universe was extremely flat.[10]

- Rapid expansion caused magnetic monopoles that may have formed earlier to become very widely dispersed and therefore undetectable today.
- Inflation created small density fluctuations, which were then the condensation nuclei for later matter structures (galaxies, galaxy clusters, ...).

In Fig. 15.15 the thermal evolution of the universe is sketched.

Even earlier than 10^{-43} s (Planck time) one needs a quantum theory of gravity. Here there are many speculations like *Superstring theories* of *Green* and *Schwartz*.

> The inflationary phase occurred due to the phase transition that led to the splitting of the strong force from the electroweak force; the universe expanded by a factor of 10^{26} from the size of a proton to the size of 10 cm.

15.6.4 String Theory

We have already discussed the problem of unification of the four fundamental forces. The so-called standard model is unsatisfactory on this point because gravity cannot be incorporated. In string theory one assumes that particles consist of strings, which do not

---

[10] For a flat universe $\Omega_0 = 1$.

Fig. 15.16 Interaction
between strings; this does not
happen at a well-defined point

Fig. 15.17 Interaction in the
Standard Model; this happens
at a point and does not make
mathematical sense for
gravitons

show any further structure, but can appear in different states of vibration. Particles are thus
vibrations of a string, and one finds:

- Particles that form matter: Electrons, muons, neutrinos and elementary quarks,
- transmitters of interactions: Photons, W and Z bosons, gluons,
- Graviton: this particle, which is supposed to transmit gravitation, is also postulated, and
  thus gravitation would be integrated by string theory (Fig. 15.16).

The four-dimensional space-time structure of the universe should naturally follow from the
string theories. Relativistic quantum theory describes the properties of elementary particles
in the presence of very weak gravity, and particle theory works only in the absence of
gravity. General relativity describes the orbits of planets, big bang and black holes, and
gravitational lensing. However, here one mostly calculates with a classical universe, i.e.
quantum effects are not considered. Therefore, string theory would fill this gap.

Gravitons are the transmitters of gravity. Suppose a graviton is placed in a quantum
field, the particle interactions happen at a point in space-time with zero distance between
the interacting particles (Fig. 15.17). This does not make sense. In string theory, strings
collide at a small finite distance, and the mathematical solutions appear to make sense.

In string theory, elementary particles are described as excitation modes of elementary
strings. The strings are free in space-time. The tension of the strings is described by:

$$\frac{1}{2\pi a'} \tag{15.125}$$

whereby a' is equal to the square of the string length scale. Strings are of size in the range of the Planck length (10^{-35} m). There are two kinds of strings: open and closed strings. In order to integrate also the fermions into the theory, one needs the so-called *Supersymmetry:* For every boson (which are the particles that transmit forces) there exists a corresponding fermion (which are the particles that make up matter). And for every fermion there should be a supersymmetric boson. These heavy supersymmetric particles should be found by particle accelerators of the coming generation.

String theories can thus be formulated in different ways:

* open or closed strings,
* boson strings only: Boson strings,
* for bosons and fermions → Super symmetry, super strings.

For a string theory of bosons one needs 26 space-time dimensions, for super strings only ten. For the former, there is a particle with an imaginary mass, the tachyon. The reduction of, say, ten space-time dimensions to our known four is called *compactification.* M-theory *(mother of all theories)* unifies the many different string theories.

> According to string theories, there must be more than the four space-time dimensions we know.

The idea of more than four dimensions is not new. *Klein* and *Kaluza* around 1921 were the first to advocate the idea that our universe could have more than three spatial dimensions. But why do we observe only the known three? Here is an analogy: If we look at a garden hose from a distance, it looks like a one-dimensional object. Only up close do we realize that it is a three-dimensional object. It could also be the case below the current measurement limits of 10^{-16} cm there are further dimensions. According to the theory of Kaluza-Klein, gravity and Maxwell's equations can be described together in a 5-dimensional universe. In addition to the three space dimensions we know, there should be one more. The problem was, however, that this theory could not be unified with quantum mechanics at all, and so it was abandoned.

String theory now says that specific particle properties such as mass or electric charge are determined by the size, number, shape of the holes, conditioned by the higher dimensions.

In *Topology* one examines the properties of geometric space, which do not change under:

* Space dilation,
* twist,
* bending.

Example

A doughnut and a sphere are topologically distinct, there is no way to create a doughnut from a sphere by twisting or bending. A doughnut and a teacup look different. However, these two bodies are topologically the same.

In general relativity, space-time is assumed to be constantly changing its size and shape (expansion of the universe). However, the topology of the universe remains the same.

Let us consider quantum mechanical processes on the smallest size scales: At smallest scales there are fluctuations, and therefore one must assume such fluctuations in the space-time structure as well. These fluctuations average out when we look at the universe on larger scales. Mathematically, this can be represented by shrinking to a singular point followed by expansion in an orthogonal direction. The singularity cannot be described by general relativity.

The boundary conditions of strings can be different: Closed strings have periodic boundary conditions, open strings have two types of boundary conditions: (a) Neumann boundary conditions, the endpoint is free to move but has no momentum, (b) Dirichlet boundary conditions, where the endpoint is fixed on a manifold, this is called a Dp-brane (p indicates the number of spatial dimensions of this manifold). In Fig. 15.18 one has open strings with one or both endpoints fixed on a D2-Brane.

Let us examine the interaction of D-branes with gravity. In Fig. 15.19 one has a graviton, represented by a closed string, interacting with a D2-brane. The string thereby becomes an open string, and at the time of the interaction the endpoints are on the D-brane.

A good application to test this concept is found in Hawking black hole radiation. If one assumes open strings, then radiation follows in the form of closed strings. The *Beckenstein-*

Fig. 15.18 D2 brane of an open string

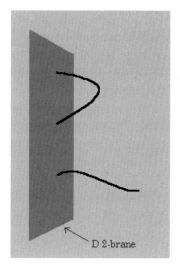

D 2-brane

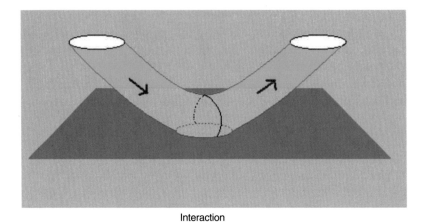

Interaction

Fig. 15.19 D-brane interaction with a graviton. Note the transition to an open string at the time of the interaction

Hawking entropy formula states that the entropy S of a black hole can be expressed by:

$$S = A/4, \tag{15.126}$$

where A is the event horizon. Once an object has penetrated the event horizon, it can never escape from it. The above formula can be explained using string theoryren.

15.6.5 Quantum Foam

The two major theories of 20th century physics are general relativity and quantum theory. If one applies these two theories to scales of Planck length, i.e., about 10^{-35} m, then bubbles of space-time are continuously formed, which decay again. Due to the uncertainty principle, the uncertainty of location (Δx) and momentum (Δp) of a particle:

$$\Delta x \Delta p \geq \hbar/2 \tag{15.127}$$

The momentum can therefore fluctuate by $\Delta p < \frac{\hbar}{2\Delta x}$ without being able to measure it. The uncertainty of the energy is:

$$mc^2 = \Delta E = \Delta pc = \frac{\hbar c}{2\Delta x} \tag{15.128}$$

A black hole is determined by the Schwarzschild radius: $r = \frac{2Gm}{c^2}$.

Objects whose extent is equal to the Planck length would have to possess a mass of 10^{-8} kg , because:

- $m < 10^{-8}$ kg: spatial uncertainty → greater expansion.
- $m > 10^{-8}$ kg: Schwarzschild radius > Planck length.

15.6.6 Quantum Vacuum

As already shown, the vacuum is not empty, but filled with virtual particles. According to the Heisenberg uncertainty principle, a particle of energy ΔE a short time

$$\Delta t < \hbar/\Delta E \qquad (15.129)$$

exist. Thus, virtual particle/antiparticle pairs are created. If the time is very short, high energy or large masses are created because $m_0 = E/c^2$. Large masses in turn bend space-time, and bubbles are created in space-time. The quantum foam is chaotic in structure.

The quantum vacuum represents a state of lowest energy → However, the energy is never exactly zero. Into the above formula enters:

- Planck length: 10^{-35} m,

$$l_P = \sqrt{\frac{\hbar G}{c^3}} \qquad (15.130)$$

- Planck time: 10^{-43} s.

$$t_P = \sqrt{\frac{\hbar G}{c^5}} \qquad (15.131)$$

Time needed for light to travel one Planck length.

The location and time of a particle can never be determined more precisely than these two quantities.

→ Space and time are not continuous, but are a composite of quanta of space-time .

15.6.7 Loop Gravity, Quantum Loop Gravity

We have used string theory as a way of to describe gravity with the other forces of nature. The singularity of the big bang could be avoided. Any object smaller than the Planck scale would immediately collapse into a black hole due to the uncertainty principle, as it has

Fig. 15.20 The MAGIC telescope at the Observatorio Roque de los Muchchos in La Palma

extremely high energy or mass. In loop gravity, one assumes (as in string theory) that there cannot be arbitrarily small structures. Space itself obeys quantum mechanics, it can be described by quantum states. Such a quantum state is described by a network of nodes connected by lines. The distances between nodes correspond to the Planck length. A cubic centimeter would thus contain 10^{99} nodes. The universe contains about 10^{85} cm^3. Thus, one could obtain much more information with a perfect microscope that could zoom into a cubic centimeter than with a perfect telescope that resolves the universe to cm-length!

Where is this mesh embedded? The mesh itself defines the space. There is nothing between the links and nodes. Analogy: consider a sand dune. It is made up of many grains of sand. Between the sand grains there is nothing. The elementary particles are then net nodes. The movement of the particles is a displacement of the nodes in the net.

Some effects can be predicted with loop gravity: long wavelength gravitational waves, Hawking radiation, black hole entropy, positive cosmological constant. Other effects would be a wavelength dependence of the speed of light. This would be especially apparent when the wavelength becomes comparable to the distances between nodes, i.e. the Planck length. If detectable at all, this could only be detected for highest energy cosmic rays, here the difference would amount to 10^{-9}. This would therefore mean differences in transit time at different wavelengths, e.g. during a gamma-ray burst. The blazar Markarjan 501, is 500 million light-years away. This object was studied with the MAGIC telescope (diameter about 15 m, Fig. 15.20)[11] and showed transit time differences.

---

[11] MAGIC stands for Major Atmospheric Gamma Imaging Cerenkov Telescopes.

15.6.8 The First Stars

We have seen that the universe consists of three components:

1. Baryonic matter: visible, observable; about 4%.
2. Dark matter: not observable, gravitational effect only! About 23%.
3. Dark energy: explains accelerated expansion of universe.

On large scales, the universe exhibits a filamentary structure. This cannot be explained by baryonic matter alone. Baryonic matter is subject to thermal pressure, i.e. gas pressure → No clumping of matter. Dark matter is subject only to gravity and not to pressure.

So dark matter explains the filamentary distribution of baryonic matter. Giant clouds formed at the network nodes and collapsed. Population III stars formed: they contained only hydrogen and helium, which affects opacity. This is why they could be more massive than the Stars.

Population III stars: some 100 solar masses
Diameter $\approx 10\,R_\odot$
Surface temperatures: some 100,000 K
"Dark" stars could have been formed during the collapse of dark matter. However, they shone due to annihilation of dark matter particles and antiparticles. Mass determined according to

$$E = mc^2 \tag{15.132}$$

is converted into energy. Dark stars were therefore extremely bright. Lifetimes were only a few million years. Now there are two scenarios:

- After this Dark Matter dispersion due to anihilation, the star shines due to normal nuclear fusion. However, because of the large mass, this means also a short lifetime, and the star explodes into a supernova. A black hole remains, and the interstellar matter is enriched with heavier elements.
- Massive dark stars do not reach normal stellar stage (i.e. thermonuclear fusion) and become heavy black holes of several thousand or more solar masses. This may have given rise to the Supermassive Black Holes (SMBH) in the nuclei of galaxies.

The successor to the Hubble telescope, the James Webb Space Telescope, will be used to search for such supernovae. Since they originate from the early phase of the universe, they show an extreme red shift and should therefore be observable in the IR.

15.6.9 Parallel Universes

There are two ways to think of the creation of parallel universes:

The universe originated from quantum foam; if our universe originated from it, then any number of universes with possibly quite different laws of nature may have originated.

Many-worlds interpretation of quantum mechanics: During observations, the world splits into multiple worlds. The totality of all parallel worlds is also called multiverse. The difference to above is, that here the same laws of nature must be valid in all universes.

15.7 Time Scale

We scale the evolutionary history of the universe from the origin to the present to one year. Then we find the data given in Table 15.6 given data.

15.8 Further Literature

We give a small selection of recommended further reading.
Bergström, L., Goobar, A., Cosmology and Particle Astrophysics (Springer Praxis Books) (Paperback) Springer, 2006

Table 15.6 History of the evolution of the universe from the origin to the present day scaled to one year

Time	Event
January 1 0^h0^m	Big bang, creation of H, He
January 1 0^h14^m	Decoupling of radiation and matter
5 January	First stars and black holes, C, N, O, ...
16 January	oldest known galaxy, quasar
9 September	Formation of solar system and earth
28 September	First life on Earth (cyanobacteria)
December 16–19	Vertebrate fossils, plants
20–24 December	Forest, fish, reptiles
December 25	Mammals
28 December	Extinction of the dinosaurs
December 31, 20^h00^m	First humans
December 31 23^h55^m	Neanderthals
December 31, $23^h55^m56^s$	Year 0
January 12	Earth becomes too hot
April 7	Sun becomes a red giant
April 16	Collision Milky Way with Andromeda Galaxy

Sexl, R.U., Urbantke, H.K.: Gravitation and Cosmology, Spektrum Akademischer Verlag, Heidelberg, 5th edition 200
Introduction to General Relativity and Cosmology, C.G. Böhmer, World Scientific, 2016.
Gravitation and Cosmology, S. Weinberg, WSE, 2008
Modern Cosmology, A. Liddle, Wiley, 2015
Extragalactic Astronomy and Cosmology, P. Schneider, Springer, 2014
Foundations of Modern Cosmology, J.F. Hawley, K.A.Holcomb, Oxford Univ. Press, 2005
Astrophysics and the Evolution of the Universe, L.S. Kisslinger, World Scientific, 2016

Tasks

15.1 Under which simplification does $\tau = 1/H$ gives the actual age of the world?

Solution
Uniform expansion.

15.2 At what distance is there a galaxy from us whose red shift is $z = 0.1$ is?

Solution
If we calculate in the units for the speed of light [km/s] or for the Hubble constant [km s^{-1} Mpc^{-1}]: $d = cz/H_0 = 3 \times 10^5 \times 0.1/(50 \times 1.1) \rightarrow d = 600$ Mpc.

15.3 When cosmic ray particles collide, muons are produced in the Earth's atmosphere. These travel at 99.4% of the speed of light. Their half-life is, for particles at rest 1.5×10^{-6} s. By what amount does its half-life increase?

Solution
One first calculates $\gamma = 1/\sqrt{1 - 0.994^2} \rightarrow \Delta t' = \gamma \Delta t = 13.7 \times 10^{-6}$ s. Fast-moving muons decay more slowly, travel preserves young.

15.4 Determine the components of the metric tensor in the case of Minkowski space!

Solution
From $ds^2 = dx^0 dx^0 g_{00} + dx^0 dx^1 g_{01} + dx^0 dx^2 g_{02} + dx^0 dx^3 g_{03}$
$+ dx^1 dx^0 g_{10} + dx^1 dx^1 g_{11} + dx^1 dx^2 g_{12} + dx^1 dx^3 g_{13}$
$+ dx^2 dx^0 g_{20} + dx^2 dx^1 g_{21} + dx^2 dx^2 g_{22} + dx^2 dx^3 g_{23}$
$+ dx^3 dx^0 g_{30} + dx^3 dx^1 g_{31} + dx^3 dx^2 g_{32} + dx^3 dx^3 g_{33}$ it follows that $g_{00} = -1$, $g_{11} = 1$, $g_{22} = 1$, $g_{33} = 1$ and all other components are zero.

15.5 Could a dark matter planet entering our solar system be observed?

Solution
No, only by its gravitational effects; you couldn't even observe a transit!

15.6 Show how to calculate from the formula 15.55 find the classical Doppler effect!

Solution
Set $c \to \infty$.

15.7 Show that, unlike the classical Doppler effect, there is a red shift in the relativistic one even when the source is moving transversely!

Solution
Set $\cos \Theta = 0$. A consequence of time dilation.

Astrobiology

16

Astrobiology investigates the possibility of the origin of life on other planets and objects (i) in the solar system, (ii) around other stars, so-called exoplanets . This branch of research was initially purely speculative. It was not until around 1990 that direct evidence of planets around other stars was obtained.

16.1 Life on Earth and in the Solar System

So far, we only know of life on Earth. It is usually assumed that the origin of life is linked to the presence of liquid water as well as complex organic compounds. We briefly describe the evolution of life on Earth, as well as the importance of Earth's atmosphere and magnetic field to life.

16.1.1 What Is Life?

The definition of life is complex. Above all, life requires a source of energy. Plants on Earth need solar energy, carbon dioxide, and water. Life grows and reproduces. Life is characterized by *metabolism*. These characteristics have been used in various experiments on the surface of Mars to search for life. Life organizes itself into complex structures and adapts to its environment.

Life on Earth has as its smallest unit the cell. There are unicellular life forms that consist of a single cell (most species of bacteria), as well as multicellular life. The human body consists of about 100 trillion cells.

Life on Earth is based primarily on the presence of *water*. Water has special properties: it stores heat, serves as a solvent, and acts as a means of transport in cells. When it evaporates, water can have a cooling effect and thus keep a complex organism at a nearly

© Springer-Verlag GmbH Germany, part of Springer Nature 2023 605
A. Hanslmeier, *Introduction to Astronomy and Astrophysics*,
https://doi.org/10.1007/978-3-662-64637-3_16

constant temperature. Ice is lighter than water, which is why a layer of ice always forms on the surface of a lake or body of water, allowing living things in the water below the ice crust to survive. However, there are some creatures that can go for long periods without water and thus survive extreme conditions: spores of bacteria dehydrate and can even withstand the low temperatures of space. Therefore, precautions are taken so that landing space probes on other objects of the solar system (Mars, Titan, Europa) does not infect them with terrestrial life.

In addition to water, carbon is also of particular importance to life on Earth. Carbon atoms can share electrons with other carbon atoms and form long chains of carbon molecules.

Both water and carbon are found relatively frequently in the universe:

- Water: Traces of water have been found on all planets in the solar system. The outer planets and their moons contain particularly large amounts of water.
- Carbon or organic compounds: were found in various planetary atmospheres such as Venus, Mars, Jupiter; but also on the surface of Titan, Europa or Pluto, and in comets and meteorites.

At present, we know of only one example of the development of life: life on Earth.

16.1.2 Life on Earth

Earth was formed about 4.6 billion years ago. It took about 10–20 million years for the Earth to form. During the early phase of the earth there was the big bombardment, there were still a lot of small bodies in the solar system, which crashed onto the earth and other planets. It is possible that much of the water on Earth came from comet impacts. Life was not possible at this early stage.

The origin of life on Earth about 3.5 billion years ago is still debated:

- Urey-Miller experiment: conducted in 1952/1953 by Urey and Miller at the University of Chicago, it has since been refined in several variations. A container of water, methane, ammonia and hydrogen was heated and subjected to electrical discharges. After a week, 10–15% organic compounds were detected in the mixture.

  Later studies showed that UV radiation from the Sun also promotes the formation of such compounds (e.g., in the dense atmosphere of Saturn's moon Titan).
- Panspermia theory: Life must have evolved only once somewhere in the universe and then spreads all by itself (cf. spores of bacteria that survive space flights). However, whether spores survive the long journey between stars, which takes tens of thousands of years cannot be answered.

- Black Smokers: The Earth is geologically active, plates are moving, and hot springs are forming on the ocean floors. Normally, the water at the ocean floors has a temperature of about 2 °C but at the hot springs there are temperatures of 60–400 °C. At these hot springs, we found *archaea* and *extremophilic bacteria* bacteria that survive in extreme conditions (e.g. very hot or very salty). If life on earth would have originated in such black smokers, then it would have been well protected by the water masses against the strong UV radiation of the sun in the early time of the earth. The UV radiation of the early Sun was much stronger than it is today. Sun-like young stars shine in the UV and X-ray range 100–1000 times brighter than today's Sun in these wavelength ranges. Moreover, this radiation reached the Earth's surface unimpeded because the Earth's atmosphere did not yet have a UV-absorbing ozone layer.

    Black Smokers could also exist on the ocean floors of Jupiter's moon Europa.

Life on Earth originated about 3.5 billion years ago *(Cyanobacteria)*. By photosynthesis

$$6CO_2 + 6H_2O \rightarrow C_6H_{12}O_6 + 6O_2 \tag{16.1}$$

the earth's atmosphere was slowly enriched with oxygen and an ozone layer was formed. Only then could living organisms evolve from water to land. About 750 million years ago, the first multicellular *eukaryotes* appeared.[1] about 600 million years ago there was the so-called *Cambrian explosion* during which life diversified extremely rapidly.

16.1.3 Protective Shields for Life on Earth

There are two important shields, without which life on the Earth's surface would not be possible:

- Earth's atmosphere: It protects life from short-wave radiation, i.e. UV and X-ray radiation. UV radiation from the sun is absorbed in the ozone layer (20–40 km altitude in the stratosphere). Ozone O_3 is formed by ($h\nu$ is a UV photon from the Sun):

$$O_2 + h\nu \rightarrow O + O \tag{16.2}$$

$$O_2 + O \rightarrow O_3 \tag{16.3}$$

X-rays and the far UV are partly absorbed at higher altitudes in the Earth's atmosphere (dissociation of O_2 and other molecules).

The Earth's ozone layer is influenced by solar activity.

---

[1] Eukaryotes have a nucleus.

- Magnetosphere: Near the Earth's surface the Earth's magnetic field is a dipole. The dipole axis is inclined by $10°$ with respect to the Earth's axis of rotation and the magnetic south pole is near the geographic north pole. The field strength varies between 0.25 Gauss (equator) and 0.60 Gauss (poles).[2] The solar wind compresses the magnetic field on the side facing the sun. The most important function of the Earth's magnetic field is to shield the Earth's surface from charged particles of the solar wind and cosmic rays. There have been repeated polarity reversals of the Earth's magnetic field, the last one 780,000 years ago. Nevertheless, a residual magnetic field remains even during these reversals, and no correlation is found between the reversals and the occurrence of mass extinctions of animal and plant species.

> Without the magnetosphere and atmosphere, life on Earth would be almost inconceivable.

16.1.4 Life in the Solar System

The for life on other solar system bodies has so far been unsuccessful. Possible candidates are:

- Mars: origin of life during earlier warmer climatic periods on Mars? Are there simple life forms below the surface or near the polar regions?
- Jupiter moons Europa, Ganymede: Both are very likely to have a saline ocean beneath the surface.
- Saturn's moon Enceladus: water geysers have been found.
- Saturn's moon Titan: Organic compounds were detected in its atmosphere (tholins) as well as small lakes of hydrocarbon compounds on the surface.

The discovery of life forms that prefer extreme environments (extremophiles) raises hope of possibly finding at least primitive life forms on some of the above objects after all.

16.2 Discovery of Extrasolar Planetary Systems

Problem: Seen from Earth, the planets are very close to their host star. The host star outshines the planets. The contrast is somewhat better in the infrared, since planets shine predominantly in the IR. Therefore a direct observation of exoplanets is very difficult.

---

[2] For comparison, field strength of a small electric motor armature: 100 gauss, 1 gauss. $= 10^{-4}$ T, T...Tesla.

Fig. 16.1 Exoplanet (red) orbiting a brown dwarf (object 2MASSWJ1207334-393254 or 2M1207); the object orbits the brown dwarf at a distance equivalent to twice the Sun-Neptune distance (Credit: ESO)

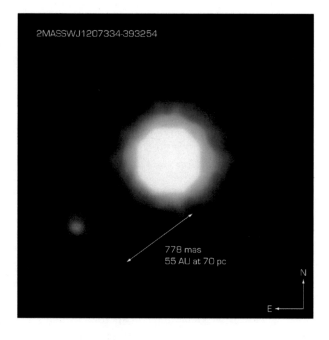

Figure 16.1 shows an exoplanet around a brown dwarf. Brown dwarfs are stars whose mass is insufficient to sustain permanent nuclear fusion at the center. Stars exist only from about 0.08 solar masses, stars with lower masses are called brown dwarfs. The brown dwarf in Fig. 16.1 has about 25 times the mass of Jupiter.[3]

16.2.1 Astrometry

Consider a planet around a star. Both move around the common center of gravity. An exoplanet with a mass of Jupiter M_J orbiting a star of one solar mass in two years causes the parent star to move by 0.3 milliarc seconds (mas) when the star is 10 pc away from us. To find exoplanets with astrometry, we need measurement accuracies in the μ as range.

16.2.2 Radial Velocity Method

The motion of the parent star around the centre of gravity of its system causes periodic variations of the radial velocity. The required precise measurements can be made with the HARPS (High Accuracy Radial velocity Planet Searcher) instruments on the 3.6 m ESO telescope or with the HIRES on the Keck telescope. The problem is when the planets are

---

[3] The brown dwarf corresponds to 1/40 $M_\odot$.

Fig. 16.2 Transit of an
exoplanet and variation of the
brightness of the parent star

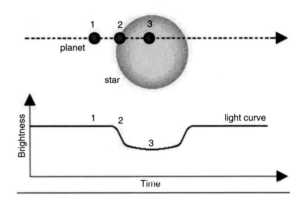

at a greater distance from the parent star: Jupiter takes about 12 years to orbit the Sun, so you have to wait 12 years to get a complete radial velocity curve. The planet 51 Pegasi b was the first found with this method. 51 Peg is a six billion year old yellow dwarf with apparent magnitude $5^m.1$ at a distance of about 50 Ly from us and of spectral type G2 V. In 1995, the first exoplanet orbiting a Sun-like star, 51 Pegasi b, was discovered by M. Mayor and D. Queloz of the Department of Astronomy at the University of Geneva. 51 Pegasi b has 0.46 Jupiter masses and orbits the star in 4.2 days at a distance of only 0.05 AU.

16.2.3 Light Curves, Transit Observations

When the Earth is practically on the orbital plane of an extra solar planet, a transit occurs as seen from Earth (cf. Venus transit, the planet moves in front of the star), as outlined in Fig. 16.2. Accurate measurements of the brightness of a star can detect such transits, but there are several other ways in which a star can change brightness, such as:

- Starspots,
- stellar variations

Satellite missions hope to find Earth-like planets through such observations.

Transit observations can also be used to infer the atmosphere of an exoplanet.

ETD is the Exoplanet Transit Database. On this web page you can get predictions about transits.

16.2.4 Microlensing

This effect has already been discussed in the previous chapter. Problem: The microlensing effect occurs only once—if at all—for an exoplanet, so the observations are not reproducible.

Fig. 16.3 The effect of
relativistic aberration

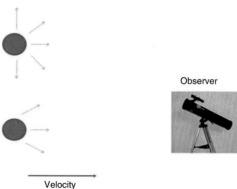

Observer

Velocity

16.2.5 Einstein-Beaming

Einstein-Beaming is a relativistic effect, just like microlensing. In a star-planet system, both bodies move around the common centre of gravity. If the star moves towards us as a result of this motion of the centre of gravity, it becomes slightly brighter and its radiation is focused according to the theory of relativity. One effect is relativistic aberration (Fig. 16.3). In this figure, the observer should be on the right. In the upper left of the figure are drawn the rays emanating from the hemisphere of an object at rest facing the observer, and in the lower left are drawn the focused rays of an object moving toward the observer.

This effect has been used to find the planet Kepler-76b ("Einstein's Planet"), which orbits its star in only 1.5 days.

16.2.6 Earth-Based Observations

There are numerous telescopes used only for transit observations:

- TrES, Trans Atlantic Exoplanet Survey; three 10-cm telescopes at Lowell Observatory, Palomar Observatory, and the Canary Islands are used.
- XO telescope in Hawaii (200-mm telephoto lens).
- HATNet: Hungarian Automated Telescope Network XO, 6 robotic telescopes in operation WASP Wide
- Super WASP Wide Angle Search for (Wide Angle HAT net Search TrES for Planets ETDB); 200 mm telescope.

Satellite Observations
- COROT (Convection Rotation et Transits): this was used to find Corot-1b in May 2007. Observations lasted until 2013. 27-cm telescope.
- Kepler: Launched in March 2009; about 145,000 stars in a preselected field of stars in the Milky Way are continuously monitored. It uses 42 CCDs, each with 2024 × 1024

pixels. The selected field is chosen so that it lies outside the ecliptic and no objects of the Kuiper belt disturb the observations. By the end of 2017, about 4500 candidate exoplanets had been found, about 10% of them Earth-sized, 30% super-Earths, about 45% Neptune-like, and the rest Jupiter-sized or larger than Jupiter. Due to problems, the mission continued as the K2 mission from 2015 to end in October 2018. In December 2017, a planetary system of 8 planets was discovered (Kepler 90).

- GAIA (Global Astrometric Interferometer for Astrophysics): Launched in December 2013. The goal of this mission is to determine positions, distances, and annual motions of about a billion stars in our galaxy. As a by product, it is hoped to find several tens of thousands of exoplanets. For now, the COROT mission is scheduled to end in 2025.

16.3 Host Stars

In this chapter we discuss host stars, the parent stars around which exoplanets orbit. The luminosity of these stars defines a circumstellar habitable zone.

16.3.1 Hertzsprung-Russell Diagram

There are three main groups of stars found in the HRD:

- Main stars,
- giants and supergiants,
- white dwarfs.

Most stars are on the main sequence because this represents the longest phase in stellar evolution. The lifetime of a star is important for finding life on planets around that star. On Earth, it took about 1 billion years before the first life forms evolved. The lifetime of a star can be defined in a good approximation as the time τ_{MS} when the star is on the main sequence, and this depends on the following factors :

$$\tau_{MS} = (1.0 \times 10^{10}) \times \frac{M}{M_\odot} \frac{L}{L_\odot} \text{ years} \tag{16.4}$$

M, L denote mass and luminosity of the parent star, τ_{MS} is the main sequence lifetime of the star in years.

Typical values for late spectral types (main sequence stars only):

- G stars: about 1 solar mass, surface temperatures about 5800 K, $\tau_{MS} = 10 \times 10^9$ years.
- K stars: about 0.6 solar masses, $T = 4000K$, $\tau_{MS} = 32 \times 10^9$ years,
- M stars: 0.22 solar masses, $T = 2800K$, $\tau_{MS} = 210 \times 10^9$ years.

Fig. 16.4 Artist's impression of exoplanet HD 189733b, its atmosphere partially vaporized by a stellar flare (Credit: NASA)

Assuming that life takes time to evolve (it took about 1 billion years on Earth), hot O, B, and A stars are ruled out as parent stars of habitable planets.

Exoplanets are influenced by the activity of their parent star. In Fig. 16.4 we sketch how the exoplanet HD 189733b loses parts of its atmosphere due to a stellar flare outburst. This exoplanet was studied with the HST in 2010 and 2011. The sudden appearance of a gas flare was observed, and estimates indicated that the planet lost about 1000 t of mass per second. The UV radiation from the parent star is about 20 times stronger than that from the Sun, and the X-ray radiation is about 1000 times stronger. The stellar flare was measured by NASA's Swift satellite. This example shows what conditions might have been like in the early solar system, when the Sun was much more active than it is today.

M stars are the most common, but the habitable zone moves closer and closer to the star because of the star's low luminosity. Stars of spectral type earlier than F have too short a main sequence lifetime (less than 10^9 years), so life on planets around these stars is very unlikely.

Therefore, only stars with a spectral type between F and K are preferred for the search for habitable exoplanets.

16.3.2 Habitable Zone

If we assume that life forms similar as on Earth, the habitable zone comprises the area around a star where water in liquid form can exist on a planet. The *circumstellar habitable zone* depends on:

- Spectral type and thus surface temperature of the host star.
- Size of the planet.
- Atmosphere of the planet, greenhouse effect, etc.

Even large moons around exoplanets could be habitable if a liquid salt ocean lies beneath the surface heated by tidal forces (cf. Jupiter's moon Europa). This is called a *circumplanetary habitable zone*.

There is also a galactic habitable zone. Too close to the galactic center, stellar density is high, supernova explosions occur more frequently, and life can be extinguished by such explosions.[4] Too far from the galactic center, the content of elements heavier than helium (metals) is too low and the formation of planetary systems is unlikely. Life, moreover, could not develop until the universe had a certain minimum content of metals. Therefore, several stellar generations had to pass before Earth-like planets could form from the interstellar nebulae.

Examples of exoplanets in habitable zones are given in Figs. 16.5 and 16.6

- Gliese 581 b: parent star about 22 light-years away from us, spectral type M5. Gliese 581 d has about 8 Earth masses, orbital period 65 days.
- Kepler 22 b: about 2.2 Earth masses, orbital period about 220 days.

Figure 16.5 shows:

- The habitable zone is farther away from the parent star for hot stars than for cool stars.
- The extent of the habitable zone is smaller for cooler stars.

---

[4] A supernova explosion less than 100 Ly away could destroy large parts of the terrestrial ozone layer.

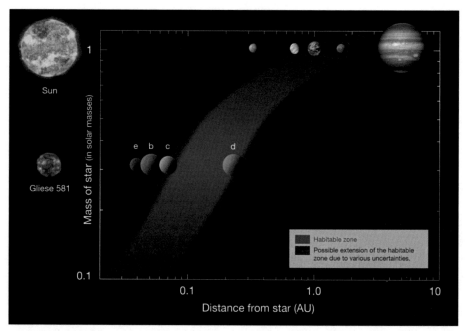

Fig. 16.5 Habitable zones: comparison: solar system (top) and system Gliese 581, where a planet could be in the habitable zone (Sketch: NASA)

- Since the habitable zone is closer to the parent star for cool stars, possible planets in this zone are very exposed to stellar winds and radiation bursts from that star. These could destroy possible life.

The planet Kepler 22b was the first exoplanet found by the Kepler mission in a habitable zone (Fig. 16.6).

16.3.3 Examples

The first detection of an extra solar planet was in 1995 at the star 51 Pegasi. For this M. Mayor and D. Queloz were awarded the . Nobel Prize in Physics.

Two Jupiter-like planets were found at the star 47 UMa by measurements of Doppler shifts in the star's spectrum. 47 UMa is thought to be 7 billion years old, and the distance is 51 light years.

Orbiting the red dwarf OGLE-05-390L is a planet with only five Earth masses. The star Gliese 876 is orbited by a planet with 7.5 Earth masses Gliese-876b in only 47 h.

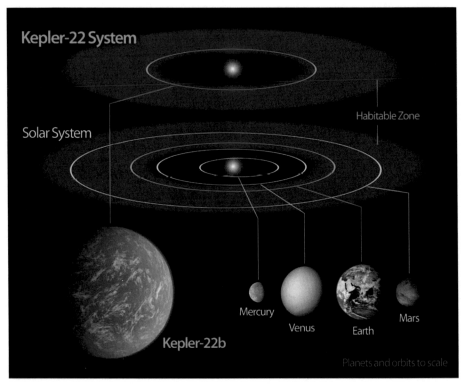

Fig. 16.6 The planet found by KEPLER with the designation 22b is located in a habitable zone around its parent star. It takes about 290 days to complete one orbit and is 2.4 times the diameter of Earth. The system is 600 light years away from us (Credit: NASA/Ames ResearchGliese-581b Center)

> Although low-mass, cool stars are more common than hotter stars, it is unlikely to find life in very cool M stars in the habitable zone because these stars are unstable.

The new satellite missions mentioned earlier will certainly lead to the discovery of numerous Earth-sized exoplanets. It will also be seen whether our solar system is the normal case or rather a special case. The exoplanets found so far (Ex. Table 16.1) mainly comprise large Jupiter-like objects that are very close to their parent star. However, these observations are selection effects. It is easier to find large planets close to the star than smaller planets. It is possible that these giant planets migrated (migrated) during the evolution of the system.

Table 16.1 Data of some extra solar planets

Designation	$M \sin i$ [M_{Jupiter}]	Orbit [d]	Major semimajor axis [AU]	Exc.
HD83.443	0.34	2.986	0.038	0.08
HD46375	0.25	3.024	0.041	0.02
TauBoo	4.14	3.313	0.047	0.02
UpsAndb	0.68	4.617	0.059	0.02
RhoCrB	0.99	39.81	0.224	0.07
UpsAndc	2.05	241.3	0.828	0.24
16CygB	1.68	796.7	1.69	0.68
47UMa b	2.54	1089.0	2.09	0.06
UpsAndd	4.29	1308.5	2.56	0.31
EpsEri	0.88	2518.0	3.36	0.60
47UMa c	0.76	2594	3.73	0.1

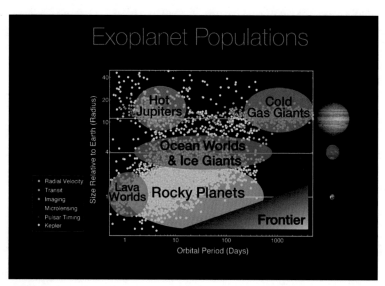

Fig. 16.7 Different types of exoplanets found so far. Close to the parent star (short orbital period) are planets with liquid lava surfaces, rocky planets are expected to be up to about 4 times the size of Earth, above that are planets composed mainly of water, and ice giants; hot Jupiters have orbital periods below 10 days, cool gas giants have orbital periods of more than 1000 days (Credit: NASA/Ames Research Center/Natalie Batalha/Wendy Stenzel)

Figure 16.7 summarizes few exoplanets found so far. One divides these into:

- Lava planets: very close to parent star, orbital periods of a few days. These planets are about the size of Earth. No analogue in the solar system.
- Rocky planets: Orbital periods from a few days to more than 100 days. They are about the size of the earth, up to a maximum of about 4 times the size of the earth. Analogue in the solar system: terrestrial planets.

- Planets of water and ice: larger than 4 Earth sizes. In our solar system they correspond approximately to the planets Uranus and Neptune.
- Hot Jupiters: 10 to 20 times Earth size; very close to parent star, orbital period a few days. No analogue in the solar system.
- Cold gas giants: 10 to 20 times Earth's size, farther from parent star. Analogue in the solar system: Jupiter, Saturn.

This figure also shows the methods used to find the objects.

16.4   Further Literature

We give a small selection of recommended further literature.
Astrobiology: Understanding Life in the Universe, C.S. Cockell, Wiley, 2015
Ice Worlds in the Solar System, M. Carroll, Springer, 2019
Astrobiology: From the Origins of Life to the Search for Extraterrestrial Intelligence, A. Yamagishi, T. Kakegawana et al, Springer, 2019
Biosignatures for Astrobiology, B. Cavalazzi, F. Westall, Springer, 2018.
Habitability and Cosmic Catastrophes, A. Hanslmeier, Springer, 2008

Tasks

16.8 Discuss what conditions prevail on a planet orbiting a brown dwarf. Can there be a habitable zone there?

Solution
Consider the luminosity of a brown dwarf and its evolution!

16.9 Why is it easier to find large Jupiter-like planets near their parent star than Earth-like planets?

Solution
Probability of a transit is greater, as is the center of mass motion of the parent star.

16.10 Jupiter produces a motion of the Sun about the common center of gravity. This is just outside the Sun. Estimate where the centre of gravity is in relation to the centre of the Sun when the Jupiter-Sun distance is reduced to 1/10th!

Solution

At about 11 $R_\odot$.

16.11 Is 51 Peg b located in a habitable zone?

Solution

No.

Mathematical Methods

17

In this chapter we give very briefly some important mathematical procedures which are frequently used in astrophysics. Special emphasis is put on the practical application. There are ready-made program packages for data evaluation. Here we will specifically discuss free and freely available packages. Python in particular has recently become popular here.

17.1 Python-a Crash Course

17.1.1 What Is Python?

Python is a general-purpose programming language that is relatively easy to learn, and it was developed by G. van Rossum in the early 1990s. The latest version is 3.7. The Python 3.0 version is not fully compatible with the 2.X versions. A big advantage is the easy integration of Python programs in other languages.

Python can be downloaded for free. It is available for most popular operating systems:

- Windows
- Linux—here it is already included in many newer distributions
- Mac OS

It is also one of the preferred languages for the Raspberry Pi and was developed as part of the project *100 Dollar Laptop* project.

We assume here that the reader has installed a Python package (e.g. Anaconda). This distribution has the advantage that you can load further program packages at any time using the command `conda install xxxx`, where xxx is the name of the program you want to load.

© Springer-Verlag GmbH Germany, part of Springer Nature 2023
A. Hanslmeier, *Introduction to Astronomy and Astrophysics*,
https://doi.org/10.1007/978-3-662-64637-3_17

There are of course countless more detailed introductions. Here we cover some basic elements of this language that allow you to perform your own data analysis.

Python consists of a number of loadable add-on libraries. A very common one is numpy, which is loaded by the command `import numpy as np` to a program you have created yourself.

For science, Python is particularly interesting because there is a huge selection of scientific libraries. The numerical calculations are mostly done with numpy and matplotlib. Anaconda SciPy bundle has many scientific libraries.

17.1.2 A First Simple Python Program

We see in Fig. 17.1 a simple Python program that is self-explanatory.

The commands used are `print` for printing, `np.arange` to define a field with defined start and end values and interval between values, and the command `np.array` to define the individual elements. To plot the data, the program package `import matplotliob.pyplot as plt`, then define a figure with `plt.figure` and execute the plot (`plt.plot`) and label the axes (e.g. `plt.xlabel`).

With this we have already made an essential step to visualize data.

```
import numpy as np
import matplotlib.pyplot as plt
# Definioeren einiger Variablen
a=4
b=2
print('a+b=',a+b)
c=a+b
print('Variable c:',c)
print('a*b=',a*b)
print('a:b=',a/b)
print('a^b=',a**b)
# Definieren eines eindimensionalen Arrays mit den Zahlen 0,...9 und dem Intervall 1
a1=np.arange(0,10,1)
#Bildung des Quadrats der Elemente des Arrays
b1=a1**2
print('Vektor:',a1)

print('Vektor ^2',b1)
#Zeichnerische Darstellung
plt.figure(1)
plt.plot(a1,b1,label='Quadrat')
plt.xlabel('Zahlen')
plt.ylabel('Funktionswerte')
plt.plot(a1,a1**3,label='Kubik')
plt.legend()
# definieren eines beliebigen arrays
a=np.array([2,4,6])
print(a)
print(a**2)
```

Fig. 17.1 A very simple Python program

```
n=int(input("Geben sie eine Zahl ein bis zu der die vorhergehenden addiert werden sollen"))

s=0
i=0
while i<=n:
    s=s+i
    print('Zwischensumme',s)
    i=i+1
print('Endergebnis=',s)


for x in range(0,10):
    print(x**4)
print("Ende der for-Schleife")
```

Fig. 17.2 Interactive data entry as well as a while loop and a for loop

In Fig. 17.2 shows the interactive data input and explains a while loop is explained. As long as a numerical value (in this case the value of the variable i is smaller than a given value (in this case n), the loop is executed. Finally a for loop is explained.

17.1.3 Example: Brightness Measurements

Measurements are always subject to errors. Therefore, when specifying a measurement, one should also specify the accuracy of the measurement. There are various methods to determine the mean value of a measurement, as well as to specify errors.

Very important for scientific computing are *arrays*. These can be one-dimensional, but also multi-dimensional; here are some examples:

- One-dimensional arrays: Arrays of numbers in a series of measurements; for example brightness measurements of a star at different times: 7.7, 7.2, 6.9, 7.3, 7.0 be the values.
- Two-dimensional array: Suppose we were taking images with a CCD. The CCD would have 1000 × 1000 Pixels (picture elements). Then each pixel has a specific brightness value and so a two-dimensional matrix of values is obtained:

$$M_{ij} = \begin{pmatrix} m_{11} & m_{12} & \cdots & m_{1j} \\ m_{21} & m_{22} & \cdots & m_{2j} \\ \vdots & \vdots & \vdots & \vdots \\ m_{i1} & m_{i2} & \cdots & m_{ij} \end{pmatrix} \tag{17.1}$$

How do we enter this in Python?
We first define the array of brightness measurements:
We start Python and first have to integrate the package numpy package with
import numpy as np. For example, if we define a measurement series as an *array* with the name brightness:

```
brightness=np.array([7.7 7.2 6.9 7.3 7.0])
```

To check, we output the vector with the brightnesses again by typing the variables:

`brightnesses`

Now we could plot these values using the plot commands explained earlier. If there are random errors, the values should not show a trend. A trend in the values might occur if we measure brightnesses over a period of several hours, as the height of the measured object above the horizon changes.

17.2   Statistics

We cover some basic elements of statistics. Important: If one gives results of measurements, it is always necessary to estimate how accurate these measurements are, i.e. a statement of error is necessary!

17.2.1 Mean Values

Measurements are always subject to error. A variable X is supposed to contain N values $X_1, X_2, \ldots, X_N$ assume.

Arithmetic Mean

$$\bar{X} = \frac{X_1 + X_2 + X_3 + \ldots + X_N}{N} = \frac{\sum_{j=1}^{N} X_j}{N} = \frac{\sum X}{N} \tag{17.2}$$

We calculate the mean value from the brightness values given above, using the function `np.mean`:
`mean=np.mean(luminosities)`. By typing again: `mean` we output the mean value.

Weighted Arithmetic Mean
You assign the numbers $X_1, \ldots, X_N$ Weighting factors (or weights) $w_1, \ldots, w_N$ to and then has:

$$\bar{X} = \frac{w_1 X_1 + w_2 X_2 + \ldots + w_k X_k}{w_1 + w_2 + \ldots + w_k} = \frac{\sum w X}{\sum w} \tag{17.3}$$

The numbers 5, 8, 6, 2 occur with frequencies 3, 2, 4 and 1 respectively, then is:

$$\bar{X} = \frac{(3)(5) + (2)(8) + (4)(6) + (1)(2)}{3 + 2 + 4 + 1} = 5,7$$

Let's put this example in Python: $z1$ denotes the measured values, zw the associated static weights.

```
z1=([5,8,6,2])
zw=([3,2,4,1])
wtmean=np.average(z1,weights=zw)
wtmean
```

Median

A set of numbers is ordered by their magnitude. If the number of values is odd, the median is equal to the value in the middle of the sequence of numbers; if the number is even, the median is equal to the arithmetic mean of the two values around the middle.

Geometric interpretation: the median is the value of the abscissa that corresponds to the perpendicular which divides a histogram into two parts with equal areas.

Geometric Mean

$$G = (X_1 \cdot X_2 \cdot X_3 \cdot \ldots \cdot X_N)^{1/N} \tag{17.4}$$

Harmonic Mean

$$H = \frac{1}{\frac{1}{N}\sum_{j=1}^{N}\frac{1}{X_j}} = \frac{N}{\sum \frac{1}{X}} \tag{17.5}$$

or also

$$\frac{1}{H} = \frac{1}{N}\sum \frac{1}{X} \tag{17.6}$$

It holds:

$$H \le G \le \bar{X} \tag{17.7}$$

Quadratic Mean

$$QM = \sqrt{\bar{X^2}} = \sqrt{\frac{\sum X^2}{N}} \tag{17.8}$$

The above discussed mean values give the most probable value of a measurement, but measurements scatter, the less accurate the scatter. The measures of scatter used are:

Span
Difference between the largest and smallest number in the set.

Mean or Average Deviation

$$MA = \frac{\sum_{j=1}^{N} |X_j - \bar{X}_j|}{N} = \overline{|X - \bar{X}|} \tag{17.9}$$

Given: 2, 3, 6, 8, 11; $\bar{X} = 6$.

$$MA = \frac{|2 - 6| + |3 - 6| + |6 - 6| + |8 - 6| + |11 - 6|}{5} = 2,8$$

In Python: `a1=np.array[2,3,6,8,11])`
`np.var(a1)/len(a1)`
With the command `np.sum(a1)` you can sum up the numerical values.

Standard Deviation

$$s = \sqrt{\frac{\sum_{j=1}^{N}(X_j - \bar{X})^2}{N}} \tag{17.10}$$

If the $X_1, X_2, \ldots, X_N$ occur with the frequencies $f_1, f_2, \ldots, f_N$ occur, then: $N = \sum_{j=1}^{N} f_j = \sum f$ and:

$$s = \sqrt{\frac{\sum_{j=1}^{N} f_j(X_j - \bar{X})^2}{N}} \tag{17.11}$$

The **Variance** is the square of the standard deviation, i.e. s^2.

In Python, you simply calculate the standard deviation from:
`np.std(z1)`
For the standard deviation of the brightness measurements, one uses:
`np.std(brightnesses)`

For a set Variance of numbers N_1, N_2 with the variances s_1^2, s_2^2 applies to the composite variance:

$$s^2 = \frac{N_1 s_1^2 + N_2 s_2^2}{N_1 + N_2} \tag{17.12}$$

For moderately skewed distributions, the empirical formulas apply:
Mean variation = 4/5 (standard deviation)
The coefficient of variation (coefficient of dispersion) is defined as:

$$V = \frac{s}{\bar{X}} \tag{17.13}$$

17.2.2 Distribution Functions

The Normal Distribution
If random errors occur in the measurements, then we have a normal distribution. Normal distributions lead to a *Gaussian bell curve*. Let μ the mean and σ the standard deviation, then:

$$f(x) = \frac{1}{\sigma \sqrt{2\pi}} \exp\left[-\frac{1}{2} \left(\frac{x - \mu}{\sigma} \right)^2 \right] \tag{17.14}$$

The normal distribution occurs with:

- random errors of observation and measurement,
- random deviations,
- for example, in the description of Brownian molecular motion.

Furthermore:

- Within standard deviation, $\pm\sigma$, 68.7% of all values of a normally distributed random variable lie to the right and left of the expected value (mean),
- Within $\pm 2\sigma$ 95.45%;
- Within $\pm 3\sigma$ 99.73% of all values.

Binomial Distribution
This is a discrete probability distribution. It describes the likely outcome of a sequence of similar trials, each of which has only two possible outcomes. The order does not matter, and the elements taken are put back.

Example Basket with N balls, of which M are red and $N - M$ Black. The probability, under n balls k red ones to be found:

$$P(k) = \binom{n}{k} \frac{M^k (N - M)^{n-k}}{N^n} = \binom{n}{k} \left(\frac{M}{N}\right)^k \left(\frac{N - M}{N}\right)^{n-k}$$

$$= \binom{n}{k} p^k (1 - p)^{n-k}$$

(17.15)

where is $p = M/N$ is the probability of finding a red ball. Note:

$$\binom{n}{k} = \frac{n!}{k! \cdot (n - k)!}$$

(17.16)

17.2.3 Moments

If $X_1, X_2, \ldots, X_N$ that of the variable X assumed N values, then we call

$$\bar{X}^r = \frac{X_1^r + X_2^r + \ldots + X_N^r}{N}$$

(17.17)

the ordinary moment r-th order. The first moment with $r = 1$ is the arithmetic mean. The r-th moment with respect to the mean is defined as:

$$m_r = \frac{\sum_{j=1}^{N} (X_j - \bar{X})^r}{N}$$

(17.18)

can be seen: $m_2 = s^2$. The r-th moment with respect to some starting point A is defined as:

$$\acute{m}_r = \frac{\sum_{j=1}^{N} (X_j - A)^r}{N}$$

(17.19)

If the $X_1, X_2, \ldots$ occur with the frequencies $f_1, f_2, \ldots$, then:

$$\bar{X}^r = \frac{\sum_{j=1}^{N} f_j X_j^r}{N}$$

(17.20)

$$m_r = \frac{\sum_{j=1}^{N} f_j (X_j - \bar{X})^r}{N}$$

(17.21)

$$\acute{m}_r = \frac{\sum_{j=1}^{N} f_j (X_j - A)^r}{N}$$

(17.22)

Skewness

It indicates the degree of asymmetry of a distribution.

- Positive skewness: distribution has a longer tail on the right side of the maximum (right skewed);
- Negative skewness: distribution has a longer tail on the left side of the maximum (left skewed).

The mode (modal value) is the most frequent value of a frequency distribution, the value with the greatest probability. Since a distribution can have multiple peaks → Several modes are also assigned to a distribution.

$$\text{Skewness} = \frac{\text{Mean Value} - \text{Modus}}{\text{Standard deviation}} = \frac{\bar{X} - \text{Modus}}{s} \tag{17.23}$$

Empirical formula:

$$\text{Skewness} = \frac{3(\bar{X} - \text{Median})}{s} \tag{17.24}$$

Moment coefficient of skewness e.g. a_3

$$a_3 = \frac{m_3}{s^3} \tag{17.25}$$

For completely symmetrical curves (normal distribution) are a_3, $b_3 = a_3^2$ equal zero.

Kurtosis

It indicates the degree of steepness of a distribution. A distinction is made between:

- leptokurtic: relatively high peak
- platykurtic: strongly flattened
- mesokurtic

The Moment coefficient the Kurtosis reads:

$$a_4 = m_4/s_4 = m_4/m_2^2 \tag{17.26}$$

17.3   Curve Fits and Correlation Calculation

Problem One has given measured values and would like to put a curve through these values, so that the measured values are represented as well as possible by the curve, which is called a *fit*.

17.3.1 Fitting Curves, Least Squares Method

We assume that there is a relation between the quantities X and Y, i.e., we have the samples (measurements) $X_1, X_2, \ldots, X_N$ and the corresponding quantities $Y_1, \ldots Y_N$.

In a *scatter diagram,* one can now plot the set of points $(X_1, Y_1), (X_2, Y_2), \ldots$ the points of the scatter plot. These points can then be represented more or less well by a curve.

Equations of Approximation Curves
The variable X is often referred to as the independent, Y as the dependent variable. Frequently used equations for approximate curves:

$$Y = a_0 + a_1 X \rightarrow \text{straight line} \tag{17.27}$$

$$Y = a_0 + a_1 X + a_2 X^2 \rightarrow \text{parabola} \tag{17.28}$$

For a curve n-th order:

$$Y = a_0 + a_1 X + a_2 X^2 + \ldots + a_n X^n \tag{17.29}$$

If one fits data with a parabola, one speaks of a Parabola fit.

Example of an application would be a parabolic fit around the minimum of the measurement points of a spectral line. This can be used to determine the Doppler shift in the center of the line.
Furthermore, one uses:
Hyperbola:

$$Y = \frac{1}{a_0 + a_1 X} \qquad \frac{1}{Y} = a_0 + a_1 X \tag{17.30}$$

Exponential Curve:

$$Y = ab^X \qquad \log Y = \log a + (\log b)X = a_0 + a_1 X \tag{17.31}$$

Geometric curve:

$$Y = aX^b \qquad \log Y = \log a + b \log X \tag{17.32}$$

Modified exponential curve

$$Y = ab^X + g \tag{17.33}$$

Modified geometric curve:

$$Y = aX^b + g \tag{17.34}$$

Gompertz-Curve:

$$Y = pq^{bX} \qquad \log Y = \log p + b^X \log q = ab^X + g \tag{17.35}$$

Modified Gompertz Curve:

$$Y = pq^{bX} + h \tag{17.36}$$

Logistic Curve:

$$Y = \frac{1}{ab^X + g} \tag{17.37}$$

Straight Line Fit
Simplest type for approximations:

$$Y = a_0 + a_1 X \tag{17.38}$$

Given any two points on the straight line, the constants. a_0, a_1 can be determined; the equation of the straight line is then:

$$Y - Y_1 = \left(\frac{Y_2 - Y_1}{X_2 - X_1} \right) (X - X_1) \tag{17.39}$$

Respectively:

$$Y - Y_1 = m(X - X_1) \tag{17.40}$$

Here one calls $m = \frac{Y_2 - Y_1}{X_2 - X_1}$ slope of the straight line. If you compare this with the formula $Y = a_0 + a_1 X$ we have:

- a_0... Value of Y if $X = 0$, ordinate intercept;
- a_1 is equal to the slope m.

Method of Least Squares
Given the pairs of values $(X_1, Y_1), \ldots$ we want to put a curve through this. To do this, we first determine the deviations $D_1, D_2, \ldots$ which are the vertical distances of the points from the curve to be determined. Now we define:

Of all the curves that fit a given set of pairs of values, a curve with the property that $D_1^2 + D_2^2 + \ldots + D_N^2 \rightarrow Min$ is called the best fitting curve. (method of least squares)

Straight Line Least Squares
Given are $(X_1, Y_1), (X_2, Y_2), \ldots, (X_N, Y_N)$. The straight line of least squares, which approximates the set of these points, has the formula: $Y = a_0 + a_1 X$. Now we have to determine the coefficients a_0, a_1.

$$\sum Y = a_0 N + a_1 \sum X \tag{17.41}$$

$$\sum XY = a_0 \sum X + a_1 \sum X^2 \tag{17.42}$$

$\sum X$ etc. denotes the sum of all values X_i etc. These equations are also called Normal equations. From this you can find the coefficients:

$$a_0 = \frac{\left(\sum Y\right)\left(\sum X^2\right) - \left(\sum X\right)\left(\sum XY\right)}{N \sum X^2 - \left(\sum X\right)^2}$$

$$a_1 = \frac{N \sum XY - \left(\sum X\right)\left(\sum Y\right)}{N \sum X^2 - \left(\sum X\right)^2} \tag{17.43}$$

One can also make the calculation easier by performing the following transformation: $x = X - \bar{X}; y = Y - \bar{Y}$; then you have the following straight line equation, which goes through the point $(\bar{X}, \bar{Y})$ which is sometimes called the center of gravity of the data.

$$y = \frac{\sum xy}{\sum x^2} x \tag{17.44}$$

Parabola of Least Squares

Again, one sets up the normal equations:

$$\sum Y = a_0 N + a_1 \sum X + a_2 \sum X^2 \tag{17.45}$$

$$\sum XY = a_0 \sum X + a_1 \sum X^2 + a_2 \sum X^3 \tag{17.46}$$

$$\sum X^2 Y = a_0 \sum X^2 + a_1 \sum X^3 + a_2 \sum X^4 \tag{17.47}$$

From this follows the parabola:

$$Y = a_0 + a_1 X + a_2 X^2 \tag{17.48}$$

Example the following brightness measurements were made of a star during the night:

Time	22.00	22.30	23.00	24.00	0.30	1.00	1.30	2.00
Bright.	11.42	11.39	11.35	11.27	11.33	11.39	11.42	11.51

Put a parabola through these points! What simple cause could the increase or decrease in brightness of the star have? Solution: Coefficients of the parabola: $a_0 = 11.4325$, $a_1 = -0.0760714$, $a_2 = 0.0125000$.

In Python, we can solve this very easily:

- We define the vector of brightness values:
  ```
  h1=np.array([11.42,11.39,11.35,11.27,11.33,11.39,11.42,
  11.51])
  ```
- For simplicity, we define the time steps of the measurement by the numbers from 0 to 7.
  ```
  ht=np.array([0,1,2,3,4,5,6,7])
  ```
- The 2nd order fit (parabolic fit) is then obtained:
  ```
  z=np.polyfit(ht, h1,2)
  ```
- We output the numerical values for the coefficients of the parabola:
  ```
  z
  ```

The function `polyfit` gives the polynomial; the parameter 2 means that we put a 2nd-order polynomial through the measured values. We get as solution:
```
array([0.0125,-0.07607143,11.4325])
```

These values are the coefficients of the second order polynomial, which is given by the value 2 in the function `polyfit`.

- The polynomial is: $a_0 + a_1 x + a_2 x^2$
- The coefficients: $a_2 = 0.0125$, $a_1 = -0.07607143$, $a_0 = 11.4325$

Now we want to plot the whole thing. For this we need the library `matplotlib.pyplot`, which can be called with the command `import matplotlib.pyplot`.

We plot only the original data:

```
plt.plot(h1)
plt.show()
```

Now we create a program that plots (i) the original data, (ii) a second order polynomial fit (iii) a third order fit and (iv) a fourth order fit. Where the third and fourth order results are shifted vertically by a factor of 0.01 for better visibility. The original data is drawn in blue, the second-order fit in red, the third-order fit shifted up by 0.01 in green, and the fourth-order fit shifted up by 0.02 in magenta.

The program is shown in Fig. 17.3. This program then yields the plot in Fig. 17.4.

17.3.2 Correlations

X, Y are the variables to be examined. A scatter plot shows the position of the points. If all points in the scatter plot appear to be close to a straight line, the correlation is said to be linear. One has:

- positive linear correlation: if Y tends to increase as X also tends to increase;
- negative linear correlation: if Y tends to fall as X tends to fall;
- non-linear correlation: when all points appear to be close to a curve;
- uncorrelated: when there is no correlation.

Consider the two data series 1, 2, 3, 4, 5, 6 and 3, 5, 6, 7, 8 The correlation coefficient between the two is calculated using Python as follows:

```
b1=([1,2,3,4,5,6])
b2=([3,5,6,7,8])
np.correlate(b1,b2)
```

One can also use autocorrelation, `acf` for a given *lag*. This calculates the correlation of a function with itself, where the function is shifted against each other by a certain amount, called a *lag*, the function is shifted against itself. Consider the example in Fig. 17.5. The vector $b1$ defines a sinusoidal function, as does the vector $b2$. The correlation between the

```
import numpy as np
import matplotlib.pyplot as plt
h1=np.array([11.42,11.39,11.35,11.27,11.33,11.39,11.42,11.51])
ht=np.array([0,1,2,3,4,5,6,7])
hp2=np.zeros(8)
hp3=np.zeros(8)
hp4=np.zeros(8)
# now starts the polynom fits
# fit n=2
z2=np.polyfit(ht,h1,2)
# fit n=3

z3=np.polyfit(ht,h1,3)
#fit n=4
z4=np.polyfit(ht,h1,4)

for i in np.arange(0,8,1):
        hp2[i]=z2[2]+z2[1]*ht[i]+z2[0]*ht[i]*ht[i]
        hp3[i]=z3[3]+z3[2]*ht[i]+z3[1]*ht[i]**2+z3[0]*ht[i]**3
        hp4[i]=z4[4]+z4[3]*ht[i]+z4[2]*ht[i]**2+z4[1]*ht[i]**3+z4[0]*ht[i]**4
print("hp2",hp2)
print("hp3",hp3)
print("hp4",hp4)

print(z2)
print(z3)
plt.xlabel('Time')
plt.ylabel('Measured magnitude')
plt.title('Original data')
plt.plot(h1,'b')
plt.plot(hp2,'r')
plt.plot(hp3+.01,'g')
plt.plot(hp3+0.02,'c')
plt.show()
```

Fig. 17.3 Program for generating the polynomial fits in Python

two vectors is given by the vector c. To determine the autocorrelation, we define the lag, e.g. length $length=5$. Thus, the result provides the correlation for lag=0, 1, 2, etc.

The results of the program are shown in Fig. 17.6.

Regression Lines

As we have already seen in the previous chapter, the regression line of Y with respect to X:

$$Y = a_0 + a_1 X \tag{17.49}$$

From this we had the normal equations (17.41).

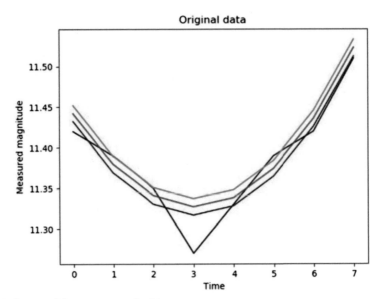

Fig. 17.4 Output of the program `polyfit`

```
# programm zur Berechnung der Korrelation zwichen zwei Datenreihen
import numpy as np
import matplotlib.pyplot as plt
b1=np.arange(0,100,1)
b1=np.sin(b1/5.)
b2=np.arange(0,100,1)
b2=np.sin(b2/.2)
# Correlation Coeffcient
c=np.corrcoef(b1,b2)
print("Correl. Coeff.:",c)

x=b1
#plt.plot(x)

def acf(x, length=5):
    return np.array([1]+[np.corrcoef(x[:-i], x[i:])[0,1]  \
        for i in range(1, length)])

cc=acf(x)

print('..........')
print(cc)
print('.-.-.-.')

plt.plot(cc)
```

Fig. 17.5 The program for calculating the correlation or autocorrelation (acf)

```
Correl. Coeff.: [[ 1.              -0.00697499]
 [-0.00697499  1.           ]]
.. .. .. .. ..
[1.           0.97916463 0.91809238 0.8204463  0.69141825]
.-.-.-.
```

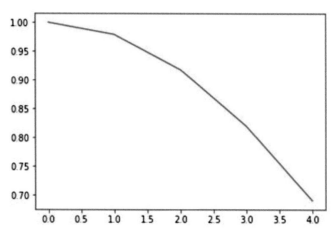

Fig. 17.6 Results of the correlation program

The regression line of X with respect to Y is given by:

$$X = b_0 + b_1 Y \tag{17.50}$$

$$\sum X = b_0 N + b_1 \sum Y \tag{17.51}$$

$$\sum XY = b_0 \sum X + b_1 \sum Y^2 \tag{17.52}$$

$$b_0 = \frac{\left(\sum X\right)\left(\sum Y^2\right) - \left(\sum Y\right)\left(\sum XY\right)}{N \sum Y^2 - \left(\sum Y\right)^2}$$

$$b_1 = \frac{N \sum XY - \left(\sum X\right)\left(\sum Y\right)}{N \sum Y^2 - \left(\sum Y\right)^2} \tag{17.53}$$

The regression lines are identical if and only if All points of the scatter plot lie on a straight line. In this case, therefore, there is a perfect linear correlation between the X and Y.

17.4   Differential Equations

Equations in which the function we are looking for occurs as derivatives are called differential equations. A simple example is the Newtonian relation between the force exerted on a mass and the acceleration caused by it:

$$ma = f(x, t)$$

$$a = \frac{d^2x}{dt^2}$$

$$m\frac{d^2x(t)}{dt^2} = f(x, t)$$

Furthermore, the following applies to gravity:

$$f(x, t) = mg$$

Another simple example is radioactive decay; the change in quantity. $m(t)$ is proportional to the quantity currently present and the time interval under consideration dt:

$$dm(t) = -\lambda m(t)dt$$

$$\frac{dm(t)}{dt} = -\lambda m(t)$$

An ordinary differential equation contains only one dependent variable, one independent variable, and their derivatives. In a partial differential equation, there are several independent variables and partial derivatives.

17.4.1 First Order Linear Differential Equations

The equation is linear in y and contains only the first derivative. The general form is:

$$y'(x) = a(x)y(x) + b(x) \tag{17.54}$$

If $a(x) = 0$, then holds:

$$\frac{dy}{dx} = b(x) \qquad \Longrightarrow y = \int b(x)dx + a \tag{17.55}$$

Let us consider the free fall:

$$\dot{v} = -g \tag{17.56}$$

$$v(t) = v(t_0) - (t - t_0)g \tag{17.57}$$

g is the acceleration due to gravity, which acts opposite to the direction in which the velocity is given. The derivative with respect to the time variable t is expressed by a point on the variable.

Since $v(t)$ is equal to the change in distance:

$$\dot{s}(t) = v(t) = v_0 + g t_0 - g t \tag{17.58}$$

And by integration one finds:

$$s(t) = s_0 + (v_0 + g t_0)(t - t_0) - \frac{1}{2} g (t^2 - t_0^2) \tag{17.59}$$

If the object at the beginning ($t_0 = 0$) at the point $s(0) = s_0$ was at rest ($v_0 = 0$), then the classical law of falling follows:

$$s(t) = s_0 - \frac{1}{2} g t^2 \tag{17.60}$$

Let be given: A first order homogeneous linear differential equation

$$y'(x) = a(x) y(x), \qquad y(x_0) = y_0, \tag{17.61}$$

then the solution is:

$$y(x) = y_0 \exp \left(\int_{x_0}^{x} dt\, a(t) \right) \tag{17.62}$$

Example The solution to the differential equation for radioactive decay:

$$\dot{m}(t) = -\lambda m(t) \qquad m(t_0) = m_0 \tag{17.63}$$

$$m(t) = m_0 \exp[-\lambda(t - t_0)] \tag{17.64}$$

17.4.2 Oscillator Equation

Many problems in physics and astrophysics can be traced back to oscillations. We briefly consider the equation of oscillation. Let there be a mass attached to a spring, the properties of the spring are described by the spring constant D. Then holds:

$$m\ddot{x} + Dx = 0 \qquad \omega_0 = \sqrt{D/m} \tag{17.65}$$

$$\ddot{x} + \omega_0^2 x = 0 \tag{17.66}$$

ω_0 is the undamped natural frequency. One solves the equation by the approach $x = \exp(\lambda t)$ and obtain:

$$\lambda^2 + \omega_0 = 0 \tag{17.67}$$

Thus results in $\lambda_{1,2} = \sqrt{-\omega_0^2} = \pm\omega_0 i$. From this one obtains the solutions:

$$x_1 = c_1 \exp(\omega_0 i t) \qquad x_2 = c_2 \exp(-\omega_0 i t) \tag{17.68}$$

This can be converted into a sine function using Euler's formula

$$\exp(i\phi) = \cos\phi + i\sin\phi \tag{17.69}$$

into a sine function.

17.4.3 Partial Differential Equations

We consider here as an example the heat conduction equation. Let $u(\mathbf{x}, t)$ the temperature at the point $\mathbf{x}$ at the time t, a is the thermal diffusivity of the medium, then the thermal equation isrme conduction equation:

$$\frac{\partial u(\mathbf{x}, t)}{\partial t} - a\triangle u(\mathbf{x}, t) = 0, \tag{17.70}$$

where $\triangle$ is the Laplace operator. If we apply the Laplace operator in Cartesian coordinates to a function $f(x, y, z)$ an:

$$\triangle f(x, y, z) = \frac{\partial^2 f}{\partial x^2} + \frac{\partial^2 f}{\partial y^2} + \frac{\partial^2 f}{\partial z^2} \tag{17.71}$$

In the stationary case (when the time derivative vanishes), the Laplace equation is obtained from the heat conduction equation:

$$\triangle u(x, y, z) = 0 \tag{17.72}$$

The solution for the one-dimensional case is:

$$u(x,t) = \frac{1}{\sqrt{4\pi a t}} \exp(-\frac{x^2}{4at})$$

(17.73)

This is also called the fundamental solution.

As another example, consider the wave equation:

$$\frac{1}{c^2} \frac{\partial^2 u}{\partial t^2} - \sum_{i=1}^{n} \left(\frac{\partial^2 u}{\partial x_i^2} \right) = 0$$

(17.74)

for a function $u(t, x_1, \ldots, x_n)$. The solution in the one-dimensional case is:

$$y(t, x) = y_{\max} \sin(2\pi (t/T - x/\lambda))$$

(17.75)

with $cT = \lambda$. Thus one can calculate the deflection y at the point x of the wave at a time t. The wave equation for an electric field is:

$$\Delta \mathbf{E} = \mu_0 \epsilon_0 \frac{\partial^2 \mathbf{E}}{\partial t^2}$$

(17.76)

and for the speed of light c one has $c = 1/(\mu_0 \epsilon_0)$.

17.5 Numerical Mathematics

17.5.1 Interpolation Polynomials

It is sometimes of great advantage, to use a polynomial $p(x)$ as an approximation for a function $y(x)$:

- Polynomials are easy to calculate, integer exponents only,
- easy to form derivatives and integrals,
- easy determination of zeros.

Error of approximation: $y(x) - p(x)$. It holds:

1. Existence and uniqueness theorem: There exists exactly one polynomial of degree at most n, so that for the arguments $x_0, \ldots, x_n$ respectively $y(x) = p(x)$.

2. Division algorithm: any polynomial can be written as:

$$p(x) = (x - r)q(x) + R \tag{17.77}$$

where r any number, $q(x)$ a polynomial of degree $n - 1$ and R is a constant.

3. Remainder: $p(r) = R$
4. Linear factor: from $p(r) = 0$ it follows that $x - r$ a factor of $p(x)$ is.
5. Zeros: A polynomial of degree n has at most n zeros, i.e. the equation $p(x) = 0$ has at most n solutions.
6. Error: Let be the product:

$$\pi(x) = (x - x_0)(x - x_1)\ldots(x - x_n) \tag{17.78}$$

For the number of interpolation points this vanishes, and one can show that the error of the interpolation polynomial is given by:

$$y(x) - p(x) = y^{(n+1)}(\xi)\pi(x)/(n + 1)! \tag{17.79}$$

Where ξ is somewhere between the outermost grid points and x.

17.5.2 Divided Differences

Consider a function $y = f(x)$ at the $n + 1$ points $x_0, x_1, \ldots, x_n$. Then let:

$$[x_0] = y_0, [x_1] = y_1, \ldots, [x_n] = y_n \tag{17.80}$$

and further the divided differences

$$[x_i x_j] = \frac{y_i - y_j}{x_i - x_j}, \text{ e. g. } [x_1 x_0] = \frac{y_1 - y_0}{x_1 - x_0} \tag{17.81}$$

and

$$[x_i x_j x_k] = \frac{[x_i x_j] - [x_j x_k]}{x_i - x_k} \tag{17.82}$$

As an example see Table 17.1.

Table 17.1 Example: scheme
of divided differences

$x_{i+2} - x_i$	$x_{i+1} - x_i$	x_i	y_i	Δ	$[x_{i+1}x_i]$	Δ	$[x_2x_1x_0]$
		1	1				
	2			26	13		
3		3	27			24	8
	1			37	37		
		4	64				

17.5.3 Newton's Interpolation Method

The Newton's Interpolation method with differences is:

$$y = f(x) = y_0 + [x_1x_0](x - x_0) + [x_2x_1x_0](x - x_0)(x - x_1) + \ldots \tag{17.83}$$

Linear Interpolation
Given are (x_0, y_0) as well as (x_1, y_1). If you are looking for a value between these two,
you can interpolate linearly as follows:

$$y(x) = y_0 + \frac{y_1 - y_0}{x_1 - x_0}(x - x_0) \tag{17.84}$$

Finite Differences
Given a discrete function whose arguments are x_k and images y_k and the arguments should
be equidistant:

$$x_{k+1} - x_k = h \tag{17.85}$$

$$\Delta y_k = y_{k+1} - y_k \tag{17.86}$$

It is best to write the so-called Difference scheme:

$$\begin{array}{llllll}
x_0y_0 & & & & & \\
 & \Delta y_0 & & & & \\
x_1y_1 & & \Delta^2 y_0 & & & \\
 & \Delta y_1 & & \Delta^3 y_0 & & \\
x_2y_2 & & \Delta^2 y_1 & & \Delta^4 y_0 & \quad (17.87)\\
 & \Delta y_2 & & \Delta^3 y_1 & & \\
x_3y_3 & & \Delta^2 y_2 & & & \\
 & \Delta y_3 & & & & \\
x_4y_4 & & \Delta^2 y_0 & & & \\
\end{array}$$

One sees immediately, for example:

$$\Delta^3 y_0 = y_3 - 3y_2 + 3y_1 - y_0 \text{ resp. in general:}$$

$$\Delta^2 y_k = \Delta(\Delta y_k) = \Delta y_{k+1} - \Delta y_k = y_{k+2} - 2y_{k+1} + y_k \tag{17.88}$$

$$\Delta^n y_k = \Delta^{n-1} y_{k+1} - \Delta^{n-1} y_k \tag{17.89}$$

$$\Delta^k y_0 = \sum_{i=0}^{k} (-1)^i \binom{k}{i} y_{k-i} \tag{17.90}$$

Newton's interpolation formula with differences taken forward is:

$$p_k = y_0 + \binom{k}{1}\Delta y_0 + \binom{k}{2}\Delta^2 y_0 + \ldots = \sum_{i=0}^{n} \binom{k}{i}\Delta^i y_0 \tag{17.91}$$

Hence:

$$p(x_k) = y_0 + \frac{\Delta y_0}{h}(x_k - x_0) + \frac{\Delta^2 y_0}{2!h^2}(x_k - x_0)(x_k - x_1) + \ldots \tag{17.92}$$

Let be given:

k	x_k	y_k	Δy_k	$\Delta^2 y_k$	$\Delta^3 y_k$
0	4	1			
			2		
1	6	3		3	
			5		4
2	8	8		7	
			12		
3	10	20			

Then one gets: $p_k(x_k) = 1 + \frac{2}{2}(x_k - 4) + \frac{3}{8}(x_k - 4)(x_k - 6) + \frac{4}{48}(x_k - 4)(x_k - 6)(x_k - 8)$

17.5.4 Interpolation with Unevenly Distributed Grid Points

Lagrange's Interpolation Formula
It holds:

$$p(x) = \sum_{i=0}^{n} L_i(x) y_i \tag{17.93}$$

Thereby $L_i(x)$ is the Lagrange' basis function:

$$L_i(x) = \frac{(x - x_0)(x - x_1)\dots(x - x_{i-1})(x - x_{i+1})\dots(x - x_n)}{(x_i - x_0)(x_i - x_1)\dots(x_i - x_{i-1})(x_i - x_{i+1})\dots(x_i - x_n)} \tag{17.94}$$

17.5.5 Numerical Differentiation

Approximations for the derivative of a function can be obtained by an interpolation polynomial by using p', $p^{(2)}$, ... as a substitute for y', $y^{(2)}$, ...:

1. Newton's interpolation formula with differences taken forward:

$$y'(x) \approx \frac{y(x + h) - y(x)}{h} \tag{17.95}$$

2. Stirling's Formula for numerical differentiation:

$$y'(x) \approx \frac{y(x + h) - y(x - h)}{2h} \tag{17.96}$$

3. Newton's formula with differences taken backwards:

$$y'(x) \approx \frac{y(x) - y(x - h)}{h} \tag{17.97}$$

4. one takes several terms from Newton's formula:

$$y'(x) \approx \frac{1}{h}\left[\Delta y_0 + \left(k - \frac{1}{2}\right)\Delta^2 y_0 + \frac{3k^2 - 6k + 2}{6}\Delta^3 y_0 + \dots\right] \tag{17.98}$$

5. From the Stirling formula:

$$y'(x) \approx \frac{1}{h}\left[\delta\mu y_0 + k\delta^2 y_0 + \frac{3k^2 - 1}{6}\delta^3\mu y_0 + \dots\right] \tag{17.99}$$

6. for the second derivative:

$$y^{(2)}(x) \approx \frac{y(x + h) - 2y(x) + y(x - h)}{h^2} \tag{17.100}$$

7. Or:

$$y^{(2)}(x) \approx \frac{y(x + h) - 2y(x) + y(x - h)}{h^2} \tag{17.101}$$

Problem Mostly $y(x) - p(x)$ very small, but $y'(x) - p('x)$ can be very large. Geometrically, this means: Two curves can run very close to each other even though the slopes are very different.

Find the derivatives up to order 4 from Newton's interpolation formula with differences taken forward!

Solution

$p_k = y_0 + \binom{k}{1}\Delta y_0 + \binom{k}{2}\Delta^2 y_0 + \ldots$

From $x = x_0 + kh$ one finds:

$$p'(x) = \frac{1}{h}\left(\Delta y_0 + (k - \frac{1}{2}\Delta^2 y_0 + \frac{3k^2 - 6k + 2}{6}\Delta^3 y_0 + \ldots\right)$$

$$p^{(2)}(x) = \frac{1}{h^2}\left(\Delta^2 y_0 + (k - 1)\Delta^3 y_0 + \frac{6k^2 - 18k + 11}{12}\Delta^4 y_0 + \ldots\right)$$

$$p^{(3)}(x) = \frac{1}{h^3}\left(\delta^3 y_0 + \frac{2k - 3}{2}\Delta^4 y_0 + \ldots\right)$$

$$p^{(4)}(x) = \frac{1}{h^4}(\Delta^4 y_0 + \ldots)$$

Apply these formulas to find the first derivative of the given function $\sqrt{x}$ at the point 1. One compares the results with the actual values!

x	$\sqrt{x}$			
1.00	1.00000			
		2470		
1.05	1.02470		−59	
		2411		5
1.10	1.04881		−54	
		2357		4
1.15	1.07238		−50	
		2307		2
1.20	1.09544		−48	
		2259		3
1.25	1.11893		−45	
		2214		
1.30	1.14017			

Solution: from the above table of differences we obtain for the 1st derivative:

$p'(1) = 20(0.02470 + 0.000295 + 0.000017) = 0.50024$

2nd derivative:
$$p^{(2)}(1) = 400(-0.00059 - 0.00005) = -0.256$$
3rd derivative:
$$p^{(3)}(1) = 8000(0.00005) = 0.4$$
Now compare with the correct results:

$$y'(1) = 1/2$$

$$y^{(2)}(1) = -1/4$$

$$y^{(3)}(1) = 3/8$$

You see: The input data were accurate to five digits, the first derivative was now accurate to only three digits, the second to two digits, and the third to one digit!

17.5.6 Numerical Integration

Integration of Newton's Interpolation Formula

We consider the Newton's interpolation formula with forward differences of degree n between x_0, x_n:

$$\int_{x_0}^{x_1} p(x)dx = \frac{h}{2}(y_0 + y_1) \tag{17.102}$$

$$\int_{x_0}^{x_2} p(x)dx = \frac{h}{3}(y_0 + 4y_1 + y_2) \tag{17.103}$$

$$\int_{x_0}^{x_3} p(x)dx = \frac{3h}{8}(y_0 + 3y_1 + 3y_2 + y_3) \tag{17.104}$$

From these simple formulas, one can determine composite formulas by repeated application of the basic formulas, i.e., several straight line pieces, parabola pieces, etc. This is usually easier than using higher degree polynomials.

Trapezoidal Rule

$$\int_{x_0}^{x_n} y(x)dx \approx \frac{1}{2}h[y_0 + 2y_1 + \ldots + 2y_{n-1} + y_n] \tag{17.105}$$

One has here straight line pieces as a substitute for the function $y(x)$. The truncation error is

$$-(x_n - x_0)h^2 y^{(2)}(\xi)/12 \qquad (17.106)$$

Simpson's Rule

Here one uses parabolic arcs Simpson's rule:

$$\int_{x_0}^{x_n} y(x)dx \approx \frac{h}{3}[y_0 + 4y_1 + 2y_2 + 4y_3 + \ldots + 2y_{n-2} + 4y_{n-1} + y_n] \qquad (17.107)$$

The truncation error is:

$$-(x_n - x_0)h^4 y^{(4)}(\xi)/180 \qquad (17.108)$$

17.5.7 Numerical Solution of Differential Equations

Very many applications lead to Differential equations, but only a few of them are analytically solvable.

Definition of a classical initial value problem: One shall solve a function $y(x)$ that satisfies the first order differential equation $y' = f(x, y)$ and takes the initial value $y(x_0) = y_0$. There are very many numerical methods, we will discuss some:

Isocline Method

Let $y'(x)$ the slope of the solution curve. Thus the function defines $f(x, y)$ be the prescribed slope at each point.

Euler Method

$$y_{k+1} = y_k + hf(x_k, y_k) \qquad (17.109)$$

Thereby $h = x_{k+1} - x_k$.

Runge-Kutta Method

It is used very frequently:

$$k_1 = hf(x, y) \qquad (17.110)$$

$$k_2 = hf(x + \frac{1}{2}h, y + \frac{1}{2}k_1) \qquad (17.111)$$

$$k_3 = hf(x + \frac{1}{2}h, y + \frac{1}{2}k_2) \tag{17.112}$$

$$k_4 = hf(x + h, y + k_3) \tag{17.113}$$

$$y(x + h) = y(x) + \frac{1}{6}(k_1 + 2k_2 + 2k_3 + k_4) \tag{17.114}$$

Predictor-Corrector Formulas

There are different Runge-Kutta possibilities:

$$y_{k+1} \approx y_k + hy'_k \tag{17.115}$$

$$y_{k+1} \approx y_k + \frac{1}{2}h(y'_k + y'_{k+1}) \tag{17.116}$$

Here the predictor is Euler's formula and the Corrector is a modified Euler formula. Because of $y'_k f(x_k, y_k)$ and $y'_{k+1} = f(x_{k+1}, y_{k+1})$ the first predictor is estimated to be y_{k+1} This estimate then leads to y'_{k+1} and to a corrected value y_{k+1}. Further corrections of y'_{k+1}, y_{k+1} can then be applied one after the other until sufficient accuracy is achieved.

Milne's Method

Here one uses the predictor-corrector pair:

$$y_{k+1} \approx y_{k-3} + (4h/3)(2y'_{k-2} - y'_{k-1} + y'_k) \tag{17.117}$$

$$y_{k+1} \approx y_{k-1} + (h/3)(y'_{k+1} + 4y'_k + y'_{k-1}) \tag{17.118}$$

Four start-up values are therefore required y_k, y_{k-1}, y_{k-2}, y_{k-3}.

Adams' Method

One uses the following predictor-corrector pair:

$$y_{k+1} \approx y_k + (h/24)(55y'_k - 59y'_{k-1} + 37y'_{k-2} - 9y'_{k-3}) \tag{17.119}$$

$$y_{k+1} \approx y_k + (h/24)(9y'_{k+1} + 19y'_k - 5y'_{k-1} + y'_{k-2}) \tag{17.120}$$

So again four start-up values.

Let us do an example to illustrate the methods. One applies Milne's method to solve $y' = -xy^2$ for $y(0) = 2$ using , $h = 0.2$.

Solution: So, in this method, one has for the predictor:

$$y_{k+1} \approx y_{k-3} + 4/3h(2y'_{k-2} - y'_{k-1} + 2y'_k) \tag{17.121}$$

as well as a corrector:

$$y_{k+1} \approx y_{k-1} + 1/3h(y'_{k+1} + 4y'_k + y'_{k-1}) \tag{17.122}$$

Since the predictor requires four given values, but we have only $y(0) = 2$ we have to calculate the others: e.g. with Runge-Kutta:

$$y(0.2) = y_1 \approx 1.92308; \; y(0.4) = y_2 \approx 1.72414; \; y(0.6) = y_3 \approx 1..47059$$

The differential equation then yields:

$$y'(0) = y'_0 = 0 \qquad y'(0.2) = y'_1 \approx -0.73964$$
$$y('0.4) = y'_2 \approx -1.18906 \qquad y'(0.6) = y'_3 \approx -1.29758$$

Thus. the Milne predictor is:

$$y_4 \approx y_0 + 4/3(0.2)(2y'_3 - y'_2 + 2y'_1) \approx 1.23056$$

respectively:

$$y'_4 \approx -(0.8)(1.23056)^2 \approx -1.21142$$

The Milne corrector then yields:

$$y_4 \approx y_2 + \frac{1}{3}(0.2)(-1.21142 + 4y'_3 + y'_2) \approx 1.21808$$

A recalculation of y' using the differential equation yields the new estimate $y'_4 \approx -1.18698$, and the new application of the corrector yields:

$$y_4 \approx y_2 + 1/3(0.2)(-1.18698 + 4y'_3 + y'_2) \approx 1.21971$$

Using the differential equation again, we have $y'_4 \approx -1.19015$ and with the corrector:

$$y_4 \approx y_2 + 1/3(0.2)(-1.19015 + 4y'_3 + y'_2) \approx 1.21950$$

The next two computations yield

$$y_4' \approx -1.18974;\ y_4 \approx 17.21953;\ y_4' \approx -1.18980;\ y_4 \approx 1.21953$$

and one can stop there.

Solving Differential Equations with Python

The program package numpy again provides some algorithms. We give as a very simple example the solution of the equation for a harmonic oscillator.

The harmonic oscillator is given by the equation:

$$\ddot{x} + \omega^2 x = 0 \tag{17.123}$$

In this case, there is an analytical solution:

$$x(t) = x_0 \cos(\omega t) + \frac{v_0}{\omega} \sin(\omega t) \tag{17.124}$$

The differential equation for the harmonic oscillator is a second order differential equation. This can be represented by a system of first order differential equations:

$$\dot{x} = y \tag{17.125}$$

$$\dot{y} = -\omega^2 x \tag{17.126}$$

The Python program is shown in Fig. 17.7 given.

Higher Order Differential Equations

One can replace a higher order differential equation with a system of first order differential equations, for example:

$$y'' = f(x, y, y') \tag{17.127}$$

can be replaced by:

$$y' = p \qquad p' = f(x, y, p) \tag{17.128}$$

We consider the Runge-Kutta method:

$$y' = f_1(x, y, p) \qquad p' = f_2(x, y, p) \tag{17.129}$$

```python
1 #!/usr/bin/env python3
2 # -*- coding: utf-8 -*-
3 """
4 Created on Mon Apr  8 06:12:54 2019
5
6 @author: arnoldhanslmeier
7 """
8
9 import numpy as np
10 import matplotlib.pyplot as plt
11 from scipy.integrate import odeint
12
13 om = 4                     # Hz
14 z0 = [1.5, -5.7]           # Anfangsbedingungen
15 t = np.linspace(0,10,400)  # 400 Zeitschritte von 0 bis 10s
16
17 # F      = (x, y)
18 # dF/dt = (y, -om^2 x)
19 def dF(z,t):
20     return[ z[1], -(om**2)*z[0] ]
21
22 # Zwei gekoppelte DGLs 1.Ordnung numerisch loesen
23 z = odeint(func=dF, y0=z0, t=t)
24
25 plt.plot(t, z0[0]*np.cos(om*t) + z0[1]/om*np.sin(om*t), label="analytisch")
26 plt.plot(t, z[:,0], "rx",                               label="odeint")
27 plt.legend()
28 plt.grid(True)
29 plt.show()
```

Fig. 17.7 Solving the oscillation equation with Python

with the initial conditions $y(x_0) = y_0$; $p(x_0) = p_0$. The formulas are then:

$$k_1 = hf_1(x_n, y_n, p_n)$$

$$l_1 = hf_2(x_n, y_n, p_n)$$

$$k_2 = hf_1(x_n + \frac{1}{2}h, y_n + \frac{1}{2}k_1, p_n + \frac{1}{2}l_1)$$

$$l_2 = hf_2(x_n + \frac{1}{2}h, y_n + \frac{1}{2}k_1, p_n + \frac{1}{2}l_1)$$

$$k_3 = hf_1(x_n + \frac{1}{2}h, y_n + \frac{1}{2}k_2, p_n + \frac{1}{2}l_2)$$

$$l_3 = hf_2(x_n + \frac{1}{2}h, y_n + \frac{1}{2}k_2, p_n + \frac{1}{2}l_2) \qquad (17.130)$$

$$k_4 = hf_1(x_n + h, y_n + k_3, p_n + l_3)$$

$$l_4 = hf_2(x_n + h, y_n + k_3, p_n + l_3)$$

$$y_{n+1} = y_n + \frac{1}{6}(k_1 + 2k_2 + 2k_3 + k_4)$$

$$p_{n+1} = p_n + \frac{1}{6}(l_1 + 2l_2 + 2l_3 + l_4)$$

17.6 Fourier Methods

17.6.1 Autocorrelation

We assume a set of measuring points. If one compares this with itself, then we speak of an autocorrelation. An unshifted sequence is most similar to the original one. However, if there is a relationship between the members of a measurement series, then the correlation of the original sequence with the shifted sequence also has a non-zero value.

Let x_k a given measurement series and x_{k+L} a series shifted by the lag L. Let the mean be $\bar{x}$ then the autocorrelation is ACF at the displacement (lag L):

$$ACF(L) = ACF(-L) = \frac{\sum_{k=0}^{N-L-1}(x_k - \bar{x})(x_{k+L} - \bar{x})}{\sum_{k=0}^{N-1}(x_k - \bar{x})^2} \tag{17.131}$$

and the autocovariance:

$$R_x(L) = R_x(-L) = \frac{1}{N}\sum_{k=0}^{N-L-1}(x_k - \bar{x})(x_{k+L} - \bar{x}) \tag{17.132}$$

One uses this for one-dimensional time series. Note: The mean is subtracted beforehand. In the case of *(image processing)* it is better to use the Fast Fourier Transform (FFT): The ACF is calculated from the FFT of a 2-D array is as follows:

$$\text{array}_1 = \text{FFT(array)} \tag{17.133}$$

$$ACF = FFT^{-1}(\text{array}_1 * conj(\text{array}_1)) \tag{17.134}$$

Where: FFT an algorithm for calculating the FFT and FFT^{-1} the inverse FFT and *conj* the conjugate complex function.

17.6.2 The Fast Fourier Transform, FFT

This is a method based on the *Cooley* and *Tukey,* algorithm (1965) for the fast computation of the Fourier transform.[1] We only briefly outline the Fourier transform.

---

[1] Also called the FFT, Fast Fourier Transform; first form as early as 1805 by Gauss to compute the orbits of Juno and Pallas.

Fourier (1768–1830) showed that periodic functions can be written as superpositions of harmonic oscillations. Let $f(x)$ is a function with the period $T = 2\pi$, then:

$$f(x) = \frac{a_0}{2} + \sum_{k=1}^{\infty} a_k \cos(kx) + b_k \sin(kx)$$

$$a_k = \frac{1}{\pi} \int_{-\infty}^{\infty} f(u) \cos(ku) du, \qquad k = 0, 1, 2, \ldots \tag{17.135}$$

$$b_k = \frac{1}{\pi} \int_{-\infty}^{\infty} f(u) \sin(ku) du, \qquad k = 0, 1, 2, \ldots$$

→ Details in spatial domain correspond to high frequencies in frequency domain. For compression methods we can neglect these high frequencies, image manipulation.

Euler's formula states:

$$e^{ikx} = \cos(kx) + i \sin(kx) \tag{17.136}$$

and therefore:

$$f(x) = \sum_{k=-\infty}^{\infty} c_k e^{ikx} \qquad c_k = \frac{1}{\pi} \int f(x) e^{-ikx} dx \tag{17.137}$$

The discrete Fourier transform maps $f(0), f(1), \ldots, F(N-1)$ to:

$$F(u) := \frac{1}{N} \sum_{x=0}^{N-1} f(x) e^{(-2\pi i u x)/N} \tag{17.138}$$

Important in astrophysics is the application to 2-D arrays (images):

$$F(u, v) := \frac{1}{MN} \sum_{x=0}^{M-1} \sum_{y=0}^{N-1} f(x, y) \exp(-[2\pi i (ux/M + vx/N)]) \tag{17.139}$$

From the Euler formula, the real or imaginary part then follows again:

$$R(u, v) := \sum_{x=0}^{M-1} \sum_{y=0}^{N-1} f(x, y) \cos(-2\pi (ux/M + vy/N)) \tag{17.140}$$

$$I(u, v) := \sum_{x=0}^{M-1} \sum_{y=0}^{N-1} f(x, y) \sin(-2\pi (ux/M + vy/N)) \tag{17.141}$$

The amplitude spectrum (power spectrum) reads:

$$|F(u, v)| := \sqrt{R^2(u, v) + I^2(u, v)} \tag{17.142}$$

The phase spectrum is:

$$\phi(u, v) := \arctan\left(\frac{I(u)}{R(u)}\right) \tag{17.143}$$

Example We define on a one-dimensional array $a1$ the function $f3$

$$f3 = \sin(a1/20) + \cos(a1/10)/2 \tag{17.144}$$

Then we compute the Fourier transform ff by means of np.fft.rfft and the power spectrum ff_{Pow}, i.e., the absolute value of the Fourier transform. Finally, we try to filter out a period from the function a from the function. For this we set a range of values of the Fourier transform ffn zero and calculate the inverse transform using np.fftz.irfft. The Python program is given in Fig. 17.8 given, and the results are shown in Fig. 17.9.

17.6.3 Digital Filters

Let us assume , we wanted to filter a signal, e.g. with a:

- High-pass filter
- low-pass filter
- band-pass filter

Imagine a CCD image of a galaxy. What would be the purpose of the three filter options mentioned here?

With the FFT this is very simple: Calculate the Fourier transform of the data, multiply the result with the filter function $H(f)$ and do an inverse FFT to restore the data in time. It should be noted:

- One must multiply the filter function $H(f)$ for positive and negative frequencies.
- If the measured data is real and the filtered output should also be real, then the following must hold for the filter function: $H(-f) = H(f)^*$.
- If $H(f)$ has sharp edges, then the impulse response will get damped frequencies at the edges. Therefore, one should consider a smooth transition.
- Before doing so, it is best to remove a trend in the data by subtracting the data from a straight line, which is determined by the first and last points.

```
#  fast fourier transform
import numpy as np
import matplotlib.pyplot as plt

a=np.arange(0,16000,.1)
a1=a/57.3
f1=np.sin(a1/(20 ))
f2=np.cos(a1/(10))/2.
f3=f1+f2
print(f3)
plt.tight_layout()
plt.subplots_adjust(left=0.1, bottom=.1, right=.9, top=None, wspace=0.5, hspace=.5)
plt.figure(0)
plt.subplot(2,2,1) # 2 Zeilem, 2 Spalte 1... erster Plot
# Plotten der Originaldaten
plt.plot(a1,f3)
plt.title("Original data")
# Berechnen der Fouriertransformation
ff=np.fft.rfft(f3)
plt.subplots_adjust(left=0.1, bottom=.1, right=.9, top=None, wspace=0.4, hspace=.4)
plt.subplot(2,2,2)
plt.xlim(left=0,right=20)# x-Wertebereich Grenze
# Plotten der Foureirtranaf.
plt.plot(ff)
plt.title("FFT")
ff_pow=abs(ff*ff) # Power Spektrum
plt.subplots_adjust(left=0.1, bottom=.1, right=.9, top=None, wspace=0.4, hspace=.4)
plt.subplot(2,2,3)
plt.xlim(left=0,right=20)
plt.plot(ff_pow)
plt.title("Power")
ffn=ff
ffn[2:15998]=0.+0.# Filtern der Daten im Fourierraum
print(ffn[0:12])
plt.subplots_adjust(left=0.1, bottom=.1, right=.9, top=None, wspace=0.4, hspace=.4)
plt.subplot(2,2,4)
plt.plot(a1,np.fft.irfft(ffn))
plt.title("Filtered function")
```

Fig. 17.8 Python program for Fourier transform and filtering

An example of filtering data is given in Fig. 17.9. By truncating the data in Fourier space, the cos function is almost eliminated. Experiment with the program and set the values from 4 to 15996 for the function ffn equal to zero.

17.6.4 Fourier Transforms in Optics

In the field of optics, it is often necessary to restore Images. Images can become degraded by various effects:

- Seeing: due to fluctuations of the refractive index in the Earth's atmosphere.
- Blurring: effect of defocusing due to air streaks that vary from place to place. The whole image loses sharpness as a result.

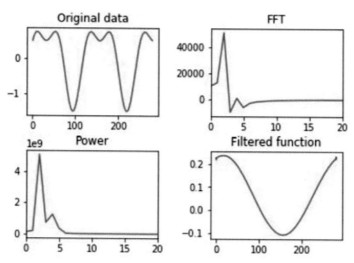

Fig. 17.9 Results for the FFT program

- Image Motion: The image remains sharp, but it moves rapidly back and forth;
- Image Distortion: Most of the image remains sharp, but smaller parts of the image are shifted against each other.

Image Motion occurs at a frequency of up to 100 Hz. So if you only use 10^{-2} s long exposure, one can reduce these interferences considerably.

A real image of, say, the sun is to be imaged in the x-y plane of a telescope. At each point the intensity $I(x, y)$ has contributions belonging to neighboring points of the imperfect image. The Point Spread Function $PSF(x, y, \xi, \eta)$ describes these influences and the intensity becomes:

$$I(x, y) = \int_{-\infty}^{\infty} I_0(\xi, \eta) PSF(x, y, \xi, \eta) d\xi d\eta \tag{17.145}$$

Short: $I = I_0 * PSF$. This can be simplified if the PSF only depends on the distances $\xi - x$ and $\eta - y$ as in the case of uniform blurring. Then one considers the FT of the image, and because of the convolution theorem, each Fourier component of I is simply multiplied by the corresponding Fourier component of the PSF. The modulus of the FT is called modulation transfer function, MTF:

$$MTF(k_x, k_y) = \left| \int \int_{-\infty}^{\infty} PSF(x, y) e^{[-2\pi i(k_x x + k_y y)]} dx dy \right| \tag{17.146}$$

What is the advantage of a Fourier transform? One can easily represent certain effects by a sequence of additive terms in the MTF:

$$MTF_{telescope} + MTF_{seeing} = MTF_{total} \tag{17.147}$$

If the PSF is rotationally symmetric in the x-y plane, i.e. $PSF(x, y) = PSF(r)$; $r = \sqrt{x^2 + y^2}$ then the MTF is rotationally symmetric in the k_x, k_y-plane and given by the Hankel TF:

$$MTF(k) = 2\pi \int_0^\infty r PSF(r) J_0(2\pi kr) dr \tag{17.148}$$

Where $k = \sqrt{k_x^2 + k_y^2}$ and J_0 is the Bessel function of order 0.

As an example, consider the Airy image of a point source given by a diffraction-limited telescope:

$$PSF_D(r) = \frac{1}{\pi}[J_1(br)/r]^2 \tag{17.149}$$

Thereby J_1 is the Bessel function is of order 1, and $b = D\pi/\lambda f$, D is the aperture, f is the focal length. The first zero of the PSF_D is of $r = r_1 = 3{,}832/b$ and is called the radius of the central Airy disk. The angle $\alpha_1 = r_1/f$ is usually called the resolution of a diffraction-limited telescope:

$$\alpha_1 = 1.22\lambda/D \tag{17.150}$$

Now substitute the PSF into the formula for the MTF and get:

$$MTF_D(k) = \frac{2}{\pi}\left[\arccos(k/k_m) - \frac{k}{k_m}\sqrt{1 - (k/k_m)^2}\right] \tag{17.151}$$

$k_m = b/\pi$ is the largest wavenumber still transmitted by the telescope. $MTF_D = 0, k \geq k_m$.

The MTF_D is therefore the factor by which a signal at the wavenumber k is reduced. The wavenumber corresponding to the resolution is then: $k = 1/r_1 = k_m/1.22$ and $MTF_D = 0.0894$ for this $k \to$ of the original signal only this amount remains.

If one averages over periods of 1 s or even longer, then the PSF of the seeing is well described by a rotationally symmetric Gaussian function:

$$PSF_S(r) = \frac{1}{2\pi s_0^2}e^{-r^2/2s_0^2} \tag{17.152}$$

Thereby the parameter s_0 is a quantitative measure for the seeing. One speaks with $s_0 = 1''$ good seeing and $s_0 = 0.5''$ excellent seeing. The FT of a Gaussian function again yields a Gaussian function. The MTF is:

$$MTF_s(k) = e^{-2\pi^2 s_0^2 k^2} \tag{17.153}$$

17.7 Vector Calculus

17.7.1 General

Vectors are quantities with magnitude and direction, e.g. the velocity **v**. *Scalars* can be described by a simple number (mass, temperature). In a three-dimensional space, one can write for the components of a vector:

$$\mathbf{A} = (A_x, A_y, A_z) = (A_1, A_2, A_3) \tag{17.154}$$

The length of the vector is given by:

$$|\mathbf{A}| = \sqrt{A_1^2 + A_2^2 + A_3^2} \tag{17.155}$$

In a Cartesian coordinate system, one uses the following unit vectors: $\mathbf{i} = (1, 0, 0)$, $\mathbf{j} = (0, 1, 0)$, $\mathbf{k} = (0, 0, 1)$, and therefore: $\mathbf{A} = (A_x\mathbf{i} + A_y\mathbf{j} + A_z\mathbf{k})$. Two vectors can be added and subtracted component wise.

The scalar product (inner product) gives a scalar quantity as the result:

$$\mathbf{A} \cdot \mathbf{B} = |\mathbf{A}||\mathbf{B}| \cos\alpha \tag{17.156}$$

where α is the angle between **A** and **B** is.

→ If the two vectors **A** and **B** are orthogonal, then the scalar product vanishes.

The vector product of two vectors is defined by

$$\mathbf{C} = \mathbf{A} \times \mathbf{B} = |\mathbf{A}||\mathbf{B}| \sin\alpha \tag{17.157}$$

The resulting vector **C** is perpendicular to **A** and **B**, and the following applies (see also Fig. 17.10).

Fig. 17.10 Right rule for the
vector product

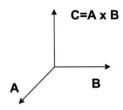

Note:

$$\mathbf{i} \times \mathbf{i} = 0 \qquad \mathbf{i} \times \mathbf{j} = \mathbf{k}$$

$$\mathbf{j} \times \mathbf{j} = 0 \qquad \mathbf{j} \times \mathbf{k} = \mathbf{i}$$

$$\mathbf{k} \times \mathbf{k} = 0 \qquad \mathbf{k} \times \mathbf{i} = \mathbf{j}$$

One can also write:

$$\mathbf{A} \times \mathbf{B} = \begin{vmatrix} \mathbf{i} & \mathbf{j} & \mathbf{k} \\ A_x & A_y & A_z \\ B_x & B_y & B_z \end{vmatrix} \tag{17.158}$$

Trajectories: when a vector depends on a parameter, it is called a vector function: $\mathbf{a}(t)$.

Let $\mathbf{r}(t)$ the trajectory of a point over time t. The velocity is $\mathbf{v} = \dot{\mathbf{r}}$ and the acceleration $\mathbf{a} = \ddot{\mathbf{r}}$. The distance travelled s is $ds = |d\mathbf{r}| = v\,dt$. We consider three unit vectors perpendicular to each other $\mathbf{e_T}, \mathbf{e_N}, \mathbf{e_B}$. The vector $\mathbf{e_T}$ points in the direction of $\mathbf{v}$.

The tangent vector is:

$$\mathbf{e_T} = \frac{d\mathbf{r}}{ds} = \frac{d\mathbf{v}}{v} \tag{17.159}$$

The curvature of the curve:

$$\kappa = \left| \frac{d\mathbf{e_T}}{ds} \right| = \left| \frac{d^2\mathbf{r}}{ds^2} \right| \tag{17.160}$$

the radius of curvature:

$$R = \frac{1}{\kappa} \tag{17.161}$$

the normal vector:

$$\mathbf{e_N} = R\frac{d\mathbf{e_T}}{ds} = R\frac{d^2\mathbf{r}}{ds^2} \tag{17.162}$$

It is perpendicular to $\mathbf{e_T}$. Thus, for the velocity and the acceleration, we find:

$$\mathbf{v} = \dot{\mathbf{r}} = v\mathbf{e_T} \qquad \mathbf{a} = \ddot{\mathbf{r}} = \dot{v}\mathbf{e_T} + \frac{v^2}{R}\mathbf{e_N} \qquad (17.163)$$

So we get tangential acceleration when the magnitude of the velocity changes, and normal acceleration when the direction changes.

Vector field: let each location have a vector associated with it. Important for field description in physics.

17.7.2 Gradient, Divergence, Curl

In this Sectionconsider we:

- Scalar functions: These assign to each point $P = (x, y, z)$ in space a scalar quantity. An example of this would be a temperature field. At every point in space, there is a temperature that is a scalar quantity, so it has no direction.
- Vector field: a vector is assigned to each point in space, e.g., a force field $\mathbf{F}(x, y, z)$.

The gradient of a scalar function ψ Is the vector field:

$$\mathrm{grad}\,\psi = \nabla\psi \qquad \nabla\psi = \mathbf{e_x}\frac{\partial\psi}{\partial x} + \mathbf{e_y}\frac{\partial\psi}{\partial y} + \mathbf{e_z}\frac{\partial\psi}{\partial z} \qquad (17.164)$$

The gradient at a point is a vector pointing in the direction of the sharpest increase.

The divergence (source density) generates from a vector field $\mathbf{A}$ the scalar field:

$$\mathrm{div}\,\mathbf{A} = \nabla\cdot\mathbf{A} = \frac{\partial A_x}{\partial x} + \frac{\partial A_y}{\partial y} + \frac{\partial A_z}{\partial z} \qquad (17.165)$$

The divergence indicates the sources ($div > 0$) or sinks of a vector field.

The curl of a vector field $\mathbf{A}$ gives the vector field:

$$\mathrm{curl}\,\mathbf{A} = \nabla\times\mathbf{A} \qquad (17.166)$$

and one determines:

$$\nabla\times\mathbf{A} = \mathbf{e_x}\left(\frac{\partial A_z}{\partial y} - \frac{\partial A_y}{\partial z}\right) + \mathbf{e_y}\left(\frac{\partial A_x}{\partial z} - \frac{\partial A_z}{\partial x}\right) + \mathbf{e_z}\left(\frac{\partial A_y}{\partial x} - \frac{\partial A_x}{\partial y}\right)$$

Another calculation of the curl:

$$\nabla \times \mathbf{A} = \begin{vmatrix} \mathbf{e_x} & \mathbf{e_y} & \mathbf{e_z} \\ \frac{\partial}{\partial x} & \frac{\partial}{\partial y} & \frac{\partial}{\partial z} \\ A_x & A_y & A_z \end{vmatrix} \tag{17.167}$$

Important theorems:

$$\nabla \cdot (\nabla \times \mathbf{A}) = \mathbf{0} \tag{17.168}$$

$$\nabla \times (\nabla \times \mathbf{A}) = \nabla(\nabla \cdot \mathbf{A}) - \nabla^2 \mathbf{A} \tag{17.169}$$

Divergence theorem:

$$\int_V \nabla \cdot \mathbf{A} d^3 x = \int_S \mathbf{A} \cdot d\mathbf{S} \tag{17.170}$$

Stokes-Theorem:

$$\int_S (\nabla \times \mathbf{A}) \cdot d\mathbf{S} = \oint_C \mathbf{A} \cdot d\mathbf{l} \tag{17.171}$$

17.7.3 Applications

Let us consider two simple applications of the divergence operator.

- Given the vector field $\mathbf{F}(x, y) = x\mathbf{i} + y\mathbf{j}$. A graph of this field shows that all vectors point away from the point (0, 0). For example, consider the point (2, 2). At this point the vector $2\mathbf{i} + 2\mathbf{j}$ is to be drawn. Analogously, this applies to all points (x, y). One finds: The point (0, 0) is a *Source* of the field. Such a field could describe the situation of matter distribution immediately after the Big Bang.
- Let be given the vector field $\mathbf{F}(x, y) = -x\mathbf{i} - y\mathbf{j}$. A graph of this field shows that all vectors point towards the point (0, 0). This describes the situation around a black hole. Matter flows into a sink.
- Let us calculate the divergence for $\mathbf{F}(x, y) = x\mathbf{i} + y\mathbf{j}$. From $\nabla \cdot \cdot \mathbf{F} = 2$, thus > 0, therefore a source, for the case $\mathbf{F}(x, y) = -x\mathbf{i} - y\mathbf{j}$ the divergence gives the value -2, therefore a sink.

17.8 Splines

With the Spline Interpolation one divides a function f in an interval into subintervals, most simply into lines with straight lines:

$$s(x) = \frac{y_2 - y_1}{x_2 - x_1} \cdot x + y_1 - \frac{y_2 - y_1}{x_2 - x_1} \cdot x_1 \qquad (17.172)$$

$$m = \frac{y_2 - y_1}{x_2 - x_1} \qquad (17.173)$$

$$b = y_1 - m \cdot x_1 \qquad (17.174)$$

Much more accurate are the so-called cubic splines. If one has an interval [a, b] with $a = x_0 < x_1 \ldots < x_{n-1} < x_n = b$ resp. x_i, $f(x_i)$ $i = 0, \ldots, n$ and on each of these subintervals we take a 3rd degree polynomial (hence cubic spline), which is supposed to be twice differentiable, then

$$S_j(x) = a_j + b_j \cdot (x - x_j) + c_j \cdot (x - x_j)^2 + d_j \cdot (x - x_j)^3 \qquad (17.175)$$

with $x_j < x < x_{j+1}$ and $j = 0, \ldots, n - 1$. We need $4n$ conditions to solve the equations:

- Interpolation conditions:

$$S_j(x_j) = f_j \qquad j = 0, 1, \ldots, n - 1 \qquad (17.176)$$

$$S_j(x_{j+1}) = f_{j+1} \qquad j = 0, 1, \ldots, n - 1 \qquad (17.177)$$

- The spline must be twice continuously differentiable:

$$S_j'(x_j) = S_{j+1}'(x_j) \qquad j = 1, \ldots, n - 1 \qquad (17.178)$$

$$S_j''(x_j) = S_{j+1}''(x_j) \qquad j = 1, \ldots, n - 1 \qquad (17.179)$$

- The other conditions can be defined differently, e.g..:

(a) free edge (natural spline) $S_0''(x_0) = 0$, $S_{n-1}''(x_n) = 0$
(b) clamped edge: $S_0'(x_0) = f_0'$, $S_{n-1}'(x_n) = f_n'$ where f_0' and f_n' are given.
In addition, you can define other splines.

Example Python and splines.
Given the declination of the moon for the following days:

Day	Declination °
15.3.2007	−22.4259
16.3.	−17.4703
17.3	−11.4218
18.3.	−4.5013
19.3.	2.2335
20.3.	9.3104
21.3.	16.0319
22.3	21.3251
23.3.	25.3624
24.3.	27.5815
25.3.	28.3307

Calculate when the moon is in the celestial equator during this period, i.e. $\delta_{moon} = 0°$. For this we use the program library $scpy.interpolate$, from which we load the program $interp1d$. The complete Python program is shown in Fig. 17.11.
The solution results in $18.670 = 18.03.2007$ at $16:05$ min.

```python
1 #!/usr/bin/env python3
2 # -*- coding: utf-8 -*-
3 """
4 Created on Mon Jul 15 07:08:50 2019
5
6 @author: arnoldhanslmeier
7 """
8
9 from scipy.interpolate import interp1d
0 # Eingeben der Datumswerte
1 y=[15,16,17,18,19,20,21,22,23,24,25]
2 # Eingeben der Deklinationswerte
3 x=[-22.4259,-17.4703,-11.4218,-4.5013,2.2335,9.3104,16.0319,21.3251,25.3624,27.5815,28.3307]
4
5 # lineare Interpolation
6 f = interp1d(x, y)
7 # kubische Interpolation
8 f2 = interp1d(x, y, kind='cubic')
9
0 xnew = np.linspace(-22, 22, num=40, endpoint=True)
1 # Definieren des Wertes 0 für die Deklination
2 xx=0.
3 # Ergebnis
4 print(f(xx),' lineare Interpolation ',f2(xx), ' kubische Interpolation')
5 import matplotlib.pyplot as plt
6 plt.plot(x, y, 'o', xnew, f(xnew), '-', xnew, f2(xnew), '--')
7 plt.legend(['data', 'linear', 'cubic'], loc='best')
8 plt.show()
```

Fig. 17.11 Python program for spline calculation

```
import ephem
graz=ephem.Observer()
graz.lat, graz.lon='47.1','15.42'
graz.pressure=0
graz.horizon='0'
mars=ephem.Mars()
mars.compute('2019/6/10')
graz.date='2019/6/10 17:00'
print (mars.name,mars.ra, mars.dec,mars.size,ephem.constellation(mars))
print('Time is given in UT')
print('Mars rises:',graz.previous_rising(ephem.Mars()))
print('Mars sets',graz.next_setting(ephem.Mars()))
print('Sun rises:',graz.previous_rising(ephem.Sun()))
print('Sun sets',graz.next_setting(ephem.Sun()))
```

Fig. 17.12 Some calculations with the program ephem

17.9   Special Software Packages

17.9.1 The Ephem Program Package

The Python program package ephem calculates ephemeris, i.e. positions of the objects, rising and setting times, etc. One installs it with the command pip install ephem.

Let's have a look at some simple examples (more details can be found in the tutorials). With the commands mars=ephem(Mars) and mars.compute('2019...') one computes the position of the planet Mars. The program is shown in Fig. 17.12. In addition, the program calculates the rising and setting times of the Sun and Mars, as well as the constellation in which Mars is located. The data are valid for the observation site Graz, which has a latitude of about 47° and a longitude of about 15.4° East. The calculated times are in UT; thus for Central Europe one hour has to be added or 2 h because of daylight saving time.

17.9.2 Calculation of the Light Curves of Exoplanet Transits

The simplest method to find exoplanets is to measure the change of the brightness of the parent star when the exoplanet passes in front of it. This is called a transit. The change in brightness depends on several factors:

- Ratio of diameters of exoplanet and parent star. The smaller the planet is compared to its parent star, the smaller the change in brightness will be.
- Orbital inclination (inclination) of the exoplanet. Only at an inclination close to 90° can a transit be observed, since we are then looking at the edge of the planet's orbit, so to speak.

```
## lightcurve simulation of transiting planets
import batman
import numpy as np
import matplotlib.pyplot as plt
import pylab
params = batman.TransitParams()
params.t0 = 0.                    #time of inferior conjunction
params.per = 1.                   #orbital period
params.rp = 0.01                   #planet radius (in units of stellar radii)
params.a = 20.                    #semi-major axis (in units of stellar radii)
params.inc = 88.                  #orbital inclination (in degrees)
params.ecc = 0.                   #eccentricity
params.w = 90.                    #longitude of periastron (in degrees)
params.u = [0.1, 0.3]             #limb darkening coefficients [u1, u2]
params.limb_dark = "quadratic"    #limb darkening model
t = np.linspace(-0.05, 0.05, 100)  # specify the times at which you want to get values
m = batman.TransitModel(params, t)   #initializes model
flux = m.light_curve(params)         #calculates light curve
plt.plot(t, flux,label='a=20')
plt.xlabel("Time from central transit")
plt.ylabel("Relative flux")
params.a=50.
flux = m.light_curve(params)
plt.plot(t,flux+.00001,label='a=50')
pylab.legend(loc='upper left')
plt.show()
```

Fig. 17.13 Calculation of a brightness variation due to a planetary transit

- Distance of the exoplanet from its parent star. The further away the planet is from its parent star, the less likely a transit becomes, since we as observers would then have to look exactly at the edge of the planet's orbit.
- The shape of the light curve is also affected by the edge-darkening of the star. This center-edge variation depends on the type of star itself.

The Python program package `batman` provides light curves of exoplanet transits, where the parameters listed above are freely selectable. An example of this is shown in Fig. 17.13. Two light curves are generated, one for $a = 20$ and one for $a = 50$. In the latter case, the exoplanet would be 50 stellar radii away from the star and you can see from the Light curve that there is no longer a transit.

The light curves are shown in Fig. 17.14.

17.9.3 Image Processing

Images of objects are nowadays almost exclusively taken Image processing with CCD arrays. The CCD array consists of a matrix with light-sensitive picture elements, pixels. One receives thus a data matrix; each point of the matrix has a certain numerical value, the amount of which reflects the intensity. The advantage of this technique is on the one hand the high light efficiency (quantum efficiency), which means short exposure times for faint objects, and on the other hand the data are available in digital form as a two-dimensional array. We will briefly show how to use Python for image analysis. In Fig. 17.15 a simple program is shown.

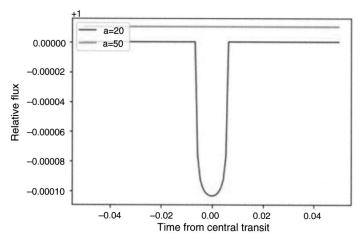

Fig. 17.14 Exoplanets: transit light curves, for $a = 20$ respectively $a = 50$ (no transit)

```
from PIL import Image
from numpy import *
from pylab import *
img = Image.open('M82_300s.jpg')
img.show()

# define a box

box=(100,100,400,400)

# define a subimage
region=img.crop(box)
# plot subimage
region.show()
# resize subimage
out=region.resize((50,50))
out.show()
# rotate image
out_rotate=img.rotate(45)
out_rotate.show()
```

Fig. 17.15 Some simple processes to manipulate images

First the necessary packages like PIL, numpy and pylab are imported. The command Image.open is used to open a sample image. The path where the image is located must be specified. In our case, the sample image is in the same directory as the program. The command img.show() shows the image. Often you are only interested in a certain area of an image. Using box this region of interest. can be defined; the coordinates of the two selection points for defining the rectangular image section are (100, 100, 400, 400). Now we define the cropped image with the command (region=img.crop(box)) and shown using region.show(). We can also resize this image with the command region.resize((50,50)). Finally we rotate the whole image with img.rotate(45) by the value of 45°.

```
from PIL import Image
from numpy import *
from pylab import *
# now read image to array
im=array(Image.open('M82_300s.jpg'))
print(im.shape, im.dtype)
imshow(im)
# mark a special area in image
x=[100,100,400,400]
y=[200,500,200,500]
# plot a se ific point in image
plot(200,200,'*')
plot(x,y,'r*')
title('M 82 Galaxy, AH')

figure()
# now we plot an intebsity profile at üposition x= 600
xv=np.arange(1,1100,1)
yv=im[1:1100,600,1:2]

plot(xv,yv)
```

Fig. 17.16 Saving an image as a data array

```
from PIL import Image
from numpy import *
from pylab import *

from PIL import Image, ImageFilter

#if __name__=="__main__":
im = Image.open("M82_300s.jpg")

kernelValues = [-31,-1,0,-1,1,1,0,1,31] #emboss
kernel = ImageFilter.Kernel((3,3), kernelValues)

im2 = im.filter(kernel)

im2.show()


from PIL import Image, ImageEnhance
im = Image.open("M82_300s.jpg")
enhancer = ImageEnhance.Sharpness(im)
enhanced_im = enhancer.enhance(20.0)
enhanced_im.show()
enhanced_im.save("enhanced.sample3.png")

from PIL import Image, ImageEnhance
im = Image.open("M82_300s.jpg")
enhancer = ImageEnhance.Contrast(im)
enhanced_im = enhancer.enhance(2.10)
enhanced_im.show()
enhanced_im.save("enhanced.sample1.png")
```

Fig. 17.17 Program for image manipulation

In Fig. 17.16 shows how to read in an image and then save it as an ARRAY. Then points are marked in the image array, first a rectangle (red asterisk), then a point (blue asterisk) and finally an intensity profile is created.

In Fig. 17.17 some examples of image manipulation are shown, such as contrast enhancement, edge enhancement, etc. The file read in is M82_300s.jpg.

```
import matplotlib.pyplot as plt
from astropy.visualization import astropy_mpl_style
plt.style.use(astropy_mpl_style)
from astropy.utils.data import get_pkg_data_filename
from astropy.io import fits
# Downlaod sample file
image_file = get_pkg_data_filename('tutorials/FITS-images/HorseHead.fits')
# print info about fits file
fits.info(image_file)
image_data = fits.getdata(image_file, ext=0)
#stores the image as 2-d numpy array

print(image_data.shape)

# display the stuff

plt.figure()
plt.imshow(image_data, cmap='gray')
plt.colorbar()
```

Fig. 17.18   Program to read a fits file and define it as a 2-D array

17.9.4 The File Format Fits

In astrophysics f the data is often stored as FITS files. FITS stands for flexible image transport system. This was developed by NASA in 1981 and is mostly used to record images, spectra or tables. It has been approved by the International Astronomical Union.

In Fig. 17.18 gives an example of reading a fits file. This file is supplied with the installation of the `astropy` package. The package `astropy` is installed, for example, with the command `conda install astropy`.

17.10 Further Literature

We give a small selection of recommended further reading.
Mathematical Methods for Physics and Engineering, K.F. Riley, M.P. Hobson, S.J. Bence, Cambridge, 2020
Principles of Digital Image Processing, W. Burger, Springer, 2011
Image Processing with Python, S. Dey, Packt Publ. 2018
Python Crash Course, E. Matthes, No Starch Press, 2019

Tasks

17.1 Measured values: 2.52 3.96 3.28 9.20 3.75. Determine the mean and the median.

Solution
Mean = $(2.52 + 3.96 + 3.28 + 9.20 + 3.75)/5 = 4.54$
Median: arrange the numbers and find the value in the middle = 3.75

17.2 Let the series of measurements be 2, 4, 8. Calculate the arithmetic mean, geometric mean and harmonic mean!

Solution
Arithmetic mean = 4.67; geometric mean = 4, harmonic mean = 3.53.

17.3 Determine the square mean of the numbers 1, 3, 4, 5, 7.

Solution
$$QM = \sqrt{\frac{1^2+3^2+4^2+5^2+7^2}{5}} = 4,47.$$

17.4 Given the sets of numbers 2, 5, 8, 11, 14 and 2, 8, 14, determine (a) the mean for each set, (b) the variance for each set, (c) the mean of the combined sets, (d) the variance of the combined sets.

Solution

(a) mean of 1st quantity = $(1/5)(2 + 5 + 8 + 11 + 14) = 8$, mean of the 2nd set = 8 (b) variance of the 1st set:
$$s_1^2 = \tfrac{1}{5}[(2-8)^2 + (5-8)^2 + (8-8)^2 + (11-8)^2 + (14-8)^2] = 18,$$
Variance of the 2nd quantity = 24
(c) Mean of the combined quantities:
$$\frac{2+5+8+11+14+2+8+14}{5+3} = 8$$
(d) Variance of the combined quantities:
$$s^2 = \frac{(2-8)^2+(5-8)^2+(8-8)^2+(11-8)^2+(14-8)^2+(2-8)^2+(8-8)^2+(14-8)^2}{5+3} = 20,25$$

17.5 Determine the first four moments with respect to the mean of the set of numbers 1, 3, 5, 6, 5.

Solution
$$m_1 = \frac{(1-4)+(3-4)+(5-4)+(6-4)+(5-4)}{5}$$
$$m_2 = \frac{(1-4)^2+(3-4)^2+(5-4)^2+(6-4)^2+(5-4)^2}{5} \ldots$$

17.6 Fit a straight line to the following data using the least squares method: $X = 1, 3, 4, 6, 8, 9, 11, 14$, $Y = 1, 2, 4, 4, 5, 7, 8, 9$. Draw the data and the straight line on a graph!

Solution

The best way to create a table is $\sum X = 56$, $\sum Y = \ldots$, $\sum X^2 = \ldots$, $\sum XY = \ldots$, $\sum Y^2 = 256$

Thus we get the normal equations to:

$8a_0 + 56a_1 = 40$

$56a_0 + 524a_1 = 364$,

and our sought straight line is:

$Y = 0.545 + 0.636X$.

17.7 Solve the previous example assuming that X is the dependent variable and Y the independent!

Solution

So we are looking for a straight line of the form: $X = b_0 + b_1 Y$ And again set up the normal equations:

$\sum X = b_0 N + b_1 \sum Y$

$\sum XY = b_0 \sum Y + b_1 \sum Y^2$

Then we get as solution:

$Y = 0.333 + 0.667X$.

17.8 Find the third degree polynomial that takes the following values: $x_k = 0, 1, 2, 4$; $y_k = 1, 1, 2, 5$

Solution

$p(x) = \frac{(x-1)(x-2)(x-4)}{(0-1)(0-2)(0-4)}1 + \frac{x(x-2)(x-4)}{1(1-2)(1-4)}1 + \frac{x(x-1)(x-4)}{2(2-1)(2-4)}2 + \frac{x(x-1)(x-2)}{4(4-1)(4-2)}5$

and hence

$p(x) = \frac{1}{12}(-x^3 + 9x^2 - 8x + 12)$

17.9 Apply Simpson's rule to obtain $\sqrt{x}$ between 1.00 and 1.30!

Solution

$\int_{1.00}^{1.30} \sqrt{x}\,dx \approx \frac{0.05}{3}[1.0000 + 4(1.02470 + 1.07238 + 1.11803) + 2(1.04881 + 1.09544) + 1.14017] = 0.32149$

17.10 One uses Euler's method to solve the differential equation.

$y' = f(x, y) = xy^{1/3}$, $y(1) = 1$

to solve it.

Solution

We calculate three values:

$y_1 \approx 1 + (..01)(1) = 1.0100$

$y_2 \approx 1.0100 + (0.01)[(1.01)(1.033)] \approx 1.0201$

$y_3 \approx 1.0201 + (0.01)[(1.02)(1.0067)] \approx 1.0304$

Compare the result with the actual values!

17.11 Apply the Runge-Kutta method to the above DE.

Solution

Let $x_0 = 1, h = 0.01$, then you have for one step:

$k_1 = (0.1) f(1, 1) = 0.1 \, k_2 = (0.1) f(1.05; 1.05) \approx 0.10672 k_3 = (0.1) f(1.05; 1.05336)$
$\approx 0.1068 k_4 = (0.1)(f(1.1; 1.10684) \approx 0.11378 y_1 = 1 + 1/6(0.1 + 0.21344 + 0.21368 +$
$0.11378) \approx 1.10682$

Calculate further steps until $x = 5$ and compare the results!

17.12 Let it be shown that the vector field $\mathbf{F}(x, y) = -y\mathbf{i} + x\mathbf{j}$ has no sources and sinks!
Sketch the vector field! What could describe such a vector field?

Solution

$\text{div}\mathbf{F} = [\partial/\partial x](-y) + [\partial/\partial y](x) = 0$ It could describe a vortex flow.

17.13 Calculate the divergence of the vector field $\mathbf{F}(x, y, z) = xz\mathbf{i} + 2xy^2\mathbf{j} + z^2\mathbf{k}$ at the
point $P = (1, -1, 1)$!

Solution

The divergence of the vector field gives: $4xy + 3z$. If we substitute for $x = 1, y = -1, z = 1$, then the result is: divergence of the vector field at the given point $= -1$, i.e., a sink.

Appendix

A

A.1 Literature

A.1.1 General

Bennet, J., Astronomy: The cosmic perspective, H. Lesch (editor), Pearson Study—Physics, Addison-Wesley Publishing; 5th edition, 2009

Encyclopedia of Astronomy and Astrophysics, Inst. of. Phys. Publ., London, 2001 (3 vols.; also available online at http://eea.crcpress.com)

Hawking, S.: A brief history of time, Rowohlt, Hamburg, 1998

Karttunen, H., Krüger, P., Oja, H., Poutanen, M, Dorner, K.J. eds, Fundamental Astronomy, Springer-Verlag, Berlin, 4th edition 2004

Lesch, H., Müller, J.: Big Bang second act. On the tracks of life in space, Goldmann Verlag 2005

Meyers Handbuch Weltall, B.I., Mannheim, 7th edition 1994

Pasachoff, J.M., Filippenko, A.: The Cosmos, Cambridge Univ. 5th ed, 2019

Sexl, R., Schmidt, H.K.: Space, Time, Relativity, Springer, Berlin, 7th ed. 2000

Sexl, R., Sexl, H.: White dwarfs – black holes, Springer, Berlin, 2nd ed. 1999

Sexl, R.U., Urabantke, H. Gravitation and Cosmology, Spektrum, 1995

Spatschek, K.H., Astrophysics, 2nd edition, Springer Sektrum, 2017.

Unsöld, A., Baschek, B.: The new cosmos, Springer-Verlag, Heidelberg, 7th edition 2015.

Voigt, H.H.: Abriss der Astronomie, H.J. Röser, W. Tscharnuter (editors), Wiley-VCH, 6th edition, 2012

Weigert, A., Wendker, H.J.: Astronomy and Astrophysics, A Basic Course, Wiley–VCH, Weinheim, 5th edition 2009

© Springer-Verlag GmbH Germany, part of Springer Nature 2023
A. Hanslmeier, *Introduction to Astronomy and Astrophysics*,
https://doi.org/10.1007/978-3-662-64637-3

A.1.2 Journals

Popular Astronomy
Sky and Telescope, Sky Publ. Coop., Cambridge/Mass.
Astronomy today, Spektrum der Wissenschaft, Heidelberg
Sterne und Weltraum, Spektrum der Wissenschaft, Heidelberg
Interstellarum, Oculum Publishing House, Erlangen, Germany
BBC Sky at night, UK
Astronomy Now, UK
Sky News, Canada
Mercury, PASP
Popular Astronomy, US

Journals
Advances in Space Research, Elsevier
Annual Review of Astronomy and Astrophysics (review article)
Astronomy and Astrophysics, EDP Sciences
Celestial Mechanics and Dynamical Astronomy, Springer
Icarus, Elsevier
Solar Physics, Springer
The Astronomical Almanac (Ephemerides), US Naval Observatory
The Astronomical Journal, The University of Chicago Press
The Astrophysical Journal, The University of Chicago Press
Astrobiology, Mary Ann Liebert Publ.

A.1.3 Important Internet Addresses

ADS Abstract Service (here you can find abstracts of scientific articles from all areas of astronomy and astrophysics; you can search for a specific topic or for a specific author):

http://adsabs.harvard.edu/
Astronomical Society:

http://www.astronomische-gesellschaft.org/de
Astronomy on the Internet (Astroweb):

http://cdsweb.u-strasbg.fr/astroweb.html
European Southern Observatory, ESO: http://www.eso.org
Hubble Space Telescope: http://www.spacetelescope.org/about
International Astronomical Union (IAU): http://www.iau.org/
NASA: http://www.nasa.gov/goddard

SOHO (solar imagery): http://sohowww.nascom.nasa.gov/
Virtual tours of observatories:
http://tdc-www.harvard.edu/mthopkins/obstours.html

A.1.4   Software (Professional)

ANA (free data analysis software): http://ana.lmsal.com/
IRAF (Image Reduction and Analysis Feature): http://iraf.noao.edu/
ESO-MIDAS (Munich Image Data Analysis System, free software for data reduction):
    http://www.eso.org/projects/esomidas/
Python: www.python.org

Email of the author:
arnold.hanslmeier(at)uni-graz.at

A.2   Test Questions

True or false? You correct the statements if necessary. You will also find some *highlights* **from exams among them!**

1. Galaxies and clusters expand as the universe expands.
2. Matter cannot be created from nothing.
3. A mirror with a focal length of 10 m produces an image of the sun with a diameter of 10 cm.
4. Hubble's law applies to quasars and galaxies.
5. Superluminal velocities can be explained as projection effects.
6. The universe is static.
7. Flares are luminous phenomena in the earth's atmosphere.
8. The universe is composed mostly of dark, invisible matter.
9. Galaxies evolve according to the Hubble sequence.
10. The star of Bethlehem was a bright comet.
11. Spiral arms are formed by density waves.
12. Black holes cannot be detected.
13. Active optics are used to improve image quality.
14. Convection only exists in stars.
15. General relativity is a theory of gravity.
16. Pulsars are pulsating stars.
17. White light has a higher intensity than red light.
18. Galaxy masses can be inferred from rotation.
19. All supernovae are exploding stars.

20. Stars shine constantly.
21. Astronomical unit is the mean distance earth-sun.
22. Venus is the brightest planet in the evening or morning sky.
23. Active galactic nuclei are caused by black holes at their centers.
24. Most stars are single stars.
25. Asteroids cannot become dangerous.
26. Flare outbursts are produced by fusion reactions inside the sun.
27. Bright stars are closer.
28. The period-luminosity relation holds for all Cepheids.
29. Many stars have measurable proper motions.
30. The color of a star is a measure of its temperature.
31. Forbidden lines occur in the spectrum of the photosphere.
32. Kepler's third law states that a planet moves around the sun more rapidly when it is near the sun than when it is far from it.
33. After five billion years, the Sun evolves into a supernova and then into a black hole.
34. You can always see sunspots.
35. Cosmic rays are created by close neutron binaries.
36. The photospheric spectrum of the sun is in absorption, that of the chromosphere and corona in emission.
37. Extrasolar planets have been photographed.
38. Star diameters are determined by interferometer.
39. The latitude of an observing site can be determined by measuring the altitude of Polaris.
40. The asteroid belt cannot be penetrated.
41. The rings of Uranus have long been known.
42. The rings of Saturn are solid.
43. A G2 star is hotter than an A1 star.
44. All giant planets have rings.
45. Sidereal time is about 4 min shorter than solar time because Earth's rotation is slowed down.
46. Mars is 200 times farther away than the moon.
47. The absolute brightness of a star can be read from the HRD.
48. Growth curves indicate how fast a star forms from a protostellar object.
49. The Doppler effect no longer applies to quasars.
50. Tides are caused by the moon.
51. Tidal forces decrease with the square of the distance.
52. The moon can only be seen at night.
53. The shape of a spectral line depends only on the temperature and abundance of the element in question.
54. Telescopes are needed for high magnifications.
55. Copernicus introduced elliptical planetary orbits.
56. A star with many spectral lines is very hot.

57. Neutrinos are formed when protons and neutrons combine.
58. In the Galaxy, dark matter consists of the dark clouds that can already be seen with the naked eye.
59. The big bang theory can only be proved by the expansion of the universe.
60. The resolving power of a telescope depends on its focal length.
61. Quasars are like pulsars.
62. Fe lines in the solar spectrum indicate a large abundance of Fe.
63. Dark energy accounts for about 70% of the energy of the universe.
64. Fusion occurs in stellar centers due to high temperatures.
65. Planetary nebulae form in late stages of red giant evolution.
66. Helioseismology uses waves propagating through the solar body.
67. Supermassive black holes at the centers of galaxies cannot be detected.
68. The extent of the Milky Way is 206,265 times the Earth-Sun distance.
69. The galactic double wave of velocities produces density waves.
70. Parallax and aberration ellipses can only be distinguished by different phases.
71. At the North Pole the Pole Star is seen at the zenith.
72. At maximum western elongation Mercury is seen in the evening sky.
73. The diameter of Betelgeuse has been determined by lunar occultations.

A.3   Tables

The following tables give an overview of important properties of bright stars (both from the northern and southern hemisphere) as well as important astrophysical constants, units in the SI system and important conversions. The CGS system is still frequently used in the field of astrophysics (Tables A.1, A.2, A.3, A.4 and A.5).

Table A.1 Bright star clusters (open clusters OC, globular clusters GC), planetary nebulae (PN) and galactic nebulae (GN)

Designation	Type	Brightness	Distance
M 2	GC	$6^{m}5$	11.3 kpc
M 3	GC	$6^{m}3$	9.9 kpc
M 13	GC	$5^{m}9$	7.2 kpc
M 15	GC	$6^{m}4$	9.4 kpc
h+χ Persei	OC	$4^{m}3$	2.2 kpc
M 45 Pleiades	OC	$2^{m}9$	125 pc
Hyades	OC	$3^{m}4$	46 pc
Praesepe	OC	$3^{m}1$	160 pc
M 11	OC	$5^{m}8$	1.7 kpc
M 57, Ring Nebula	PN	$9^{m}7$	2000 Ly
M 27, Dumbbell Nebula	PN	$7^{m}6$	1000 Ly
M 1, Crab Nebula	PN	$9^{m}0$	6000 Ly
M 97, Owl Nebula	PN	$12^{m}0$	1300 Ly
M 42, Orion Nebula	GN	$4^{m}0$	1500 Ly
M 8, Lagoon Nebula	GN	$6^{m}0$	5200 Ljy
M 17, Omega Nebula	GN	$7^{m}0$	5000 Ly
M 20, Trifid Nebula	GN	$9^{m}0$	5200 Ly

Table A.2 Bright Stars

Star	Name	Semblance bright.	Spectrum	Abs. bright.	Distance pc
Sirius	α CMa A	−1.46	A1 V	1.4	2.6
Canopus	α Car	−0.72	F0 Ia	−8.5	360
Arcturus	α Boo	−0.04	K2 III	−0.2	280
Rigil Cen	α Cen A	0.00	G2 V	4.4	1.3
Wega	α Lyr	0.03	A0 V	0.5	8.1
Capella	α Aur	0.08	G8 III	0.3	13
Rigel	β Ori A	0.12	B8 Ia	−7.1	280
Procyon	α CMi A	0.38	F5 IV	2.6	3.5
Betelgeuse	α Ori	0.50	M2 Iab	−5.6	95
Achernar	α Eri	0.46	B5 IV	−1.6	26
Hadar	β Cen AB	0.61	B1 II	−5.1	140
Altair	α Aql	0.77	A7 IV	2.2	5.1
Aldebaran	α Tau A	0.85	K5 III	−0.3	21
Spica	α Vir	0.98	B1 V	−3.5	79
Antares	α Sco A	0.96	M1 Ib	4.7	100
Fomalhaut	α PsA	1.16	A3 V	2.0	6.7
Pollux	β Gem	1.14	K0 III	0.2	11
Deneb	α Cyg	1.25	A2 Ia	−7.5	560

Table A.3 Astronomical and physical constants

Gravitational constant	G	$6{,}6726 \times 10^{-11}\,\text{N}\,\text{m}^2\,\text{kg}^{-2}$
Speed of light in a vacuum Planck's constant	c	299,792.458 km/s
Quantum of action	$h\,(\hbar = h/2\pi)$	$6.626076 \times 10^{-34}\,\text{J}\,\text{s}$
Boltzmann constant	k	$1.380662 \times 10^{-23}\,\text{J/K}$
Stefan Boltzmann constant	σ	$5.67032 \times 10^{-8}\,\text{W}\,\text{m}^{-2}\,\text{K}^{-4}$
Elementary charge	e	$1.6021773 \times 10^{-19}\,\text{C}$
Rest mass of the electron	m_e	$9.109390 \times 10^{-31}\,\text{kg}$
Rest mass of the proton	m_p	$1.672623 \times 10^{-27}\,\text{kg}$
Atomic mass unit	AMU	$1.660539 \times 10^{-27}\,\text{kg}$
Gas constant	$\mathfrak{R}$	$8.31451\,\text{J/(mol K)}$
Permeability of the vacuum Dielectric const.	μ_0	$1.256637 \times 10^{-6}\,\text{V}\,\text{s}\,\text{A}^{-1}\,\text{m}^{-1}$
d. vacuum	ϵ_0	$8.854188 \times 10^{-12}\,\text{A}\,\text{s}\,\text{V}^{-1}\,\text{m}^{-1}$
Electronvolt	eV	$1.62 \times 10^{-19}\,\text{J}$
Light year	Ly	$9.4608 \times 10^{12}\,\text{km}$
Parsec	pc	$3.08568 \times 10^{13}\,\text{km}$
Astronomical unit	AU, AE	$149.5979 \times 10^6\,\text{km}$
Earth radius	R_E	6378 km
Earth mass	R_{Erde}	$5.98 \times 10^{24}\,\text{kg}$
Solar radius	$R_\odot$	695,980 km
Solar mass	$M_\odot$	$1.99 \times 10^{30}\,\text{kg}$
Solar luminosity	$L_\odot$	$3.9 \times 10^{26}\,\text{W}$

Table A.4 SI units and derived units

Length	Metre (m)
Time	Second (s)
Mass	Kilogram (kg)
Amount of substance	Mol (mole)
Current	Ampère (A)
Temperature	Kelvin (K)
Brightness	Candela (cd)
Force Newton (N)	$1\,\text{N} = 1\,\text{kg}\,\text{m/s}^2$
Energy, work: Joule (J)	$1\,\text{J} = 1\,\text{N}\,\text{m}$
Power: Watt (W)	$1\,\text{W} = 1\,\text{J/s}$
Frequency: Hertz (Hz)	$1\,\text{Hz} = 1/\text{s}$
Charge: Coulomb (C)	$1\,\text{C} = 1\,\text{A}\,\text{s}$
Magnetic induction: Tesla (T)	$1\,\text{T} = 1\,\text{N/A}\,\text{m}$
Pressure: Pascal (Pa)	$1\,\text{Pa} = 1\,\text{N/m}^2$

Table A.5 Conversions

1 km = 0.6215 mi
1 in = 2.54 cm
1 Å= 0.1 nm
1 gal = 3.76 L
1 km/h = 0.2778 m/s
1 rad = 57.3 = 206 265 $''$
1 t = 1000 kg
1 g/cm^3 = 1000 kg/m^3
1 N = 10^5 dyn
1 atm = 101.325 kPa = 1.01325 bar
1 bar = 100 kPa
1 kW h = 3.6 MJ
1 eV = 1.602 × 10^{-19} J
1 erg = 10^{-7} J
1 W = 1 J/s = 10^7 erg/s
1 Gauss (G) = 10^{-4} T
1 kT (TNT) = 4.18 × 10^{-12} J

Printed in the United States
by Baker & Taylor Publisher Services